Biosimilar Drug Product Development

DRUGS AND THE PHARMACEUTICAL SCIENCES
A Series of Textbooks and Monographs

Series Executive Editor
James Swarbrick
PharmaceuTech, Inc.
Pinehurst, North Carolina

Recent Titles in Series

Biosimilar Drug Product Development, *Laszlo Endrenyi, Paul Declerck, and Shein-Chung Chow*

High Throughput Screening in Drug Discovery, *Amancio Carnero*

Generic Drug Product Development: International Regulatory Requirements for Bioequivalence, Second Edition, *Isadore Kanfer and Leon Shargel*

Aqueous Polymeric Coatings for Pharmaceutical Dosage Forms, Fourth Edition, *Linda A. Felton*

Good Design Practices for GMP Pharmaceutical Facilities, Second Edition, *Terry Jacobs and Andrew A. Signore*

Handbook of Bioequivalence Testing, Second Edition, *Sarfaraz K. Niazi*

Generic Drug Product Development: Solid Oral Dosage Forms, Second Edition, *edited by Leon Shargel and Isadore Kanfer*

Drug Stereochemistry: Analytical Methods and Pharmacology, Third Edition, *edited by Krzysztof Jozwiak, W. J. Lough, and Irving W. Wainer*

Pharmaceutical Powder Compaction Technology, Second Edition, *edited by Metin Çelik*

Pharmaceutical Stress Testing: Predicting Drug Degradation, Second Edition, *edited by Steven W. Baertschi, Karen M. Alsante, and Robert A. Reed*

Pharmaceutical Process Scale-Up, Third Edition, *edited by Michael Levin*

Sterile Drug Products: Formulation, Packaging, Manufacturing and Quality, *Michael J. Akers*

Freeze-Drying/Lyophilization of Pharmaceutical and Biological Products, Third Edition, *edited by Louis Rey and Joan C. May*

Oral Drug Absorption: Prediction and Assessment, *edited by Jennifer B. Dressman and Christos Reppas*

Generic Drug Product Development: Specialty Dosage Forms, *edited by Leon Shargel and Isadore Kanfer*

Generic Drug Product Development: International Regulatory Requirements for Bioequivalence, *edited by Isadore Kanfer and Leon Shargel*

Active Pharmaceutical Ingredients: Development, Manufacturing, and Regulation, Second Edition, *edited by Stanley Nusim*

A complete listing of all volumes in this series can be found at ***www.crcpress.com***

Biosimilar Drug Product Development

Edited by
Laszlo Endrenyi
University of Toronto, Canada
Paul Jules Declerck
KU Leuven, Leuven, Belgium
Shein-Chung Chow
Duke University School of Medicine, Durham, USA

CRC Press
Taylor & Francis Group
Boca Raton London New York

CRC Press is an imprint of the
Taylor & Francis Group, an **informa** business

CRC Press
Taylor & Francis Group
6000 Broken Sound Parkway NW, Suite 300
Boca Raton, FL 33487-2742

First issued in paperback 2020

ISBN-13: 978-1-4987-1879-0 (hbk)
ISBN-13: 978-0-367-55249-7 (pbk)

Library of Congress Cataloging-in-Publication Data

Names: Endrenyi, Laszlo, editor. | Declerck, Paul (Professor in
pharmaceutical sciences), editor. | Chow, Shein-Chung, 1955- editor.
Title: Biosimilar drug product development / [edited by] Endrenyi, Laszlo,
Declerck, Dr. Paul, Chow, Shein-Chung.
Other titles: Drugs and the pharmaceutical sciences.
Description: Boca Raton : CRC Press, [2016] | Series: Drugs and the
pharmaceutical sciences | Includes bibliographical references and index.
Identifiers: LCCN 2016038570| ISBN 9781498718790 (hardback : alk. paper) |
ISBN 9781498718806 (e-book)
Subjects: | MESH: Biosimilar Pharmaceuticals | Drug Evaluation | Drug Approval
Classification: LCC RM301.25 | NLM QV 241 | DDC 615.1/9--dc23
LC record available at https://lccn.loc.gov/2016038570

Visit the Taylor & Francis Web site at
http://www.taylorandfrancis.com

and the CRC Press Web site at
http://www.crcpress.com

Contents

Preface

More and more innovative biological products will lose their patents in the coming decade. Therefore, in order to reduce costs, attempts will continue to be made to establish an abbreviated regulatory pathway for the approval of biosimilar drug products of the innovator's biological products. However, owing to the complexity of the structures of biosimilar products and the nature of the manufacturing process, biological products differ from the traditional small-molecule (chemical) drug products. Many scientific challenges remain for establishing an abbreviated regulatory pathway for the approval of biosimilar products due to their unique characteristics.

This book is devoted entirely to the development of biosimilar drug products. It covers the scientific factors and/or practical issues that are commonly encountered at various stages of research and development of biosimilar products. It is our goal to provide a useful desk reference to scientists and researchers engaged in pharmaceutical/clinical research and the development of biosimilar drug products, those in the regulatory agencies who have to make decisions in the review and approval process of biological regulatory submissions. We hope that this book can serve as a bridge among the pharmaceutical/biotechnology industry, government regulatory agencies, and academia.

This book follows the FDA's and EMA's proposed stepwise approach for evaluation and approval of the development of biosimilar products. The stepwise approach starts with analytical similarity assessment for functional and structural characterization of critical quality attributes that are relevant to clinical outcomes at various stages of the manufacturing process, *in vitro* studies for pharmacological activities, additional nonclinical studies if needed, and clinical studies for pharmacokinetic and immunogenicity assessment and efficacy confirmation. Thus, this book consists of 17 chapters. These chapters cover analytical similarity assessment (Chapters 2 through 4), manufacturing process control (Chapter 5), nonclinical studies (Chapter 6), design and analysis for assessing biosimilarity and drug interchangeability (Chapters 8, 10, and 11), pharmacovigilance (Chapter 13) and immunogenicity (Chapter 12), clinical development (Chapter 7), patent exclusivities (Chapter 14), extrapolation of indications for biosimilars (Chapter 9), and other issues (Chapters 15 through 17). The chapters intend to illuminate the many current issues and future directions of the development of biosimilars. They at times contain repetition of material, which may shed light on some topics from different directions.

From Taylor & Francis, we would like to thank Barbara Norwitz and Hilary Lafoe for giving us the opportunity to work on this book and to the support and production teams for their excellent assistance. We would like to thank colleagues and friends from academia, the pharmaceutical industry, and regulatory agencies for their support and discussions during the preparation of this book. We are, in particular, grateful to Dr. Agnes V. Klein and Dr. Jian Wang, Health Canada, not only for their own chapter contributions but also for helping to acquire and oversee other chapters.

Finally, the views expressed are those of the authors and not necessarily those of the University of Toronto, Toronto, Canada, the University of Leuven, Leuven, Belgium, and Duke University School of Medicine, Durham, North Carolina. We are solely responsible for the contents and any possible errors of this book. Any comments and suggestions will be much appreciated.

Laszlo Endrenyi, PhD, Paul Declerck, PhD, and Shein-Chung Chow, PhD

Contributors

Lyudmil Antonov
Bulgarian Drug Agency
Sofia, Bulgaria

Sigrid Balser
bioeq GmbH
Holzkirchen, Germany

Steven A. Berkowitz
Consultant
Sudbury, Massachusetts

Erwin A. Blackstone
Department of Economics
Temple University
Philadelphia, Pennsylvania

Paul Chamberlain
NDA Advisory Services Ltd.
Leatherhead, United Kingdom

Shein-Chung Chow
Department of Biostatistics & Bioinformatics
Duke University School of Medicine
Durham, North Carolina

Noel Courage
Bereskin & Parr LLP
Toronto, Ontario, Canada

Paul Declerck
Department of Therapeutic and Diagnostic Antibodies
University of Leuven
Leuven, Belgium

Karen De Smet
Federal Agency for Medicines and Health Products
Brussels, Belgium

Laszlo Endrenyi
Department of Pharmacology and Toxicology
University of Toronto
Toronto, Ontario, Canada

Alan Fauconnier
Federal Agency for Medicines and Health Products
Brussels, Belgium
and
Culture in vivo ASBL
Nivelles, Belgium

Joseph P. Fuhr Jr.
College of Population Health
Thomas Jefferson University
Philadelphia, Pennsylvania

J. Christopher Hall
School of Environmental Sciences
University of Guelph
Guelph, Ontario, Canada
and
PlantForm Corporation
Toronto, Ontario, Canada

Shehla Hashim
Marketed Biologicals, Biotechnology and Natural Health Products Bureau
Health Canada
Ottawa, Ontario, Canada

Roy Jefferis
The Division of Immunity & Infection
University of Birmingham
Birmingham, United Kingdom

Agnes V. Klein
Centre for the Evaluation of Radiopharmaceuticals and Biotherapeutics
Health Canada
Ottawa, Ontario, Canada

Wallace Lauzon
Centre for the Evaluation of Radiopharmaceuticals and Biotherapeutics
Health Canada
Ottawa, Ontario, Canada

Li Liu
Department of Biostatistics & Bioinformatics
Duke University School of Medicine
Durham, North Carolina

Michael R. Marit
PlantForm Corporation
Toronto, Ontario, Canada

Mark McCamish
Forty Seven, Inc.
Menlo Park, California

Michael D. McLean
PlantForm Corporation
Toronto, Ontario, Canada

Catherine Njue
Centre for the Evaluation of Radiopharmaceuticals and Biotherapeutics
Health Canada
Ottawa, Ontario, Canada

Felix Omara
Marketed Biologicals, Biotechnology and Natural Health Products Bureau
Health Canada
Ottawa, Ontario, Canada

Sol Ruiz
Spanish Medicines Agency
Madrid, Spain

Souleh Semalulu
Marketed Biologicals, Biotechnology and Natural Health Products Bureau
Health Canada
Ottawa, Ontario, Canada

Kenny K. Y. So
School of Environmental Sciences
University of Guelph
Guelph, Ontario, Canada
and
Department of Plant Science
University of Manitoba
Winnipeg, Manitoba, Canada

Fuyu Song
Center for Food and Drug Inspection
China Food and Drug Administration
Beijing, People's Republic of China

Lynn C. Tyler
Barnes & Thornburg LLP
Indianapolis, Indiana

Leon AGJM van Aerts
Medicines Evaluation Board
Utrecht, the Netherlands

Duc Vu
Marketed Biologicals, Biotechnology and Natural Health Products Bureau
Health Canada
Ottawa, Ontario, Canada

Jian Wang
Centre for the Evaluation of Radiopharmaceuticals and Biotherapeutics
Health Canada
Ottawa, Ontario, Canada

Gillian Woollett
Avalere Health
Washington, DC

1 Introduction
Scientific Factors in Biosimilar Product Development

Laszlo Endrenyi
University of Toronto

Paul Declerck
University of Leuven

Shein-Chung Chow
Duke University School of Medicine

CONTENTS

1.1 BACKGROUND

When an innovative drug product is going off patent, generic companies may file an abbreviated new drug application (ANDA) for the approval of the generic copies (with an identical active ingredient) of the innovative drug product under the Hatch–Waxman Act. For approval of generic drug products, the United States Food and Drug Administration (FDA) as well as other regulatory agencies require that evidence in average bioavailability be provided through the conduct of pharmacokinetic (PK) bioequivalence (in terms of rate and extent of drug absorption) studies. The assessment of bioequivalence as a surrogate endpoint for the evaluation of drug safety and efficacy is based on the *Fundamental Bioequivalence Assumption*. It states that if two drug products are shown to be bioequivalent in average bioavailability, it is assumed that they are therapeutically equivalent and can be used interchangeably.

Unlike drug products with identical active ingredients, the concept for the development of copies of biological products is different because they are made of living cells. The copies of biological products are referred to as biosimilars by the European Medicines Agency (EMA), similar biotherapeutic products (SBPs) by the World Health Organization (WHO), and subsequent-entry biologics (SEB) by Health Canada.

Biosimilars are fundamentally different from generic (chemical) drugs. Important differences include the size and complexity of the active substance and the nature of the manufacturing process. Because biosimilars are not exact copies of their originator products, different criteria for regulatory approval are required. This is partly a reflection of the complexities of manufacturing and the safety and efficacy controls of biosimilars when compared to their small-molecule generic counterparts (see, e.g., Chirino and Mire-Sluis, 2004; Crommelin et al., 2005; Roger and Mikhail, 2007; Schellekens, 2005). Since biological products are (recombinant) proteins produced by living cells, manufacturing processes for biological products are highly complex and require hundreds of specific isolation and purification steps. In practice, it is impossible to produce an identical copy of a biological product, as changes to the structure of the molecule can occur with changes in the production process. Since a protein can be modified during the process (e.g., different sugar chains may be added, the structure may have changed due to protein misfolding and so on), different manufacturing processes may lead to structural differences in the final product, which may result in differences in efficacy and safety, and may have an impact on the immune responses of patients. In some cases, these issues also occur during the postapproval changes of the innovator's biological products.

Since 2006, the EMA has provided several guidelines for the development of biosimilars. These have been followed by guidelines established by other regulatory agencies (Australia, Japan, South Korea, Canada) and the WHO. In 2015, the FDA published several guidances on the development of biosimilar products (FDA, 2015a–c). The guidance entitled *Scientific Considerations in Demonstrating Biosimilarity to a Reference Product* recommends a stepwise approach for obtaining the totality of the evidence for assessing biosimilarity between a proposed biosimilar product and its corresponding innovative biological drug product. The stepwise approach starts with analytical similarity assessment for functional and structural characterization of critical quality attributes (CQAs) that are relevant to clinical outcomes at various stages of the manufacturing process; animal studies for toxicity; pharmacokinetics and pharmacodynamics for pharmacological activities; clinical studies for efficacy confirmation; immunogenicity for safety and tolerability; and pharmacovigilance for long-term safety. Accordingly, the purpose of this chapter is to outline scientific factors and practical issues that are commonly encountered in the development of biosimilar products.

Section 1.2 describes fundamental differences and assumptions between conventional drug products and follow-on biologics. Section 1.3 presents scientific factors and practical issues that are commonly encountered in the development of biosimilar products. The aim and scope of the book are provided in Section 1.4.

1.2 FUNDAMENTAL DIFFERENCES FROM GENERICS AND ASSUMPTIONS FOR BIOSIMILARS

1.2.1 Fundamental Differences from Generics

In comparison with conventional drug products, the concept for the development of follow-on biologics is very different. Webber (2007) defines follow-on (protein) biologics as products that are intended to be sufficiently similar to an approved product to permit the applicant to rely on existing scientific knowledge about the safety and efficacy of the approved reference product. Under this definition, follow-on products are intended not only to be similar to the reference product, but also to be therapeutically equivalent with the reference product. As a number of biological products patents have expired and many more are due to expire in the next few years, the subsequent follow-on products have generated considerable interest within the pharmaceutical/biotechnological industry as biosimilar manufacturers strive to obtain part of an already large and rapidly growing market. The potential opportunity for price reductions versus the innovator biologic products remains to be determined, as the advantage of a cheaper price may be outweighed by the potential increased risk of side-effects from biosimilar molecules that are not exact copies of their innovators. In this chapter, we focus on issues surrounding biosimilars, including manufacturing, quality control, clinical efficacy, side-effects (safety), and immunogenicity. In addition, we attempt to address the challenges in imposing regulations that deal with these issues.

1.2.2 Fundamental Assumptions

As indicated by Chow and Liu (2008), bioequivalence studies are performed under the so-called Fundamental Bioequivalence Assumption, which constitutes the legal basis for the regulatory approval of generic drug products. As noted earlier, the Fundamental Bioequivalence Assumption states:

> If two drug products are shown to be bioequivalent, it is assumed that they will reach the same therapeutic effect or they are therapeutically equivalent and hence can be used interchangeably.

Note that this statement can be interpreted to mean that the confidence interval for the ratio of geometric means is between 80% and 125%. An alternative would be to show that the tolerance intervals (or a distribution-free model) overlap sufficiently.

To protect the exclusivity of a brand-name drug product, the sponsors of the innovator drug products will make every attempt to prevent generic drug products from being approved by regulatory agencies such as the FDA. One strategy used in the United States is to challenge the Fundamental Bioequivalence Assumption by filing a *citizen petition* with scientific/clinical justification. Upon receipt of a citizen petition, the FDA has the legal obligation to respond within 180 days. It should be noted, however, that the FDA will not suspend the review/approval process of a generic submission of a given brand-name drug even if a citizen petition is under review within the FDA.

In spite of the Fundamental Bioequivalence Assumption, one of the controversial issues that has arisen is that bioequivalence may not necessarily imply therapeutic equivalence and therapeutic equivalence does not guarantee bioequivalence either. One criticism lodged in the assessment of average bioequivalence for generic approval is that it is based on legal/political considerations rather than scientific arguments. In the past several decades, many sponsors/researchers have attempted to challenge this assumption but without success.

In practice, verification of the Fundamental Bioequivalence Assumption is often difficult, if not impossible, without conducting clinical trials. Notably, the Fundamental Bioequivalence Assumption applies to drug products with identical active ingredient(s). Whether the Fundamental Bioequivalence Assumption is applicable to drug products with similar but different active ingredient(s), as in the case of biosimilars, becomes an interesting but controversial question.

Similar to the Fundamental Bioequivalence Assumption described above, it has been suggested that a Fundamental Biosimilarity Assumption be developed. The following statement could be considered:

> When a follow-on biological product is claimed to be biosimilar to an innovator product in some well-defined study endpoints, it is assumed that they will reach similar therapeutic effect or they are therapeutically equivalent.

Some well-defined study endpoints are those from different functional areas such as certain physicochemical characteristics, biological activities, pharmacokinetics/pharmacodynamics (PK/PD), and immunogenicity.

1.3 SCIENTIFIC FACTORS AND PRACTICAL ISSUES

1.3.1 Criteria for Biosimilarity

For the comparison between drug products, some criteria for the assessment of bioequivalence, similarity (e.g., comparison of dissolution profiles), and consistency (e.g., comparisons between manufacturing processes) are available in either regulatory guidelines/guidances and/or the literature. These criteria, however, can be classified as (1) absolute change versus relative change, (2) aggregated versus disaggregated, or (3) moment-based versus probability-based. In this section, we briefly review different categories of criteria.

1.3.1.1 Absolute Change versus Relative Change

In clinical research and development, for a given study endpoint, either posttreatment absolute change from a baseline or posttreatment relative change from a baseline is usually considered for making comparisons between treatment groups. A typical example would be the study of weight reduction in an obese patient population. In practice, it is not clear whether a clinically meaningful difference in terms of an absolute change from a baseline can be translated to a clinically meaningful difference in terms of a relative change from the baseline. Sample-size calculations based on power analysis in terms of an absolute change from a baseline or a relative change from a baseline could lead to a very different result.

Current regulations for the assessment of bioequivalence between drug products in terms of average bioavailability are based on relative change. In other words, we conclude (average) bioequivalence between a test product and a reference product if the 90% confidence interval for the ratio of geometric means of the primary pharmacokinetic response, such as area under the blood or plasma-concentration versus time curve (AUC) between the two drug products, is (in %) totally within 80% and 125%. Note that regulatory agencies suggest that a log-transformation be performed before data analysis for the assessment of bioequivalence.

1.3.1.2 Aggregated versus Disaggregated Criteria

As indicated by Chow and Liu (2008), bioequivalence can be assessed by evaluating *separately* differences in averages, intrasubject variabilities, and the variance due to subject-by-formulation interaction between drug products. Individual criteria for the assessment of differences in averages, intrasubject variabilities, and the variance due to subject-by-formulation interaction between drug products are referred to as disaggregated criteria. If the criterion is a single summary measure composed of these individual criteria, it is called an aggregated criterion.

For the assessment of average bioequivalence (ABE), most regulatory agencies, including the FDA, recommend the use of a disaggregate criterion based on average bioavailability. In other words, bioequivalence is concluded if the average bioavailability of the test formulation is between 80% and 125% of the ABE for the reference formulation, with a certain assurance. Note that EMA disaggregated (2010) and WHO (2005) use the same equivalence criterion of 80%–125% for the log-transformed pharmacokinetic responses such as AUC.

For assessment of population bioequivalence (PBE) and individual bioequivalence (IBE), however, the following aggregated criteria have been considered. For the assessment of IBE, a criterion proposed in the FDA guidance (FDA, 2001) can be expressed as:

$$\theta_I = \frac{\left(\delta^2 + \sigma_D^2 + \sigma_{WT}^2 - \sigma_{WR}^2\right)}{\max\left\{\sigma_{W0}^2, \sigma_{WR}^2\right\}}, \tag{1.1}$$

where

$\delta = \mu_T - \mu_R$, σ_{WT}^2, σ_{WR}^2, σ_D^2 are the true difference between means, the intrasubject variabilities of the test product and the reference product, and the variance component due to subject-by-formulation interaction, respectively.

σ_{W0}^2 = a scale parameter specified by the regulatory agency.

Similarly, the criterion for the assessment of population bioequivalence suggested in the FDA guidance (FDA, 2001) is given by:

$$\theta_P = \frac{\left(\delta^2 + \sigma_{TT}^2 - \sigma_{TR}^2\right)}{\max\left\{\sigma_{T0}^2, \sigma_{TR}^2\right\}}, \quad (1.2)$$

where

$\sigma_{TT}^2, \sigma_{TR}^2$ = the total variances for the test product and the reference product, respectively.

σ_{T0}^2 = a scale parameter specified by the regulatory agency.

Population and individual bioequivalence are discussed in greater detail in Chapter 8 of this book.

A typical approach is to construct a one-sided 95% confidence interval for $\theta_I(\theta_P)$ for the assessment of individual (population) bioequivalence. If the one-sided 95% upper confidence limit is less than the bioequivalence limit of $\theta_I(\theta_P)$, then we conclude that the test product is bioequivalent to that of the reference product in terms of individual (population) bioequivalence. More details regarding individual and population bioequivalence can be found in Chow and Liu (2008).

Note that although individual bioequivalence has been discussed extensively in the past, it has been dropped by the FDA and is no longer used or considered.

1.3.1.3 Moment-Based versus Probability-Based Criteria

Schall and Luus (1993) proposed moment-based and probability-based measures for the expected discrepancy in pharmacokinetic responses between drug products. The moment-based measure suggested by Schall and Luus (1993) is based on the following expected mean-squared differences:

$$d\left(Y_j;Y_{j'}\right) = \begin{cases} E(Y_T - Y_R)^2 & \text{if } j = T \text{ and } j' = R \\ E\left(Y_R - Y_R'\right)^2 & \text{if } j = R \text{ and } j' = R \end{cases}. \quad (1.3)$$

For some prespecified positive number R, one of the probability-based measures for the expected discrepancy is given as (Schall and Luus, 1993):

$$d\left(Y_j;Y_{j'}\right) = \begin{cases} P\left\{\left|Y_T - Y_R\right| < r\right\} & \text{if } j = T \text{ and } j' = R \\ P\left\{\left|Y_R - Y_R'\right| < r\right\} & \text{if } j = R \text{ and } j' = R \end{cases}. \quad (1.4)$$

$d(Y_T;Y_R)$ measures the expected discrepancy for some pharmacokinetic metric between test and reference formulations, and $d\left(Y_R;Y_R'\right)$ provides the expected discrepancy between the repeated administrations of the reference formulation. The role of $d\left(Y_R;Y_R'\right)$ in the formulation of bioequivalence criteria is to serve as a control. The rationale is that the reference formulation should be bioequivalent to itself.

Therefore, for the moment-based measures, if the test formulation is indeed bioequivalent to the reference formulation, then $d(Y_T;Y_R)$ should be very close to $d\left(Y_R;Y_R'\right)$. It follows that if the criteria are functions of the difference (or ratio) between $d(Y_T;Y_R)$ and $d\left(Y_R;Y_R'\right)$, bioequivalence is concluded if they are smaller than some prespecified limit. For probability-based measures, however, if the test formulation is indeed bioequivalent to the reference formulation, as compared with $d\left(Y_R;Y_R'\right)$, $d(Y_T;Y_R)$ should be relatively large. As a result, bioequivalence, or biosimilarity, is concluded if the criterion based on the probability-based measure is higher than some prespecified limit (Chow et al., 2010).

1.3.1.4 Remarks

Although several criteria for similarity are available in both regulatory guidelines/guidances and the literature, these criteria do not translate each other. In other words, one may pass one criterion but fail to pass others. Moreover, these criteria do not address the following critical questions: (1) how similar is considered to be similar? and (2) what is the impact of the level of similarity on drug interchangeability.

1.3.2 Statistical Methods

Since there are many critical attributes of a potential patient's response in follow-on biologics, for a given critical attribute, valid statistical methods need to be developed under a valid study design and a given set of criteria for similarity, as described in Section 1.3.1. Several areas can be identified for developing appropriate statistical methodologies for assessment of the biosimilarity of follow-on biologics. These areas include, but are not limited to:

1. Consistency in manufacturing processes
   Since changes in the manufacturing process could have a significant impact on the clinical outcome of follow-on biologics, tests for consistency in manufacturing processes are critical in the assessment of biosimilarity.
2. Stability testing (multiple labs, multiple lots)
   Since biological products are sensitive to environmental factors such as light and temperature, we suggest that stability testing be conducted under study designs that can account for these environmental factors following both the International Conference on Harmonization (ICH) and FDA guidelines for determination of shelf life.
3. Comparability in quality attributes of structural and functional characterization
   As indicated by the FDA, test for comparability in quality attributes in structural and functional characterization is essential for assessment of the biosimilarity of follow-on biologics. Valid statistical methods are necessarily developed with respect to the study design, endpoints, and criteria employed.
4. Sequential testing procedures
   Due to the complexity of the manufacturing process of biological products, sequential testing procedures for statistical quality control may be useful to ensure biosimilarity.

5. Criteria for biosimilarity (in terms of average, variability, or distribution) to address the question of "how similar is similar"
   We suggest establishing criteria for biosimilarity in terms of average, variability, and/or distribution.
6. Criteria for interchangeability
   In practice, it is recognized that drug interchangeability is related to the variability due to subject-by-drug interaction. However, it is not clear whether a criterion for interchangeability should be based on the variability due to subject-by-drug interaction or on the variability due to subject-by-drug interaction adjusted for intrasubject variability of the reference drug.
7. Bridging studies for assessing biosimilarity
   Because most biosimilar studies are conducted using a parallel design rather than a replicated crossover design, independent estimates of variance components such as the intrasubject and the variability due to subject-by-drug interaction are not possible. In this case, bridging studies may be considered.
8. Use of a percentile method for the assessment of variability
   In addition to classical F-type test statistics for assessment of variability, use of a percentile method may be useful.
9. Assessment of immunogenicity
   As indicated by the FDA, assessment of immunogenicity is important for assessment of biosimilarity. Appropriate statistical methods should be developed according to study endpoints and criteria employed.
10. Multiple testing procedures for global assessment of biosimilarity
    Since the assessment of biosimilarity of follow-on biologics comprises different properties such as biological activities, PK/PD, immunogenicity, and clinical response, multiple testing procedures should be considered for the assessment of global biosimilarity.

In addition to PK/PD, biomarkers such as genomic data could serve as surrogate endpoints for the assessment of biosimilarity of follow-on biologics if they are predictive of clinical responses.

1.3.3 The Manufacturing Process

Unlike small-molecule drug products, biological products are made of living cells. Thus, the manufacturing of biological products is a very complicated process that involves (1) cell expansion, (2) cell production (in bioreactors), (3) recovery (through filtration or centrifugation), (4) purification (through chromatography), and (5) formulation. A small discrepancy at each step (e.g., purification) could lead to a significant difference in the final product, which might cause a difference in clinical outcomes. Thus, process control and validation plays an important role in the success of the manufacturing of biological products. In addition, since at each step (e.g., purification), different methods may be used for different biological manufacturing processes (within the same company or at different biotech companies), tests for consistency are necessarily performed. Note that at the purification step, the following chromatography media or resins are commonly considered: (1) gel filtration, (2) ion exchange, (3) hydrophobic interaction, (4) reversed phase/normal phase, and (5) affinity. Thus, at each step of the manufacturing process, primary performance characteristics should be identified, controlled, and tested for consistency of process control and validation.

Issues of manufacturing and process control in the development of biosimilars are discussed in Chapter 5 of this book.

1.3.4 Similarity in Size and Structure

In practice, sponsors perform various *in vitro* tests such as the assessments of the primary amino acid sequence, charges, and hydrophobic properties to compare the structural aspects of biosimilars with their originator molecules. However, whether *in vitro* tests can be predictive of biological activity *in vivo* is a concern inasmuch as there may be significant differences in biological activity despite similarities in size and structure. Besides, it is difficult to assess biological activity adequately as few animal models can provide the data needed to extrapolate for an accurate and reliable prediction of biological activity in humans. Thus, controlled clinical trials remain often necessary for confirming the similarity between a biosimilar molecule and the originator product.

The FDA has proposed a tiered approach for analytical similarity assessment of CQAs relevant to clinical outcomes at various stages of the manufacturing process. The FDA suggests first classifying the identified CQAs into three tiers depending on their criticality or risk ranking relevant to clinical outcomes. The FDA then recommends using equivalence tests for CQAs in Tier 1 that are considered most relevant to clinical outcomes; a quality range approach for CQAs in Tier 2 that are considered mild to moderate relevant to clinical outcomes; and raw data and graphical comparisons for CQAs in Tier 3 that are considered least relevant to clinical outcomes (see, e.g., Chow et al., 2016). Analytical similarity assessment is discussed in greater detail in Chapter 3 of this book.

1.3.5 Biosimilarity in Biological Activity

Pharmacological or biological activity is an expression describing the beneficial or adverse effects of a drug on living matter. When the drug is a complex chemical mixture, this activity is exerted by the substance's active ingredient or pharmacophore but can be modified by the other constituents. A crucial component of biological activity is a substance's toxicity. Activity is generally dosage dependent, and it is not uncommon to have effects ranging from beneficial to adverse for one substance when going from low to high doses. Activity depends critically on the fulfillment of the absorption, distribution, metabolism, and excretion (ADME) criteria.

Note that the EU *Pharmaceutical Review* legislation published on April 30, 2004, amended the EU community code on medicinal products to provide for the approval of biosimilars based on fewer preclinical and clinical data than had been required for the original reference product. The complexity of the protein and knowledge of its structure–function relationships determine the types of information needed to establish similarity.

1.3.6 The Problem of Immunogenicity

Since all biological products are biologically active molecules derived from living cells and have the potential to evoke an immune response, immunogenicity is probably the most critical safety "uncertainty" for the assessment of biosimilarity of follow-on biologics. The commonly seen possible causes of immunogenicity include, but are not limited to: (1) sequence differences between a therapeutic protein and endogenous proteins, (2) nonhuman sequences or epitopes, (3) structural alterations, (4) storage conditions, (5) purification during the manufacturing process, (6) formulation (e.g., surfactants), (7) route, dose, and frequency of administration, (8) patient status such as concomitant therapy (e.g., immunosuppressants) or genetic background. Thus, the following questions should be asked when assessing biosimilarity between biological products: (1) What is the immunogenic potential of the therapeutic protein? (2) What is the impact of the generating antibodies to the self-protein or to the therapeutic drug? (3) What is the impact of immunogenicity on preclinical toxicity (e.g., pharmacokinetic levels and dose-limiting toxicity)? (4) What is the impact of immunogenicity of the therapeutic protein on safety? (5) What are the risk evaluation and mitigation strategy processes required by regulatory agencies?

The immune responses to biological products can lead to: (1) anaphylaxis, (2) injection site reactions, (3) flu-like syndromes, and (4) allergic responses. Note that one of the most serious adverse events occurs when neutralizing antibodies to the drug cross-react with endogenous proteins that have a unique physiological role. The risk of immunogenicity can be reduced through stringent testing of the product during its development. It should be noted, however, that immunogenicity in animals does not predict immunogenicity in clinical trials, and analytical techniques may not detect differences that may impact immunogenicity. Therefore, the immunogenicity of a biological product depends heavily on the attributes of product quality such as the physical, structural, and functional properties of the active pharmaceutical ingredients; as well as excipients, container closure, and delivery system. It turns out that similarity of the acceptable ranges of these quality attributes is crucial to the evaluation of similarity between the biosimilar and the reference product.

Problems of immunogenicity in the development of biosimilars are discussed in Chapter 12 of this book.

1.3.7 Drug Interchangeability

Basically, drug interchangeability can be classified either as drug prescribability or drug switchability. Drug prescribability is defined as the physician's choice for prescribing an appropriate drug product for his or her new patients between a brand-name drug product and a (number of) biosimilar drug product(s) that have been shown to be bioequivalent/biosimilar to the brand-name drug product. The underlying assumption of drug prescribability is that the brand-name drug product and its biosimilars can be used alternatively in terms of the efficacy and safety of the drug product. Drug switchability, in contrast, is related to the switch from a drug product (e.g., a brand-name drug product) to an alternative drug product (i.e., a biosimilar of the brand-name drug product) within the same subject, whose concentration of the drug product has been titrated to a steady, efficacious, and safe level. As a result, drug switchability is considered more critical than drug prescribability in the study of drug interchangeability for patients who have been on medication for a while. Drug switchability, therefore, is exchangeability within the same subject.

Issues of interchangeability, switchability, and substitution of biosimilars are discussed in Chapter 10 of this book.

1.3.8 Development of the Biosimilarity Index

Chow (2011) proposed the development of a composite index for assessing biosimilarity based on the facts that (1) the concept of biosimilarity for biological products (made of living cells) is very different from that of bioequivalence for chemical drug products and (2) critical quality attributes of biological products are dependent on the manufacturing process. Although some research on the comparison of moment-based criteria and probability-based criteria for the assessment of (1) average biosimilarity and (2) variability of biosimilarity for some given study endpoints by applying the criteria for bioequivalence are available in the literature (see, e.g., Chow and Liu, 2010; Hsieh et al., 2010), universally acceptable criteria for biosimilarity are not available in the regulatory guidelines/guidances. Thus, Chow (2011) proposed a biosimilarity index based on the concept of the probability of reproducibility as follows:

Step 1: Assess the average biosimilarity based on bioequivalence criteria—that is, biosimilarity is claimed if the 90% confidence interval of the ratio of means of a given study endpoint falls within the biosimilarity limit of (80%, 125%) based on log-transformed data.

Step 2: Once the product passes the test for biosimilarity in Step 1, calculate the probability of reproducibility (Shao and Chow, 2002) based on the observed ratio and variability. The primary reason for this is to take the variability and the sensitivity of heterogeneity in variances into consideration.

Step 3: We shall claim biosimilarity if the probability of reproducibility is larger than a prespecified number p_0, which can be obtained based on the comparison of a "reference product" to the "reference product." For example, if the R-R comparison suggests a reproducibility of 60%, then p_0 could be chosen as 80% of the 60%, which is 48%.

As indicated by Chow (2011), the above proposal has the advantages that (1) we still follow the well-established criterion for the assessment of bioequivalence, which has been used for decades and (2) the probability of reproducibility will reflect the sensitivity of heterogeneity in variance. Note that the proposed biosimilarity index is developed based on the probability of reproducibility. It can be applied to different functional areas (domains) of biological products such as PK, biological activities, biomarkers (e.g., pharmacodynamics), immunogenicity, the manufacturing process, and efficacy. For a given domain such as PK, the proposal is briefly described as follows:

Step 1: Assess biosimilarity based on a prespecified criterion;
Step 2: Calculate the probability of reproducibility;
Step 3: Claim "success" if the probability of reproducibility is larger than p_0.

As a result, an overall biosimilarity index across domains can be developed as follows:

Step 1: Obtain p_i, the probability of reproducibility for the ith domain, $i = 1, \dots, K$;
Step 2: Define the biosimilarity index $p = \sum_{i=1}^{K} w_i p_i$ where w_i is the weight for the ith domain;
Step 3: Claim global biosimilarity if $p > p_0$ where p_0 is a prespecified value.

The statistical properties of Chow's proposed biosimilarity index are currently evaluated through simulations.

1.3.9 Remarks

Current methods for the assessment of bioequivalence for drug products with identical active ingredients are not applicable to biosimilars due to fundamental differences. The assessment of biosimilarity between a biosimilar and the reference product in terms of surrogate endpoints (e.g., pharmacokinetic parameters and/or pharmacodynamic responses) requires the establishment of the Fundamental Biosimilarity Assumption in order to bridge the surrogate endpoints and/or biomarker data to clinical safety and efficacy.

Under the established Fundamental Biosimilarity Assumption and the selected biosimilarity criteria, it is also recommended that appropriate statistical methods (e.g., comparing distributions and the development of the biosimilarity index) be developed under valid study designs for achieving the study objectives (e.g., the establishment of biosimilarity at specific domains or drug interchangeability) with a desired statistical inference (e.g., power or confidence interval). To ensure the success of studies conducted for the assessment of biosimilarity, regulatory guidelines/guidances need to be developed. Product-specific guidelines/guidances published by the EMA have been criticized for not having standards. Although product-specific guidelines/guidances do not help to establish standards for the assessment of biosimilarity of biosimilars, they do provide the opportunity for accumulating valuable experience/information for establishing standards in the future. Thus, several numerical studies could be pursued, including simulations, meta-analysis, and/or sensitivity analysis, in order to (1) provide a better understanding of these product-specific guidelines/guidances and (2) check the validity of the established Fundamental Biosimilarity Assumption, which is the legal basis for assessing biosimilarity.

1.4 AIM AND SCOPE OF THE BOOK

This book is devoted to the development of biosimilar products. It covers the scientific factors and/or practical issues that are commonly encountered at various stages of research and development of biosimilar products. Our goal is to provide a useful desk reference for scientists and researchers engaged in pharmaceutical/clinical research and the development of biosimilar products, and for those in the regulatory agencies who have to make decisions in the review and approval process of biological regulatory submissions. We hope that this book can serve as a bridge among the pharmaceutical/biotechnology industry, government regulatory agencies, and academia.

This book follows the FDA's proposed stepwise approach as well as the approach of the EMA and other regulatory agencies such as China Food and Drug Administration (CFDA) of China for evaluation and approval of the development of biosimilar products. The stepwise approach starts with analytical similarity assessment for functional and structural characterization of critical quality attributes that are relevant to clinical outcomes at various stages of the manufacturing process, pharmacological activities, additional nonclinical studies if needed, and clinical studies for pharmacokinetic and immunogenicity assessment and efficacy confirmation. This book consists of 17 chapters. These chapters cover analytical similarity assessment (Chapters 2 through 4), manufacturing process control (Chapter 5), nonclinical studies (Chapter 6), clinical development (Chapter 7), extrapolation of indications for biosimilars (Chapter 9), design and analysis for assessing biosimilarity and drug interchangeability (Chapters 8, 10, and 11), pharmacovigilance (Chapter 13) and immunogenicity (Chapter 12), patent exclusivities (Chapter 14), and other issues (Chapters 15 through 17).

Discussions of these topics cover a broad range of issues that are important for the development of biosimilar drug products. It is hoped that the explorations will benefit researchers and practitioners involved in such investigations and applications.

REFERENCES

Chirino AJ, Mire-Sluis A. (2004) Characterizing biological products and assessing comparability following manufacturing changes. *Nature Biotechnology* **22**, 1383–1391.

Chow SC. (2011) Quantitative evaluation of bioequivalence/biosimilarity. *Journal of Bioequivalence & Bioavailability* **S1:002**, 1–8.

Chow SC, Hsieh TC, Chi E, Yang J. (2010) A comparison of moment-based and probability-based criteria for assessment of follow-on biologics. *Journal of Biopharmaceutical Statistics* **20**, 31–45.

Chow SC, Liu JP. (2008) *Design and Analysis of Bioavailability and Bioequivalence Studies*, 3rd edition. Chapman Hall/CRC Press, Taylor & Francis, New York.

Chow SC, Liu JP. (2010) Statistical assessment of biosimilar products. *Journal of Biopharmaceutical Statistics* **20**, 10–30.

Chow SC, Song FY, Bai H. (2016) Analytical similarity assessment in biosimilar studies. *AAPS Journal* **18(3)**, 670–677.

Crommelin D, Bermejo, T, Bissig M., et al. (2005) Biosimilars, generic versions of the first generation of therapeutic proteins: do they exist? *Contributions to Nephrology* **149**, 287–294.

EMA. (2010) Guideline on the investigation of bioequivalence. European Medicines Agency, London, UK.

FDA. (2001) Guidance on statistical approaches to establishing bioequivalence. Food and Drug Administration, Center for Drug Evaluation and Research, Rockville, MD.

FDA. (2015a) Scientific considerations in demonstrating biosimilarity to a reference product. Food and Drug Administration, Silver Spring, MD.

FDA. (2015b) Quality considerations in demonstrating biosimilarity to a reference protein product. Food and Drug Administration, Silver Spring, MD.

FDA. (2015c) Biosimilars: questions and answers regarding implementation of the Biologics Price Competition and Innovation Act of 2009. Food and Drug Administration, Silver Spring, MD.

Hsieh TC, Chow SC, Liu JP, et al. (2010) Statistical test for evaluation of biosimilarity of follow-on biologics. *Journal of Biopharmaceutical Statistics* **20**, 75–89.

Roger SD, Mikhail A. (2007) Biosimilars: opportunity or cause for concern? *Journal of Pharmaceutical Science* **10**, 405–410.

Schall R, Luus H. (1993) On population and individual bioequivalence. *Statistics in Medicine* **12**, 1109–1124.

Schellekens H. (2005) Follow-on biologics: challenges of the 'next generation'. *Nephrology, Dialysis, Transplantation* **20**, 31–36.

Shao J, Chow SC. (2002) Reproducibility probability in clinical trials. *Statistics in Medicine* **21**, 1727–1742.

Webber KO. (2007) Biosimilars: are we there yet? Presented at Biosimilars 2007, George Washington University, Washington, DC.

WHO. (2005) Multisource (generic) pharmaceutical products: guidelines on registration requirements to establish interchangeability (draft revision). World Health Organization, Geneva, Switzerland.

2 Analytical Characterization

Structural Assessment of Biosimilarity

Steven A. Berkowitz
Consultant

CONTENTS

2.1 INTRODUCTION

In general, once the patent protection of a product expires, others have the legal opportunity to freely copy (manufacture) and sell the same product that previously was only allowed to be made and sold by its innovator. However, in the case of making and selling a copy of an innovator's (original) drug, a regulated approval process is required. In the United States (US), until the mid-1980s, the approval process required was effectively the same lengthy and costly regulated approval process needed to obtain the approval of the original drug (see Figure 2.1A), but without the need to conduct drug discovery activities. This situation was a significant impediment to making a copy of a drug that could be sold at a much lower price. However, in 1984 with passage of the Hatch–Waxman Act (also known more formally as the Drug Price Competition and Patent Term Restoration Act of 1984), the requirements for obtaining regulatory approval for making and selling a copy of an innovator's drug, specifically for innovative small-molecule drugs called pharmaceuticals, greatly changed (Frank, 2007). Of great importance in passage of this Act is the availability of a significantly abbreviated approval pathway that greatly reduces the time and cost involved in making a copy of a drug by reducing or eliminating the need for clinical work (see Figure 2.1A vs. 2.1B). Key to being able to use this abbreviated approval pathway is the critical outcome of demonstrating that the structure of the drug copy (called a generic) is an *identical* copy of the structure of the innovator's drug. At the heart of this work are the analytical physical and chemical (physicochemical) data submitted to regulatory agencies to support this *structural identity*, along with data showing the comparability of these materials in terms of purity, bioequivalence, and stability.

For pharmaceuticals whose size in terms of molecular weight (MW) is typically only a few hundred Daltons (Da), the availability of appropriate analytical physicochemical tools [such as nuclear magnetic resonance (NMR), mass spectrometry (MS), Fourier transform infrared (FTIR), spectroscopy, chromatography, and electrophoresis] can, with very high fidelity, confirm the structural identity between an

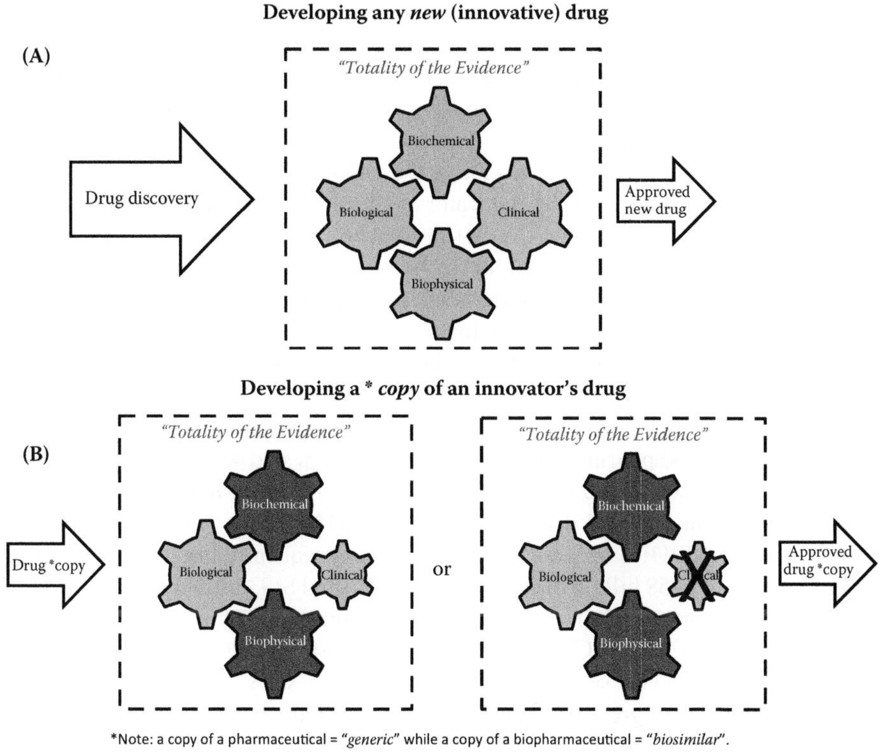

FIGURE 2.1 A simple pictorial view of the key development areas responsible for generating the *Totality of the Evidence* (a phrase, introduced by the FDA that has become associated with the underlying approach for evaluating and approving biosimilars, which, in reality should be applied more universally, as indicated here, to the evaluation and approval of *all* drugs) data package for supporting a drug's regulatory approval: (A) for a new or innovative drug (pharmaceutical or biopharmaceutical) relative to (B) a highly similar or identical copy of an innovator's drug, where the time and cost intensive search of drug discovery is replaced with the less intensive task of demonstrating that an appropriate drug candidate is an adequate copy of an innovative drug resulting in a significant reduction or elimination of clinical data to obtain its regulatory approval. (Reprinted from Houde JD and Berkowitz SA eds., 2014a. Biophysical characterization: an integral part of the "Totality of the Evidence" concept. In: *Biophysical Characterization of Proteins in Developing Biopharmaceuticals*, 385–396. Elsevier, Amsterdam. With permission from Elsevier with minor modifications.)

innovator's pharmaceutical and its generic. In addition, the nature and amount of impurities in the innovator's pharmaceutical and its generic can also be accurately assessed and compared typically using the same mentioned analytical physicochemical tools used to demonstrate structural identity. Once a generic's structural identity to the innovator's drug is established along with a comparable level of purity, bioequivalence, and stability, the generic manufacturer is in a position to take advantage of the long established clinical history of efficacy and safety of the innovator's drug to support the clinical performance of its generic. As a result, clinical development activity for a generic is greatly reduced or potentially even eliminated (see Figure 2.1B).

This reduction or elimination of clinical work, coupled with the elimination of the normal long and precarious drug discovery process, greatly reduces the time and cost of developing and commercializing a generic. The success of this activity enables the generic manufacturer to charge the end user and/or payer of its generic a much lower price than what the innovator charges, while still being able to make a reasonable profit. The end result is that those who use and/or pay for the original drug could now pay the lower price for effectively the same drug (generic version) in preference to the more costly innovator's drug. As a result, the business of making generic drugs has been an overwhelming success not only as a new business, but more importantly as a way to provide a major cost saving to those who need and pay for these drugs.

This success is validated in the US by the fact that in 2009 (Kozlowski et al., 2011) over 75% of the pharmaceutical prescription drugs sold were generic versions, which increased to 86% in 2014 (Hirsch et al., 2014) resulting in drug price reductions relative to the original drug of 77%–90% (FDA, 2015a; Hirsch et al., 2014; Kozlowski et al., 2011; Sarpatwari et al., 2015). Consequently, the US health care system saved over $158 billion in 2010 (FDA, 2015a) and $193 billion in 2011, which, when summed over various time periods, have been reported to total savings of about $1.1 trillion for 2002–2011 (GPhA, 2012) or $1.5 trillion for 2005–2015 (Sarpatwari et al., 2015). It is this cost saving that is the real success story and intended purpose of generics, which has greatly reduced the economic burden placed on government agencies and other third-party payers of health insurance who are the major payers, along with the patients who use these drugs. Indeed, by achieving this remarkable cost reduction, these life-saving and life-enhancing drugs have become more readily available to a much wider population of individuals who would not have otherwise enjoyed their benefit.

Nevertheless, the landscape of drug development in the last three decades has undergone great changes in terms of the nature of the type of drugs that are now being developed to achieve even greater life-improving outcomes. This major paradigm change has been fueled by the applied application of knowledge gained through the scientific breakthroughs in the various areas of molecular biology, biochemistry, and other related and peripheral areas of science that have led to development of an entirely new class of drugs. These new drugs are called biopharmaceuticals and are predominately proteins (note: although biopharmaceuticals can be composed of other biological materials, e.g., nucleic acids and carbohydrates, all further discussions in this chapter referencing biopharmaceuticals will be concerned specifically with protein biopharmaceuticals).

Although biopharmaceuticals represent a great improvement in our ability to provide much more effective and highly targeted therapeutic agents, unfortunately they are generally felt to come associated with a much higher cost in their development and manufacturing (Blackstone and Fuhr, 2012; Ventola, 2013). This higher cost has been associated with their much greater complexity and consequently the much more challenging task of producing and characterizing them, especially from a physicochemical prospective. Properties of these new drugs that create this complexity and challenge include the following: (1) their very high MW, (2) intricate and dynamic three-dimensional (3D) structure, conformation or higher-order structure (HOS), and (3) novel mode of production that involves the use of an enormous number of specially designed microscale factories called "living cells" instead of a collection of sequentially chemical reactions conducted in large, well-controlled chemical reactors, as is done in the case of pharmaceuticals (see Figure 2.2 as well as Table 2.1,

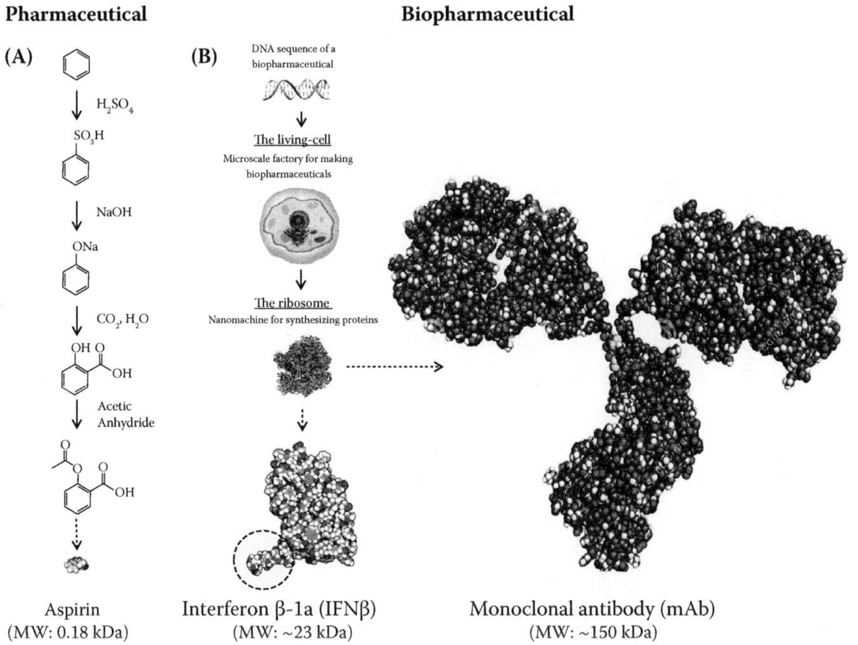

FIGURE 2.2 **(See color insert.)** A simple comparison illustrating differences in the process of making a pharmaceutical versus making a biopharmaceutical: (A) Coarse outline of the sequential chemical reactions for making a pharmaceutical, using aspirin as an example and (B) a coarse outline of the basic steps for making a biopharmaceutical, which consists of first synthesizing a piece of DNA containing the correct nucleotide sequence code for making the desired biopharmaceutical's polypeptide chain(s), the insertion of this DNA into an initial small collection of cells (the microscale factories for making the biopharmaceutical) using recombinant DNA technology, the large-scale growth of these cells during which the cell's internal protein synthesizing nanomachine (the ribosome, a complex cellular organelle composed of many proteins and several pieces of RNA) are directed to synthesize the target biopharmaceutical, illustrated here as either interferon beta-la (IFNβ) or a monoclonal antibody (mAb). Note that the space-filling molecular models of aspirin, IFNβ and mAb have all been displayed roughly on the same arbitrary scale to help provide the reader with an approximate perspective on how they would relatively compare to each other on the basis of size. The dashed circle highlighting part of the structure of IFNβ corresponds to the carbohydrate-containing portion of this biopharmaceutical that plays a dominant role in giving rise to its microheterogeneity shown in Figures 2.4D and 2.9C through E due predominantly to the complexity of the different carbohydrate structures that are found attached to the biopharmaceutical (shown in Figure 2.10) when coupled with other post-translational modifications (PTMs).

TABLE 2.1
Comparison of Key Attributes of a Pharmaceutical (Generic) and Biopharmaceutical (Biosimilar) That Make Them Different in Terms of Physicochemical Analysis

Attribute	Pharmaceutical (Generic)	Biopharmaceutical (Biosimilar)
Size and chemical complexity (primary structure)	Small drugs having a single unique structure with a small number of different chemical targets capable of giving rise typically to a small number of covalent altered forms of the drug with no or altered drug activity.	Very large drugs made of a large number of different chemical targets capable of giving rise to many different covalent forms of the drug. These drugs also require the intricate folding of their main structural element, the polypeptide chain(s), which is stabilized via a collection of weak noncovalent bonds (or weak secondary bonds). Alteration of these noncovalent bonds can give rise to alterations in this folding providing the opportunity for generating additional drug forms with different chemical properties and biological activity. Consequently these drugs tend to be a very heterogeneous collection of highly similar drugs having the same or similar drug activity, but some forms may have altered or no drug activity.
Mode of synthesis	Direct chemical synthesis using simple well controlled raw materials that requires a limited number of chemical reaction steps that can overall be well monitored and controlled.	A synthesis process that is totally dependent on the use of living cells resulting in the following: (1) exposure of the drug to an enormous array of enzymatic and chemical reactions called posttranslational modifications (PTMs) within the cell resulting in a very heterogeneous drug product with many variant forms and (2) to a very difficult drug manufacturing problem due to the intrinsic complexity of cells requiring a very wide range of factors to be controlled that are highly sensitive to small changes (e.g., raw materials—complex growth media, growth conditions, container surfaces).

(Continued)

TABLE 2.1 (*Continued*)
Comparison of Key Attributes of a Pharmaceutical (Generic) and Biopharmaceutical (Biosimilar) That Make Them Different in Terms of Physicochemical Analysis

Attribute	Pharmaceutical (Generic)	Biopharmaceutical (Biosimilar)
HOS (secondary, tertiary, and quaternary structure)	Overall structure is fairly rigid involving only covalent (primary) bonds. As a result, these drugs tend to be stable molecules that are more capable of tolerating a wide range of physical conditions.	In addition to the primary structure of these drugs, consisting of a long linear string (polypeptide chain) of covalent linked units (amino acids) plus PTMs, their structure also involves the intricate folding of the polypeptide (secondary and tertiary structure) that can involve more than one polypeptide chain (quaternary structure) that is all held together predominately by a large array of weak noncovalent bonds. The weakness of these bonds allow some to be easily broken providing these drugs with dynamic properties that enables them to interconvert into a large number (ensemble) of uniquely folded structures important to their function and potential modes of degradation. This higher order structure (HOS), important for drug function, makes these molecules fairly unstable, resulting in significant constraints on storage (requiring low temperatures) and handling conditions.
Aggregation and immunogenicity	This typically is not a factor observed with this class of drugs.	The self-association or aggregation, typically of the monomeric form of these drugs, represents a major unique form of degradation for this class of drugs, which along with the other changes in chemical (primary structure) and HOS can lead to adverse immune responses that is also a unique problem associated with these drugs.

which provides more information about the attributes that make the process of producing biopharmaceuticals very different and more difficult from the process of making pharmaceuticals). Although it would appear that these and other differentiating attributes would greatly impact the R&D cost involved in making biopharmaceuticals versus pharmaceuticals and would thus be good reasons for the higher cost of biopharmaceuticals, economic and financial assessments have not seemed to have borne this out as of yet (DiMasi and Grabowski, 2007; DiMasi et al., 2016); see Chapter 16 for further discussion on this topic. Irrespective of what may be causing the high cost associated with biopharmaceuticals, which needs to be paid to acquire these drugs, this high cost is again placing a very heavy economic burden on those who must pay for them (Hirsch et al., 2014; McCamish and Woollett, 2011) and thus is creating an economic roadblock in getting these drugs to those who need them.

As these biopharmaceuticals start to lose their patent protection, it is hoped that the same approach of making copies (generic versions) of these drugs will provide the same economic benefits as has been achieved in the case of pharmaceuticals. Unfortunately, the attributes of biopharmaceuticals and the manner in which they are made present the manufacturer trying to make a generic (identical) copy of a biopharmaceutical with an impossible task. The reality of this situation is made apparent when it is realized that even the innovator of a biopharmaceutical cannot make its own biopharmaceutical so that every biopharmaceutical molecule in a given lot is identical. Nor can the innovator make the actual collection or distribution of different biopharmaceutical molecules that is present in a given lot identical on a lot-to-lot basis. Rather, the variation in the different forms of the biopharmaceutical that are present in a given production lot is restricted by limits or specifications associated with a collection of critical quality attributes (CQAs). These limits or specifications (which were established by the innovator in collaboration with regulators who approved them) define the allowable distribution and variation of different forms of the biopharmaceutical that can be present in any lot, so that each manufactured biopharmaceutical lot is "comparable or highly similar" on a lot-to-lot basis (Schneider, 2013). Consequently, a manufacturer trying to make a copy of a biopharmaceutical cannot possibly make an identical copy of something that itself was never identical to begin with. Thus, a copy of a biopharmaceutical cannot be called a generic (an identical copy of an innovative biopharmaceutical). Rather, it is a highly similar copy of the different forms of a biopharmaceutical that are found in the innovator's biopharmaceutical that are allowed to vary within the same (highly similar) range of limits or specifications for the same CQAs that characterize the innovator's biopharmaceutical. Nomenclature used to describe a copy of a biopharmaceutical includes several different terms such as a biosimilar, follow-on biologic, and subsequent-entry biologic (Rader, 2007), with biosimilar appearing to be by far the most common name used to describe this type of drug product.

Given the inability to make identical copies of biopharmaceuticals, in the US the availability of the abbreviated approval pathway offered via the Hatch–Waxman Act is not applicable to biopharmaceuticals. Consequently, the development of new legal regulatory legislation and scientific thinking has been required in the US (and other countries and geographical regions) to provide a process for developing copies of biopharmaceuticals through an abbreviated pathway similar to generics, as shown in Figure 2.1B. However, in so doing significant debate and concern has been raised and to some extent still exists over the clarity as to what exactly is needed to establish and acquire regulatory approval of biosimilars, and about the potential problems or issues that biosimilars may create. Nevertheless, many biopharmaceutical and pharmaceutical companies are actively pursuing their development (Calo-Fernández and Martínez-Hurtado, 2012; Thayer, 2013) with the hope that it will bring the same success story (both as a new business opportunity and as an effective cost-reduction route to achieve better and increased access to these life-changing drugs) as that achieved by implementation of generics in the case of pharmaceuticals.

This chapter takes a close look at the scientific process of assessing the high comparability or similarity of a biosimilar to its corresponding innovator's biopharmaceutical from a structural perspective (which is associated with the biochemical and biophysical areas of the drug development process, highlighted as the darker areas in Figure 2.1B). In so doing, the following two areas will be emphasized: (1) the challenges that need to be overcome to successfully assess and characterize the physicochemical structure and properties of these complex drugs to obtain regulatory approval, which the FDA has already pointed out is dominated by the important issues and specific challenges concerning structural heterogeneity, HOS, and aggregation (FDA, 2009a) and (2) the analytical capabilities (in terms of tools and methods) that are available to establish adequate physicochemical structural comparability or similarity.

Before embarking on this discussion, it should again be pointed out that the analytical characterization of the structural or physicochemical component of the comparability or similarity process (which is the first and most fundamental step needed to be successfully executed in obtaining regulatory approval of a biosimilar) is only a part of a much broader process (which is largely the subject matter covered in this book). A process that has come to be referred to as the assessment of *biosimilarity*, which is assessed using the general concept referred to by the FDA as the *Totality of the Evidence* (Kozlowski et al., 2011; Woodcock et al., 2007), which in reality is a more general concept implemented in assessing and approving *all* drugs (Houde and Berkowitz, 2014a), as indicated in Figure 2.1A and B.

2.2 BIOSIMILARITY: AN EXTENSION OF THE CONCEPT OF COMPARABILITY

The underlying concept of biosimilarity is directly linked to what is now considered a common activity often carried out in developing and commercializing biopharmaceuticals called comparability studies; this activity was initiated by the FDA in 1996 (FDA, 1996) and formalized into the International Conference on Harmonisation (ICH) document designated as Q5C (ICH, 2004). Comparability studies are specifically associated with two important tasks: (1) assessing the ability of a biopharmaceutical manufacturer to be able to consistently make a biopharmaceutical on a lot-to-lot basis (which amounts to saying all lots are comparable in terms of meeting a set of established limits or specifications) and (2) enabling a biopharmaceutical manufacturer to introduce a change(s) into the process of making its biopharmaceutical (e.g., change in raw material, process step, site of manufacturing, etc.), as long as the pre- and postchange biopharmaceuticals are comparable to each other (i.e., they meet the same set of limits or specifications). The former task is an important requirement that is conducted at various stages of development and commercialization of a biopharmaceutical to demonstrate to regulators that the biopharmaceutical manufacturer can successfully control its drug's manufacturing process to make a consistent drug product. The latter task is important in enabling a biopharmaceutical manufacturer to introduce a manufacturing change(s) during its development and still be allowed to proceed with filing a new investigational drug application (IND) to conduct clinical trials without the need to repeat earlier clinical work or to allow a manufacturer of a commercial biopharmaceutical product to introduce a change(s) into its drug's manufacturing process without having to provide additional clinical data.

In the situation where a *different* manufacturer (biosimilar manufacturer) is trying to make a copy of a biopharmaceutical, one could also consider this a process or manufacturing change in making the *same* biopharmaceutical. In this case, however, the *change* in the process is in the *company* that is making the biopharmaceutical. As a result, in this comparability exercise (which would be performed by the biosimilar manufacturer, to show regulators that both forms of the same biopharmaceutical are in fact comparable or highly similar) the innovator's biopharmaceutical would be the prechange biopharmaceutical and the biopharmaceutical copy of the innovator drug (biosimilar) would be the postchange biopharmaceutical. In principle this type of comparability could be conducted using the same science and regulatory concept as that employed in any normal comparability study. In fact this is the approach taken by the European Medicines Agency (EMA) in assessing biosimilarity (Berkowitz et al., 2012; McCamish and Woollett, 2013). As a result, the process of assessing biosimilarity can also be regarded as a comparability study. Nevertheless, a critically important difference exists when a different manufacturer (biosimilar manufacturer) attempts to make a copy of an innovator's biopharmaceutical. The important difference is that all of the information concerning the prechange biopharmaceutical (the innovator's biopharmaceutical) exists *external* to the biosimilar manufacturer making the postchange biopharmaceutical (biosimilar), which differs from the situation when an innovator conducts a comparability study, where all the information about the prechange and postchange forms of the same biopharmaceutical exist *internal* within the same drug company (the innovator). In such a situation one could consider the comparability conducted by an innovator as an *internal comparability* and comparability conducted by a biosimilar manufacturer as an *external comparability*. However, the ramifications of this difference are unique and important (Declerck, 2016), leading the FDA to see comparability or internal comparability as uniquely different from external comparability, and thus refer to the former as a comparability study, process or exercise and the latter as abiosimilarity study, process or exercise, see Figure 2.3.

In considering the concepts of comparability and biosimilarity, it is helpful to note a rare situation that arose in the early 1990s that in some ways links or bridges these two concepts. This rare situation came about when Rentschler in Germany contracted or partnered with another company, Biogen, in the US to develop and produce a potential innovative biopharmaceutical called interferon beta-1a (IFNβ)

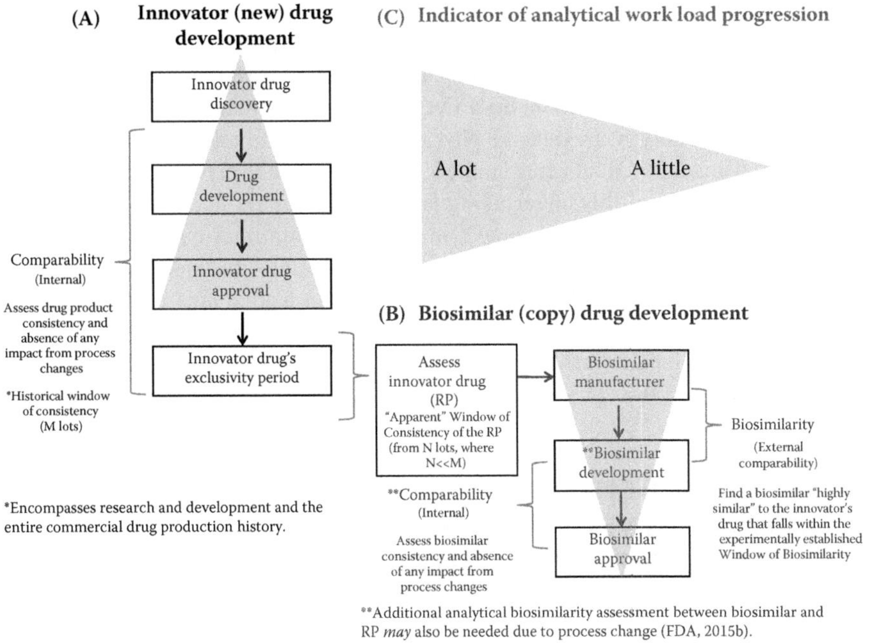

FIGURE 2.3 Comparability and biosimilarity, temporal intensity of analytical workload, and the relationship of the innovator's lot production history relative to the assessment of this lot production history by a biosimilar manufacturer when comparing (A), the development of a new (innovative) biopharmaceutical, to (B), the development of its corresponding biosimilar. (C) Triangle on the top right side of this figure shows how this triangle is used to indicate the critical difference in the time course of the intensity of the analytical physicochemical workload between developing an innovative biopharmaceutical in "A," where the analytical physicochemical workload is much lower at the beginning of drug development, but increases with time relative to developing a biosimilar in "B," where the analytical physicochemical workload is much greater in the beginning, but decreases with time.

to help patients with multiple sclerosis. In 1984, both companies formed a company called Bioferon in Germany to undertake this task. In this partnership, Rentschler had the responsibility to develop the cell line for the production and manufacture of this biopharmaceutical, while Biogen handled the clinical development side. During the pivotal phase III clinical trials, the Rentschler–Biogen partnership failed. As a result, Bioferon went into bankruptcy, with Rentschler retaining the rights to the cell clone (or cell line) used to produce IFNβ and the production process, while Biogen retained the rights to the clinical data.

Fortunately, Biogen had enough of the Rentschler clinical interferon material to continue and finish the pivotal phase III clinical trials. With favorable clinical results, Biogen was left in the precarious position of having an actual biopharmaceutical that it could potentially get approved, but it could not use the cell line developed by Rentschler to generate the IFNβ material used in the pivotal phase III clinical trials to commercialize it. Given this situation, Biogen took the bold step of developing its own cell line in an attempt to make an appropriate copy of the IFNβ drug used in the pivotal phase III clinical trials. By so doing, Biogen hoped it could make use of the prior phase III clinical trial data in filing the IFNβ drug copy for approval without the need to conduct much additional clinical work. In the end, Biogen was successful in showing and convincing FDA regulators that their copy of IFNβ was adequately comparable (via extensive physicochemical and biological work) to the IFNβ material made by Rentschler and used in the original pivotal clinical phase III trials. The success of this undertaking led to the landmark approval of Biogen's first biopharmaceutical, IFNβ (Avonex), in 1996 with minimal additional clinical work, which ushered in the concept of comparability into the biopharmaceutical industry (Blaich et al., 2007). Today this achievement also stands as a milestone event in fostering and bridging the ideas of comparability (or internal comparability) with the concept of biosimilarity (or external comparability) (Kozlowski et al., 2011; Woodcock et al., 2007).

2.3 THE UNIQUE CHALLENGES IN MAKING BIOSIMILARS VERSUS GENERICS FROM A PHYSICOCHEMICAL PERSPECTIVE

The introduction to this chapter presented the main features of biopharmaceuticals (including biosimilars) that make them different from pharmaceuticals (including generics). These differences prevent a biopharmaceutical manufacturer from making its biopharmaceutical identical and the task of making it adequately comparable on a lot-to-lot basis very difficult (Geigert, 2004). As a consequence, it creates an even greater challenge for the biosimilar manufacturer trying to make a biosimilar of an innovator's biopharmaceutical. This is because the underlying additional fine detail of physicochemical structural heterogeneity of a biopharmaceutical, which a biosimilar manufacturer is trying to copy, goes beyond the commonly known main structural element of the innovator's biopharmaceutical—the linear sequential ordering of the amino acids in its polypeptide chain(s). It is this additional fine detail of physicochemical structural information and its heterogeneous distribution among the biopharmaceutical molecules in a given lot and the variation of this distribution on a lot-to-lot basis (which is known only to the innovator and to the regulators who evaluated and approved the innovator's biopharmaceutical) that makes the process of producing a biosimilar even more challenging than making a generic of a pharmaceutical.

Indeed, in the case of making a generic drug, one only needs to know the *unique documented and publicly known chemical structure* of the pharmaceutical (e.g., aspirin). Once that structure is known, one can then simply design the chemical synthetic route to make the *identical* molecule (see Figure 2.2A). In contrast, in the case of attempting to make a biosimilar of a biopharmaceutical, knowing the chemical structure of an innovator's biopharmaceutical's amino acid sequence is unfortunately not its complete final structure. From a physicochemical point of view, a biopharmaceutical's structure also involves a collection of additional chemical modifications to its polypeptide chain(s) (involving primary or covalent bonds) and physical structural changes to its folded state (resulting from chemical modifications or/and physical changes that alter the noncovalent secondary bonds responsible for the 3D folding and stabilization of a biopharmaceutical). These additional physicochemical changes, along with the sequence of amino acids, constitute a biopharmaceutical's final *total* structure. However, the presence of these additional physicochemical structural changes does not necessarily occur in every molecule of a biopharmaceutical. In fact, each structural change is typically distributed independently among the biopharmaceutical molecules, such that the percentage of biopharmaceutical molecules with each of these additional changes can vary greatly, ranging from every biopharmaceutical molecule having all of the physicochemical changes observed in a biopharmaceutical sample to some biopharmaceutical molecules having only a partial fraction of the total observed changes and others having none of the observed changes.

In addition, a number of chemical structure modifications observed on biopharmaceuticals can occur at multiple sites on a given biopharmaceutical molecule, offering another route for distributing these structural alterations. Coupling all of this with a certain level of variability in these distributions of structural changes in biopharmaceutical molecules on a lot-to-lot basis, and with the multitude of different types of changes that can occur to a biopharmaceutical, gives rise to many different subpopulations or "drug-product variants" (as simply illustrated in Figure 2.4 A through C). The end result is that a biopharmaceutical can be considered to be a heterogeneous mixture of protein molecules (or proteoforms; Smith and Kelleher, 2013) in comparison to a pharmaceutical, as shown experimentally in Figure 2.4D using a modern separation technology called capillary electrophoresis (CE). In many cases, this heterogeneity of biopharmaceuticals is frequently referred to as microheterogeneity owing to the relatively small size of these additional chemical changes or the area of a biopharmaceutical that undergoes a physicochemical change, especially in comparison to the entire size of the biopharmaceutical molecule.

This heterogeneity or wide collection of drug-product variant forms of a biopharmaceutical can be broken into two general major classes called *drug-product-related substances* (which are comparable to the main form of the drug in terms of potency and safety) and *drug-product-related impurities* (which are not comparable to the main form of the drug in terms of potency and safety). It is thus this additional fine-detail structural complexity and its variability on a lot-to-lot basis (which gives rise to the structural heterogeneity of an innovator's biopharmaceutical) that a biosimilar manufacturer must *uncover*, and then *duplicate* and *control*, to successfully develop a biosimilar.

2.4 POSTTRANSLATIONAL MODIFICATIONS

In the previous section, the discussion focused on the additional structural changes that can occur to the polypeptide chain(s) of a biopharmaceutical during its production. Most of these changes are due to covalent (chemical or primary structural) modifications that occur *in vivo* (inside the cell) after the polypeptide chain is synthesized and released from the ribosome, and they are referred to as posttranslational modifications (PTMs). It should be noted, however, that some of these covalent modifications can actually occur while the polypeptide chain is still attached to the

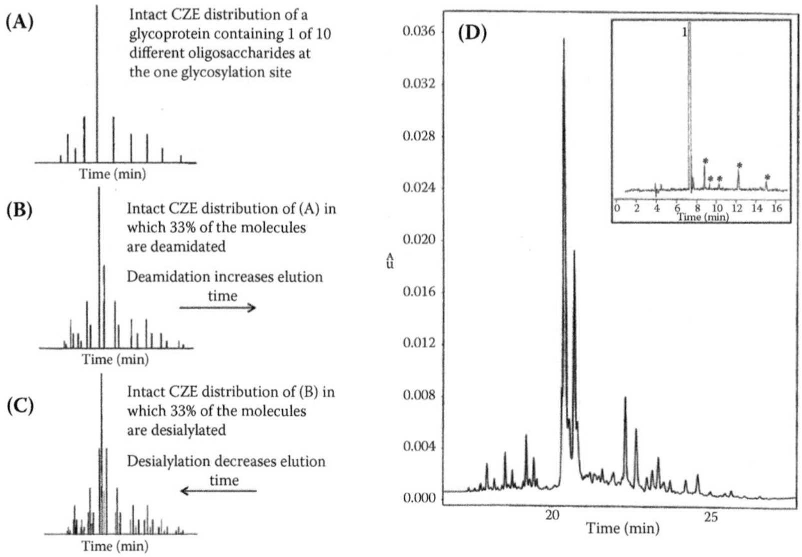

FIGURE 2.4 How PTMs greatly increase the heterogeneity of biopharmaceuticals: (A) a hypothetical capillary zone electrophoresis (CZE) electropherogram showing the resulting separation of the different (variant) forms of an intact glycosylated biopharmaceutical that has only one glycosylation site that can contain 1 of 10 different fully sialylated oligosaccharides. These different variant forms of the intact biopharmaceutical are called glycoforms. For illustration purposes, each glycoform is shown simply as the spike peak whose peak height corresponds to the relative amount of that glycoform that is present. As a result, the complete set of peaks in this electropherogram corresponds to the distribution of glycoforms that is present in the biopharmaceutical sample injected into the capillary. (B) The same CZE separation of the glycosylated biopharmaceutical shown in part "A," but in this case 33% of each glycoform has the same asparagine amino acid deamidated, which causes a reduction in the mobility of each glycoform containing the deamidation. (C) The same CZE separation of the glycosylated and deamidated biopharmaceutical shown in part "B," but in this case 33% of each glycoform has incurred some form of desialylation, which causes an increase in the mobility of each glycoform that has been desialylated. (D) A real CZE electropherogram obtained using UV detection of an intact biopharmaceutical, IFNβ (which has only one glycosylation site, whose space-filling structure is shown in Figure 2B) revealing its underlying microheterogeneity [of about 80 peaks in this case, while more recent CZE-MS work on this biopharmaceutical has revealed 138 peaks (Bush et al., 2016)] resulting from this biopharmaceutical's one glycosylation site (which is occupied by a wide range of different sialylated oligosaccharide structures), deamidation, and other PTMs. The insert electropherogram shown in part "D" corresponds to the capillary electrophoretic purity assessment of the pharmaceutical naproxen indicated by the off-scale main peak labeled (1), which is separated from several very minor impurity peaks labeled (*) that are present in this sample. Note the simplicity of the pharmaceutical's (naproxen) electropherogram relative to the biopharmaceutical's (IFNβ) electropherogram. (Figure D is reprinted from Berkowitz SA et al. 2005. Rapid quantitative capillary zone electrophoresis method for monitoring the micro-heterogeneity of an intact recombinant glycoprotein. *Journal of Chromatography A* 1079(1–2), 254–265. With permission from Elsevier with minor modifications; The insert figure was reprinted from Grossman PD, Colburn JC. 1992. *Capillary Electrophoresis: Theory and Practice*, 331–345. Academic Press, San Diego, CA. With permission from Elsevier with minor modifications.)

ribosome and is being synthesized (or translated). In this latter case, the chemical modifications are referred to as cotranslational modifications (Fedorov and Baldwin, 1997). However, for the purpose of this chapter, we will simply consider all types of chemical changes to a biopharmaceutical's polypeptide chain(s) inside the cell (whether they occur during the synthesis or after the complete synthesis of that polypeptide) and even outside the cell as PTMs.

Most of the *in vivo* covalent PTMs found on a biopharmaceutical result from enzymatic reactions that usually add a simple chemical group to different amino acid side chains in the polypeptide chain (e.g., phosphate, sulfate; see Figure 2.5A). Since the diversity of different simple chemical groups that can be added is rather large (Walsh, 2006a; Walsh and Jefferis, 2006; Walsh et al., 2005), this is a significant source for generating heterogeneous biopharmaceuticals. In some cases, however, the type of chemical group that is added constitutes a basic collection of somewhat similar building blocks (e.g., monosaccharides) that can be covalently strung together in different ways to generate their own unique source of complexity. This type of PTM is exemplified by a process called glycosylation where a wide range of uniquely linked monosaccharides can give rise to structures called oligosaccharides (which corresponds to a collection of typically 3–9 monosaccharides chemical linked in varying complex configurations) that can be found attached to a given site(s) on the polypeptide chain(s) of a biopharmaceutical during its production inside a cell (Kyte, 1995; see Figure 2.5B and Section 2.6.1.2 for further discussion on this topic). In other cases, covalent PTMs can involve the cleavage of the polypeptide (Walsh, 2006b; see Figure 2.5C) or the formation and cleavage of intrachain disulfide bonds involving two cysteine amino acids within the same polypeptide chain (see Figure 2.5D) or interchain disulfide bonds also involving two cysteine amino acids located in a different polypeptide chain of the biopharmaceutical. In the case of forming/cleaving disulfide bonds, more complex situations can arise that involve a process called disulfide scrambling (Lu and May, 2012; Wang et al., 2011), as illustrated in Figure 2.5D where the disulfide scrambling between two intrachain disulfide bonds in a biopharmaceutical form two different intrachain disulfide bonds (see Section 2.6.1.3 for further discussion of this topic).

A critical feature of these covalent PTMs is that they can be accompanied by changes in the HOS and/or surface properties of the biopharmaceutical. Such changes can serve important functional roles in controlling the biological activity and therefore the therapeutic activity of the biopharmaceutical via interactions that the biopharmaceutical will have with other biological molecules when injected into a patient (see Chapter 7 in this book). In other cases, however, PTMs simply constitute a form of degradation (especially those PTMs that occur outside the cell). In this latter form of PTMs, the resulting changes in the HOS and/or surface properties of the biopharmaceutical serve no useful function. More importantly such PTMs could cause adverse effects (e.g., aggregation), which can induce life-threatening immunogenicity issues (Filipe et al., 2010; Rosenberg, 2006) or alter cellular function that can lead to fatal disease states (Bucciantini et al., 2002; Dobson, 2001) (also see Section 2.9).

Once a biopharmaceutical makes its way outside the cell, it is still susceptible to chemical (covalent) and physical (noncovalent) modifications. These extracellular or *in vitro* PTMs correspond to a range of degradation processes that are predominately

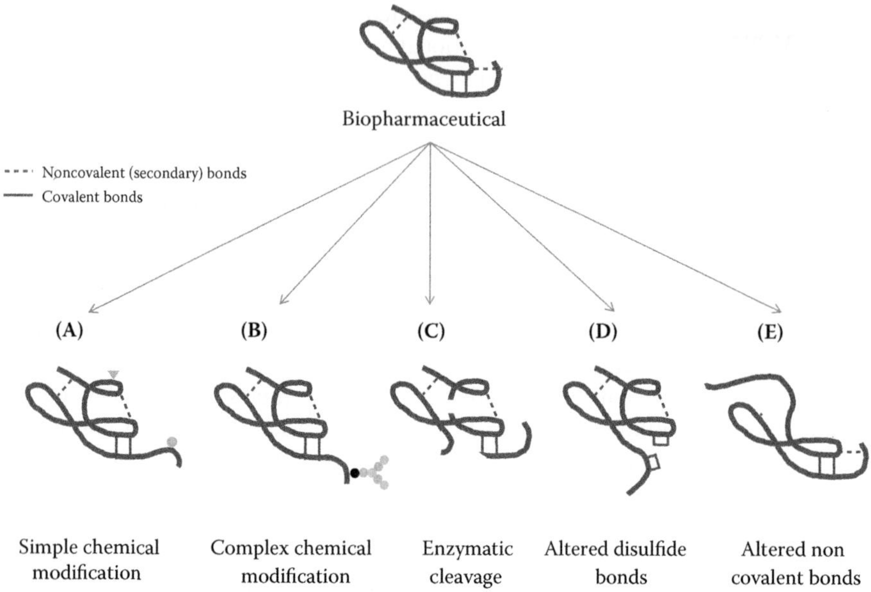

FIGURE 2.5 Common forms of PTMs found in biopharmaceuticals that can change a biopharmaceutical's HOS and/or surface properties: (A) simple covalent addition of a small chemical group, frequently via the alteration of the chemical side chain of an amino acid (e.g., phosphate, deamidation, indicated in the figure by a small triangle and sphere), (B) the covalent addition of a complex collection of covalent linked similar chemical building blocks to an amino acid's chemical side chain (e.g., glycosylation as indicated here by the "Y"-shaped array of small spheres), (C) covalent cleavage of the polypeptide chain, (D) altered disulfide bonds, which in the case shown involves intrachain disulfide scrambling, and (E) breaking of noncovalent secondary bonds.

due to environmental conditions and/or the presence of reactive process-related impurities (e.g., leachables from an array of container surfaces) that the biopharmaceutical is exposed to during cellular production in a bioreactor and/or during its purification, formulation, filling, and storage (which in some cases may still be present in the final biopharmaceutical product, resulting in the biopharmaceutical product's degradation over time or possibly to artifacts during some forms of analytical analysis). In most cases, the chemical modifications that occur in these situations are nonenzymatic in nature, although enzymatic degradation reactions due to contamination (e.g., host cell proteins), especially during the earlier stages of a biopharmaceutical's purification, are possible.

In addition to the above-mentioned chemical PTMs, biopharmaceuticals can also undergo *in vivo* and *in vitro* noncovalent (physical) PTMs (see Figure 2.5E). What is unique about physical PTMs is that, unlike chemical PTMs, no change occurs to the primary structure or chemical composition of biopharmaceuticals. As a result, physical PTMs are virtually invisible to all of the common approaches used for assessing the presence of chemical (covalent) PTMs, which frequently involve MS as the key detection tool (see Section 2.6.1.1). Consequently physical PTMs are a real challenge to detect since there is no change in the biopharmaceutical's MW. Changes that do occur in a biopharmaceutical, as a result of physical PTMs, may typically involve just a small portion of a biopharmaceutical's large collection of noncovalent weak secondary bonds, the dominant element in a biopharmaceutical that is responsible for its HOS. Such changes can arise from a range of different conditions that a biopharmaceutical experiences during its production that can physically stress the biopharmaceutical altering or breaking some of its weak secondary bonds (e.g., low pH viral inactivation). When the stress is removed, these weak secondary bonds may not be regenerated properly, resulting in small or potentially even large permanent change(s) in a given biopharmaceutical's HOS and causing changes also in the surface properties of these drugs.

Collectively, the total collection of all the covalent and noncovalent PTMs found on a given biopharmaceutical, their distribution among drug molecules, and variability on a lot-to-lot basis that a biopharmaceutical can experience is the key factor for their high heterogeneity. The unique features of this structural heterogeneity and its associated variability, as evaluated by the innovator over many years of work, defines what will be referred in this chapter as the physicochemical "window of consistency" of an innovator's biopharmaceutical (see Figure 2.3A). This experimental window comprises a large collection of physicochemical characteristics associated with CQAs of a biopharmaceutical, which its innovator used in collaboration with regulators to define the complex boundaries of limits or specifications that each innovator's biopharmaceutical lot must fall within in order to be released for commercial sale. It's this window of consistency that a biosimilar manufacturer must try to successfully uncover about the biopharmaceutical it is trying to copy through its own experimental work on the innovator's commercial biopharmaceutical. Once this task is achieved, the biosimilar manufacturer will use this window of consistency to define the window of biosimilarity that its biosimilar will need to fall within in order to achieve the necessary biosimilarity to the innovator's biopharmaceutical to obtain regulatory approval. The experimental physicochemical window of consistency assessed by the biosimilar manufacturer should ideally be identical to the physicochemical window of consistency that the innovator established and filed with regulators to obtain their drug's approval. In reality, however, small differences may exist, which will be explained in more detail below in various subheadings listed under Section 2.5.1, making this the window of consistency assessed by a biosimilar manufacturer as an *apparent* window of consistency.

2.5 THE PHYSICOCHEMICAL "WINDOW OF CONSISTENCY" OF THE REFERENCE PRODUCT (RP) AND ASSESSING BIOSIMILARITY

The process of assessing a biopharmaceutical's physicochemical window of consistency in building the totality of the evidence data package for a biosimilar is the first major step a biosimilar manufacturer will need to undertake in initiating its biosimilar program (see Figure 2.3B) (Holzmann et al., 2016). It involves the critical analytical process of establishing the fine detail of biochemical and biophysical PTMs present in the innovator's biopharmaceutical or reference product (RP; Federal Register, 2009) and the range of variability of these PTMs on a lot-to-lot basis. Once assessed, the biosimilar manufacturer will then need to find the best cell line in combination with appropriate bioreactor growth conditions that will produce a biosimilar that adequately matches the RP's entire physicochemical structure and its structural heterogeneity when both the biosimilar and RP are analytically characterized. This package of analytical physicochemical information will include common analytical release testing data [required for all biopharmaceuticals submitted for regulatory approval, e.g., size-exclusion chromatography (SEC), sodium dodecyl sulfate polyacrylamide gel electrophoresis (SDS-PAGE)], as well as additional characterization testing. In many cases, this extended characterization work will involve the use of advanced state-of-the-art analytical methods, potentially capable of delivering detailed fingerprint physicochemical structural information that was very likely not even available to the innovator when their drug was first approved. In addition, the redundant assessment of some key physicochemical parameters will also very likely be needed using orthogonal analytical methods to assure confidence in the resulting data.

Although the fine-detail structural information on an innovator's biopharmaceutical readily exists with the innovator and to a large extent with the regulators who reviewed the chemistry, manufacturing, and control (CMC) section of the biopharmaceutical's biological licensing application (BLA) and gave their approval to the innovator to allow its commercial marketing, this information (as previously noted) is not publicly available, nor are the regulators legally allowed to reveal it. The only route for a biosimilar manufacturer to obtain this information is to access it through its own analytical work that it conducts on RP material available commercially during the time period in which a biosimilar manufacturer is committed to developing its biosimilar. Hence, a biosimilar company must go out on the open market and purchase an appropriate amount of innovator drug material from a sufficient number of different innovator lots, potentially resulting in a significant cost to the biosimilar company (amounting to millions of dollars to cover biological, physicochemical, and limited clinical testing). These lots should preferably span the widest range of manufacturing time in terms of years and lot age as possible to adequately assess the historical and intrinsic manufacturing variability of the RP. Indeed, such variability will likely include the occasional manufacturing process changes made by the innovator (Crommelin et al., 2015; Schiestl et al., 2011; Schneider, 2013; Tebbey et al., 2015) and possibly unknown small deviations in the manufacturing process of the RP that can accumulate over time to give rise to what is called *drift* in some physicochemical attributes of the RP (Ramanan and Grampp, 2014; Ventola, 2013). In many of these measurements, head-to-head comparisons between representative lots of the biosimilar and RP should be conducted at the same time on the same instrument. Such head-to-head comparisons minimize the analytical method's day-to-day variability and even instrument to instrument variability (Ghirlando et al., 2013; Zhao et al., 2015) that could obscure and even bias the analytical data generated.

Once the physicochemical window of consistency for the RP is established (which in principle should be considered an "apparent" physicochemical window of consistency, owing to the inherent limitations a biosimilar manufacturer will face in securing the complete variable history of the RP; see Sections 2.5.1.3 through 2.5.1.5), it will then be used to develop the physicochemical window of biosimilarity. With this window of biosimilarity in hand, the biosimilar manufacturer will use it to find the best cell line and experimental growth conditions that will produce a biosimilar that will best match the known physicochemical structural attributes of the RP (see Figure 2.3B).

An important element in assessing an RP's window of consistency and in establishing the window of biosimilarity of a biosimilar is being able to achieve a clear understanding of the challenges and uncertainties via statistical assessments associated with the physicochemical measurements used to generate data to set the boundaries of these windows (Chow, 2015). Without such an assessment, our ability to conduct meaningful biosimilarity evaluations would not be feasible. Consequently, several topics related to the issues that impact the overall uncertainty associated with assessing and establishing these windows will be discussed below under Section 2.5.1 and its subheadings.

2.5.1 Challenges in Assessing the Physicochemical Window of Consistency of an RP in Establishing Biosimilarity

Assessing the physicochemical window of consistency of an RP and using it to develop the physicochemical window of biosimilarity to assess a biosimilar's biosimilarity to its RP would appear to be a fairly straightforward task. Just conduct the same array of appropriate biochemical and biophysical measurements on a number of different lots of the RP and biosimilar and compare the data! However, several important issues that can challenge this process should be considered and evaluated to achieve an accurate and meaningful biosimilar assessment. They include the following:

1. An understanding of each physicochemical method's limit of detection and quantitation used in a biosimilarity assessment and the factors that influence these limits on order to detect and remove any potentially significant bias between the RP and biosimilar owing to a difference(s) in their matrix that includes the sample formulation differences (e.g., buffers, excipients, pH) and residual process impurities differences (e.g., due to the use of different sources of raw materials or different contact surfaces due to the use of different container closures, holding vessels, etc.).
2. An understanding of the potential impact of implementing sample preparation (handling and processing) steps in order to be able to remove bias or interference effects in conducting analytical measurements using a given physicochemical method.
3. An understanding of the impact in only being able to acquire a certain limiting fraction of the total number of (historical) different lots of the RP produced by its innovator and the impact of RP lot age.
4. An understanding of the impact concerning the possibility that a number of the different uniquely labeled RP lots acquired and analyzed by a biosimilar manufacturer were in fact derived from the same bioreactor run rather than from a different bioreactor run.

5. An understanding of the regulatory constraints concerning the source of RP that can be used in conducting biosimilarity comparisons within a given regulatory jurisdiction and its potential impact.

2.5.1.1 Potential Analytical Method Bias Effects due to Sample Matrix Differences between the RP and Its Biosimilar

In applying an analytical method to measure any physicochemical attribute [parameter *X*, e.g., the sedimentation coefficient, or collection of "*i*" data points (x_i) that gives rise to a graphical pattern or plot of data points; e.g., circular dichroism (CD) spectrum] that characterizes a biopharmaceutical, two basic and important statistical values are extracted as a result of making measurements on a number of different aliquots (*n*) of the same sample to assess the parameter or plot. These statistical values include the *mean* value of the parameter, $<X>$, or collection of "*i*" mean values, $<x_i>$, involved in generating a given plot, and their associated *uncertainty* (error or variability), σ, or "*i*" uncertainties, σ_i, respectively [which are obtained from standard deviation, SD, calculations computed from a limited population of individual experimental measurements (Beers, 1957; Mandel, 1964)]. For chemical or biochemical (primary structure) analysis methods, the value of these means and their associated uncertainties are likely to be invariant to the nature of a biopharmaceutical's sample matrix. However, for analytical methods that assess or depend on the HOS of a biopharmaceutical (e.g., biophysical and functional analysis methods), a sample matrix difference between the RP and the biosimilar typically have significant impact on the measured mean value of a parameter or plot and may even impact their associated uncertainties (Holzmann et al., 2016; Panjwani et al., 2010) (see Section 2.5.1.1.2 for further discussion of this topic).

2.5.1.1.1 Detecting the Possible Effects of Sample Matrix Differences in Biochemical Methods Used for Assessing Primary Structure

Although differences in a biopharmaceutical sample's matrix typically do not present a major problem of introducing sample bias in biochemical methods used in primary structure characterization, some of these analysis methods could. Consequently, an observed difference(s) in the biochemical primary structure between the biosimilar and RP may not really be due to an actual true chemical difference between the active pharmaceutical ingredient (API, the protein molecule) in the RP and biosimilar. Rather, the difference recorded could be due to some difference between the matrix (formulation or process-related impurities that are present) of the biopharmaceutical samples being compared that influences the data output from a biochemical method that leads to an apparent difference between the RP and biosimilar. To assess this potential problem, especially when samples being compared are known to be in different formulations, the approach discussed below, as well as in Figure 2.6, might prove helpful.

An appropriate amount of sample (which will need to be precalculated) should be taken from a container (e.g., vial, syringe) of the RP and biosimilar and in each case split into two equal parts (note: in some cases, it may be necessary to pool the contents from more than one container from the same lot of RP material and similarly from more than one container from the same lot of biosimilar material if a

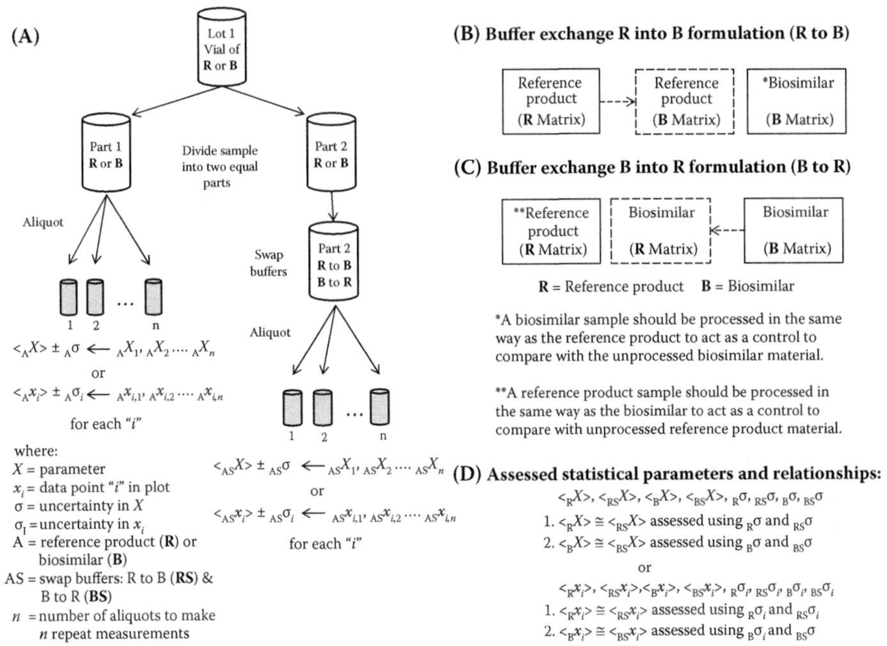

FIGURE 2.6 A schematic presentation of the sample processing procedure outlined in Section 2.5.1.1.1 for generating test samples to detect the presence of matrix bias effects when conducting physicochemical measurements in assessing biosimilarity: (A) Illustrates the fractionation of the reference product, RP or R, and biosimilar, B, into part 1, which is than aliquoted, and part 2, which is first buffer exchanged (as indicated in "B" and "C" below) and then aliquoted in the same manner as part 1, (B) shows the specific buffer exchange of the RP into the biosimilar formulation buffer (with the recommended processing of the biosimilar in the same way as the RP to act as a control to help detect sample processing and handling effects), (C) shows the specific buffer exchange of the biosimilar into the RP formulation buffer (with the recommended processing of the RP in the same way as the biosimilar to act as a control to help detect processing and sample handling effects), and (D) summarizes the key statistical parameters and relationships assessed to make overall assessment.

single container does not provide enough material to carry out the experiment to be described). One part of the RP sample is then aliquoted into small "*n*" size aliquots (note: "*n*" should at least be equal or greater than three), as shown in Figure 2.6A, and analyzed to obtain $\langle {}_{R}X \rangle$ or "*i*" $\langle {}_{R}x_i \rangle$ values along with their associated ${}_{R}\sigma$ or "*i*" ${}_{R}\sigma_i$ values (where "R" refers to the RP). The same procedure and analysis is then also carried out on the biosimilar sample to generate $\langle {}_{B}X \rangle$, ${}_{B}\sigma$ or $\langle {}_{B}x_i \rangle$, ${}_{B}\sigma_i$ values (where B refers to the biosimilar). The second part of the RP sample should then be buffered exchanged into the biosimilar's formulation, as indicated in Figure 2.6B, and the second part of the biosimilar sample is buffered exchanged into RP's formulation, as indicated in Figure 2.6C. These two buffered exchanged (or swapped formulation) samples should then be aliquoted and analyzed, as done for the first part of RP and biosimilar samples, to obtain the corresponding $\langle X \rangle$ or "*i*" $\langle x_i \rangle$ values and their associated σ and "i" σ_i values for their swapped formulations (symbolized as $\langle {}_{RS}X \rangle$, ${}_{RS}\sigma$ or $\langle {}_{RS}x_i \rangle$, ${}_{RS}\sigma_i$ and $\langle {}_{BS}X \rangle$, ${}_{BS}\sigma$ or $\langle {}_{BS}x_i \rangle$, ${}_{BS}\sigma_i$, where RS refers to the RP sample swapped or exchanged into the biosimilar formulation and correspondingly BS refers to a biosimilar sample swapped or exchanged into the RP formulation).

Using the above experimental calculated information, if $\langle {}_{R}X \rangle \approx \langle {}_{RS}X \rangle$ or all "i" $\langle {}_{R}x_i \rangle \approx \langle {}_{RS}x_i \rangle$ are found to be true via statistical testing (e.g., t-test) using an appropriate statistical criteria (e.g., 95% or 99% confidence limit, CL, computed from the corresponding experimentally measured uncertainties ${}_{R}\sigma$ and ${}_{RS}\sigma$ or "i" ${}_{R}\sigma_i$ and ${}_{RS}\sigma_i$ values that go with each appropriate mean comparison) and the same situation holds true for the biosimilar [$\langle {}_{B}X \rangle \approx \langle {}_{BS}X \rangle$ or all "i" $\langle {}_{B}x_i \rangle \approx \langle {}_{BS}x_i \rangle$], then there is a high level of confidence that method bias due to any matrix differences between the RP and biosimilar is absent. However, if any of these effective equalities between the two different formulations for the RP or the biosimilar is determined to be statistically not true, a possible method bias may exist due to differences in the matrix of the samples being compared. As a result, appropriate steps will need to be taken to overcome this bias in order to achieve a meaningful biosimilarity assessment between the RP and biosimilar.

2.5.1.1.2 *Detecting the Possible Effects of Sample Matrix Differences in Biophysical Methods Used for Assessing HOS*

The approach outlined in Section 2.5.1.1.1 can also be applied to biophysical methods to assess matrix bias effects between the RP and its biosimilar. However, as already mentioned in Section 2.5.1.1, biophysical measurements display a high sensitivity to a sample's matrix. This is due to the dominant role that the weak secondary chemical bonds play in establishing and maintaining the HOS of biopharmaceuticals and the high sensitivity of these bonds to be altered simply by changing their chemical and physical environment (e.g., buffer, excipients, and pH) of the biopharmaceutical. Hence, in the biophysical assessment of a biosimilar to its corresponding RP, any known difference in the formulation between these two materials will likely lead to a difference in the biophysical measurements between these samples. Consequently, known formulation differences between samples need to be removed in order to conduct valid biophysical biosimilarity assessments. Needless to say, any steps taken to achieve this must be carefully assessed to make sure no sample bias is introduced; see the next section.

2.5.1.2 The Impact of Sample Handling and Processing Steps in Conducting Physicochemical Measurements

As noted previously, for a biosimilar manufacturer the only source of RP material to use to experimentally assess the window of consistency of an RP are the various commercial innovator lots of the RP that are available on the open market. Such material, however, may actually be compounded into a formulation matrix that is different in composition from that used for the biosimilar (e.g., due to patent issues surrounding the RP formulation). This difference in formulation may require the introduction of sample preparation (handling and processing) steps to make sure that all samples are in the same formulation to avoid issues concerning matrix bias effects, especially in the case for biophysical methods, and remove method-specific interfering excipients. In general, such sample-handling and processing steps can create their own potential problems, especially when not applied in an identical manner to all of the samples being compared. As a result, any asymmetry in the way the RP and biosimilar are handled and processed can lead to erroneous conclusions concerning the presence or absence of biosimilarity and to erroneous data for setting target specification values and/or range limits for physicochemical parameters used in biosimilarity assessments. As a result, it will be important for the biosimilar manufacturer to validate any sample handling and processing used during any physicochemical measurement to assure that it does not alter the RP and/or biosimilar (Heavner et al., 2007; Panjwani et al., 2010).

2.5.1.3 Potential Impact of Investigating a Limited Number of RP Lots

A challenging characteristic of the process of assessing biosimilarity relative to the traditional exercise of (internal) comparability is due to the inability of the biosimilar manufacturer to have access to the complete historical knowledge base of information about the RP it is trying to copy (which encompasses lots made during research and development, e.g., toxicology and IND studies, and the complete commercial lot history of an innovator's biopharmaceutical). This is indicated in Figure 2.3 by the fact that the total number of all RP lots produced by an innovator, represented by *M*, is always going to be much greater relative to the total number of lots that can be secured by a biosimilar manufacturer for its characterization work on the RP, represented by *N*.

Over the time span in which an innovator develops and commercializes those *M* lots, a number of intentional changes in the production of the innovator material are very likely to have occurred (Crommelin et al., 2015; McCamish and Woollett, 2012; Schneider, 2013; Tebbey et al., 2015), along with some unintentional small level of drift in its manufacturing (Ramanan and Grampp, 2014; Ventola, 2013) process. If lot samples of the RP material obtained by a biosimilar manufacturer do not span the entire range of this variability, the physicochemical window of consistency assessed by the biosimilar manufacturer will likely be different and show a biased reduced variability in comparison to what the innovator has historically experienced, established, and employed in dealing with its regulatory filings. This is a likely outcome simply because biopharmaceutical lots typically have a commercial expiry (shelf life) of only about 2 years. Coupling this with the fact that a drug's patent protection can cover over a decade and possibly more of commercial lot production, a biosimilar company is only likely to be able to get access to a fairly limited fraction of the total number of commercial lots made (that have not expired). In addition, this limited fraction of RP lots will be dominated by lots corresponding to those made only in the latter years of production of an RP unless the biosimilar manufacturer had the foresight and willingness to start collecting samples of innovator RP lots a number of years before it actually started its biosimilar development program. Furthermore, those lots will also need to be analyzed before they expire, since RP lots analyzed after their expiration date may well raise questions as to the validity of the information they contribute to biosimilarity studies.

2.5.1.3.1 RP Lot Age

A source for introducing variability into an RP concerns its level of stability in its container closure during its approved shelf life (when stored within stated storage conditions indicated by the innovator). Although an appropriately formulated biopharmaceutical can prevent or greatly reduce the rate of its degradation so that they never reach the limits for lot rejection (defined by its window of consistency), some level of age-related degradation may occur. Unfortunately, these aging effects do not happen uniformly in all biopharmaceutical container units (e.g., vials, prefilled syringe, etc.). As a result, over time an RP lot's uniformity will diverge from its initial low degradation state when the biopharmaceutical lot was initially released, reaching higher degradation state values, but still within specification values or limits, by the time they reach expiry (typically 2 years later). Consequently, it is important that the biosimilar manufacturer sample RP lots that encompass as wide a range of an RP's shelf life (or lot age) as possible to help assess the RP's full range of variability in terms of its CQAs.

2.5.1.4 Impact of Different RP Lots Derived from the Same Bioreactor Run versus Different Bioreactor Runs

A challenging factor that may lead a biosimilar manufacturer to generate a different window of consistency, relative to that generated by the innovator, has to do with the overestimation of the actual number of uniquely different lots of RP that a biosimilar manufacturer thinks it used to characterize the RP. This arises, as illustrated in Figure 2.7A, when the number of uniquely different-labeled commercial RP lots a biosimilar manufacturer secures are actually derived from the same innovator bioreactor run (see scenario #1 in Figure 2.7A) versus the situation where each commercial RP lot is associated with a unique bioreactor run (see scenario #2 in Figure 2.7A). Since the most significant source for introducing variations into the physicochemical attributes of a biopharmaceutical (especially in terms of PTMs) arises during its biological production (Moroco and Engen, 2015), all lots generated via scenario #1 (Figure 2.7A) will show very similar values for their physicochemical attributes relative to the situation illustrated in scenario #2 (Figure 2.7A), where all lots will show a much greater variability in the value of their physicochemical attributes. This occurs due to the complex and sensitive linkage of cell growth to its physical and chemical environment, which makes the task of replicating the physical and chemical environment inside a bioreactor on a run-to-run basis very difficult. The end result is that the biological production phase of producing a biopharmaceutical is the key source for introducing variability into the biopharmaceutical product relative to the other steps associated with a biopharmaceutical's production (purification, formulation, vialing, and storage; see Figure 2.7B). Consequently, if the manufacturer of an innovative biopharmaceutical uses scenario #1 to make "m" lots of its innovative biopharmaceutical that are each uniquely labeled with a different lot identification number rather than scenario #2, then a biosimilar manufacturer analyzing more than one of those "m" uniquely labeled commercial RP lots would effectively be analyzing the *same* RP lot more than once. As a result, the mean value measured for any attribute (parameter or plot) that characterizes the RP will be falsely skewed (to some extent)

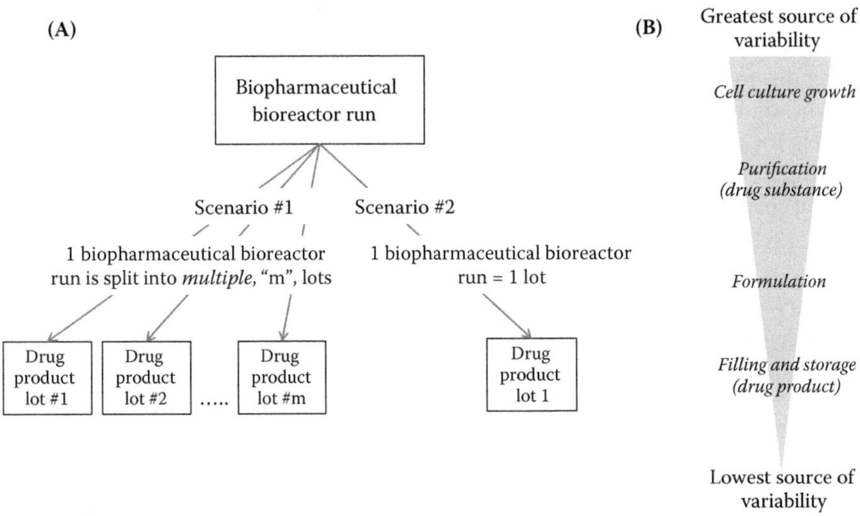

FIGURE 2.7 Understanding the potential relationships of uniquely identified biopharmaceutical drug product lots that are released by an innovator for commercial use to the bioreactor run from which they are derived. (A) Scenario #1: one large bioreactor run is used to make multiple commercial biopharmaceutical drug product lots that are each labeled by a uniquely different identification number versus scenario #2: one bioreactor run is used to only make one uniquely identified and labeled commercial biopharmaceutical drug product lot. (B) A rough rank ordering of the factors that contribute to the physicochemical variability in the production of a biopharmaceutical drug product lot, listed in order where the top factor typically introduces the most variability, while the bottom factor typically introduces the least variability.

by the inappropriate weighting (bias) created by those repeated measurements on effectively the same RP lot. In addition, the resulting measured lot-to-lot variability of any physicochemical attribute evaluated in such a situation will also be biased to a lower value.

2.5.1.5 RP Sourcing Limitations for Different Regulatory Regions

The final challenging complication in the biosimilarity process is associated with the issues surrounding the sourcing requirements (or limitations) of the RP material that can be used in a biosimilar filing in each country or geographical region that a specific regulatory agency oversees. In most cases, the RP lots that can be used by a biosimilar company must be from lots that were specifically approved in the regulatory region where the biosimilar manufacturer will be filing its biosimilar. Since innovator biopharmaceuticals are commonly made at more than one site, which only service certain specific regulatory regions, this can create problems for the biosimilar manufacturer who wants to file its biosimilar globally (Greer, 2012). In this case, the RP physicochemical data package used for one regulatory region may not be able to readily be applied to another regulatory region. However, some regulatory agencies will allow their inclusion if appropriate bridging data can be provided, e.g., FDA (2015b). As a result, this can lead to a physicochemical window of consistency for the same RP that may vary somewhat from one regulatory region to another due to slight variations in manufacturing at different sites of production and/or to the significant differences in the number of commercially available RP lots available for analysis, coupled with the resulting inherent nature of statistical variations, especially when the total RP lot population is low for one or more of these different specific regulatory regions.

2.6 ANALYTICAL PHYSICOCHEMICAL CHARACTERIZATION CAPABILITIES IN ASSESSING BIOPHARMACEUTICAL CONSISTENCY AND BIOSIMILARITY

An important point to realize when making copies of biopharmaceuticals that have or will be coming off patent protection in the next few years is that these biopharmaceuticals were developed and approved using analytical technologies that existed two to three decades ago. Since that time, an explosion of improvements in analytical instrumentation and techniques, especially in the areas of biochemical and biophysical analysis, has occurred that has greatly enhanced our ability to characterize complex biomacromolecules such as biopharmaceuticals (Berkowitz et al., 2012; Marino et al., 2015). One area that has facilitated many of these achievements has been associated with developments in digital electronics and computers (in terms of both hardware and software). In addition, advances in other scientific and technical areas (e.g., material science, higher sensitivity detectors), as well as advances in many unrelated scientific areas, coupled with the continuing ingenuity of scientists and engineers, have contributed greatly to the development of new and improved technologies and analytical instruments with ever greater analytical power. With this greater analytical capability we are now able to analyze the earlier approved commercial biopharmaceuticals in much greater detail with much greater resolution, sensitivity, precision, and accuracy (frequently in much less time and using much less material) relative to when they were first approved. As a result, when an RP is characterized by a biosimilar manufacturer today, more detailed information will likely be uncovered about that RP than what was known when it was initially approved. It must also be realized that these earlier approved biopharmaceuticals are more likely to display less consistent manufacturing in comparison to today's biopharmaceuticals (which includes biosimilars) owing to the more recent advances and improvements in today's manufacturing resulting from implementing process analytical technology (PAT) (FDA, 2004; Glassey et al., 2011; Kozlowski and Swann, 2006; Rathore et al., 2010) and quality by design (QbD) concepts (FDA, 2009b; ICH, 2005; Kozlowski and Swann, 2006; Rathore and Mhatre, 2009).

Nevertheless, even with the significant advances and improvements in our present manufacturing and analytical instrumentation, our present limited knowledge, capability, and know-how as to what exactly needs to be altered to produce a biosimilar so that it adequately displays the specific physicochemical characteristics that will make it highly similar to the RP to receive regulatory approval, is just as big a problem and challenge as the analytical assessment part. Analytics can establish the target window of biosimilarity and tell a biosimilar manufacturer what physicochemical attributes need to be changed and how close or far away those attributes are from that target window (thus guiding the biosimilar manufacturer in its attempts to find and produce a biosimilar with specific physicochemical attributes that will be highly similar to its RP). Unfortunately, however, analytics cannot tell us how to make those changes or what exactly needs to be changed in the manufacturing process so that the resulting biosimilar produced will be highly similar to the RP. That is a totally different problem and challenge.

2.6.1 Analytics at the Biochemical Level: Primary Structure

In dealing with the daunting challenges of characterizing the biochemical heterogeneity of a biopharmaceutical at the level of its primary structure, today's biopharmaceutical scientists are heavily dependent on separation technologies that are combined with various types of physicochemical detectors. The most prominent of these separation techniques include liquid chromatography (LC) (Fekete and Guillarme, 2014; Fekete et al., 2012, 2015; Haverick et al., 2014; Reusch et al., 2015a; Sandra et al., 2014) (which can separate biopharmaceuticals based on their chemical composition and spatial arrangement of those chemical groups on the biopharmaceutical's exposed surface in combination with the biopharmaceutical's overall size and shape), electrophoresis (Anderson et al., 2012; Berkowitz et al., 2005; Fekete et al., 2013; Moritz et al., 2015; Rustandi et al., 2008; Tamizi and Jouyban, 2015) (which can separate biopharmaceuticals on the basis of their net charge, given their physicochemical environment, in combination with the biopharmaceutical's overall size and shape), and MS (Kaltashov et al., 2012; Leurs et al., 2015; Zhang et al., 2009) (which separates biopharmaceuticals simply, in most cases, on the basis of their mass-to-charge ratio, m/z). Using these separation modes independently or in appropriate combinations with various detectors offers significant capabilities in revealing the underlying fingerprint heterogeneity in the physicochemical structure and properties of biopharmaceuticals. Such a situation is shown in Figure 2.4D where capillary zone electrophoresis (CZE, a mode of CE) in combination with simple UV detection was used to fractionate an intact biopharmaceutical (IFNβ) under conditions capable of maintaining the native-like structure of this biopharmaceutical to reveal its underlying microheterogeneity. The complex electropherogram shown in Figure 2.4D is due predominantly to the primary structural differences (in terms of charge) present and distributed among the biopharmaceutical molecules; however, additional separation is achieved through the contributing factors arising from a biopharmaceutical's HOS, for example, size and shape (or hydrodynamic properties).

In the case of MS, the separation and analysis capability that it can provide are tightly linked. This unique feature of MS allows it to stand on its own as an especially powerful independent analytical separation-analysis tool. However, when MS is coupled to LC or CE to add multidimensional separation capability to enhance its ability to fractionate complex mixtures, some of the most powerful separation-analysis tools feasible for characterizing and analyzing biopharmaceuticals are created (see Figure 2.8). Such a multidimensional arrangement of hyphenated separations and analysis capabilities now dominates the landscape of biopharmaceutical primary structure characterization for revealing and quantifying the underlying complex chemical heterogeneity of biopharmaceuticals (Ayoub et al., 2013; Beck et al., 2012, 2013, 2015; Chen et al., 2013; Dotz et al., 2015; Fekete et al., 2015; Gahoual et al., 2014a,b; Klepárník, 2015; Reusch et al., 2015b; Sandra et al., 2014; Xie et al., 2010).

This enhanced ability to separate a complex mixture of variant forms of a biopharmaceutical into its individual constituent components, or into far less complex mixtures of variant forms of the biopharmaceutical, can go a long way in enhancing the characterization process by providing a more detailed view of the biochemical fingerprint of the heterogeneity of these drugs to provide a more meaningful biosimilarity assessment. In some cases, even higher multidimensional separation-analysis approaches beyond LC-MS and CE-MS have been employed. Often, this is achieved by carrying out appropriate fractionation of a sample during a given separation (either manually or via automated fraction collection) to create a number of fractions that are further processed individually offline using a second different separation mode that frequently incorporate the additional unique separation and analysis capabilities of MS (Biacchi et al., 2015; Neill et al., 2015). However, preferable automated online coupling of these multiple separations using special automated coupled hardware that ends with the employment of one of several general MS analysis approaches outlined in Figure 2.8 are becoming feasible and more attractive for carrying out higher-level primary structure characterization work (Mellors et al., 2013; Sandra and Sandra, 2015; Stoll, 2015; Stoll et al., 2015; Vanhoenacker et al., 2015).

By using these separation- and multidimensional separation-analysis approaches, complex fingerprint data patterns are frequently generated (see Figure 2.9A and B). The creation of such complex fingerprint patterns, without knowing exactly what each observed peak corresponds to (initially), can in themselves allow for making useful detailed empirical biosimilarity comparisons and consistency assessments between RP lots and between biosimilar lots. Indeed, since a biosimilarity project is a multistep exercise (consisting effectively of the following steps)—(1) experimentally determine and define the RP's window of consistency, (2) find a biosimilar that will adequately match the window of consistency of the RP, (3) tweak various manufacturing conditions to maximize biosimilarity of the biosimilar to its RP and experimentally establish and document this biosimilarity, and (4) assess the consistency of the biosimilar's manufacturing process to make sure that biosimilar lots can adequately meet all the established biosimilarity and consistency criteria—these initial acquired empirical fingerprint patterns can at each of these steps help to make rapid and meaningful assessments to speed up a biosimilar's development. For example, initial empirical fingerprints may just serve as a rapid comparison template for scanning and honing in on finding a feasible collection of biosimilar candidates that could best match the same fingerprint profile observed for the RP, for example, as is commonly done using intact MS (Beck et al., 2015; Xie et al., 2010; see Figure 2.9B). In later work, these same fingerprint assessments can also serve as a key starting point for much more intense, but focused, investigational work into understanding the exact nature of any observed differences between RP lots and biosimilar lots, and eventually between RP and biosimilar lots.

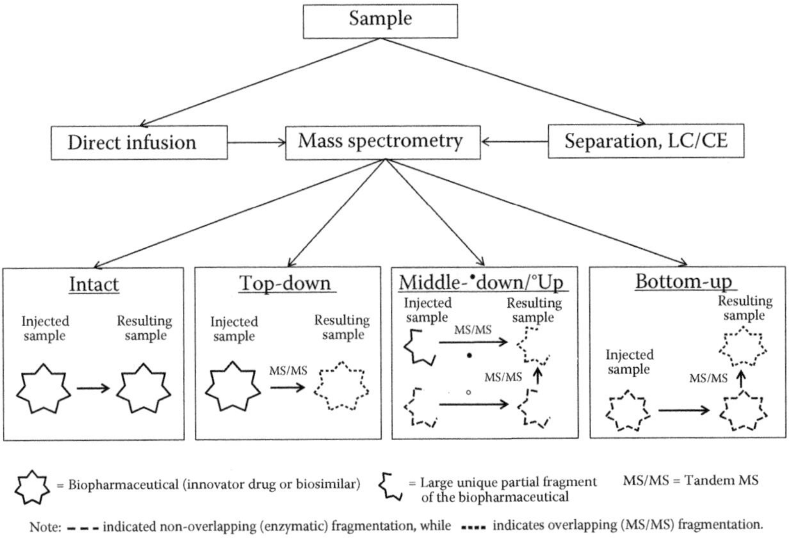

FIGURE 2.8 A pictorial scheme highlighting MS approaches commonly used to extract primary structure information about a biopharmaceutical. They include the following: *intact MS*, which involves the direct injection of the intact biopharmaceutical into the mass spectrometer, *top-down MS* (via tandem MS, MS/MS), which also involves the direct injection of the intact biopharmaceutical into the mass spectrometer where it is then randomly fragmented into overlapping peptides, *middle-down MS* (via MS/MS), which involves the injection of a specific large partial fragment of the biopharmaceutical (generated enzymatically outside the mass spectrometer) into the mass spectrometer where it is then randomly fragmented into overlapping peptide fragments, *middle-up MS*, where the specific large partial fragment of the biopharmaceutical used in middle-down MS is digested into a complex mixture of unique nonoverlapping peptides (using a specific enzyme) that is then injected into the mass spectrometer (*note*: these specific nonoverlapping peptides can be further characterized in a down mode via MS/MS), and *bottom-up MS*, where the intact biopharmaceutical is digested into a complex mixture of unique nonoverlapping peptides (using a specific enzyme) that is then injected into the mass spectrometer (*note*: these specific nonoverlapping peptides can be further characterized in a down mode via MS/MS). The implementation of these MS approaches can be carried out via the simple direct infusion of the biopharmaceutical samples into the mass spectrometer or, alternatively, via the application of a prior online separation (using LC or CE) to significantly improve the mass spectrometer's ability to better characterize a biopharmaceutical's heterogeneity. Note that all samples entering a mass spectrometer must have their noncompatible MS buffer components either adequately minimized or removed via appropriate dilution or solid-phase extraction. In the case of online LC and CE, the mobile phase or electrophoretic buffer system must be MS compatible.

2.6.1.1 MS: A Key Primary Structure Tool for Assessing Biopharmaceutical Consistency, Comparability, and Biosimilarity

Since the successful development of methods for getting proteins into mass spectrometers using the soft ionization modes [such as electrospray ionization (ESI) (Meng et al., 1988) and matrix-assisted laser desorption/ionization (MALDI) (Karas and Hillenkamp, 1988)] in the late 1980s, the role of MS in characterizing the biochemical primary structure of biopharmaceuticals has grown spectacularly (Beck et al., 2012, 2013, 2015; Dotz et al., 2015; Kaltashov et al., 2012; Leurs et al., 2015; Reusch et al., 2015b; Rogstad et al., 2016;Xie et al., 2010; Zhang et al., 2009). This growth, as we will see in Section 2.6.2.2.1, is extending its reach into areas of biophysical analysis (Beck et al., 2012, 2013, 2015; Chen et al., 2011; Leurs et al., 2015; Mo et al., 2012), which is an area of particular importance in biopharmaceutical development that needs significant improvement (Challner, 2014).

Today, the availability of different types of mass spectrometers allows this analytical separation-analysis tool to take on a broad range of characterization problems to provide a rich source of information that is unmatched by any other class of analytical instrumentation within the biopharmaceutical industry. The combined use of different mass spectrometers along with the different MS approaches, as outlined in Figure 2.8, plays an important role in primary structure elucidation by providing both complementary and redundant information. Although the latter may initially sound wasteful, this redundant information is generated via orthogonally different approaches requiring different sample processing. Hence, when agreement between these various MS methods is obtained, significant confidence in the validity of the experimentally acquired data is achieved. For example, in bottom-up reversed phase LC-MS (a common approach used routinely to assess the amino acid sequence of biopharmaceuticals; see Figure 2.9A), it is known that this form of analysis is prone to potential method artifacts (Leurs et al., 2015; Zhang et al., 2009) that could lead to erroneous conclusions. By obtaining results via top-down MS and/or middle-down MS that agrees with the bottom-up data, strong support for the validity of the bottom-up approach is generated, building significant confidence in the total data package provided by MS.

It should be pointed out that what makes MS's ability to separate and assess the different molecular species present in a sample, in terms of their *m/z* ratio (and

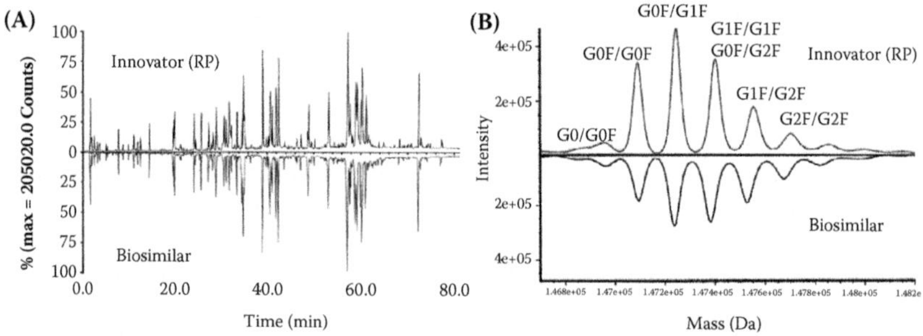

FIGURE 2.9 **(See color insert.)** Some examples of MS being used to characterize biopharmaceuticals. (A) Peptide map comparison of an innovator monoclonal antibody (RP) with its corresponding biosimilar using online LC-MS to separate and detect the eluting peptides. The resulting fingerprint elution profiles are shown in a mirror (or butterfly) format plot to visually show a high level of similarity between the RP and its corresponding biosimilar. (B) Simple intact MS comparison of an innovator monoclonal antibody (RP) to its corresponding biosimilar conducted via the direct infusion. Resulting MS spectrum displayed in a mirror plot format show some corresponding inconsistency in the higher MW variant glycoforms between the RP and the biosimilar.

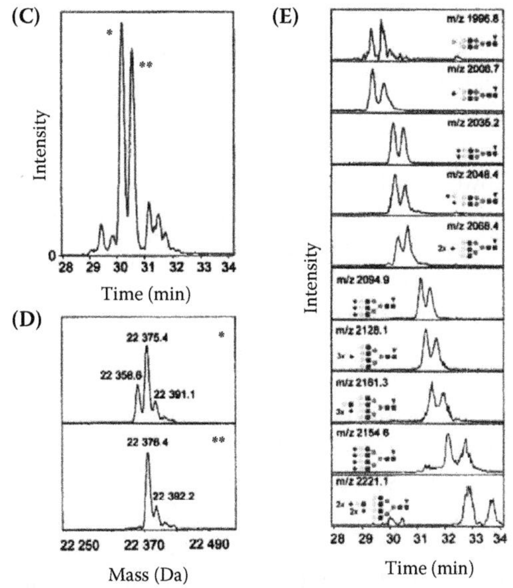

FIGURE 2.9 (CONTINUED) (See color insert.) (C) Online CZE-MS characterization (Haselberg et al., 2013) of the same intact IFNβ shown in Figure 2.4D (under similar low pH conditions, but with significantly lower resolution). The base peak electropherogram resulting from the online MS reveals the IFNβ sample's underlying heterogeneity. (D) MS analysis of the two major peaks in the resulting electropherogram labeled (*) and (**) in "C" (which are also seen in Figure 2.4D) corresponds to the same major glycoform known to be present in this biopharmaceutical (a disialylated bianternnary oligosaccharide); however, the deconvoluted mass spectra of the major species present in each of these two peaks differ from each other by a mass of 1Da, due to the presence of a deamidation site in the protein portion of the intact IFNβ material in the peak labeled (**). This deamidation increases the negative charge on this glycoform of IFNβ, resulting in a net increase in its elution time from the capillary, giving rise to the observed two different IFNβ proteoforms. (E) Overall, the underlying heterogeneity of IFNβ is due predominantly to the numerous glycoforms and their associated deamidated products present in this biopharmaceutical. This is supported by the resulting extracted ion electropherograms generated from the various fractions obtained from this CZE separation at different elution times which show the deduced different oligosaccharide structures attached to the same single glycosylation site on the nondeamidated and deamidated forms of IFNβ that were electropherotically separated. (Figure 2.9A reprinted from Xie H et al. 2010. Rapid comparison of a candidate biosimilar to an innovator monoclonal antibody with advanced liquid chromatography and mass spectrometry technologies. *MAbs* 2(4), 379–394. With permission from Taylor & Francis with minor modifications; Figure 2.9B reprinted from Ivleva VB et al. 2012. Structural comparability assessment of innovator and biosimilar Rituximab using the biopharmaceutical system solution with INIFI, Waters Application note. Available from: http://www.waters.com/webassets/cms/library/docs/720004445en.pdf. With permission from Waters Corp. with minor modifications; Figure 2.9C through E reprinted from Haselberg R et al. 2013. Low-flow sheathless capillary electrophoresis-mass spectrometry for sensitive glycoform profiling of intact pharmaceutical proteins. *Analytical Chemistry* 85(4), 2289–2296. With permission from American Chemical Society with minor modifications. Copyright 2013.)

therefore MW), so powerful is that it does this with very high accuracy due to the technique's high level of resolution that can yield MW accuracy approaching a few parts per million. In addition, this is achieved with very high sensitivity (requiring a very small sample, a few picomoles or less) and with reasonably good quantitation in terms of the relative amounts of the different ionized molecular species that are observed. However, MS has another attribute that is also very important, making MS even more powerful. This additional attribute concerns the ability of mass spectrometers to fragment specifically selected molecular ions using methods such as collision-induced dissociation (CID), electron transfer dissociation (ETD), or electron capture dissociation (ECD) (Leurs et al., 2015). The ability of MS to pick ionized species (precursor ions) with specific *m/z* values and then fragment them at limited cleavage sites, but in a random manner to create a collection of overlapping smaller fragments that can be further separated and analyzed in terms of very accurate *m/z* values and relative amounts, is referred to as tandem MS, MS/MS or MS^2 (Siuzdak, 2003), or through a somewhat similar process developed by Waters called MS^e (Plumb et al., 2006; Waters Corporation, 2011).

The application of tandem MS can also be extended to the resulting smaller product fragments created from the first fragmentation process within the same run. Such a repetitive process of applying tandem MS to prior generated fragments is referred to more generally as MS^n, multiple stage, or sequential MS, and is an additional powerful attribute of MS. Although historically the utility of MS fragmentation has played an important role in the use of MS in general structural characterization, in the case of characterizing biopharmaceuticals its full capability (along with common enzymatic fragmentation outside the mass spectrometer) is realized when the fragmentation information generated is combined with the following list of prior key knowledge:

1. The known amino acid sequence of the biopharmaceutical's polypeptide chain(s).
2. The known corresponding MWs of the amino acid building blocks.
3. The known MW changes that will occur when specific PTMs are encountered.
4. The known specific points of cleavage when using various proteases or the nature of the limited number of specific cleavage points encountered when physicochemical fragmentation procedures inside a mass spectrometer are used.

Using this prior knowledge, the biopharmaceutical scientist can create a constraining database of possible MW fragments that could be generated during these fragmentation processes. This theoretical information is then employed in analyzing the actual experimentally measured array of MS-MW data generated from various fragmentation approaches. To facilitate these analytical comparisons, appropriate computer software is used to conduct informatic searches to help reconstruct and identify the initial species that were present in the injected sample. By so doing, very accurate and knowledgeable qualitative and quantitative conclusions about the primary structure of a biopharmaceutical and its heterogeneity can be achieved. Structural information that can be extracted from these MS studies include the following:

1. A confirmation of the biopharmaceutical's amino acid sequence.
2. The detection of PTMs.
3. Identification of the specific nature of those PTMs.
4. The localization as to where those PTMs occur (what exact amino acid is modified) within the biopharmaceutical's unique sequence of amino acids.
5. The percent level to which these specific amino acids are modified.

This ability to provide such an array of information leaves little doubt that MS can provide the biopharmaceutical scientist with powerful capabilities for attacking the challenges of characterizing the chemical heterogeneity of a biopharmaceutical.

In addition, as mentioned in Section 2.6.1, in many cases the resulting raw fingerprint output from these MS experiments themselves or in combination with a prior separation method can provide an informative empirical approach for conducting a range of powerful comparisons, without even initially understanding the nature of what each peak corresponds to. Indeed, in head-to-head comparisons conducted in a biosimilarity study, the resulting MS spectrum data from the biosimilar and its corresponding RP can be plotted on the same graph but in opposite directions on the *y*-axis. Such a plot should ideally appear as a highly similar mirror or butterfly plot of the fingerprint pattern of MS peaks having peak areas, *m/z* or MW and even retention or elution time values (generated from the two samples being compared) that are highly similar to each other (see Figure 2.9A and B). Any qualitative observed nonmirrored areas would indicate a difference that should readily be recognized visually. Furthermore, such collected data could also be analyzed quantitatively, if so desired, to achieve a higher level of assessment of biosimilarity using various approaches (e.g., by simply conducting difference measurements between the two plots coupled with appropriate statistical assessment—for example, using the *t*-test with appropriate CL; see section 2.5.1.1.1) to help assess whether the observed difference constitutes a real difference as in the case of analyzing hydrogen/deuterium exchange-MS (H/DX-MS) data (Houde et al., 2011), shown in Section 2.6.2.2.1, or by using local correlation calculation as is done for the case of proton nuclear magnetic resonance (^{1}H NMR) spectral data comparisons (Amezcua and Szabo, 2013; Mei-feng et al., 2015).

2.6.1.2 Structural Analysis of Glycosylation

Of all the different types of PTMs identified on biopharmaceuticals [which total well over several hundred (Doll and Burlingame, 2015; Walsh, 2006a)], none can compare in terms of importance, complexity, and level to which they have and are being researched as glycosylation (Moremen et al., 2012; Solá et al., 2007) (see Chapter 4 for a more in-depth discussion). In this type of PTM, a collection of monosaccharides are linked together through a complex sequence of enzymatic reactions (within the cell) to form a structure called a glycan, which is covalently coupled to the polypeptide chain of a biopharmaceutical to yield a glycosylated biopharmaceutical. Although the size of the glycan that can be coupled to a protein can be very large, glycans that consist of only a relatively small number of monosaccharides (e.g., 3–9 monosaccharide units) are referred to as an oligosaccharide (as mentioned earlier in this chapter). Oligosaccharides attached to biopharmaceuticals can in general be put into two general classes that differ in terms of the specific amino acid they are linked to in the biopharmaceutical's polypeptide chain. These two classes include oligosaccharides linked to the amide nitrogen side chain of an asparagine, called *N*-linked oligosaccharides, and those linked to the hydroxyl side chain on serine or threonine, called *O*-linked oligosaccharides. In the physicochemical characterization of a glycosylated biopharmaceutical, the following are key points of interest concerning these attached oligosaccharides:

1. The monosaccharide composition of each oligosaccharide.
2. The specifics of the covalent linkage of these monosaccharides to each other and their sequential arrangement in each oligosaccharide.
3. The characterization of additional chemical modifications that might be present on a monosaccharide and their specific location on the monosaccharide structure.
4. The number and location of the oligosaccharide's glycosylation site(s) on the polypeptide chain(s) of a given biopharmaceutical.
5. The percent of biopharmaceutical molecules in a given sample that have an oligosaccharide present at each glycosylation site (commonly referred to as the percent occupancy of each glycosylation site) on a biopharmaceutical.
6. The distribution profile of different oligosaccharides' structures that occupy each glycosylation site.

To obtain all this information requires similar analytical approaches used to characterize the protein part of a biopharmaceutical. This includes the collaboration of information extracted from the analysis of the intact biopharmaceutical along with peptide mapping, where separation science together with MS characterization again plays a dominant role in revealing and quantitating the detailed characteristics of the oligosaccharide(s) and their location on the biopharmaceutical (Ayoub et al., 2013; Beck et al., 2012, 2013, 2015; Biacchi et al., 2015; Chen et al., 2013; Dotz et al., 2015; Gahoual et al., 2014a,b; Kaltashov et al., 2012; Klepárník, 2015; Leurs et al., 2015; Reusch et al., 2015b; Sandra et al., 2014; Xie et al., 2010; Zhang et al., 2009). However, an additional mode of oligosaccharide characterization also exists that involves the unique ability of the biopharmaceutical investigator to release quantitatively the oligosaccharides from a biopharmaceutical to assess the average global distribution of all the different oligosaccharides present in a given biopharmaceutical sample. The ability to release these oligosaccharides and analyze them independent of a biopharmaceutical's protein structure can provide a detailed fingerprint of the glycosylation pattern (oligosaccharide profile) of the biopharmaceutical using various types of LC and CE separations in combination with any one of a number of various detectors [e.g., pulse amperometric detection (PAD), fluorescent, or MS]. Such fingerprint information can be used empirically to qualitatively, as shown in Figure 2.10, or quantitatively assess glycosylation lot-to-lot variability of the RP and biosimilar lots to establish their consistency of manufacturing and in the glycosylation biosimilarity assessment of a biosimilar to its RP (Beck et al., 2013; Reusch et al., 2015a,b; Sandra et al., 2014; Suzuki, 2013).

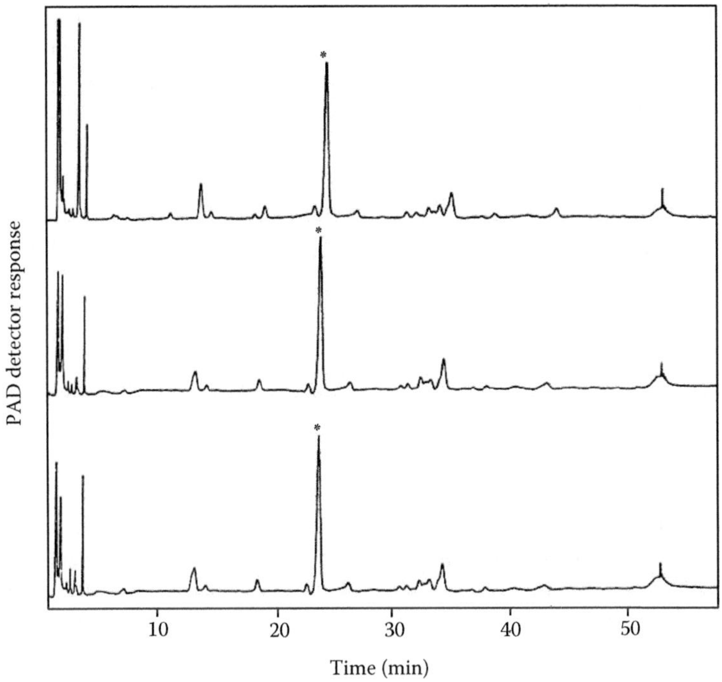

FIGURE 2.10 Assessing the lot-to-lot reproducibility of the glycosylation profile (or oligosaccharide profile) of a glycosylated biopharmaceutical (IFNβ). In this example, three different IFNβ lots were used to conduct oligosaccharide profiling. The *N*-linked oligosaccharides from IFNβ's one glycosylation site were first released using PNGase (a specific enzyme that cleaves *N*-linked oligosaccharides from proteins) and then separated and detected using liquid chromatography with pulse-amperometric detection (PAD). The resulting chromatography data from each lot shows a high level of comparability in the elution profile of the different oligosaccharides (which elute within the time window of 10-50 min) present on IFNβ on a lot-to-lot basis. Note that the single major peak (*) in each of these electropherograms corresponds to the known major glycoform found on this biopharmaceutical, which is a disialylated bianternnary oligosaccharide. However, when the oligosaccharides are left attached to the IFNβ protein and CZE analysis is conducted on the intact IFNβ, two closely spaced major peaks are observed; see Figures 2.4D and 2.9C. This splitting of the major disialylated bianternnary glycoform is due to the impact of deamidation as speculated by Berkowitz et al. (2005) and confirmed by Haselberg et al. (2013) and is one of the other major factors that gives rise to the heterogeneity observed in this biopharmaceutical as illustrated in Figure 2.4A and B.

In terms of assessing the details of an oligosaccharide's structure, much of this information is acquired via the collaboration of MS with the use of various glycolytic enzymes, which are capable of specifically cleaving various monosaccharides off the oligosaccharide (depending on the specific nature of the monosaccharide and its specific linkage to the other monosaccharides) outside the MS before the sample is injected and analyzed by MS or tandem MS. Significant progress in unraveling the intricacies of glycosylation has been made through utilization of such enzymatic cleavage/fragmentation outside the mass spectrometer, in combination with fragmentation and accurate experimental MW information obtained inside the mass spectrometer, coupled again with prior knowledge concerning the known MW of the monosaccharides, the various known observed oligosaccharide structures, and the known various adducts found on monosaccharides (Moremen et al., 2012).

However, when it comes to the task of linking the diversity of monosaccharide and oligosaccharide structures to biopharmaceutical function, much remains to be learned (Moremen et al., 2012). Nevertheless, some aspects of glycosylation functionality have already been realized (e.g., its role in the drug clearance; Solá and Griebenow, 2010), binding to certain biological receptors (Arnold et al., 2007; Shibata-Koyama et al., 2009), along with its structural impact on the HOS of biopharmaceuticals (Solá and Griebenow, 2009; Solá et al., 2007). As a result, the detailed characterization of the glycosylated structures on biopharmaceuticals will continue to receive a great deal of attention, especially when assessing biosimilarity.

2.6.1.3 Structural Analysis of Cysteine and Cystine

The presence of the amino acid cysteine [the most reactive of all the common amino acids found in proteins (Walsh, 2007)] and its chemical status in the linear sequence of amino acids of a biopharmaceutical can play a particularly important role in the structure of these drug molecules. This unique role of cysteine is due to the high reactivity of its side chain sulfhydryl group, which can readily react with another cysteine sulfhydryl group that might be present in the same or different polypeptide chain of a biopharmaceutical to form a cystine residue via an intrachain or interchain disulfide bond, respectively. This cross-linking reaction between two cysteine residues in the same biopharmaceutical molecule allows the polypeptide chain to take on a unique stable folded structure that will significantly influence the biopharmaceutical's HOS and stability. In addition, the presence of more than one cystine or the presence of an additional free cysteine in the same biopharmaceutical also offers the opportunity for the miscoupling of these disulfides or the scrambling of the different cysteines involved in a disulfide bond. As a result, different cystine cross links may arise due to the differences in the actual two cysteines involved in forming the disulfide bond to make a cystine (see Figure 2.5E).

The result of such miscoupling or scrambling events will create variants or subpopulations of a biopharmaceutical with different HOS(s). Hence, an important task in the primary structure (and HOS) assessment between the biosimilar and its RP, as well as in the structural consistency assessments between lots of RP and lots of biosimilar, not only involves assessing the agreement in the number of free cysteine and cystine residues that are present, but also in the case of cystine demonstrating that the correct specific two cysteine residues are cross-linked in forming each cystine in a biopharmaceutical. Such analysis typically involves conducting nonreducing and reducing peptide mapping using MS detection via a collection of LC-MS techniques. Although this task can be very challenging, a number of approaches to tackle this problem have been documented in the literature to help facilitate this work (Wang et al., 2011; Wiesner et al., 2015).

Another point of concern associated with the chemical status of cysteine and cystine residues in the RP and biosimilar is related to the possibility of intermolecular cross-linking between two or more independent monomeric biopharmaceutical molecules, resulting in the formation of covalent aggregates (Costantino et al., 1994; Lu and May,

2012). The task of detecting and quantitating the presence of such material is commonly the responsibility of analytical aggregation tools used under nonreducing and reducing conditions, for example, SDS-PAGE (den Engelsman et al., 2011) as well as MS.

2.6.2 Analytics at the Biophysical Level: HOS Including Secondary, Tertiary, and Quaternary Structure

In assessing the biophysical consistency of a biopharmaceutical, in terms of its HOS, the approach that is used is significantly different from that taken in assessing the biochemical primary structure of a biopharmaceutical, where one analytical tool, the mass spectrometer (in combination with the supporting role provided by various

TABLE 2.2
Low-Resolution Biophysical Method Being Used and High-Resolution Methods Being Investigated in Characterizing Biopharmaceuticals

Method	Key Attributes of HOS that are Assessed
	Low-Resolution
UV spectroscopy	Aromatic amino acid physicochemical environment
Fluorescence spectroscopy	Aromatic amino acid physicochemical environment
Circular dichroism	Polypeptide (secondary) structure and aromatic amino acid physicochemical environment (tertiary structure)
Fourier transform infrared spectroscopy	Polypeptide (secondary) structure via information acquired about the amide bond
Differential scanning calorimetry	Global and domain structure
Size-exclusion chromatography	Quantitative aggregation and aggregate size distribution
Analytical ultracentrifugation	Quantitative aggregation and aggregate size distribution, and shape, physicochemical properties
Asymmetric flow field-flow fractionation	Quantitative aggregation and aggregate size distribution
Light scattering (static and dynamic)	Qualitative or semi-quantitative information about aggregation and aggregate size distribution
	High-Resolution
Mass spectrometry	
Footprinting	Indirect solution structure information (approaching amino acid resolution) via chemical reporter's ability to react with the biopharmaceutical
Native	Global structure information in the gas phase via charge state distribution
Ion-mobility	Global structure information via size, shape, and charge in the gas phase
Antibody arrays	Indirect solution structure information (at the peptide level) via epitope reporter accessibility
Nuclear magnetic resonance	Structure information (approaching the atomic level) by studying the physicochemical environment of active NMR nuclei (^{1}H, ^{13}C, ^{15}N)

separation technologies), is the key analysis tool. In the case of biophysical characterization, the approach that has evolved has resulted in the use of a number of different low-resolution biophysical analytical tools (see Table 2.2). However, the poor resolution of these tools coupled with the driving need to obtain more detailed information about a biopharmaceutical's HOS has led to the search for more informative, yet still practical, biophysical tools. In the last decade, this situation is beginning to show significant signs of being realized as a growing number of more advanced high-resolution biophysical tools (again see Table 2.2) are being investigated and developed in an attempt to achieve a more detailed characterization of a biopharmaceutical's overall HOS (Berkowitz et al., 2012; Marino et al., 2015).

2.6.2.1 Commonly Used Low-Resolution Biophysical Tools to Assess HOS

The most common and readily available, and therefore frequently employed, biophysical tools used in the biopharmaceutical industry assess the HOS of biopharmaceuticals from a global perspective. Such biophysical tools probe and collect the signal output from a number of physicochemical structural elements on a biopharmaceutical that may be different or identical (but located in different physicochemical environments) that nevertheless emit a highly similar signal output that changes little from their unaltered state even when the physicochemical environment of these structural elements is changed. In so doing, they frequently have little capability to detect small subtle changes in the HOS of a biopharmaceutical, especially when one includes their inherent low signal-to-noise ratio.

This difficulty is illustrated by a hypothetical case where the CD measurements are conducted on an RP and its biosimilar that consist of a number of separate α-helix segments that are spread across different locations of the entire biopharmaceutical molecule. Since CD measurements from virtually all α-helices produce effectively the same CD spectrum (signal output), if only a small part of one α-helix segment in the biosimilar is altered (e.g., converted to a random coil) relative to the RP, that altered region of α-helix segment will yield a different CD output spectrum that still overlaps extensively with the normal CD outputs from the rest of the unaltered α-helix material still present in the biosimilar sample. As a result, the small altered CD signal output between the RP and biosimilar will need to be extracted from nearly the same large normal background CD spectrum generated from the intact (unaltered) α-helix material (along with the measurement's associated noise) still present in both samples. This large background of normal α-helix CD signals will significantly reduce one's ability to see the actual small difference that exists between the RP and the biosimilar relative to an ideal hypothetical case where each individual α-helix segment generated its own unique CD spectrum that didn't overlap. In the latter case increasing the concentration of both samples would improve one's ability to assess the difference in the altered α-helix CD spectrum between the RP and the biosimilar over the noise that is present in the physicochemical measurements.

However, in the real case where the CD spectrum from the normal and altered α-helix material present overlap significantly, the approach of increasing concentration would not be very effective since the large normal CD background spectrum from the unaltered α-helix material would also increase. In this situation, one is faced with the typically difficult problem of trying to extract a small signal difference between two very large and highly similar signals that also contain signal noise. In such cases, the resulting small differences will likely be buried in the noise of the two biophysical measurements unless one can find ways to improve the signal-to-noise ratio in these measurements (for a graphical illustration of this problem, see Figure 3.1 in Houde and Berkowitz, 2014b). Nevertheless, the collective nature of the information obtained from low-resolution biophysical techniques does offer, as a pooled dataset, some underpinning of a fingerprint of the HOS of the RP and biosimilar; it's just that in most cases the ability of this fingerprint to reveal a small difference in HOS is going to be significantly limited.

One approach that might help improve the capability of some of these low-resolution biophysical methods to better detect the presence of small differences in the HOS between biopharmaceutical samples is to apply identical amounts of a physical or/and chemical stress to the samples being compared. The premise here is that if an inherent small difference in the HOS actually exists and makes that biopharmaceutical slightly less stable, applying an appropriate amount of stress over time may make that biopharmaceutical preferentially undergo some further structural change and/or a structural change at a different rate relative to the unaltered form of the same biopharmaceutical, as illustrated in Figure 2.11. Hence, for the sample having the slightly altered HOS, the introduction of stress might now provide a significantly large enough signal output in response to this stress, so that the actual difference present between both samples can be detected. Consequently, if one can introduce in an identical manner the same form and amount of an appropriate stress to biopharmaceutical samples being compared, this approach could be effective in assessing the biophysical biosimilarity using these low-resolution biophysical tools. Indeed, it is exactly this type of an approach that is employed in differential scanning calorimetry (DSC) that tends to make this particular biophysical tool fairly useful in detecting HOS differences between highly similar biopharmaceuticals (Demarest and Frasca, 2014). In this case, the controlled reproducible application of an increasing amount of heat (stress, which eventually denatures the biopharmaceutical) is used to reveal differences in the HOS of biopharmaceutical samples.

This general concept of applying stress to help facilitate the detection of underlying structural difference(s) between two or more samples is the basis for revealing the underlying stability issues when developing any biopharmaceutical or pharmaceutical via the use of accelerated stability studies. However, in these stability studies, the level of stress is greatly reduced, resulting in the amount of time to potentially reveal a structural problem that is very long (months). Consequently, the application of stress being called for here to potentially help low-resolution biophysical tools to be more useful in revealing possible structural differences in biopharmaceutical samples is significantly higher in order to carry out these measurements in a much shorter time scale (minutes).

Overall low-resolution biophysical tools provide only a coarse footing for assessing biosimilarity from a biophysical perspective. To improve upon this situation, one will need to use much higher resolution biophysical tools to generate more sensitive assessments of consistency and biosimilarity in terms of HOS. As a result, as mentioned at the end of Section 2.6.2, a growing interest in finding and developing such analytical tools for the biopharmaceutical area has attracted a fair amount of interest in recent years (Berkowitz et al., 2012; Marino et al., 2015).

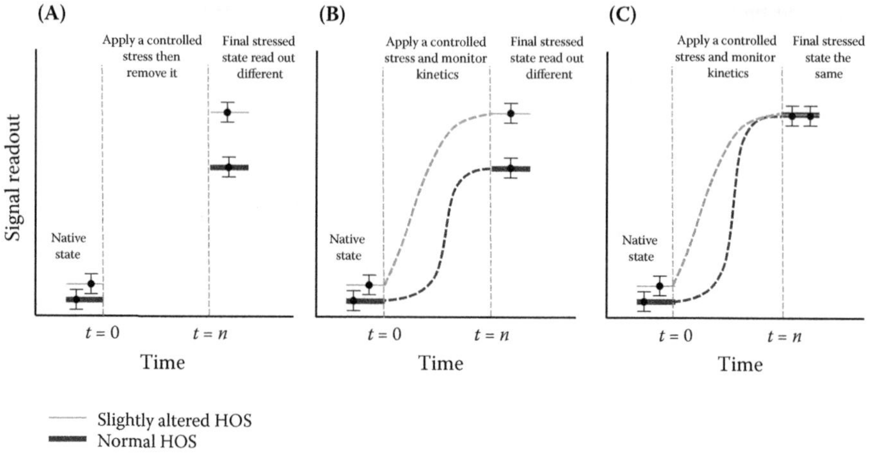

FIGURE 2.11 **(See color insert.)** The utility of stress studies to facilitate the detection of differences in the HOS between two biopharmaceuticals using low-resolution biophysical methods. (A) In this example, the signal readout from a low-resolution biophysical method for two different samples of the same biopharmaceutical that differ slightly in their HOS under normal stable conditions is too small to detect (the difference is within the inherent statistical variability of the biophysical method). However, if an appropriate identical stress is applied to both samples (at $t = 0$) and is then removed from both samples at the same later time point (at $t = n$), the resulting difference in HOS between the two protein samples may potentially now be revealed (*note*: depending on circumstances, the actual stress may need to be still in place while the biophysical measurement is being made (at $t = n$) in order to reveal the difference in HOS between the two samples). (B) Same situation as in "A"; however, the stress applied allows the biophysical method to detect the initial difference in the HOS between the samples by also studying the kinetic of the change in the HOS between the samples as a function of time. (C) Same situation as "B" but in this case the final state of both samples is effectively the same, but the differences in the HOS between the two samples can only be detected by monitoring the kinetics of the change in the HOS between the samples as a function of time. (Reprinted from Houde JD, Berkowitz SA eds. 2014b. Biophysical industry's biophysical toolbox. In: *Biophysical Characterization of Proteins in Developing Biopharmaceuticals*, 46–78. Elsevier, Amsterdam. With permission from Elsevier with minor modifications.)

2.6.2.2 Higher Resolution Biophysical Tools to Assess HOS

Until recently, the approach of using higher resolution biophysical tools for characterizing the consistency, comparability, and biosimilarity of biopharmaceuticals has not been fruitful since the only biophysical techniques that seem to be available to provide higher resolution characterization involved techniques that actually determined the 3D structure of these molecules, such as X-ray crystallography and NMR. Although these biophysical tools are very powerful, they are far too complex and time consuming for practical use in a routine development environment and come with their own set of limitations that have been enumerated numerous times in the scientific literature. However, interest in finding and using more practical high-resolution biophysical tools in the last decade has started to look more promising.

Several MS-based footprinting approaches using reversible hydrogen/deuterium exchange (H/DX-MS) or irreversible covalent labeling (e.g., with hydroxyl radicals), along with native MS, are showing signs of being capable of making their way into more critical assessments of comparability and biosimilarity of biopharmaceuticals (Beck et al., 2012, 2013; Bobst and Kaltashov, 2011; Edgeworth et al., 2015; Houde and Berkowitz, 2016; Houde et al., 2011; Zhang et al., 2014). In addition, advancements in NMR are attracting attention (Arbogast et al., 2015a,b; Aubin et al., 2008, 2014; Poppe et al., 2013, 2015; Wishart, 2013), and even new tools such as antibody arrays (Davies et al., 2015; Wang et al., 2013) are showing signs that they can effectively be used to interrogate large portions or even the entire biopharmaceutical molecules to assess the HOS of these molecules with high resolution that can detect small changes in a biopharmaceutical's HOS on a somewhat more practical level than what was possible a few years ago.

Although these analytical biophysical tools in most cases require a high level of expertise, very expensive instrumentation, or critical reagents that are only commercially available (via the vulnerable situation) from a single source, the fact remains that in assessing biosimilarity, the need to perform these types of high-resolution measurements can be contained to a fairly limited number of specific key samples and experiments that when performed can make a significant contribution in successfully demonstrating biosimilarity. As a result, the investment of resources into conducting such analytical biophysical measurements can pay big dividends in facilitating efforts in picking the best biosimilar to work on and/or in acquiring critical data for a BLA filing of a biosimilar to support its approval. Given the potential limited needs for using these advanced high-resolution technologies, it may not be necessary to bring these capabilities in-house (especially in the early phase of biosimilar work). Rather, key collaborations with specific facilities (analytical contract research organization, CRO, or cutting-edge academic laboratories) can offer opportunities for undertaking these types of measurements. Such an approach would allow for the initial evaluation of these tools to determine whether efforts should be made for their acquisition and development in-house, or access to these techniques can should be better pursued scientifically and/or economically by appropriate contract or collaboration approaches.

2.6.2.2.1 MS Footprinting to Assess HOS: H/DX-MS and Covalent Labeling MS

Although H/DX-MS has been around for more than two decades, only in the last decade has the biopharmaceutical industry begun to realize the potential of this technique to greatly improve the industry's ability to provide useful indirect detailed information about the HOS of biopharmaceuticals and its dynamics in a very practical way (Engen and Wales, 2015; Houde and Berkowitz, 2016; Houde et al., 2011; Iacob and Engen, 2012; Kaltashov et al., 2010; Wei et al., 2014). Much of the delay in realizing H/DX-MS's capabilities can be traced to the intense amount of time and effort initially required to obtain useful data from these measurements and the availability of only home-built instrumentation to carry out these measurements. However, with the great advances in MS instrumentation, which includes the commercialization of a turnkey H/DX-MS system by Waters

Associates (Rouhi, 2012), coupled with the ever present and ubiquitous developments in computer hardware/software, this technique is now seen as an attractive tool to improve our ability to conduct HOS comparability and biosimilarity assessments on biopharmaceuticals (Houde and Berkowitz, 2016; Houde et al., 2011; Kaltashov et al., 2010; Majumdar et al., 2015; Mo et al., 2012; Pirrone et al., 2015; Wei et al., 2014).

What is particularly attractive about this technique is that the basic reporting element, the amide hydrogen atom associated with each peptide bond and therefore every amino acid (with the exception of proline, which does not have an amide hydrogen), is intrinsically built into the biopharmaceutical along the biopharmaceutical's polypeptide backbone. What is particularly important about the hydrogen amide is its ability to exchange with the hydrogen atoms in the bulk aqueous solution that is critically dependent on its local physicochemical environment or structure within the biopharmaceutical. This property makes it a natural reporter for detecting small changes in the HOS of a biopharmaceutical. Hence, in conducting H/DX-MS measurements, all that is needed to carry out an experiment is to dilute a sample directly into the sample's normal formulation solution made with deuterated water (D_2O) instead of normal H_2O and to monitor the change (increase in this case) in mass of the intact biopharmaceutical with time using MS (under appropriate well-controlled experimental conditions). Conducting such an experiment is called *global* H/DX-MS.

A far more popular, more informative and much higher resolution mode of H/DX-MS [which can reach a level of resolution of one amino acid (Fajer et al., 2012; Pan and Borchers, 2014; Pan et al., 2014; Rand et al., 2009)] involves the introduction into the H/DX process of a proteolytic step before the sample is injected into the mass spectrometer. Such H/DX-MS experiments are called *local* (or bottom-up) H/DX-MS. This mode of H/DX-MS can not only detect changes in HOS, it can also provide fairly specific information about where these changes in HOS are occurring in the biopharmaceutical samples being compared. Recent developments in local H/DX-MS have shown that this mode of H/DX-MS can be further simplified and improved by replacing the proteolytic step conducted outside the MS with direct fragmentation inside the mass spectrometer using ETD and ECD fragmentation to conduct what is called top-down H/DX-MS (Pan and Borchers, 2014; Pan et al., 2014; Rand et al., 2009).

By using local H/DX-MS, the HOS of a biosimilar can be compared to its RP by monitoring the extent and rate of hydrogen exchange of the same structural fragments in these molecules with time under the same experimental conditions. Such capability can uncover potentially subtle differences in the HOS structure and its structural dynamics between these two samples, by conducting a number of such studies over a range of different experimental conditions. What is also important about H/DX-MS is its ability to probe virtually the entire structure of a biopharmaceutical at the same time. Such capability is demonstrated for two specific cases of the same biopharmaceutical, IFNβ, in Figure 2.12. In the first case, a significant change in the cell culture media used to make this biopharmaceutical showed no significant change in H/DX properties of IFNβ when about 95% of the total sequence of amino acids in this molecule was probed (see Figure 2.12A and B). In the second case, the chemical addition of a single small chemical group to the side chain sulfhydryl group on the only free cysteine residue in the IFNβ molecule results in a significant increase in the extent and rate of H/DX in a number of specific regions along the polypeptide chain of IFNβ (see Figure 2.12C and D). Such results imply that a significant change to the IFNβ's HOS (which includes structural dynamics) had occurred as a result of this single simple chemical modification (allowing more amide hydrogen atoms to be exposed to the bulk water so that they can more readily exchange). Such results clearly demonstrate the utility of H/DX-MS's in generating detailed indirect HOS fingerprint profiles of biopharmaceuticals, which can be used for various qualitative or quantitative comparison purposes (Houde and Berkowitz, 2016; Houde et al., 2011) that can include biosimilarity studies.

In the case of conducting covalent labeling MS to assess HOS biopharmaceutical comparisons, one is effectively using an approach that is very similar to H/DX-MS. Here, instead of using the reversible exchange reaction of the amide hydrogen atoms (intrinsically) present on a biopharmaceutical with the deuterium atoms from the deuterated water added to a biopharmaceutical sample, a reactive chemical reagent is added to the biopharmaceutical sample that can covalently react with either specific amino acids (selective labeling, e.g., using acetic anhydride) or a wide arrangement of amino acids (nonspecific labeling, e.g., using hydroxyl radicals)

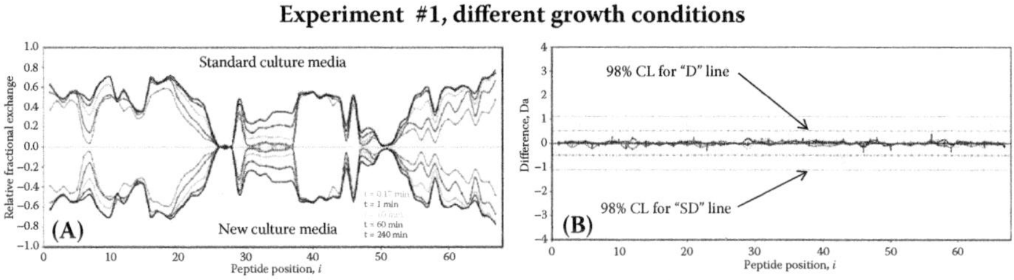

FIGURE 2.12 **(See color insert.)** Two different local H/DX-MS comparison experiments conducted on a biopharmaceutical (IFNβ). In experiment #1, the impact of a significant change in the cell culture media used to produce IFNβ (New Culture Media) was compared to the IFNβ produced using the normal culture media (Standard Culture Media). Results from this H/DX-MS experiment are shown in (A) as a mirror (or butterfly) format plot of all the deuterium uptake data for each peptide [plotted along the x-axis starting from the peptide nearest the N-terminus ($X = 1$) to the peptide nearest the C-terminus ($X = 67$) covering about 95% of the amino acid sequence of IFNβ] at each time point (ranging from 0.17 to 240 min., where the same time point for each peptide is connected by the same colored line) for IFNβ made using Standard Culture Media (control), which is displayed in the positive y-axis direction, while all the corresponding deuterium uptake data for IFNβ made using New Culture Media (experimental) is displayed in the negative y-axis direction. Data in "A" are also shown in (B) as a difference plot, which corresponds to the difference data computed as Standard Culture Media data minus New Culture Media data for each peptide at each time point (*note*: for each X peptide the sum of hydrogen exchange difference data for all time points is presented as a bar plot). In experiment #2, the impact of a chemical modification to the only free cysteine amino acid in IFNβ using N-ethylmaleimide, NEM, (IFNβ-NEM) is compared with unmodified IFNβ (IFNβ).

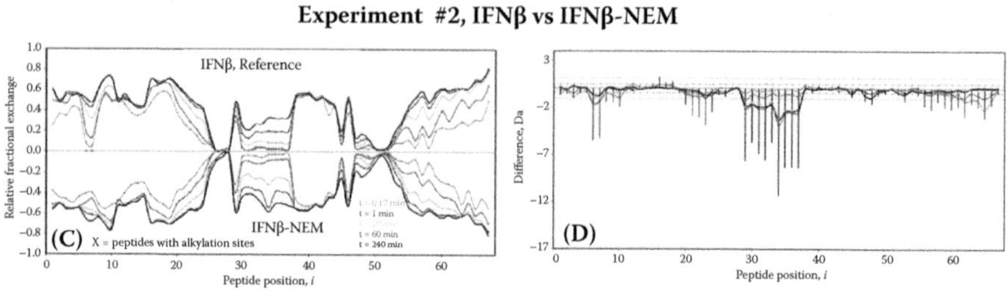

FIGURE 2.12 (CONTINUED) **(See color insert.)** Results from this experiment are shown in (C), again as a mirror plot where the deuterium uptake data for the unmodified IFNβ Reference (control) are displayed in the positive *y*-axis direction and the deuterium uptake data for IFNβ-NEM (experimental) are displayed in the negative *y*-axis direction. Data in "C" are also shown in (D) as a difference plot as described in "C" as described in "B". In H/XD-MS experiment #1, data show no significant indication of difference in the H/DX-MS data between the two IFNβs, as indicated by the absence of any qualitative visual differences in the mirror plot "A" and the absence of quantitative difference data in plot "B" exceeding its 98% confidence limit (CL) D line (blue) and sum of difference data exceeding its 98% CL SD line (black). These results support the comparability of the HOS of the two IFNβ indicating the absence of any effect on the biophysical properties of IFNβ as a result of the change made in its production cell culture media. In H/XD-MS experiment #2, however, data show that the simple chemical modification of IFNβ's sole free cysteine has a significant impact on its HOS, as indicated by the qualitative visual differences seen in the mirror plot "C" and numerous quantitative difference data exceeding its 98% CL D line (blue) and sum of difference data exceeding its 98% CL SD line (black). These results support the noncomparability of the HOS of these two INFβ. (Figure 2.12A and B reprinted from Houde D et al. 2011. The utility of hydrogen/deuterium exchange mass spectrometry in biopharmaceutical comparability studies. *Journal of Pharmaceutical Sciences* 100(6), 2071–2086. With permission from Elsevier with minor modifications.)

(Kaltashov and Eyles, 2005a; Kaur et al., 2015). As in H/DX-MS, the extent and/or rate of these reactions can be followed and assessed by MS in collaboration with fragmentation methods inside or outside the mass spectrometer to assess changes in the biopharmaceutical HOS and identify which structural elements on the biopharmaceutical are being altered (Deperalta et al., 2013; Madsen et al., 2016). One important difference between covalent labeling and H/DX-MS is the absence of the reversibility of the labeling process in the case of the former, which removes some of the experimental constraints that must be adhered to when doing H/DX-MS, which could offer a significant advantage when using covalent footprinting techniques.

2.6.2.2.2 Using Native MS to Assess HOS

In conducting native MS, one effectively uses the resulting charge state distribution of the native or native-like structure of the injected biopharmaceutical retained under the gas-phase conditions inside the mass spectrometer as an indirect indicator of the biopharmaceutical's HOS (Huber, 2015; Rosati et al., 2012). Since the charge state distribution of a biopharmaceutical is highly dependent on its HOS, samples of the same biopharmaceutical in the same matrix with the same HOS should yield the same charge state distribution when compared under identical experimental conditions. Those samples showing the biopharmaceutical in higher charge states would imply that these biopharmaceuticals have differences in their HOS resulting from a more exposed or open conformation (which would enable them to take on the observed additional charges).

Such a native MS comparison is demonstrated in Figure 2.13. In this figure, the same comparison of native IFNβ versus a chemically modified IFNβ (IFNβ-NEM) as shown in Figure 2.12C and D was investigated. Native MS results show that the modified IFNβ has a significantly higher charge state distribution than the native form of IFNβ. Such results are consistent with a more open conformation for the chemically modified IFNβ. This conclusion is independently supported by the HOS information obtained by H/DX-MS (an orthogonal biophysical method) shown earlier in Figure 2.12C and D, which indicated greater H/DX (in terms of both the amount and rate of H/DX) for the same chemically modified form of IFNβ relative to the native form of IFNβ. An important difference between H/DX-MS and native MS, however, is that H/DX-MS data actually show where in the IFNβ molecule HOS changes are occurring.

In conducting native MS comparisons, an important required element is that all samples must be in the same exact MS compatible matrix that can maintain the HOS of a biopharmaceutical. Hence, only volatile buffers and excipients can be used (e.g., acetate, formate, ammonium). This will likely require some form of sample preparation (e.g., buffer exchange to be carried out) in which all sample handling and processing must be identically applied to achieve meaningful comparisons.

Further enhancements to native MS, designed to improve its ability to characterize and compare the HOS of biopharmaceutical samples, can also be achieved by combining it with other online separation techniques. Such separations include ion-exchange chromatography (IEC) (Muneeruddin et al., 2015) and SEC (Haberger et al., 2016; Muneeruddin et al., 2014) using an MS-compatible buffer that maintains the native structure of the separated biopharmaceutical material entering the mass spectrometer. In addition, the incorporation of ion-mobility (IM) within the mass spectrometer to conduct native IM-MS (Campuzano et al., 2015; Kaltashov and Eyles, 2005b) also offers an additional separation mode for enhancing native MS. This latter enhancement enables the native charged biopharmaceutical species that enter the mass spectrometer to be further separated in terms of both their native or native-like charge state and collisional cross-sectional area or overall size and shape inside the mass spectrometer. By adding these additional online modes of separation to native MS, significant additional capability is introduced for improving the ability of native MS to characterize a biopharmaceutical's HOS and structural heterogeneity by analyzing individual components or much simpler mixtures, in comparison to trying to extract the same information from the total overlapping global collection of components present in the unfractionated sample when conducting normal native MS (where the native biopharmaceutical is simply infused into the mass spectrometer).

Although native MS (and native IM-MS) characterization does not allow the HOS analysis of biopharmaceuticals to be conducted in their formulation buffer, whatever

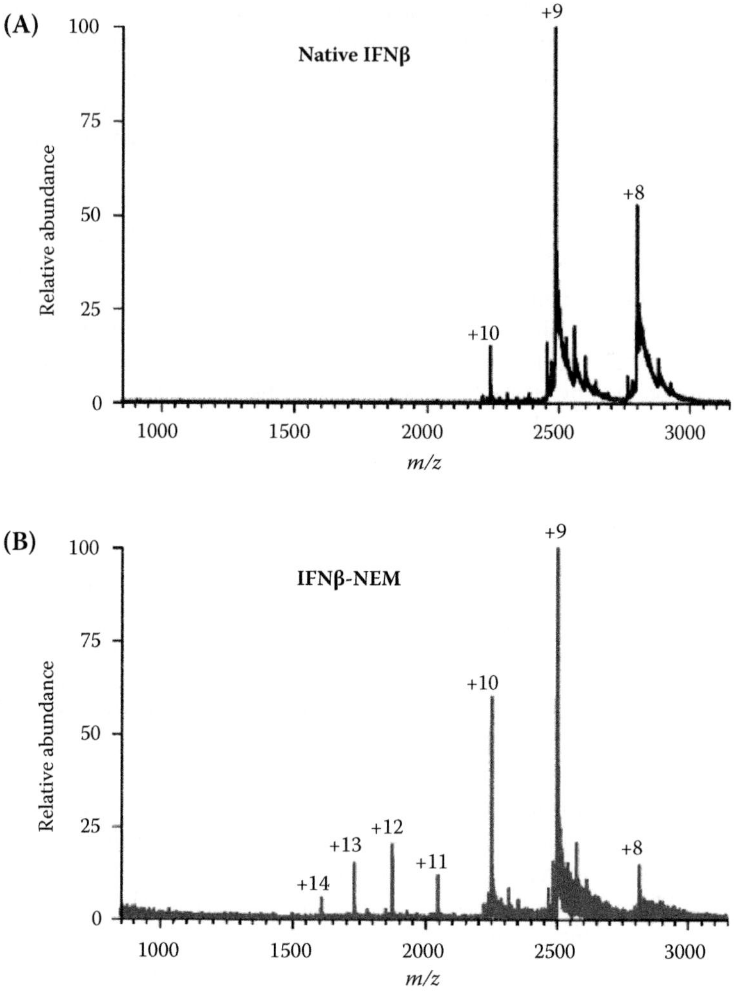

FIGURE 2.13 (A) An example of native MS where the resulting charge state distribution of an unmodified intact biopharmaceutical (IFNβ) is compared with (B), a modified form of IFNβ, where IFNβ's only free cysteine was modified with *N*-ethylmaleimide (NEM). Both samples were placed into the same MS compatible formulation buffer before being analyzed by MS. Results show that the modified IFNβ has a higher charge state distribution profile resulting from an alteration of its HOS, due to the partial unfolding of the protein structure, as a result of this single amino acid modification. Note that this conclusion is consistent with the results shown in Figure 2.12C and D for the same comparison using H/DX-MS. (Reprinted from Bobst CE et al. 2008. Detection and characterization of altered conformations of protein pharmaceuticals using complementary mass spectrometry-based approaches. *Analytical Chemistry* 80(19), 7473–7481. With permission from American Chemical Society with minor modifications. Copyright 2008.)

alteration that does occur to the HOS of the biopharmaceutical samples when they are in mass spectrometer's gas phase, the resulting alterations to these samples should be the same. Consequently, the resulting charge state distribution (and gas-phase separation profile in the case of IM-MS) of the samples being compared should be highly similar if the HOS of the biopharmaceutical samples entering the mass spectrometer were highly similar to each other to begin with. Any difference that is discerned between the MS outputs from the different samples being compared would likely be indicative of a difference in the starting HOS of these samples (again assuming that all samples are in the same matrix, are processed under identical conditions and that MS experimental conditions used in these assessments are also the same). Consequently, the act of conducting these types of comparisons in a gas-phase environment of the mass spectrometer effectively implies the application of an approach similar to that discussed in Section 2.6.2.1 and illustrated in Figure 2.11, where a stress is applied to help facilitate a comparison. Since the temperature and voltage of the native MS can be controlled, additional variables for applying different levels of stress can readily be changed to fine tune this approach in challenging the samples being compared to help reveal their potential underlying HOS differences.

2.6.2.2.3 Antibody Array Footprinting to Assess HOS

In Section 2.6.2.2.1, the discussion on footprinting techniques called H/DX-MS and covalent labeling MS made use of MS to monitor the labeling process and assess what parts of the biopharmaceutical's HOS were potentially different. Recently, an interesting alternate form of protein footprinting has been introduced that avoids the need of MS. This relatively new footprinting technique relies on using specifically designed collections (kits) of antibodies, referred to as antibody arrays (Wang et al., 2013) or protein conformational arrays (Davies et al., 2015), as a labeling reagent. Each kit of antibodies is unique to a specific biopharmaceutical. This form of footprinting offers an interesting and simple approach for allowing anyone to conduct high-resolution footprinting without the complexity and high cost of acquiring a mass spectrometer or other sophisticated instrumentation and the use of special fragmentation/cleavage procedures to conduct HOS comparisons on a biopharmaceutical.

What is involved in antibody array footprinting is the synthesis of a set of short overlapping peptides (about 30) that collectively correspond to the entire amino acid sequence of a particular target biopharmaceutical. These peptides are then used to develop a set of unique polyclonal antibodies (anti-peptide antibodies) that only bind to their respective targeted peptide (which corresponds predominantly to only a small specific linear amino acid sequence region in the polypeptide chain(s) of the targeted biopharmaceutical called a linear epitope). Since many of these peptides will normally be buried within the interior of the biopharmaceutical, they will not be accessible to bind to their probing antibodies. Hence, when each of these antipeptide antibodies is separately added to a different sample of the same biopharmaceutical (using specially designed 96-well plates), it generates a unique binding profile of the antipeptide antibodies that bind to that biopharmaceutical using the sensitive and common detection capability of enzyme-linked immunosorbent assay (ELISA) technology. If another sample of the same biopharmaceutical has incurred a change in its HOS, it may trigger a change(s) in the profile of the peptides that are exposed on the biopharmaceutical's surface in that particular sample. This HOS change would be reflected in the change in the binding profile of the antipeptide antibodies (specific to that biopharmaceutical) relative to that obtained normally from the same unaltered (native) biopharmaceutical.

This approach to assessing the HOS comparability of different samples of the same biopharmaceutical takes advantage of the unique high fidelity of antibodies to recognize specific binding sites of a protein that can be as small as 3–6 amino acids (Davies et al., 2015; Wang et al., 2013). Hence, by generating a unique set (kit) of antibodies for a particular biopharmaceutical, one can use this approach as an analytical tool to detect changes in the presence or absence of linear epitopes (and therefore HOS) on the surface of biopharmaceuticals that may exist between two or more different samples of the same biopharmaceutical (e.g., RP and its biosimilar). Since biopharmaceuticals are effectively a collection of many different epitopes, one needs to generate an array of uniquely different antibodies, where each antibody is specific to one of these epitopes on the biopharmaceutical. At present, dividing the target biopharmaceutical into 30 or so different epitopes (peptides) that cover the entire structure of the biopharmaceutical is being used to monitor the HOS of these drug molecules with reasonably high resolution. Thus, this technique can serve as a useful orthogonal approach to complement the more common biophysical toolbox to assess more quantitatively and with higher resolution the HOS biosimilarity of a biosimilar to its corresponding RP using a fingerprint-like picture of the resulting antibody binding profile (Davies et al., 2015; Wang et al., 2013). Although this analytical technique has only recently appeared, its use in biosimilarity work has already been realized in the case of the EMA filing and approval of the biosimilar Remsima® to Remicade® (Jung et al., 2014). In addition, its use in routine biosimilar candidate selection and optimization has also recently been suggested (Davies et al., 2016). At present, twelve different antibody array kits are commercially available for twelve biopharmaceuticals that have or will be losing patent coverage. As a result the opportunity now exists to use this tool to be used in the development of biosimilars to these nine commercial biopharmaceuticals.

2.6.2.2.4 NMR

Of all the biophysical tools available to study the HOS of biopharmaceuticals, NMR is the richest in terms of information content. Areas where this technique can provide physicochemical information include the following: (1) chemical identification and quantitation, (2) spatial location of atoms (3D structure), and (3) even the temporal behavior of the structural elements within a biopharmaceutical as well as the entire biopharmaceutical itself. The impressive ability of NMR to characterize the HOS of a biopharmaceutical is demonstrated graphically by visually looking at one form of data output from this technique shown in Figure 2.14. Here the native (properly folded) two-dimensional (2D) $^{1}H^{N}$-^{15}N heteronuclear single quantum coherence (HSQC) NMR spectrum of the N^{15} labeled biopharmaceutical methionyl-granulocyte colony stimulating factor (met-G-CSF), shown in Figure 2.14A, is compared to the extreme case where this same biopharmaceutical is denatured (unfolded), shown in Figure 2.14B. In the former 2D NMR spectrum, the unique and distinct features of the fold state of this biopharmaceutical enables the H-N signal generated from each

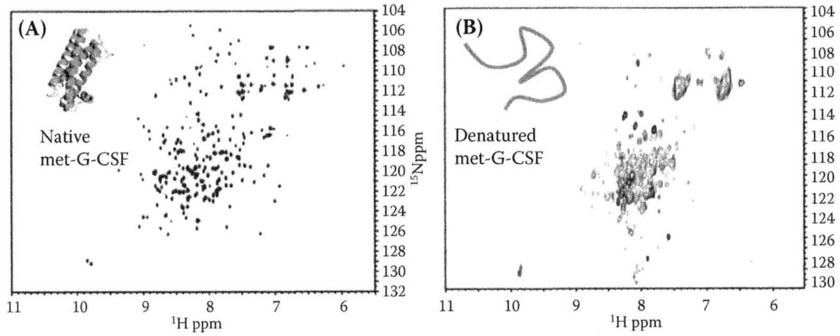

FIGURE 2.14 Comparison of the 2D ($^1H^N$-^{15}N) heteronuclear single quantum coherence (HSQC) NMR spectrum of N^{15} labeled methionyl-granulocyte colony stimulating factor (met-G-CSF) in its (A) native folded state (as indicated by the molecular structure model shown in the upper left corner of this figure) to its (B) denatured state (as indicated by the hypothetical molecular structure model shown in the upper left corner of this figure). (Reprinted from the unpublished work generated by Drs. Robert Brinson, Luke Arbogast, and John Marino at the National Institute of Standards and Technology, with minor modifications.)

peptide bond in the polypeptide chain of the biopharmaceutical (which are each located in highly specific chemical environments) to be resolved into a unique pattern of 2D contour peaks (which look like spots) in the 2D NMR plot. These unique defined spectral features are due to the limited structural variation in the biopharmaceutical's conformations resulting from the highly more rigid state of the biopharmaceutical molecules in its native conformation. In comparison the denatured unfolded state of the same biopharmaceutical displays a much greater ensemble of assessable conformations leading to 2D NMR spectrum that is more diffuse and greatly lacking in distinct features relative to the 2D NMR spectrum for native met-G-CSF molecules. Such dramatic changes in the detail fingerprint of the NMR spectrum offer significant opportunities for detecting subtle change(s) in the HOS of a biopharmaceutical (Aubin et al., 2008; Panjwani et al., 2010). Somewhat similar NMR opportunities have also been reported on larger biopharmaceuticals, monoclonal antibodies (mAbs) using ^1H-^{13}C and ^1H-^{15}N linkage (Arbogast et al., 2015a,b).

In pursuing the use of NMR to monitor the HOS of biopharmaceuticals in comparability and biosimilarity studies, a significant restriction is imposed of not being able to use isotope labeling approaches to enhance the concentration of the naturally low abundant levels of the active NMR nuclei ^{15}N and ^{13}C, especially for conducting 2D NMR. This restriction causes NMR to be a very low-sensitivity technique leading to a very long data collection times per sample (e.g., approaching times as long as 24 h or longer). This makes the use of NMR very challenging from a practical perspective. Nevertheless, several recent approaches have been investigated in an attempt to overcome this problem, with the hope of making NMR more amenable for application in the biopharmaceutical industry, especially for comparability and biosimilarity studies (Arbogast et al., 2015a; Chen et al., 2015; Frank et al., 2015; Poppe et al., 2013, 2015). In some of these cases, NMR approaches have involved use of the highly abundant NMR active ^{1}H nuclei to conduct various forms of one-dimensional (1D) proton (^{1}H) NMR, that employs new NMR procedures to extract HOS information. Such NMR procedures have been able to significantly reduce data acquisition times, while still providing the investigator with a significant amount of HOS information (Frank et al., 2015; Poppe et al., 2013, 2015). Such approaches may create significant opportunities for engaging the use of NMR in assessing HOS characterization information in the biopharmaceutical industry, especially for use in comparability and biosimilarity studies.

2.7 PHYSICOCHEMICAL METHODS THAT CAN PROVIDE BOTH PRIMARY AND HOS FINGERPRINT INFORMATION TO ASSESS BIOSIMILARITY

In characterizing the biochemical primary structure of biopharmaceuticals, it has been noted that the key analytical tool is MS. This dominant role of MS in primary structure analysis is achieved through its ability to assess the nature of the primary structure alterations present in the variant forms of a biopharmaceutical solely on the basis of accurate MW measurements. However, as pointed out in Section 2.6.1, much of the primary structure characterization work on biopharmaceuticals is facilitated by the implementation of additional prior separation techniques (see Figure 2.9A, C, and D). In many cases, these separations are conducted under denaturing conditions (e.g., using reversed phase LC). In other cases, however, separations may be conducted under experimental conditions where the native or native-like HOS of the biopharmaceutical is maintained. Under the latter conditions, the HOS of the different variant forms of the biopharmaceutical can also play an important role in the separations that are achieved. In some cases, the observed separations may not actually be associated with any primary chemical change in the biopharmaceutical's structure, but rather with the change in its HOS (e.g., see Figure 2.5E). As a result, separation methods such as IEC, isoelectric focusing (IEF) electrophoresis, CZE, or hydrophobic interaction chromatography by themselves offer the ability to empirically detect both primary structure and/or HOS differences between samples when these separations are conducted native or native-like experimental conditions (without knowing the details about the exact nature of the change in the biopharmaceutical). Indeed these separation methods on their own offer great opportunities to rapidly generate detailed empirical physicochemical fingerprint information, as shown in Figure 2.4D, to help in assessing consistency, comparability, and biosimilarity studies.

2.8 ESTABLISHING BIOSIMILARITY IN TERMS OF CONCENTRATION AND POTENCY

In establishing dosing equivalence in assessing biosimilarity, two important attributes must be assessed: (1) the physical mass of the API that is present in a biopharmaceutical's commercial product (e.g., vial, syringe) and (2) the amount of biological or functional activity measurement associated with that specific amount of physical mass of the biopharmaceutical's API. In terms of function or biological activity, a range of pertinent *in vitro* biological or binding assays or *in vivo* testing via toxicological or clinical activity such as PK, PD, or other relevant clinical indicator will need to be assessed (see Chapters 4 and 7). However, the aspect of physical mass falls under the jurisdiction of physicochemical analysis and will be the topic of this section.

In virtually all cases, the physical mass of the API present in a biopharmaceutical's commercial product is based on the total amount of the unmodified polypeptide mass of the biopharmaceutical, which is usually expressed on the basis of the molar theoretical amino acid sequence MW of the biopharmaceutical. This simple approach provides for a reasonably more accurate assessment of the biopharmaceutical's physical mass without the need to deal with the impact of mass differences due to the complexity of PTMs and their variability.

A common approach used to assess the physical unmodified polypeptide mass in a biopharmaceutical sample involves taking a UV measurement at a wavelength that is usually at or very near 280 nm of a sample and coupling it with information about the biopharmaceutical's theoretically estimated extinction coefficient at that wavelength based on its amino acid sequence. Since the only amino acids that absorb light at this wavelength are tyrosine (Tyr), tryptophan (Trp), and cysteine (Cys), knowing the amino acid sequence of the biopharmaceutical allows one to calculate the number of Tyr, Trp, and Cys moles present per mole of unmodified biopharmaceutical (note: this approach assumes that any PTM present does not contribute to the absorptivity at the wavelength employed, which is usually the case). Combining this information with the known molar extinction coefficient (at 280 nm) for Tyr, Trp, and Cys enables one to calculate a theoretical (molar or specific) extinction coefficient for a particular biopharmaceutical to provide a moderately accurate assessment of the unmodified polypeptide mass of the biopharmaceutical that is in a given solution by simply measuring its UV absorptivity at 280 nm (Gill and von Hippel, 1989; Mach et al., 1992; Pace et al., 1995) (note: in using this approach one assumes the extinction coefficient for Tyr, Trp, or Cys by themselves and in the biopharmaceutical is the same). However, in the case of a biosimilarity assessment, the FDA has stated that the extinction coefficient used to calculate the concentration of a biosimilar should be determined experimentally to be the same as the RP (FDA, 2015c).

To experimentally assess the biosimilarity of the extinction coefficient of a biosimilar to its RP, a relative experimental approach that compares the extinction coefficient of the polypeptide mass of a biosimilar to that of the RP should be just as valid an approach as their independent absolute extinction coefficient assessment and comparison. In fact, the relative approach is most likely to be more accurate. Although several simple relative approaches can be employed, probably the simplest and most accurate involves the use of a multidetector SEC system consisting of UV, light scattering (LS), and differential refractive index (DRI) detectors (Wen et al., 1996). In this case, the simple injection of the highly purified biosimilar and RP into this SEC system, integrating the total well-resolved monomer peak area resulting from the chromatogram output from each of the three detectors, and combining this data with the rearranged form of Equation 9 from the published work of Wen et al. (1996) will yield a parameter that is experimentally directly proportional to the biopharmaceutical's (unmodified) protein extinction coefficient of the injected sample (see Equation 7.9 in Houde and Berkowitz, 2014c) (note: in conducting these measurements, information about the amount of sample injected into the SEC system is not needed, which significantly improves the accuracy of the extinction coefficient assessment). By taking the ratio of Equation 7.9 in Houde and Berkowitz (2014c) containing the experiment parameters (monomer peak areas from each detector) determined for the biosimilar and the same equation containing the experiment parameters for the RP, the same collection of proportionality constants present in each equation (which includes the instrument constant and the known unmodified amino acid sequence MW of the target biopharmaceutical that converts these values to an absolute protein extinction coefficients) will cancel, further simplifying and improving the accuracy of the comparison measurements. As a result, if the biosimilar's extinction coefficient is highly similar to the RP's extinction coefficient, this experimental ratio should be very close to a value of 1.00. Since a number of experiments where repetitive measurements on two different samples of the same biopharmaceutical, conducted by the author, have yielded mean ratio values that have shown variabilities from 1.00 of only about ±1%–2% at a 95% CL (unpublished data), experimental ratios between a biosimilar and RP that range between 0.98 and 1.02 would likely establish a high level of confidence in the biosimilarity in their extinction coefficients.

2.9 THE UNIQUE PROBLEM OF AGGREGATION IN ASSESSING BIOSIMILARITY

One of the major differences indicated in Table 2.1 between pharmaceuticals (including generics) and biopharmaceuticals (including biosimilars) concerns the propensity of biopharmaceuticals to form a unique class of high MW product-related impurities due their ability to self-associate or aggregate. This form of degradation typically refers to the physical association of monomeric units of a biopharmaceutical to form higher MW material due to their unique physicochemical surface properties and surface topology. Although some biopharmaceuticals may intrinsically display concentration-dependent self-association, which is generally reversible (and may or may not be an issue), the more common and concerning forms of stable aggregates are those generated from altered biopharmaceutical material that arises from the biopharmaceutical's exposure to a range of physical and chemical environmental conditions that stress the biopharmaceutical during its production. These stress conditions may lead to a chemical or physical PTM of the biopharmaceutical that alters its HOS and/or surface properties making it prone to aggregate. Such aggregates can span an enormous range of sizes (nanometers to microns) and display different physicochemical properties, as indicated in Figure 2.15.

Although the biopharmaceutical industry depends heavily on the use of SEC as the key analytical tool for detecting and quantitating aggregation, there is now a clear understanding of SEC's potential limitations and associated artifacts (Arakawa et al., 2010; Berkowitz, 2006; Carpenter et al., 2010). In recent years, other biophysical tools, which have existed for a fairly long time, have been greatly improved and are now also capable of providing useful quantitative physicochemical characterization information about the aggregation that is present in a biopharmaceutical

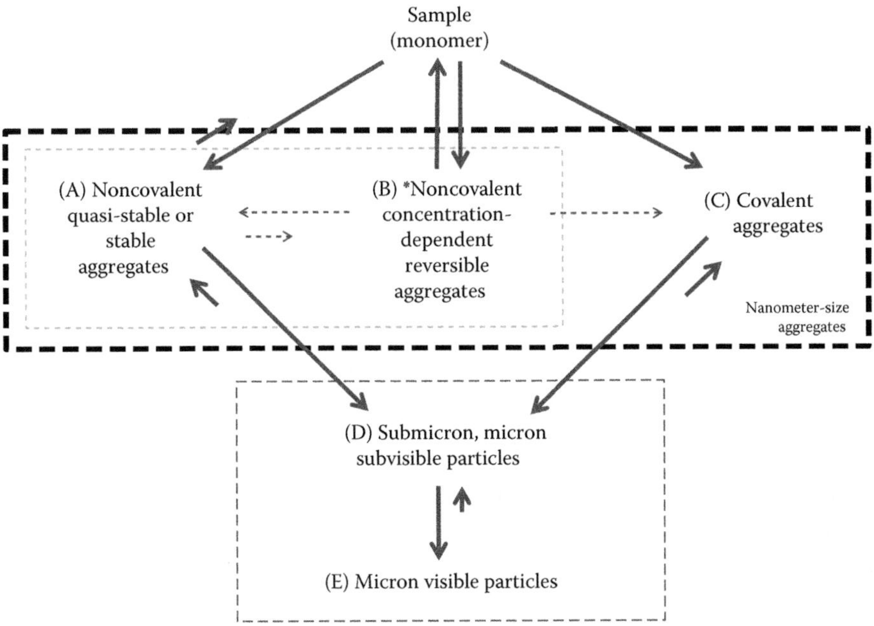

FIGURE 2.15 A simple schematic presentation of biopharmaceutical aggregates in terms of their most basic physicochemical properties and interrelationships. (Reprinted from Berkowitz S et al. 2015. Challenges in the determination of protein aggregates, part II. *LC/GC North America* 33(7), 478–489, with minor modifications.)

based on orthogonal biophysical principles. Two such important tools are analytical ultracentrifugation (AUC) and asymmetric flow field-flow fractionation (AF^4). Today these tools are used as orthogonal analytical methods to help monitor and support SEC data (along with, to some extent, light scattering (LS) tools, e.g., static-LS and dynamic-LS) in an attempt to overcome SEC's potential limitations for assessing aggregation. They therefore play an important role in helping to develop both innovator biopharmaceuticals and biosimilars.

What is particularly unique and important about aggregation, besides their potential impact of reducing the amount of active therapeutic activity present in a given biopharmaceutical sample, has been their association with adverse immunogenicity issues (Filipe et al., 2010; Moussa et al., 2016; Rosenberg, 2006) and/or the possibility that aggregates can give rise to toxicological effects that can go beyond immunogenicity (Bucciantini et al., 2002). In the case of immunogenicity, some forms of aggregation, owing to their unique physicochemical structure (e.g., size and physicochemical properties), appear to be capable to elicit immunogenicity effects with only a small amount of aggregated material being present (Ahmadi et al., 2015; Rosenberg, 2006). As a result, in the case of aggregate size, it has been realized that our dependence on using SEC to assess the relatively low MW aggregates (e.g., dimer, trimer, etc.), light obscuration methodology to assess near visual particles (e.g., 10–50 μm particles), and direct visual observation methods to assess particles

>100 μm has left a large gap in our ability to assess the level and distribution of very low micron and submicron-size aggregate particles that are felt to be an important size range of aggregates responsible for generating an immunogenic response (Carpenter et al., 2009). Hence, a significant effort has been under way for almost a decade to develop new and improved analytical tools capable of detecting and quantifying aggregates in this size range to fill this gap. In so doing, significant progress is being made in this area. This progress is supported by the development of such techniques as flow imaging microscopy, nanoparticle tracking analysis using light scattering, and other more exotic physical techniques (e.g., resonant mass measurements, Archimedes) (Hawe et al., 2014).

Clearly, in assessing biosimilarity in terms of aggregation, the level of aggregates present is critically important. Here the goal is not to match aggregate levels determined in the RP, but rather not to exceed an upper aggregation limit based on experimental aggregation information obtained on the RP. In addition, given the importance of the physicochemical properties of the aggregates, biosimilar manufacturers should also consider conducting some appropriate level of comparability work focused on characterizing the structural nature of the aggregates found in the RP to those found in its biosimilar using various physicochemical approaches that have been reported recently in the scientific literature (Iacob et al., 2013; Iwura et al., 2014; Remmele et al., 2006). Such work would be focused on detecting the presence of significantly new forms of aggregates in the biosimilar relative to the RP that might be associated with different properties and therefore different immunogenicity and toxicological effects.

2.10 THE ROLE OF PHYSICOCHEMICAL MEASUREMENTS IN ASSESSING THE IMMUNOGENICITY OF BIOPHARMACEUTICALS

In generating physicochemical characterization data to adequately support the biosimilarity of a biosimilar to its RP, there is one area where present physicochemical analytical technology still has significant shortcomings in adequately assuring that biosimilarity exists. This gap in assessing biosimilarity concerns the area of immunogenicity. At present, much of the physicochemical investigation effort in this area has been focused on trying to improve the detection and quantification of aggregation, as discussed in Section 2.9. Unfortunately, little in terms of definitive useful information has been gathered about what specific attributes of a biopharmaceutical aggregate's structure one should be concerned with in terms of immunogenicity [other than some ideas concerning size and the presence of native-like repetitive structural elements (Rosenberg, 2006)].

Although great strides have been made in establishing analytical physicochemical biosimilarity (as a way to avoid the need to conduct costly and lengthy clinical trials), at present the complexity of factors that can play a role in immunogenicity (Mukovozov et al., 2008) makes it unavoidable to realize that physicochemical or biological data cannot adequately override the need for some clinical work to establish the absence of any immunogenicity concerns with a given biosimilar. Although future research in this area may eventually allow us to overcome this problem, at present the subtle and unclear knowledge as to what it is that is responsible for immunogenicity effects, pitted against its potential life-threatening repercussions, leaves little justification for taking imprudent risks. Hence, some limited form of clinical assessment will likely continue to be a requirement for some time to deal with this issue (see Chapter 7) until better knowledge is gained concerning what physicochemical attributes of a biopharmaceutical and its aggregates are of key importance in eliciting this type of response in humans (Moussa et al., 2016).

2.11 DETECTING DIFFERENCES DURING BIOSIMILARITY ASSESSMENT: WHAT'S IMPORTANT, WHAT'S NOT IMPORTANT?

The FDA has pointed out in general terms that adequate biosimilarity of a biosimilar to its RP is reached when a biosimilar is found to be "highly similar to the reference product not withstanding minor differences in clinically inactive components and that there are no clinically meaningful difference between the biological product and the reference product in terms of safety, purity and potency of the product" (FDA, 2015b). Since a biosimilar cannot be made identical to its RP, the big question becomes: "How highly similar is highly similar enough?" Given the enormous range of potential structural alterations that can occur to a biopharmaceutical, coupled with the ever-increasing capability of our analytical tools to assess and reveal physicochemical differences (e.g., again see Figure 2.4D), "What's important and what's not important in terms of safety, potency and purity when a physicochemical difference between a biosimilar and its RP is observed?" is unfortunately a very challenging question that needs to be addressed, which Miller (2011) noted would seem to require clinical data to properly answer. Indeed, a small change such as a single amino acid substitution [as seen in the case of hemoglobin (Murayama, 1967)] or the alteration of a PTM on a single amino acid [as seen in the case of the fucosylation of the oligosaccharide on the Fc region of an antibody (Houde et al., 2010)] can alter the structure or/and physicochemical properties of a biopharmaceutical that can translate into an altered biological function. Such a change in the behavior of a protein drug in response to such subtle changes in its structure certainly presents challenges in attempting to answer our posed question. To address this problem, to some extent, the FDA has categorized biosimilarity into several coarse levels: not similar, similar, highly similar, and highly similar with fingerprint-like similarity (FDA, 2014). Nevertheless, we might draw some better insights in attempting to answer our posed question by considering the following two interesting real case situations.

The first case concerns the comparability study carried out by Genzyme in 2008 to scale-up the production of its commercial biopharmaceutical Myozyme (a drug for Pompe disease) from a 160 L reactor to a 2000 L reactor to meet the growing commercial demands for this drug. In carrying out the scale-up, analytical results showed that the levels of an important key phosphorylated monosaccharide [mannose-6-phosphate, which Genzyme researchers and their collaborators had shown earlier to play an important role in the activity of this biopharmaceutical

(McVie-Wylie et al., 2008)] on the glycans that are chemically attached to several sites on Myozyme, were observed to differ between the two reactors. Although the biopharmaceutical from the 2000 L was already approved and licensed in a number of other countries, when Genzyme filed for a supplemental license for approval in the US to sell the biopharmaceutical from its 2000 L reactor under Myozyme's original BLA, the FDA would not grant Genzyme approval owing to the lack of comparability of the phosphorylated glycans and recommended that Genzyme apply for a new BLA license for the 2000 L drug (Foumier, 2015; Mack, 2008).

The second case concerns the FDA approval in 2010 of the first generic version of low molecular weight heparin (LMWH) called Enoxaparin, to an original LMWH RP called Lovenex (FDA, 2010a,b). Although LMWH is classified as a pharmaceutical and is not a protein, the complex nature of this drug (a heterogeneous collection of different-sized oligosaccharides that is derived from a biological source, e.g., porcine intestinal mucosa) presented challenges in developing a copy of this drug that were much more similar to what a biosimilar manufacturer faces in attempting to make a copy of a biopharmaceutical than what a generic manufacturer faces in attempting to make an identical copy of a small organic molecule. Indeed, Enoxaparin is not an identical copy of Lovenex; it is a "highly similar" copy (Guerrini et al., 2015; Mourier et al., 2015). Nevertheless, the FDA was able to develop a rigorous scientific approach (based on five critical criteria) to assess the sameness of Enoxaparin to Lovenex, which was heavily dependent on analytical, functional, and pharmacodynamics data to allow the approval of Enoxaparin without the need to conduct clinical trials (Lee et al., 2013).

These two cases should make it apparent that, even though they represent very different scenarios and do not uniquely answer our question, the key factors in the critical decision-making process of granting or denying these approvals rested on the combination of analytics with whatever underpinning knowledge was known about how these drugs worked at the time in which regulatory action was taken. Both examples highlight the totality of the evidence mantra and bring into focus the realization that each drug approval or denial stands to a large extent on its own merit, spawning the need to use the infamous "case-by-case" phrase in formulating an answer to the question "What's important and what's not important?"

2.12 CONCLUSION

In the years following the approval of those biopharmaceuticals that have or are now nearing their patent expiration date, great strides have been made in improving our ability to analytically characterize these complex drugs to better support their development and approval. These new analytical capabilities are now being called upon to play an even more intensive and critical role in the approval process of a form of biopharmaceutical that has over the last decade attracted significant attention from those who make, use, and regulate biopharmaceuticals, called biosimilars. This form of biopharmaceutical is not a new or improved version of the RP with better therapeutic performance attributes. Rather, they are simply a "highly similar" copy of the RP that can be made and sold at a much lower price. The approval of biosimilars should thus play an important role in reducing the financial burden that the costly RP has unfortunately placed on those who must pay for them, which, if achieved, should also increase the distribution of these drugs to those who need them but cannot afford them.

On the one hand, the key to a biosimilar's regulatory approval is dependent on the successful demonstration of biosimilarity—that the biosimilar is a highly similar copy of its corresponding RP with no meaningful difference in terms of safety, purity, and potency. On the other hand, the key to a biosimilar's commercial success can only be realized by achieving the following: (1) the ability to make biosimilars at a low enough cost so that they can be commercially offered to those who need them and must pay for them at a price significantly low enough in comparison to the corresponding RP (the subject matter of Chapter 16), so that these biosimilars will be purchased over the more costly RP (and over any residual lingering concerns about biosimilar's potential greater risk relative to that of the RP) and (2) the ability to offer the biosimilar at a price that will also provides a profit to the biosimilar manufacturer to justify its investment in this endeavor. Clearly, if *both* of these critical milestones are not achieved, we will have accomplished little and wasted much in diverting and squandering precious resources from the opportunities of finding new and improved drugs (especially for unmet health needs). At the same time, we will have lost a great opportunity for lowering the cost of health care and improving our ability to get these drugs to those who need them.

In tackling the subject matter of biosimilarity in this chapter, the author has dealt solely with the part of the biosimilarity exercise that is concerned with demonstrating biosimilarity from a structural perspective. In taking on this task, the author has discussed the associated scientific challenges and the impressive array of analytical tools and their capabilities that are available to assess the scientific attributes required to establish biosimilarity. Nevertheless, as noted by the FDA "Despite improvements in the analytical techniques, current analytical methodology may not be able to detect or characterize all relevant structural and functional differences between two products" (FDA, 2015d). This potential shortcoming is due to the small and subtle nature of structural differences in biopharmaceuticals that can make a significant difference in the performance of this class of drugs (as mentioned in Section 2.11) that can vary in importance from one biopharmaceutical to another (e.g., deamination for one biopharmaceutical may be an issue, while deamination in an entirely different biopharmaceutical may not). Hence, any structural difference observed through the use of our powerful analytical toolbox can be a potential source of concern. Unfortunately, as we increase our analytical power, we will likely also increase our ability to detect differences that may or may not be important. Coupling all this with the very subjective terminology concerning a biosimilar's successful milestone marker that is defined by the phrase "highly similar" only brings more uncertainty into play as to what a biosimilar manufacturer exactly needs to assure regulators that any observed difference between its biosimilar and the corresponding RP does not constitute a point of uncertainty (concerning a clinical meaningful difference) that will stand in the way of the biosimilar's approval. As a result, a biosimilar manufacturer needs to work closely with regulators early on in its biosimilar development program to get a clear picture as to what structural, functional, and even clinical data regulators will want to see (especially after the biosimilar manufacturer presents them with their preliminary window of consistency data package for the RP and the window of biosimilarity data package for its biosimilar), to avoid any surprises during the final biosimilarity assessment process.

In such an environment, the best we can do is try to convert the uncertainties we encountered into a situation of understanding the nature and extent of the risks we take, which we hope we can minimize, manage, and eventually remove through continual improvement of our scientific knowledge gained through the continual improvement of our analytical capabilities. By applying this mantra of maintaining our willingness to learn and apply our ever-increasing knowledge in the best way that we can, we can move forward to reap the opportunities and benefits of that increased but imperfect knowledge to improve the quality and longevity of our lives while minimizing the risks we unfortunately cannot totally avoid.

REFERENCES

Ahmadi M, Bryson CJ, Cloake EA, et al. (2015) Small amounts of sub-visible aggregates enhance the immunogenic potential of monoclonal antibody therapeutics. *Pharmaceutical Research* **32(4)**, 1383–1394.

Amezcua CA, Szabo CM. (2013) Assessment of higher order structure comparability in therapeutic proteins using nuclear magnetic resonance spectroscopy. *Journal of Pharmaceutical Sciences* **102(6)**, 1724–1733.

Anderson CL, Wang Y, Rustandi RR. (2012) Applications of imaged capillary isoelectric focussing technique in development of biopharmaceutical glycoprotein-based products. *Electrophoresis* **33(11)**, 1538–1544.

Arakawa T, Ejima D, Li T, Philo JS. (2010) The critical role of mobile phase composition in size exclusion chromatography of protein pharmaceuticals. *Journal of Pharmaceutical Sciences* **99(4)**, 1674–1692.

Arbogast LW, Brinson RG, Formolo T, et al. (2015a) 2D $^{1}H^{N}$, ^{15}N Correlated NMR methods at natural abundance for obtaining structural maps and statistical comparability of monoclonal antibodies. *Pharmaceutical Research* **33(2)**, 462–475.

Arbogast LW, Brinson RG, Marino JP. (2015b) Mapping monoclonal antibody structure by 2D ^{13}C NMR at natural abundance. *Analytical Chemistry* **87(7)**, 3556–3561.

Arnold JN, Wormald MR, Sim RB, et al. (2007) The impact of glycosylation on the biological function and structure of human immunoglobulins. *Annual Review of Immunology* **25**, 21–50.

Aubin Y, Freedberg DI, Keire DA. (2014) One- and two-dimensional NMR techniques for biopharmaceuticals. In: *Biophysical Characterization of Proteins in Developing Biopharmaceuticals*, Houde JD, Berkowitz SA, eds., 341–383. Elsevier, Amsterdam.

Aubin Y, Gingras G, Sauvé S. (2008) Assessment of the three-dimensional structure of recombinant protein therapeutics by NMR fingerprinting: demonstration on recombinant human granulocyte macrophage-colony stimulation factor. *Analytical Chemistry* **80(7)**, 2623–2627.

Ayoub D, Jabs W, Resemann A, et al. (2013) Correct primary structure assessment and extensive glyco-profiling of cetuximab by a combination of intact, middle-up, middle-down and bottom-up ESI and MALDI mass spectrometry techniques. *MAbs* **5(5)**, 699–710.

Beck A, Debaene F, Diemer H, et al. (2015) Cutting-edge mass spectrometry characterization of originator, biosimilar and biobetter antibodies. *Journal of Mass Spectrometry* **50(2)**, 285–297.

Beck A, Sanglier-Cianférani S, Van Dorsselaer A. (2012) Biosimilar, biobetter, and next generation antibody characterization by mass spectrometry. *Analytical Chemistry* **84(11)**, 4637–4646.

Beck A, Wagner-Rousset E, Ayoub D, et al. (2013) Characterization of therapeutic antibodies and related products. *Analytical Chemistry* **85(2)**, 715–736.

Beers Y. (1957) *Introduction to the Theory of Errors*. Addison-Wesley, Massachusetts.

Berkowitz SA. (2006) Role of analytical ultracentrifugation in assessing the aggregation of protein biopharmaceuticals. *American Association of Pharmaceutical Scientists Journal* **8(3)**, E590–605.

Berkowitz SA, Engen JR, Mazzeo JR, Jones GB. (2012) Analytical tools for characterizing biopharmaceuticals and the implications for biosimilars. *Nature Review Drug Discovery* **11(7)**, 527–540.

Berkowitz SA, Krull I, Rathore A. (2015) Challenges in the determination of protein aggregates, part II. *LC/GC North America* **33(7)**, 478–489.

Berkowitz SA, Zhong H, Berardino M, et al. (2005) Rapid quantitative capillary zone electrophoresis method for monitoring the micro-heterogeneity of an intact recombinant glycoprotein. *Journal of Chromatography A* **1079(1–2)**, 254–265.

Biacchi M, Gahoual R, Said N, et al. (2015) Glycoform separation and characterization of cetuximab variants by middle-up off-line capillary zone electrophoresis-UV/electrospray ionization-MS. *Analytical Chemistry* **87(12)**, 6240–6250.

Blackstone EA, Fuhr JP Jr. (2012) Innovation and competition: will biosimilars succeed? *Biotechnology Healthcare* **9(1)**, 24–27.

Blaich G, Janssen B, Roth G, Salfeld J. (2007) Overview: differentiating issues in the development of macromolecules compared with small molecules, In: *Handbook of Pharmaceutical Biotechnology*, Gad SC, ed., 89–123. John Wiley & Sons, Hoboken, NJ.

Bobst CE, Abzalimov RR, Houde D, et al. (2008) Detection and characterization of altered conformations of protein pharmaceuticals using complementary mass spectrometry-based approaches. *Analytical Chemistry* **80(19)**, 7473–7481.

Bobst CE, Kaltashov IA. (2011) Advanced mass spectrometry-based methods for the analysis of conformational integrity of biopharmaceutical products. *Current Pharmaceutical Biotechnology* **12(10)**, 1517–1529.

Bucciantini M, Giannoni E, Chiti F, et al. (2002) Inherent toxicity of aggregates implies a common mechanism for protein misfolding diseases. *Nature* **416**, 507–511.

Bush DR, Zang L, Belov AM, et al. (2016) High resolution CZE-MS quantitative characterization of intact biopharmaceutical proteins: proteoforms of interferon-β1. *Analytical Chemistry* **88(2)**, 1138–1146.

Calo-Fernández B, Martínez-Hurtado JL. (2012) Biosimilars: company strategies to capture value from the biologics market. *Pharmaceuticals* **5(12)**, 1393–1408.

Campuzano IDG, Larriba C, Bagal D, Schnier PD. (2015) Ion mobility and mass spectrometry measurements of the humanized IgGk NIST monoclonal antibody. In: *State-of-the-Art and Emerging Technologies for Therapeutic Monoclonal Antibody Characterization Volume 3. Defining the Next Generation of Analytical and Biophysical Techniques*, Schiel JE, Davis DL, Borisov OV, eds., ACS Symposium Series Volume 1202, 75–112. doi:10.1021/bk-2015-1202.ch004.

Carpenter JF, Randolph TW, Jiskoot W, et al. (2009) Overlooking subvisible particles in therapeutic protein products: gaps that may compromise product quality. *Journal of Pharmaceutical Sciences* **98(4)**, 1201–1205.

Carpenter JF, Randolph TW, Jiskoot W, et al. (2010) Potential inaccurate quantitation and sizing of protein aggregates by size exclusion chromatography: essential need to use orthogonal methods to assure the quality of therapeutic protein products. *Journal of Pharmaceutical Sciences* **99(5)**, 2200–2208.

Challner C. (2014) Tackling the challenges of HOS determination. *BioPharm International* **27(3)**. Available from: http://www.biopharminternational.com/tackling-challenge-hos-determination (Accessed December 27, 2015).

Chen G, Warrack BM, Goodenough AK, et al. (2011) Characterization of protein therapeutics by mass spectrometry: recent developments and future directions. *Drug Discovery Today* **16(1–2)**, 58–64.

Chen K, Freedberg DI, Keire DA. (2015) NMR profiling of biomolecules at natural abundance using 2D ^{1}H-^{15}N and ^{1}H-^{13}C multiplicity-separated (MS) HSQC spectra. *Journal of Magnetic Resonance* **251**, 65–70.

Chen SL, Wu SL, Huang LJ, et al. (2013) A global comparability approach for biosimilar monoclonal antibodies using LC-tandem MS based proteomics. *Journal of Pharmaceutical and Biomedical Analysis* **80**, 126–135.

Chow S-C. (2015) Challenging issues in assessing analytical similarity in biosimilar studies. *Biosimilars* **15**, 33–39.

Costantino HR, Langer R, Klibanov AM. (1994) Moisture-induced aggregation of lyophilized insulin. *Pharmaceutical Research* **11(1)**, 21–29.

Crommelin DJ, Shah VP, Klebovich I, et al. (2015) The similarity question for biologicals and non-biological complex drugs. *European Journal of Pharmaceutical Sciences* **76**, 10–17.

Davies M, Wang G, Fu G, Wang X. (2015) mAb higher order structure analysis with protein conformational array ELISA. *British Journal of Pharmaceutical Research* **7(6)**, 401–412.

Davies M, Wang G, Gong J, et al. (2016) Biosimilar mAb in-process sample higher order structure analysis with protein conformational array ELISA. *British Journal of Pharmaceutical Research* **9(3)**, 1–11.

Declerck P, Farouk-Rezk, Rudd PM. (2016) Biosimilarity versus manufacturing change: two distinct concepts. *Pharmaceutical Research* **33**, 261–268.

Demarest SJ, Frasca V. (2014) Differential scanning calorimetry in the biopharmaceutical sciences. In: *Biophysical Characterization of Proteins in Developing Biopharmaceuticals*, Houde JD, Berkowitz SA, eds., 287–306. Elsevier, Amsterdam.

den Engelsman J, Garidel P, Smulders R, et al. (2011) Strategies for the assessment of protein aggregates in pharmaceutical biotech product development. *Pharmaceutical Research* **28(4)**, 920–933.

Deperalta G, Alvarez M, Bechtel C, et al. (2013) Structural analysis of a therapeutic monoclonal antibody dimer by hydroxyl radical footprinting. *MAbs* **5(1)**, 86–101.

DiMasi JA, Grabowski, HG. (2007) The cost of biopharmaceutical R&D: is biotech different? *Managerial and Decision Economics* **28**, 467–479.

DiMasi JA, Grabowski HG, Hansen RW. (2016) Innovation in the pharmaceutical industry: new estimates of R&D. *Journal of Health Economics* **47**, 20–33, Supplemental material B1–B4.

Dobson CM. (2001) Protein folding and its links with human disease. *Biochemical Society Symposium* **68**, 1–28.

Doll S, Burlingame AL. (2015) Mass spectrometry-based detection and assignment of protein posttranslational modifications. *ACS Chemical Biology* **10(1)**, 63–71.

Dotz V, Haselberg R, Shubhakar A, et al. (2015) Mass spectrometry for glycosylation analysis of biopharmaceuticals. *TrAC Trends in Analytical Chemistry* **73**, 1–9.

Edgeworth MJ, Phillips JJ, Lowe DC, et al. (2015) Global and local conformation of human IgG antibody variants rationalizes loss of thermodynamic stability. *Angewandte Chemie International Edition in English* **54(50)**, 15156–15159.

Engen JR, Wales TE. (2015) Analytical aspects of hydrogen exchange mass spectrometry. *Annual Review Analytical Chemistry* **8**, 127–148.

Fajer PG, Bou-Assaf GM, Marshall AG. (2012) Improved sequence resolution by global analysis of overlapped peptides in hydrogen/deuterium exchange mass spectrometry. *Journal of American Society for Mass Spectrometry* **23(7)**, 1202–1208.

FDA. (1996) Guidance concerning demonstration of comparability of human biological products, including therapeutic biotechnology-derived products. Available from: http://www.fda.gov/Drugs/GuidanceComplianceRegulatoryInformation/Guidances/ucm122879.htm (Accessed December 27, 2015).

FDA. (2004) Guidance for Industry: PAT—a framework for innovative pharmaceutical development, manufacturing and quality assurance. Available from: http://www.fda.gov/downloads/Drugs/Guidances/ucm070305.pdf (Accessed December 27, 2015).

FDA. (2009a) Potential need for measurement standards to facilitate R&D of biologic drugs. Statement of Steven Kozlowski, M.D. before the U.S. House of Representatives. Available from: http://www.fda.gov/NewsEvents/Testimony/ucm183596.htm (Accessed December 27, 2015).

FDA. (2009b) Guidance for Industry: Q8(R2) Pharmaceutical development. Available from: http://www.fda.gov/downloads/Drugs/.../Guidances/ucm073507.pdf (Accessed December 27, 2015).

FDA. (2010a) FDA approves first generic enoxaparin sodium injection. News & Events, FDA News Release, July 23, 2010. Available from: http://www.fda.gov/NewsEvents/Newsroom/PressAnnouncements/ucm220092.htm (Accessed December 27, 2015).

FDA. (2010b) Establishing active ingredient sameness for a generic enoxaparin sodium, a low molecular weight heparin. Available from: http://www.fda.gov/Drugs/DrugSafety/PostmarketDrugSafetyInformationforPatientsandProviders/ucm220023.htm (Accessed December 27, 2015).

FDA. (2014) Draft Guidance for Industry: clinical pharmacology data to support a demonstration of biosimilarity to a reference product. Available from: http://www.fda.gov/downloads/drugs/guidancecomplianceregulatoryinformation/guidances/ucm397017.pdf (Accessed February 2, 2016).

FDA. (2015a) Drugs: facts about generic drugs. FDA, Silver Spring, MD. Last updated: June 19, 2015. Available from: http://www.fda.gov/Drugs/ResourcesForYou/Consumers/BuyingUsingMedicineSafely/UnderstandingGenericDrugs/ucm167991.htm (Accessed November 27, 2015).

FDA. (2015b) Scientific considerations in demonstrating biosimilarity to a reference product. Available from: http://www.fda.gov/downloads/Drugs/GuidanceComplianceRegulatoryInformation/Guidances/UCM291128.pdf (Accessed December 27, 2015).

FDA. (2015c) Biosimilars: questions and answers regarding implementation of the Biologics Price Competition and Innovation Act of 2009. Available from: http://www.fda.gov/downloads/Drugs/GuidanceComplianceRegulatoryInformation/Guidances/UCM444661.pdf (Accessed December 27, 2015).

FDA. (2015d) Guidance for Industry: quality considerations in demonstrating biosimilarity of a therapeutic protein to a reference product. Available from: http://www.fda.gov/downloads/Drugs/GuidanceComplianceRegulatoryInformation/Guidances/UCM291134.pdf (Accessed December 27, 2015).

Federal Register. (2009) Biologics Price Competition and Innovation Act, §7002(b)(3). Available from: http://www.fda.gov/downloads/Drugs/GuidanceComplianceRegulatoryInformation/ucm216146.pdf (Accessed December 27, 2015).

Fedorov AN, Baldwin TO. (1997) Cotranslational protein folding. *Journal of Biological Chemistry* **272(52)**, 32715–32718.

Fekete S, Beck A, Veuthey JL, Guillarme D. (2015) Ion-exchange chromatography for the characterization of biopharmaceuticals. *Journal of Pharmaceutical and Biomedical Analysis* **113**, 43–55.

Fekete S, Gassner A-L, Rudaz S, et al. (2013) Analytical strategies for the characterization of therapeutic monoclonal antibodies. *Trends in Analytical Chemistry* **42**, 74–83.

Fekete S, Guillarme D. (2014) Ultra-high-performance liquid chromatography for the characterization of therapeutic proteins. *Trends in Analytical Chemistry* **63**, 76–84.

Fekete S, Guillarme D, Sandra P, Sandra K. (2015) Chromatographic, electrophoretic and mass spectrometric methods for the analytical characterization of protein biopharmaceuticals. *Analytical Chemistry* **88(1)**, 480–507.

Fekete S, Veuthey JL, Guillarme D. (2012) New trends in reversed-phase liquid chromatographic separations of therapeutic peptides and proteins: theory and applications. *Journal of Pharmaceutical and Biomedical Analysis* **69**, 9–27.

Filipe V, Hawe A, Schellekens H, Jiskoot W. (2010) Aggregation and immunogenicity of therapeutic proteins. In: *Aggregation of Therapeutic Proteins*, Wang W, Roberts CJ, eds., 400–433. John Wiley & Sons, Hoboken, NJ.

Foumier J. (2015) A review of glycan analysis requirements. *BioPharm International* **28(10)**. Available from: http://www.biopharminternational.com/review-glycan-analysis-requirements (Accessed December 27, 2015).

Frank RG. (2007) The ongoing regulation of generic drugs. *The New England Journal of Medicine* **357(20)**, 1993–1996.

Franks J, Glushka JN, Jones MT, et al. (2015) Spin diffusion editing for structural fingerprints of therapeutic antibodies. *Analytical Chemistry* **88(2)**, 1320–1327.

Gahoual R, Biacchi M, Chicher J, et al. (2014b) Monoclonal antibodies biosimilarity assessment using transient isotachophoresis capillary zone electrophoresis-tandem mass spectrometry. *MAbs* **6(6)**, 1464–1473.

Gahoual R, Busnel JM, Beck A, François YN, Leize-Wagner E. (2014a) Full antibody primary structure and microvariant characterization in a single injection using transient isotachophoresis and sheathless capillary electrophoresis-tandem mass spectrometry. *Analytical Chemistry* **86(18)**, 9074–9081.

Geigert J. (2004) *The Challenge of CMC Regulatory Compliance for Biopharmaceuticals*, 17–34. Kluwer Academic/Plenum, New York.

Ghirlando R, Balbo A, Piszczek G, et al. (2013) Improving the thermal, radial, and temporal accuracy of the analytical ultracentrifuge through external references. *Analytical Biochemistry* **440(1)**, 81–95.

Gill SC, von Hippel PH. (1989) Calculation of protein extinction coefficients from amino acid sequence data. *Analytical Biochemistry* **182(2)**, 319–326.

Glassey J, Gernaey KV, Clemens C, et al. (2011) Process analytical technology (PAT) for biopharmaceuticals. *Biotechnology Journal* **6**, 369–377.

GPhA. (2012) Generic drug savings in the US. Report from the Generic Pharmaceutical Association, 4th annual edition. Available from: http://www.gphaonline.org/media/cms/IMSStudyAug2012WEB.pdf (Accessed December 27, 2015).

Greer F. (2012) Biosimilar developers face a reference-product dilemma. *BioPharm International* **25(3)**. Available from: http://www.biopharminternational.com/biosimilar-developers-face-reference-product-dilemma (Accessed December 27, 2015).

Grossman PD, Colburn JC. (1992) *Capillary Electrophoresis: Theory and Practice*, 331–345. Academic Press, San Diego, CA.

Guerrini M, Rudd TR, Mauri L, et al. (2015) Differentiation of generic enoxaparins marketed in the United States by employing NMR and multivariate analysis. *Analytical Chemistry* **87(16)**, 8275–8283.

Haberger M, Leiss M, Heidenreich AK, et al. (2016) Rapid characterization of biotherapeutic proteins by size-exclusion chromatography coupled to native mass spectrometry. *MAbs* **8(2)**, 331–339.

Haselberg R, de Jong GJ, Somsen GW. (2013) Low-flow sheathless capillary electrophoresis-mass spectrometry for sensitive glycoform profiling of intact pharmaceutical proteins. *Analytical Chemistry* **85(4)**, 2289–2296.

Haverick M, Mengisen S, Shameem M, Ambrogelly A. (2014) Separation of mAbs molecular variants by analytical hydrophobic interaction chromatography HPLC: overview and applications. *MAbs* **6(4)**, 852–858.

Hawe A, Zolls S, Freitag A, Carpenter JF. (2014) Subvisible and visible particle analysis in biopharmaceutical research and development. In: *Biophysical Characterization of Proteins in Developing Biopharmaceuticals*, Houde JD, Berkowitz SA, eds., 261–286. Elsevier, Amsterdam.

Heavner GA, Arakawa T, Philo JS, et al. (2007) Protein isolated from biopharmaceutical formulations cannot be used for comparative studies: follow-up to "a case study using Epoetin Alfa from Epogen and EPREX". *Journal of Pharmaceutical Sciences* **96(12)**, 3214–3225.

Hirsch BR, Balu S, Schulman KA. (2014) The impact of specialty pharmaceuticals as drivers of health care costs. *Health Affairs* **33(10)**, 1714–1720.

Holzmann J, Balser S, Windisch J. (2016) Totality of the evidence at work: the first U.S. biosimilar. *Expert Opinion on Biological Therapy* **16(2)**, 137–142. Available from: http://www.tandfonline.com/doi/pdf/10.1517/14712598.2016.1128410 (Accessed February 2, 2016).

Houde JD, Berkowitz SA. (2014a) Biophysical characterization: an integral part of the "totality of the evidence" concept. In: *Biophysical Characterization of Proteins in Developing Biopharmaceuticals*, Houde JD, Berkowitz SA, eds., 385–396. Elsevier, Amsterdam.

Houde JD, Berkowitz SA. (2014b) Biophysical industry's biophysical toolbox. In: *Biophysical Characterization of Proteins in Developing Biopharmaceuticals*, Houde JD, Berkowitz SA, eds., 46–78. Elsevier, Amsterdam.

Houde JD, Berkowitz SA. (2014c) Size-exclusion chromatography in biopharmaceutical process development. In: *Biophysical Characterization of Proteins in Developing Biopharmaceuticals*, Houde JD, Berkowitz SA, eds., 139–169. Elsevier, Amsterdam.

Houde D, Berkowitz SA. (2016) The role of hydrogen exchange mass spectroscopy in assessing the consistency and comparability of the higher-order structure of protein biopharmaceuticals. In: *Hydrogen Exchange Mass Spectrometry of Proteins: Fundamentals, Methods and Applications*, Weis DD, ed., 225–246. John Wiley & Sons, Chichester, UK.

Houde D, Berkowitz SA, Engen JR. (2011) The utility of hydrogen/deuterium exchange mass spectrometry in biopharmaceutical comparability studies. *Journal of Pharmaceutical Sciences* **100(6)**, 2071–2086.

Houde D, Peng Y, Berkowitz SA, Engen JR. (2010) Post-translational modifications differentially affect IgG1 conformation and receptor binding. *Molecular Cell Proteomics* **9(8)**, 1716–1728.

Huber C. (2015) Higher order mass spectrometry techniques applied to biopharmaceuticals. *LC/GC*, Oct 02, Available from: http://www.chromatographyonline.com/higher-order-mass-spectrometry-techniques-applied-biopharmaceuticals (Accessed February 2, 2016).

Iacob RE, Bou-Assaf GM, Makowski L, et al. (2013) Investigating monoclonal antibody aggregation using a combination of H/DX-MS and other biophysical measurements. *Journal of Pharmaceutical Sciences* **102(12)**, 4315–4329.

Iacob RE, Engen JR. (2012) Hydrogen exchange mass spectrometry: are we out of the quicksand? *Journal of American Society for Mass Spectrometry* **23(6)**, 1003–1010.

ICH. (2004) Comparability of biotechnological/biological products subject to changes in their manufacturing process. ICH Harmonised Tripartite Guideline Q5E. Available from: http://www.ich.org/fileadmin/Public_Web_Site/ICH_Products/Guidelines/Quality/Q5E/Step4/Q5E_Guideline.pdf (Accessed December 27, 2015).

ICH. (2005) Q8: pharmaceutical development. Available from: http://www.ich.org/products/guidelines/quality/quality-single/article/pharmaceutical-development.html (Accessed December 27, 2015).

Ivleva VB, Yu YQ, Scott Berger B, Chen W. (2012) Structural comparability assessment of innovator and biosimilar Rituximab using the biopharmaceutical system solution with INIFI, Waters Application note. Available from: http://www.waters.com/webassets/cms/library/docs/720004445en.pdf (Accessed December 27, 2015).

Iwura T, Fukuda J, Yamazaki K, et al. (2014) Intermolecular interactions and conformation of antibody dimers present in IgG1 biopharmaceuticals. *Journal of Biochemistry* **155(1)**, 63–71.

Jung SK, Lee KH, Jeon JW, et al. (2014) Physicochemical characterization of Remsima. *MAbs* **6(5)**, 1163–1177.

Kaltashov IA, Bobst CE, Abzalimov RR, et al. (2010) Conformation and dynamics of biopharmaceuticals: transition of mass spectrometry-based tools from academe to industry. *Journal of American Society for Mass Spectrometry* **21(3)**, 323–337.

Kaltashov IA, Bobst CE, Abzalimov RR, et al. (2012) Advances and challenges in analytical characterization of biotechnology products: mass spectrometry-based approaches to study properties and behavior of protein therapeutics. *Biotechnology Advances* **30(1)**, 210–222.

Kaltashov IA, Eyles SJ. (2005a) *Mass Spectrometry in Biophysics: Conformation and Dynamics of Biomolecules*, 157–163. John Wiley & Sons, Hoboken, NJ.

Kaltashov IA, Eyles SJ. (2005b) *Mass Spectrometry in Biophysics: Conformation and Dynamics of Biomolecules*, 365–367. John Wiley & Sons, Hoboken, NJ.

Karas M, Hillenkamp F. (1988) Laser desorption of proteins with molecular masses exceeding 10,000 daltons. *Analytical Chemistry* **60**, 2299–2301.

Kaur P, Kiselar J, Shi W, et al. (2015) Covalent labeling techniques for characterizing higher order structure of monoclonal antibodies. In: *State-of-the-Art and Emerging Technologies for Therapeutic Monoclonal Antibody Characterization Volume 3. Defining the Next Generation of Analytical and Biophysical Techniques*, Schiel JE, Davis DL, Borisov OV, eds., Vol. 1202, 45–73, doi:10.1021/bk-2015-1202.ch003.

Klepárník K. (2015) Recent advances in combination of capillary electrophoresis with mass spectrometry: methodology and theory. *Electrophoresis* **36(1)**, 159–178.

Kozlowski S, Swann P. (2006) Current and future issues in the manufacturing and development of monoclonal antibodies. *Advanced Drug Delivery Reviews* **58(5–6)**, 707–722.

Kozlowski S, Woodcock J, Midthun K, Sherman RB. (2011) Developing the nation's biosimilars program. *New England Journal of Medicine* **365(5)**, 385–388.

Kyte J. (1995) *Structure in Protein Chemistry*, 102–110. Garland Publishing, Inc., New York.

Lee S, Raw A, Yu L, et al. (2013) Scientific considerations in the review and approval of generic enoxaparin in the United States. *Nature Biotechnology* **31(3)**, 220–226.

Leurs U, Mistarz UH, Rand KD. (2015) Getting to the core of protein pharmaceuticals—Comprehensive structure analysis by mass spectrometry. *European Journal of Pharmaceutics and Biopharmaceuticals* **93**, 95–109.

Lu H, May K. (2012) Disulfide bond structures of IgG molecules: structural variations, chemical modifications and possible impacts to stability and biological function. *MAbs* **4(1)**, 17–23.

Mach H, Middaugh CR, Lewis RV. (1992) Detection of proteins and phenol in DNA samples with second-derivative absorption spectroscopy. *Analytical Biochemistry* **200(1)**, 74–80.

Mack G. (2008) FDA balks at Myozyme scale-up. *Nature Biotechnology* **26(6)**, 592.

Madsen JA, Yin Y, Qiao J, et al. (2016) Covalent labeling denaturation mass spectrometry for sensitive localized higher order structure comparisons. *Analytical Chemistry* **88(4)**, 2478–2488.

Majumdar R, Middaugh CR, Weis DD, Volkin DB. (2015) Hydrogen-deuterium exchange mass spectrometry as an emerging analytical tool for stabilization and formulation development of therapeutic monoclonal antibodies. *Journal of Pharmaceutical Sciences* **104(2)**, 327–345.

Mandel J. (1964) *The Statistical Analysis of Experimental Data*. John Wiley & Sons, New York.

Marino JP, Brinson RG, Hudgens JW, et al. (2015) Emerging technologies to assess the higher order structure of monoclonal antibodies. In: *State-of-the-Art and Emerging Technologies for Therapeutic Monoclonal Antibody Characterization Volume 3. Defining the Next Generation of Analytical and Biophysical Techniques*, Schiel JE, Davis DL, Borisov OV, eds., ACS Symposium Series Volume 1202, 17–43. American Chemical Society, Washington, DC.

McCamish M, Woollett G. (2011) Worldwide experience with biosimilar development. *MAbs* **3(2)**, 209–217.

McCamish M, Woollett G. (2012) The state of the art in the development of biosimilars. *Clinical Pharmacology and Therapeutics* **91(3)**, 405–417.

McCamish M, Woollett G. (2013) The continuum of comparability extends to biosimilarity: how much is enough and what clinical data are necessary? *Clinical Pharmacology and Therapeutics* **93(4)**, 315–317.

McVie-Wylie AJ, Lee KL, Qiu H, et al. (2008) Biochemical and pharmacological characterization of different recombinant acid alpha-glucosidase preparations evaluated for the treatment of Pompe disease. *Molecular Genetics and Metabolism* **94(4)**, 448–455.

Mei-feng X, Jun-jian F, Fang-ting D, Xian-zhong Y. (2015) Higher order structure assessment of biosimilars based on the correlation of NMR spectral fingerprints. *Chinese Journal of Magnetic Resonance* **32(2)**, 342–353.

Mellors JS, Black WA, Chambers AG, et al. (2013) Hybrid capillary/microfluidic system for comprehensive online liquid chromatography-capillary electrophoresis-electrospray ionization-mass spectrometry. *Analytical Chemistry* **85(8)**, 4100–4106.

Meng CK, Mann M, Fenn JB. (1988) Of protons or proteins. *Zeitschrift für Physik D* **10**, 361–368.

Miller HI. (2011) Why an abbreviated FDA pathway for biosimilars is overhyped. *Nature Biotechnology* **29(9)**, 794–795.

Mo J, Tymiak AA, Chen G. (2012) Structural mass spectrometry in biologics discovery: advances and future trends. *Drug Discovery Today* **17(23–24)**, 1323–1330.

Moremen KW, Tiemeyer M, Nairn AV. (2012) Vertebrate protein glycosylation: diversity, synthesis and function. *Nature Reviews Molecular Cell Biology* **13(7)**, 448–462.

Moritz B, Schnaible V, Kiessig S, et al. (2015) Evaluation of capillary zone electrophoresis for charge heterogeneity testing of monoclonal antibodies. *Journal Chromatography B* **983–984**, 101–110.

Moroco JA, Engen JR. (2015) Replication in bioanalytical studies with HDX MS: aim as high as possible. *Bioanalysis* **7(9)**, 1065–1067.

Mourier PA, Agut C, Souaifi-Amara H, et al. (2015) Analytical and statistical comparability of generic enoxaparins from the US market with the originator product. *Journal of Pharmaceutical and Biomedical Analysis* **115**, 431–442.

Moussa EM, Panchal JP, Moorthy BS, et al. (2016) Immunogenicity of therapeutic protein aggregates. *Journal of Pharmaceutical Sciences* **105**, 417–430.

Mukovozov I, Sabljic T, Hortelano G, Ofosu FA. (2008) Factors that contribute to the immmunogenicity of therapeutic recombinant human proteins. *Thrombosis Haemostasis* **99(5)**, 874–882.

Muneeruddin K, Nazzaro M, Kaltashov IA. (2015) Characterization of intact protein conjugates and biopharmaceuticals using ion-exchange chromatography with online detection by native electrospray ionization mass spectrometry and top-down tandem mass spectrometry. *Analytical Chemistry* **87(19)**, 10138–10145.

Muneeruddin K, Thomas JJ, Salinas PA, Kaltashov IA. (2014) Characterization of small protein aggregates and oligomers using size exclusion chromatography with online detection by native electrospray ionization mass spectrometry. *Analytical Chemistry* **86(21)**, 10692–10699.

Murayama M. (1967) Structure of sickle cell hemoglobin and molecular mechanism of the sickling phenomenon. *Clinical Chemistry* **13(7)**, 578–588.

Neill A, Nowak C, Patel R, et al. (2015) Characterization of recombinant monoclonal antibody charge variants using OFFGEL fractionation, weak anion exchange chromatography, and mass spectrometry. *Analytical Chemistry* **87(12)**, 6204–6211.

Pace CN, Vajdos F, Fee L, Grimsley G, Gray T. (1995) How to measure and predict the molar absorption coefficient of a protein. *Protein Science* **4(11)**, 2411–2423.

Pan J, Borchers CH. (2014) Top-down mass spectrometry and hydrogen/deuterium exchange for comprehensive structural characterization of interferons: implications for biosimilars. *Proteomics* **14(10)**, 1249–1258.

Pan J, Zhang S, Parker CE, Borchers CH. (2014) Subzero temperature chromatography and top-down mass spectrometry for protein higher-order structure characterization: method validation and application to therapeutic antibodies. *Journal of the American Chemical Society* **136(37)**, 13065–13071.

Panjwani N, Hodgson DJ, Sauvé S, Aubin Y. (2010) Assessment of the effects of pH, formulation and deformulation on the conformation of interferon alpha-2 by NMR. *Journal of Pharmaceutical Sciences* **99(8)**, 3334–3342.

Pirrone GF, Iacob RE, Engen JR. (2015) Applications of hydrogen/deuterium exchange MS from 2012 to 2014. *Analytical Chemistry* **87(1)**, 99–118.

Plumb RS, Johnson KA, Rainville P, et al. (2006) UPLC/MS^E: a new approach for generating molecular fragment information for biomarker structure elucidation. *Rapid Communication in Mass Spectroscopy* **20(13)**, 1989–1994.

Poppe L, Jordan JB, Lawson K, et al. (2013) Profiling formulated monoclonal antibodies by (1)H NMR spectroscopy. *Analytical Chemistry* **85(20)**, 9623–9629.

Poppe L, Jordan JB, Rogers G, Schnier PD. (2015) On the analytical superiority of 1D NMR for fingerprinting the higher order structure of protein therapeutics compared to multidimensional NMR methods. *Analytical Chemistry* **87(11)**, 5539–5545.

Rader RA. (2007) What is a generic biopharmaceutical? Biogeneric? Follow-on protein? Biosimilar? Follow-on biologic? Part 1: introduction and basic paradigms. *BioProcess International* **5(3)**, 28–38.

Ramanan S, Grampp G. (2014) Drift, evolution, and divergence in biologics and biosimilars manufacturing. *BioDrugs* **28(4)**, 363–372.

Rand KD, Zehl M, Jensen ON, Jørgensen TJ. (2009) Protein hydrogen exchange measured at single-residue resolution by electron transfer dissociation mass spectrometry. *Analytical Chemistry* **81(14)**, 5577–5584.

Rathore AS, Bhambure R, Ghare V. (2010) Process analytical technology (PAT) for biopharmaceutical products. *Analytical and Bioanalytical Chemistry* **398(1)**, 137–154.

Rathore AS, Mhatre R. (2009) *Quality by Design for Biopharmaceuticals: Principles and Case Studies*. John Wiley & Sons, Hoboken, NJ.

Remmele RL Jr., Callahan WJ, Krishnan S, et al. (2006) Active dimer of Epratuzumab provides insight into the complex nature of an antibody aggregate. *Journal of Pharmaceutical Sciences* **95(1)**, 126–145.

Reusch D, Haberger M, Falck D, et al. (2015a) Comparison of methods for the analysis of therapeutic immunoglobulin G Fc-glycosylation profiles—Part 2: mass spectrometric methods. *MAbs* **7(4)**, 732–742.

Reusch D, Haberger M, Maier B, et al. (2015b) Comparison of methods for the analysis of therapeutic immunoglobulin G Fc-glycosylation profiles-Part 1: separation-based methods. *MAbs* **7(1)**, 167–179.

Rogstad, S, Faustino A, Ruth A, et al. (2016) A retrospective evaluation of the use of mass spectrometry in FDA biological license applications. *Journal American Society Mass Spectrometry*. doi:10.1007/s13361-016-1531-9

Rosati S, Thompson NJ, Barendregt A, et al. (2012) Qualitative and semiquantitative analysis of composite mixtures of antibodies by native mass spectrometry. *Analytical Chemistry* **84(16)**, 7227–7232.

Rosenberg AS. (2006) Effects of protein aggregates: an immunologic perspective. *American Association of Pharmaceutical Scientists Journal* **8**, E501–E507.

Rouhi, AM. (2012) Easing the toil of hydrogen exchange. *Chemical & Engineering News* **90(17)**, 16–17.

Rustandi RR, Washabaugh MW, Wang Y. (2008) Applications of CE SDS gel in development of biopharmaceutical antibody-based products. *Electrophoresis* **29(17)**, 3612–3620.

Sandra K, Sandra P. (2015) The opportunities of 2D-LC in the analysis of monoclonal antibodies. *Bioanalysis* **7(22)**, 2843–2847.

Sandra K, Vandenheede I, Sandra P. (2014) Modern chromatographic and mass spectrometric techniques for protein biopharmaceutical characterization. *Journal of Chromatography A* **1335**, 81–103.

Sarpatwari A, Avorn J, Kesselheim, AS. (2015) Progress and hurdles for follow-on biologics. *New England Journal of Medicine* **372(25)**, 2380–2382.

Schiestl M, Stangler T, Torella C, et al. (2011) Acceptable changes in quality attributes of glycosylated biopharmaceuticals. *Nature Biotechnol*ogy **29**, 310–312.

Schneider SK. (2013) Biosimilars in rheumatology: the wind of change. *Annals of the Rheumatic Diseases* **72(3)**, 315–318.

Shibata-Koyama M, Iida S, Okazaki A, et al. (2009) The *N*-linked oligosaccharide at Fc gamma RIIIa Asn-45: an inhibitory element for high Fc gamma RIIIa binding affinity to IgG glycoforms lacking core fucosylation. *Glycobiology* **19(2)**, 126–134.

Siuzdak G. (2003) *The Expanding Role of Mass Spectrometry in Biotechnology*, 45–68. MCC Press, San Diego, CA.

Smith LM, Kelleher NL. (2013) Proteoform: a single term describing protein complexity. *Nature Methods* **10(3)**, 186–187.

Solá RJ, Griebenow K. (2009) Effects of glycosylation on the stability of protein pharmaceuticals. *Journal Pharmaceutical Sciences* **98(4)**, 1223–1245.

Solá RJ, Griebenow K. (2010) Glycosylation of therapeutic proteins: an effective strategy to optimize efficacy. *BioDrugs* **24(1)**, 9–21.

Solá RJ, Rodríguez-Martínez JA, Griebenow K. (2007) Modulation of protein biophysical properties by chemical glycosylation: biochemical insights and biomedical implications. *Cell and Molecular Life Sciences* **64(16)**, 2133–2152.

Stoll DR. (2015) Recent advances in 2D-LC for bioanalysis. *Bioanalysis* **7(24)**, 3125–3142.

Stoll DR, Harmes DC, Danforth J, et al. (2015) Direct identification of rituximab main isoforms and subunit analysis by online selective comprehensive two-dimensional liquid chromatography-mass spectrometry. *Analytical Chemistry* **87(16)**, 8307–8315.

Suzuki S. (2013) Recent developments in liquid chromatography and capillary electrophoresis for the analysis of glycoprotein glycans. *Analytical Sciences* **29(12)**, 1117–1128.

Tamizi E, Jouyban A. (2015) The potential of the capillary electrophoresis techniques for quality control of biopharmaceuticals. *Electrophoresis* **36(6)**, 831–858.

Tebbey PW, Varga A, Naill M, et al. (2015) Consistency of quality attributes for the glycosylated monoclonal antibody Humira® (adalimumab). *MAbs* **7(5)**, 805–811.

Thayer AM. (2013) The new copycats. *Chemical & Engineering News* **91(40)**, 15–23.

Vanhoenacker G, Vandenheede I, David F, et al. (2015) Comprehensive two-dimensional liquid chromatography of therapeutic monoclonal antibody digests. *Analytical and Bioanalytical Chemistry* **407(1)**, 355–366.

Ventola CL. (2013) Biosimilars Part 1: proposed regulatory criteria for FDA approval. *Pharmacy and Therapeutics* **38(5)**, 270–287.

Walsh CT. (2006a) *Posttranslational Modifications of Proteins: Expanding Nature's Inventory*. Roberts & Company Publishers, Greenwood Village, CO.

Walsh CT. (2006b) *Posttranslational Modifications of Proteins: Expanding Nature's Inventory*, 203–242. Roberts & Company Publishers, Greenwood Village, CO.

Walsh CT, Garneau-Tsodikova S, Gatto GJ Jr. (2005) Protein posttranslational modifications: the chemistry of proteome diversifications. *Angewandte Chemie International Edition in English* **44(45)**, 7342–7372.

Walsh G. (2007) *Pharmaceutical Biotechnology*, 17. John Wiley & Sons, West Sussex, UK.

Walsh G, Jefferis R. (2006) Post-translational modifications in the context of therapeutic proteins. *Nature Biotechnology* **24**, 1241–1252.

Wang X, Kumar S, Singh, SK. (2011) Disulfide scrambling in IgG2 monoclonal antibodies: insights from molecular dynamics simulations. *Pharmaceutical Research* **28(12)**, 3128–3144.

Wang X, Li Q, Davies M. (2013) Development of antibody arrays for monoclonal antibody higher order structure analysis. *Frontiers in Pharmacology* **4**, 103.

Wang Y, Lu Q, Wu SL, et al. (2011) Characterization and comparison of disulfide linkages and scrambling patterns in therapeutic monoclonal antibodies—using LC-MS with electron transfer dissociation. *Analytical Chemistry* **83(8)**, 3133–3140.

Waters Corporation. (2011). White paper: an overview of the principles of MS^E, the engine that drives MS performance. Available from: http://www.waters.com/webassets/cms/library/docs/720004036en.pdf (Accessed December 27, 2015).

Wei H, Mo J, Tao L, et al. (2014) Hydrogen/deuterium exchange mass spectrometry for probing higher order structure of protein therapeutics: methodology and applications. *Drug Discovery Today* **19(1)**, 95–102.

Wen J, Arakawa T, Philo JS. (1996) Size-exclusion chromatography with on-line light-scattering, absorbance, and refractive index detectors for studying proteins and their interactions. *Analytical Biochemistry* **240(2)**, 155–166.

Wiesner J, Resemann A, Evans C, et al. (2015) Advanced mass spectrometry workflows for analyzing disulfide bonds in biologics. *Expert Review Proteomics* **12(2)**, 115–123.

Wishart DS. (2013) Characterization of biopharmaceuticals by NMR spectroscopy. *Trends in Analytical Chemistry* **48**, 96–111.

Woodcock J, Griffin J, Behrman R, et al. (2007) The FDA's assessment of follow-on protein products: a historical perspective. *Nature Review Drug Discovery* **6(6)**, 437–442.

Xie H, Chakraborty A, Ahn J, et al. (2010) Rapid comparison of a candidate biosimilar to an innovator monoclonal antibody with advanced liquid chromatography and mass spectrometry technologies. *MAbs* **2(4)**, 379–394.

Zhang H, Cui W, Gross ML. (2014) Mass spectrometry for the biophysical characterization of therapeutic monoclonal antibodies. *FEBS Letters* **588(2)**, 308–317.

Zhang Z, Pan H, Chen X. (2009) Mass spectrometry for structural characterization of therapeutic antibodies. *Mass Spectrometry Reviews* **28(1)**, 147–176.

Zhao H, Ghirlando R, Alfonso C, et al. (2015) A multilaboratory comparison of calibration accuracy and the performance of external references in analytical ultracentrifugation. *PLoS One* **10(5)**, e0126420.

3 Analytical Similarity Assessment

Shein-Chung Chow and Li Liu
Duke University School of Medicine

CONTENTS

3.1 BACKGROUND

Following passage of the Biologics Price Competition and Innovation (BPCI) Act in 2009, the FDA circulated three guidances on the demonstration of biosimilarity of biosimilar products for public comments in April 2015 (FDA, 2015a,b,c). These guidances are intended not only (1) to assist sponsors to demonstrate that a proposed therapeutic protein product is biosimilar to a reference product for the purpose of submitting a marketing application under Section 351(k) of the Public Health Service (PHS) Act, but also (2) to describe the FDA's current thinking on factors demonstrating that a proposed protein product is highly similar to a reference product, which was licensed under Section 351(a) of the PHS Act. In the guidance on *Scientific Considerations in Demonstrating Biosimilarity to a Reference Product*, the FDA introduces the concept of a stepwise approach to obtaining "Totality of the Evidence" for the regulatory review and approval of biosimilar applications (FDA, 2015a).

The stepwise approach starts with the assessment of analytical similarity of critical quality attributes (CQAs) for structural and functional characterization in the manufacturing process of biosimilar products that may have an impact on the assessment of similarity. In practice, often a large number of CQAs may be relevant to clinical outcomes. Thus, it is almost impossible to assess analytical similarity for all of these CQAs individually. As a result, the FDA suggests that the sponsors identify CQAs that are relevant to clinical outcomes and classify them into three tiers depending on their criticality risk ranking—most relevant (Tier 1), mild to moderately relevant (Tier 2), and least relevant (Tier 3) to clinical outcomes. To assist the sponsors, the FDA also proposes some statistical approaches for the assessment of analytical similarity for CQAs from different tiers. For example, the FDA recommends an equivalence test for CQAs from Tier 1, a quality range approach for CQAs from Tier 2, and descriptive raw data and graphical presentation for CQAs from Tier 3 (see, e.g., Chow, 2013, 2014, 2015; Christl, 2015; Tsong, 2015).

This chapter not only provides a close look at these approaches by providing interpretation and/or statistical justification whenever possible, but also discusses some challenging issues to the FDA's proposed approach (mainly on the equivalence test for Tier 1 CQAs). In addition, some recommendations and alternative methods are proposed.

In the next section, the stepwise approach for demonstrating biosimilarity as suggested by the FDA draft guidance is briefly outlined. Assessment of quality attributes is given in Section 3.3. Section 3.4 provides brief descriptions of the equivalence test, the quality range approach, and the method of descriptive raw data and graphical comparison. Some challenging issues to the FDA's proposed approaches are discussed in Section 3.5. Section 3.6 provides recommendations and alternative methods for the assessment of analytical similarity in CQAs from different tiers. Some concluding remarks are given in the last section of this chapter.

3.2 STEPWISE APPROACH FOR DEMONSTRATING BIOSIMILARITY

As defined in the BPCI Act, a biosimilar product is a product that is *highly similar* to the reference product notwithstanding minor differences in clinically inactive components, and there are no clinically meaningful differences in terms of safety, purity, and potency. Based on the definition of the BPCI Act, biosimilarity requires that there are no *clinically meaningful differences* in terms of *safety, purity* and *potency.* Safety could include pharmacokinetics and pharmacodynamics (PK/PD), safety and tolerability, and immunogenicity studies. Purity includes all CQAs during the manufacturing process. Potency is referred to as efficacy studies. As indicated earlier, in the 2012 FDA draft guidance on scientific considerations, the FDA recommends that a stepwise approach be considered for providing the totality of the evidence to demonstrate the biosimilarity of a proposed biosimilar product as compared to a reference product (FDA, 2015a).

The stepwise approach is briefly summarized by a pyramid illustrated in Figure 3.1. The process starts with analytical studies for structural and functional characterization. The stepwise approach continues with animal studies for toxicity, clinical pharmacology studies such as PK/PD studies, followed by investigations of immunogenicity and clinical studies for safety/tolerability and efficacy.

The sponsors are encouraged to consult with medical/statistical reviewers of the FDA with the proposed plan or strategy of the stepwise approach for regulatory agreement and acceptance. This is to make sure that the information provided is sufficient to fulfill the FDA's requirement for providing totality of the evidence for the demonstration of biosimilarity of the proposed biosimilar product as compared to the reference product. As an example, more specifically, the analytical studies are to assess similarity in CQAs at various stages of the manufacturing process of the biosimilar product as compared to those of the reference product. To assist the sponsors to fulfill the regulatory requirement for providing totality of the evidence of analytical similarity, the FDA suggests several approaches depending on the criticality of the identified quality attributes relevant to the clinical outcomes.

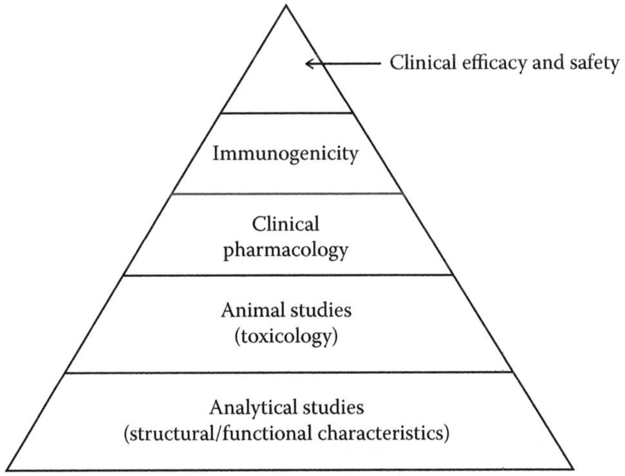

FIGURE 3.1 A stepwise approach to demonstrate biosimilarity.

3.3 TIER ASSIGNMENT FOR CRITICAL QUALITY ATTRIBUTES

Tsong (2015) indicated that CQAs are necessarily tested for the functional, structural, and physicochemical characterization of the proposed biosimilar product as compared to a reference product (either a US-licensed product or an EU-approved reference product) for analytical similarity assessment. Analytical similarity assessment is considered to be the foundation of the stepwise approach for obtaining the totality of the evidence for demonstrating biosimilarity between the proposed biosimilar product and the reference product. Gutierrez-Lugo (2015) provided a list of CQAs and methods used to evaluate the analytical similarity of the proposed biosimilar product (EP2006) as compared to a US-licensed Neupogen and EU-approved Neupogen (see Table 3.1). These CQAs are assessed for analytical similarity by means of the so-called tier approach.

3.3.1 Criticality Risk Ranking

The tier approach first assesses the criticality risk ranking of the CQAs relevant to clinical outcome and classifies these CQAs to appropriate tiers depending on their impact (degree of criticality risk ranking) on clinical outcomes. The CQAs with most relevance to clinical outcomes will be assigned to Tier 1, while the CQAs with mild to moderate relevance to clinical outcomes will be classified as Tier 2. Tier 3 will contain those CQAs with the least relevance to clinical outcomes. In practice, it is believed that biological activity assays are the best representation available to test the clinically relevant mechanism of action (MOA) and therefore should be assigned to Tier 1. Other CQAs that are tested in comparative physicochemical and functional assessment (outside of those relevant to MOA) are of potential relevance to similarity, which are considered most appropriate for Tier 2 or Tier 3.

The FDA, however, has suggested a critical risk ranking of quality attributes with regard to their potential impact on activity, PK/PD, safety, and immunogenicity, with quality attributes being assigned to tiers commensurate with their risk. As a result, it is suggested that a statistical approach should be considered to serve as a decision tool for certain CQAs that are relevant to the demonstration of similarity. In other words, it is suggested that an appropriate statistical model should be used not only to determine the relevance or association between CQAs and clinical outcomes but also to assess the criticality risk ranking of the CQAs relevant to clinical outcome by establishing a predictive model. The established predictive model can then be used to determine the degree of criticality risk ranking for assignment of the identified CQAs to appropriate tiers.

3.3.2 Statistical Model

According to the *United States Pharmacopeia* (USP), *in vitro* and *in vivo* correlation (IVIVC) is referred to as the establishment of a relationship between a biological property, or a parameter derived from a biological property produced from a dosage form, and a physicochemical property of the same dosage form (USP/NF, 2000; Chow and Liu, 2008). Typically, AUC (area under a blood or plasma concentration-time curve) or peak concentration (C_{max}) is considered the parameter derived from the biological property, while the physicochemical property is the *in vitro* dissolution profile.

TABLE 3.1
Example of Critical Quality Attributes for NEUPOGEN (Filgrastim)

Quality Attribute	Methods
Primary structure	N-terminal sequencing
	Peptide mapping with UV and MS detection
	Protein molecular mass by ESI MS
	Protein molecular mass MALDI-TOF MS
	DNA sequencing of construct cassette
	Peptide mapping coupled with MS/MS
Bioactivity	Proliferation of murine myelogenous leukemia cells (NFS-60)
Receptor binding	Surface plasmon resonance
Protein content	RP-HPLC
Clarity	Nephelometry
Subvisible particles	Microflow imaging
Higher order structure	Far and near UV circular dichroism
	^{1}H nuclear magnetic resonance
	^{1}H-^{15}N heteronuclear single quantum coherence spectroscopy
	LC-MS (disulfide bond)
High-molecular-weight variants/ aggregates	Size exclusion chromatography
	Reduced and nonreduced SDS-PAGE
Oxidized species	RP-HPLC
	LC/MS
Covalent dimers	LC/MS
Partially reduced species	LC/MS
Sequence variants:	RP-HPLC
His → Gln	LC/MS
Asp → Glu	
Thr → Asp	
fMet1 species	RP-HPLC
	LC/MS
Succinimide species	RP-HPLC
	LC/MS
Phosphoglucunoylation	LC/MS
Acetylated species	LC/MS
N-terminal truncated variants	LC-MS/MS
Norleucine species	RP-HPLC
	LC/MS
Deamidated species	RP-HPLC
	LC/MS
	IEF
	CEX

Basically, under the Fundamental Bioequivalence Assumption, the IVIVC is to use the dissolution test as a surrogate for human studies (i.e., when the drug absorption profiles in terms of AUC or C_{max} are similar, it is assumed that they are therapeutically equivalent). In addition, one of IVIVC's main roles is to assist in the quality control of functional and/or structural characteristics during the manufacturing process.

For simplicity and illustration purposes, we will consider the case where the relationship between CQA and clinical outcome is linear. The nonlinear case can be similarly treated. Let x and y be the response of a CQA and the clinical outcome, respectively. In practice, if the CQA is relevant to clinical outcome, it is assumed that the clinical outcome can be predicted by the CQA accurately and reliably with some statistical assurance. One of the statistical criteria is to examine the degree of closeness (or the degree of relevance) between the observed response y and the predicted response $\hat{y}$ through an established statistical model. To perform this examination, we will first study the association between x and y and build up a model. Then, we will validate the model based on some criteria. For simplicity, we assume that x and y can be described by the following linear model:

$$y = \beta_0 + \beta_1 x + \varepsilon \tag{3.1}$$

where ε follows a normal distribution with a mean of 0 and a variance of σ_e^2. Suppose that n pairs of observations $(x_1, y_1), \ldots, (x_n, y_n)$ are observed in a translation process. To define the notation, let

$$X^T = \begin{pmatrix} 1 & 1 & \ldots & 1 \\ x_1 & x_2 & \ldots & x_n \end{pmatrix}$$

and

$$Y^T = \begin{pmatrix} y_1 & y_2 & \ldots & y_n \end{pmatrix}.$$

Then, under model 3.1, the maximum likelihood estimates of the parameters β_0 and β_1 are:

$$\begin{pmatrix} \hat{\beta}_0 \\ \hat{\beta}_1 \end{pmatrix} = (X^T X)^{-1} X^T Y$$

with

$$\text{var}\begin{pmatrix} \hat{\beta}_0 \\ \hat{\beta}_1 \end{pmatrix} = (X^T X)^{-1} \sigma_e^2.$$

Furthermore, σ_e^2 can be estimated by the mean squared error (MSE), which is given by

$$\hat{\sigma}_e^2 = \frac{1}{n-2}\sum_{i=1}^{n}\left(y_i - \hat{y}_i\right)^2.$$

Thus, we have established the following relationship:

$$\hat{y} = \hat{\beta}_0 + \hat{\beta}_1 x. \tag{3.2}$$

For a given $x = x_0$, suppose that the corresponding observed value is given by y; however, using Equation 3.2, the corresponding fitted value is $\hat{y} = \hat{\beta}_0 + \hat{\beta}_1 x_0$. Note that $E(\hat{y}) = \beta_0 + \beta_1 x_0 = \mu_0$ and

$$\text{var}(\hat{y}) = \begin{pmatrix}1 & x_0\end{pmatrix}(X^T X)^{-1}\begin{pmatrix}1\\ x_0\end{pmatrix}\sigma_e^2 = c\sigma_e^2,$$

where

$$c = \begin{pmatrix}1 & x_0\end{pmatrix}(X^T X)^{-1}\begin{pmatrix}1\\ x_0\end{pmatrix}.$$

Furthermore, $\hat{y}$ is normally distributed with mean μ_0 and variance $c\sigma_e^2$, that is,

$$\hat{y} \sim N\left(\mu_0, c\sigma_e^2\right).$$

We may validate the translation model by considering how close an observed y is to its predicted value $\hat{y}$, which is fitted to the regression model 3.2. To assess the closeness, we propose the following two measures, which are based either on the absolute difference or the relative difference between y and $\hat{y}$:

$$\text{Criterion I.} \quad p_1 = P\left\{\left|y - \hat{y}\right| < \delta\right\},$$

$$\text{Criterion II.} \quad p_2 = P\left\{\left|\frac{y - \hat{y}}{y}\right| < \delta\right\}.$$

In other words, it is desirable to have a high probability that the difference or the relative difference between y and $\hat{y}$, given by p_1 and p_2, respectively, is less than a clinically or scientifically meaningful difference δ. Then, for either $i = 1$ or 2, it is of interest to test the following hypotheses:

$$H_0 : p_i \le p_0 \text{ versus } H_a : p_i > p_0, \tag{3.3}$$

where p_0 is some prespecified constant. The idea is to reject H_0 in favor of H_a. In other words, we would like to reject the null hypothesis H_0 and conclude H_a, which implies that the established model is considered validated.

3.3.2.1 Measure of Closeness Based on the Absolute Difference

It should be noted that we have

$$(y-\hat{y}) \sim N\left(0,(1+c)\sigma_e^2\right).$$

Therefore, p_1 can be estimated by

$$\hat{p}_1 = \Phi\left(\frac{\delta}{\sqrt{(1+c)\hat{\sigma}_e^2}}\right) - \Phi\left(\frac{-\delta}{\sqrt{(1+c)\hat{\sigma}_e^2}}\right).$$

Using the delta method through a Taylor expansion, for a sufficiently large sample size n,

$$\operatorname{var}\left(\hat{p}_1\right) \approx \left(\phi\left(\frac{\delta}{\sqrt{(1+c)\sigma_e^2}}\right) - \phi\left(\frac{-\delta}{\sqrt{(1+c)\sigma_e^2}}\right)\right)^2 \frac{\delta}{2(1-\delta)(n-2)\sigma_e^2},$$

where $\phi(z)$ is the probability density function of a standard normal distribution. Furthermore, $\operatorname{var}\left(\hat{p}_1\right)$ can be estimated by V_1, where V_1 is given by

$$V_1 = \frac{2\delta^2}{(1+c)(n-2)\hat{\sigma}_e^2}\phi^2\left(\frac{\delta}{\sqrt{(1+c)\hat{\sigma}_e^2}}\right).$$

By Slutsky's theorem, $\hat{p}_1 - p_0/\sqrt{V_1}$ can be approximated by a standard normal distribution. For the testing of the hypotheses H_0: $p_1 \leq p_0$ versus H_a: $p_1 > p_0$, we would reject the null hypothesis H_0 if

$$\frac{\hat{p}_1 - p_0}{\sqrt{V_1}} > z_{1-\alpha},$$

where $z_{1-\alpha}$ is the $100(1-\alpha)$th percentile of a standard normal distribution.

3.3.2.2 Measure of Closeness Based on the Relative Difference

In other words, for evaluation of p_2, we note that y^2 and $\hat{y}^2$ follow a noncentral χ_1^2 distribution with noncentrality parameter μ_0^2/σ_e^2 and $\mu_0^2/c\sigma_e^2$, respectively, where $\mu_0 = \hat{\beta}_0 + \hat{\beta}_1 x$. Hence, $c\hat{y}^2/y^2$ is doubly noncentral F distributed with $\upsilon_1 = 1$ and $\upsilon_2 = 1$ degrees of freedom and noncentrality parameters $\lambda_1 = \mu_0^2/c\sigma_e^2$ and $\lambda_2 = \mu_0^2/\sigma_e^2$. According to Johnson and Kotz (1970), a noncentral F distribution can be approximated by

$$\frac{1+\lambda_1\upsilon_1^{-1}}{1+\lambda_2\upsilon_2^{-1}}F_{\upsilon,\upsilon'}$$

where $F_{\upsilon,\upsilon'}$ is a central F distribution with degrees of freedom

$$\upsilon = \frac{\left(\upsilon_1 + \lambda_1\right)^2}{\upsilon_1 + 2\lambda_1} = \frac{\left(\frac{1+\mu_0^2}{c\sigma_e^2}\right)^2}{\frac{1+2\mu_0^2}{c\sigma_e^2}}$$

and

$$\upsilon' = \frac{\left(\upsilon_2 + \lambda_2\right)^2}{\upsilon_2 + 2\lambda_2} = \frac{\left(\frac{1+\mu_0^2}{\sigma_e^2}\right)^2}{\frac{1+2\mu_0^2}{\sigma_e^2}}.$$

Thus,

$$\begin{aligned} p_2 &= P\left\{\left|\frac{y-\hat{y}}{y}\right| < \delta\right\} \\ &= P\left\{(1-\delta)^2 < c\left(\frac{\hat{y}}{y}\right)^2 < (1+\delta)^2\right\} \\ &= P\left\{\frac{(1-\delta)^2}{c} < \frac{1+\lambda_1}{1+\lambda_2} F_{v,v'} < \frac{(1+\delta)^2}{c}\right\} \\ &= P\left\{\frac{(1-\delta)^2}{c}\frac{1+\lambda_2}{1+\lambda_1} < F_{v,v'} < \frac{1+\lambda_2}{1+\lambda_1}\frac{(1+\delta)^2}{c}\right\} \end{aligned}$$

Thus, p_2 can be estimated by

$$\hat{p}_2 = P\left\{\frac{(1-\delta)^2}{c}\frac{1+\hat{\lambda}_2}{1+\hat{\lambda}_1} < F_{\hat{v},\hat{v}'} < \frac{1+\hat{\lambda}_2}{1+\hat{\lambda}_1}\frac{(1+\delta)^2}{c}\right\} = P\left\{u_1 < F_{\hat{v},\hat{v}'} < u_2\right\},$$

where

$$u_1 = \frac{\left(1+\hat{\lambda}_2\right)}{c\left(1+\hat{\lambda}_1\right)}(1-\delta)^2,$$

$$u_2 = \frac{\left(1+\hat{\lambda}_2\right)}{c\left(1+\hat{\lambda}_1\right)}(1+\delta)^2$$

and $\left(\hat{\lambda}_1, \hat{\lambda}_2, \hat{\upsilon}, \hat{\upsilon}'\right)$ are the corresponding maximum likelihood estimates of $(\lambda_1, \lambda_2, \upsilon, \upsilon')$.

For a sufficiently large sample size, by Slutsky's theorem, $\hat{p}_2$ can be approximated by a normal distribution with mean p_2 and variance V_2, where

$$V_2 = \left(\frac{\partial \hat{p}_2}{\partial \beta_0}, \frac{\partial \hat{p}_2}{\partial \beta_1}, \frac{\partial \hat{p}_2}{\partial \sigma_e^2} \right) \begin{pmatrix} \left(X^T X\right)^{-1} \hat{\sigma}_e^2 & 0 \\ 0' & \dfrac{2\hat{\sigma}_e^4}{n-2} \end{pmatrix} \begin{pmatrix} \dfrac{\partial \hat{p}_2}{\partial \beta_0} \\ \dfrac{\partial \hat{p}_2}{\partial \beta_1} \\ \dfrac{\partial \hat{p}_2}{\partial \sigma_e^2} \end{pmatrix};$$

with

$$\frac{\partial \hat{p}_2}{\partial \beta_0} = \frac{2(c-1)\hat{\mu}_0}{c^2 \hat{\sigma}_e^2 \left(1+\hat{\lambda}_1\right)^2} \left[(1+\delta)^2 f(u_2) - (1-\delta)^2 f(u_1)\right]$$

$$\frac{\partial \hat{p}_2}{\partial \beta_1} = \frac{2(c-1)x_0\hat{\mu}_0}{c^2 \hat{\sigma}_e^2 \left(1+\hat{\lambda}_1\right)^2} \left[(1+\delta)^2 f(u_2) - (1-\delta)^2 f(u_1)\right]$$

$$\frac{\partial \hat{p}_2}{\partial \sigma_e^2} = \frac{\hat{\lambda}_1 - \hat{\lambda}_2}{c \hat{\sigma}_e^2 \left(1+\hat{\lambda}_1\right)^2} \left[(1+\delta)^2 f(u_2) - (1-\delta)^2 f(u_1)\right]$$

where $f(u)$ is the probability density function of an F distribution with degrees of freedom $\hat{\upsilon}$ and $\hat{\upsilon}'$. Thus, the hypotheses given in Equation 3.3 for one-way translation based on probability of relative difference can be tested. In particular, H_0 is rejected if

$$Z = \frac{\hat{p}_2 - p_0}{\sqrt{V_2}} > z_{1-\alpha},$$

where $z_{1-\alpha}$ is the $100(1-\alpha)$th percentile of a standard normal distribution. Note that V_2 is an estimate of $\mathrm{var}\left(\hat{p}_2\right)$, which is obtained by simply replacing the parameters with their corresponding estimates of the parameters.

3.3.2.3 An Example

For the two measures proposed in Section 3.3.2, p_1 is based on the absolute difference between y and $\hat{y}$. To test the hypothesis H_0: $p_1 \le p_0$ versus H_a: $p_1 > p_0$ or the given α, p_0, the set of data (x_i, y_i), and the selected observation (x_0, y_0), we have to compute the value of $\hat{p}_1$. If

$$Z = \frac{\hat{p}_1 - p_0}{\sqrt{V_1}} > z_{1-\alpha},$$

the null hypothesis is rejected. Note that the value of $\hat{p}_1$ depends on the value of δ. Furthermore, it can be shown that $\left(\hat{p}_1 - p_0 - z_{1-\alpha}\sqrt{V_1}\right)$ is an increasing function of δ over (0, +∞). That is, $\left(\hat{p}_1 - p_0 - z_{1-\alpha}\sqrt{V_1}\right) > 0$ is equivalent to $\delta > \delta_0$. Thus, the hypothesis can be tested based on δ_0 instead of $\hat{p}_1$ as long as we can find the value of δ_0 for the given dataset (x_i, y_i) and the selected observation (x_0, y_0). An example is given for illustration.

Suppose that the following data (Table 3.2) are obtained from an IVIVC study, where x is a given dose level and y is the associated toxicity measure.

This set of data is fitted to model 3.1. The estimates of the model parameters are given by

$$\hat{\beta}_0 = -0.704,\ \hat{\beta}_1 = 1.851,\ \hat{\sigma}^2 = 0.431.$$

Based on this model, given $x = x_0$, the fitted value is given by $\hat{y} = -0.704 + 1.851x_0$. Given $\alpha = 0.05$, $p_0 = 0.8$, $(x_0, y_0) = (1.0, 1.2)$, and (5.2, 9.0), we obtain $\delta_0 = 1.27$ and 1.35, respectively. If the required difference is $\delta > \delta_0$, the null hypothesis will be rejected. We conclude that the probability that the difference between y and $\hat{y}$ is less than δ is larger than 0.8. Note that δ_0 changes for different selected observations (x_0, y_0).

For practical purposes, p_2 may be more intuitive because it is based on the relative difference, which is equivalent to measuring the percentage difference relative to the observed y, and δ can be viewed as the upper bound of the percentage error. However, the method applied to test the hypothesis of p_1 cannot be used to test the hypothesis about p_2 owing to the complexity of $\hat{p}_2$. Thus, we have to test the hypothesis through computing the value of

$$Z = \frac{\hat{p}_2 - p_0}{\sqrt{V_2}}$$

and compare to $z_{1-\alpha}$ for different values of δ.

Given two values of x, estimates of p_2 are given in Table 3.3 for various choices of X and Y.

The above example illustrates that $\hat{p}_2$ is very sensitive to the choice of x_0 and y_0.

3.3.3 Remarks

As indicated in the previous section, an appropriate statistical model can be used to describe a given CQA and the corresponding clinical outcome (if available). Under the fitted model, the criticality risk ranking of the CQA with respect to the impact

TABLE 3.2
Data to Establish a Predictive Model

X	0.9	1.1	1.3	1.5	2.2	2.0	3.1	4.0	4.9	5.6
Y	0.9	0.8	1.9	2.1	2.3	4.1	5.6	6.5	8.8	9.2

TABLE 3.3
Estimates of p_2 for Various Choices of δ

				$\hat{p}_2$			
x_0	$\hat{y}$	ν	ν′	δ = 0.05	δ = 0.01	δ = 0.02	δ = 0.5
1.0	1.147	8	2	0.064	0.129	0.258	0.616
5.2	8.921	259	93	0.441	0.757	0.977	1.000

on the corresponding clinical outcome can be assessed by means of the closeness of the predicted value and the observed response in terms of either criterion based on absolute change or relative change. The degree of criticality risk ranking can be determined based on some prespecified p_0. For example, if p is larger than 80%, we may assign the CQA to Tier 1; if p is within 60% and 80%, we may consider that the CQA is mild to moderate relevant to clinical outcome and hence the quality attribute should be assigned to Tier 2. Those quality attributes whose p values are <60% will be assigned to Tier 3, which is least relevant to clinical outcome.

The above process for the development of the predictive model is usually referred to as a *one-way translational process* in translational research/medicine. That is, the information observed at basic research discoveries is translated to the clinic. As indicated by Pizzo (2006), the translational process should be a *two-way translational process*. In other words, we can exchange x and y in Equation 3.1

$$x = \gamma_0 + \gamma_1 y + \varepsilon$$

and come up with another predictive model $\hat{x} = \hat{\gamma}_0 + \hat{\gamma}_1 y$. The idea for the validation of a two-way translational process can be summarized by the following steps:

Step 1: For a given set of data (x, y), establish a predictive model, say, $y = f(x)$.

Step 2: Evaluate $\hat{p}_1 = P\{|y - \hat{y}| < \delta_1\}$ and assess the one-way closeness between y and $\hat{y}$ based on a test for hypotheses 3.3. Proceed to the next step if the one-way translation process is validated.

Step 3: Consider x as dependent variable and y as independent variable and set up the regression model. Predict x at the selected observation y_0, denoted by $\hat{x}$, based on the established model between x and y (i.e., $x = g(y)$). Note that in the above example,

$$\hat{x} = g(y) = \hat{\gamma}_0 + \hat{\gamma}_1 y$$

Step 4: Evaluate the closeness between x and $\hat{x}$ based on a test for the following hypotheses

$$H_0 : p_1 \le p_0 \text{ versus } H_a : p_1 > p_0, \tag{3.4}$$

where $p_1 = P\{|x - \hat{x}| < \delta\}$.

The above test can be referred to as a test for the two-way translational process. The idea is to reject H_0 in favor of H_a. In other words, we would like to reject the null hypothesis and conclude the alternative hypothesis that there is a two-way translation between x and y (i.e., the established predictive model is validated). Under the null hypothesis of Equation 3.4, a test based on an estimate of δ_0 can be similarly derived. We use the previous example for illustration.

Given the dataset, we set up the regression model by using y as the independent variable and x as the dependent variable. The estimates of the model parameters are $\hat{\gamma}_0 = 0.468$, $\hat{\gamma}_1 = 0.519$, and $\hat{\sigma}^2 = 0.121$. Based on this model, for the same α and p_0, given (x_0, y_0) = (1.0, 1.2) and (5.2, 9.0), the fitted values are given by $\hat{x} = 0.468 + 0.519 y_0$. We obtain δ_0 = 0.67 and 0.71, respectively. If the required difference $\delta > \delta_0$, the null hypothesis will be rejected. We conclude that the probability that the difference between x and $\hat{x}$ is less than δ is larger than 0.8.

3.4 FDA'S APPROACHES FOR TIER ANALYSIS

Analytical similarity assessment is referred to as the comparisons of functional and structural characterizations between a proposed biosimilar product and a reference product in terms of CQAs that are relevant to clinical outcomes. The FDA suggests that the sponsors identify CQAs that are relevant to clinical outcomes and classify them into three tiers depending on the criticality or risk ranking (e.g., most, mild to moderate, and least) relevant to clinical outcomes. At the same time, the FDA also recommends some statistical approaches for the assessment of analytical similarity for CQAs from different tiers. The FDA recommends an equivalence test for CQAs from Tier 1, a quality range approach for CQAs from Tier 2, and a descriptive raw data and graphical presentation for CQAs from Tier 3 (see, e.g., Chow, 2015; Christl, 2015; Tsong, 2015). They are briefly outlined in the subsequent subsections.

3.4.1 Equivalence Test for Tier 1

For Tier 1, the FDA recommends that an equivalency test be performed for the assessment of analytical similarity. As indicated by the FDA, a potential approach could be a similar approach to bioequivalence testing for generic drug products (FDA, 2003; Chow, 2015). In other words, for a given critical attribute, we may test for equivalence by the following interval (null) hypothesis:

$$H_0: \mu_T - \mu_R \le -\delta \text{ or } \mu_T - \mu_R \ge \delta, \tag{3.5}$$

where $\delta > 0$ is the equivalence limit (or similarity margin), and μ_T and μ_R are the mean responses of the test (the proposed biosimilar) product and the reference product lots, respectively. Analytical equivalence (similarity) is concluded if the null hypothesis of nonequivalence (*dis*-similarity) is rejected. Note that Yu (2004) defined inequivalence as occurring when the confidence interval falls entirely outside the equivalence limits. Similarly to the confidence interval approach for bioequivalence testing under the raw data model, analytical similarity will be accepted for a quality attribute if the $(1 - 2\alpha)100\%$ two-sided confidence interval of the mean difference is within $(-\delta, \delta)$.

Under the null hypothesis 3.5, the FDA indicates that the equivalence limit (similarity margin), δ, would be a function of the variability of the reference product, denoted by σ_R. It should be noted that each lot contributes one test value for each attribute being assessed. Thus, σ_R is the population standard deviation of the lot values of the reference product. Ideally, the reference variability, σ_R, should be estimated based on some sampled lots randomly selected from a pool of reference lots for the statistical equivalence test. In practice, it may be a challenge when there is a limited number of available lots. Thus, the FDA suggests that the sponsor provide a plan on how the reference variability, σ_R, will be estimated with a justification.

3.4.1.1 An Example

As discussed above, a statistical test for the assessment of analytical similarity for CQAs from Tier 1 is the most rigorous. Following the concept of the FDA's recommended testing procedures for CQAs from Tier 1, for illustration purposes, consider the following example.

Suppose there are k lots of a reference product (RP) and n lots of a test product (TP) available for analytical similarity assessment, where $k > n$. For a given CQA, the FDA's recommended procedure can be summarized in the following steps:

Step 1. Match the number of RP lots to TP lots.

Since $k > n$, there are more reference lots than test lots. The first step is to match the number of RP lots to TP lots for a head-to-head comparison. To *match* RP lots to TP lots, the FDA suggests *randomly* selecting n lots out of the k RP lots. If the n lots are not randomly selected from the k RP lots, justification needs to be provided to prevent *selection bias*.

Step 2. Use the remaining independent RP lots for estimating $\sigma_{R.}$

After the matching, the remaining $k - n$ lots are then used to *estimate* σ_R in order to set up the equivalency acceptance criterion (EAC). It should be noted that if $k - n \leq 2$, all RP lots should be used to estimate σ_R.

Step 3. Calculate the EAC: $\text{EAC} = 1.5 \times \hat{\sigma}_R$.

Based on the estimate of, denoted by $\hat{\sigma}_R$, the FDA recommends that the EAC be set as $1.5 \times \hat{\sigma}_R$, where $c = 1.5$ is considered a regulatory standard.

Step 4. Based on c (regulatory standard), $\hat{\sigma}_R$, and $\Delta = \mu_T - \mu_R$, an appropriate sample size can be chosen for the analytical similarity assessment

As an example, suppose that there are 21 RP lots and 7 TP lots. We first randomly select 7 out of the 21 RP lots to match the 7 TP lots. Suppose that based on the remaining 14 lots, an estimate of σ_R is given by $\hat{\sigma}_R = 1.039$. Also, suppose that the true difference between the biosimilar product and the reference product is proportional to σ_R, say $\Delta = \sigma_R/8$. Then, Table 3.4 with various sample sizes (the number of TP lots available and the corresponding test size) and statistical power for detecting the difference of $\sigma_R/8$ is helpful for the assessment of analytical similarity.

TABLE 3.4
Assessment of Analytical Similarity for CQAs from Tier 1

Number of RP Lots	Number of TP Lots	Selection of c	Test Size (Confidence Interval)	Statistical Power at (1/8) × RP SD (%)
6	6	1.5	9% (82% CI)	74
7	7	1.5	8% (84% CI)	79
8	8	1.5	7% (86% CI)	83
9	9	1.5	6% (88% CI)	86
10	10	1.5	5% (90% CI)	87

According to Table 3.4, there is 79% power for 84% CI of $\hat{\Delta} = \hat{\mu}_T - \hat{\mu}_R$ to fall within ±EAC, assuming that the number of lots is 7 and true difference between TP and RP is $\sigma_R/8$.

This approach has inflated alpha from 5% to 16%. Note that for a fixed regulatory standard c, the sponsor may appropriately select sample size (the number of lots) for achieving a desired power (for detecting a $\sigma_R/8$ difference) and significance level for analytical similarity assessment. As can be seen from the above, if one wishes to reduce the test size (i.e., α level) from 8% to 5%, 10 TP lots need to be tested. Testing 10 TP lots will give an 87% power for detecting a $\sigma_R/8$ difference.

3.4.2 Quality Range Approach for Tier 2

For Tier 2, the FDA suggests that analytical similarity be assessed on the basis of the concept of quality ranges, that is, $\pm x\sigma$, where σ is the standard deviation of the reference product and x should be appropriately justified. Thus, the quality range of the reference product for a specific quality attribute is defined as $(\hat{\mu}_R - x\hat{\sigma}_R, \ \hat{\mu}_R + x\hat{\sigma}_R)$. Analytical similarity would be accepted for the quality attribute if a sufficient percentage of test lot values (e.g., 90%) falls within the quality range.

For a given critical attribute, the quality range is set based on test results of available reference lots. If $x = 1.645$, we would expect 90% of the test results from reference lots to lie within the quality range. If x is chosen to be 1.96, we would expect that about 95% test results of reference lots will fall within the quality range. As a result, the selection of x could impact the quality range and consequently the percentage of test lot values that will fall within the quality range. Thus, the FDA indicates that the standard deviation multiplier (x) should be appropriately justified.

The quality range approach for comparing populations between a proposed biosimilar product and a reference product is a reasonable approach under the assumption that $\mu_T = \mu_R$ and $\sigma_T = \sigma_R$. Under this assumption, we expect that a high percentage (say 90%) of test values of the test product will fall within the quality range obtained based on the test values of the reference product. Thus, one of the major criticisms of the quality range approach is that it ignores the fact that there are differences in population mean and population standard deviation between the proposed biosimilar product and the reference product (i.e., $\mu_T \neq \mu_R$ and $\sigma_T \neq \sigma_R$). In practice, it is recognized that biosimilarity between a proposed biosimilar product and a reference product could be established even under the assumption that $\mu_T \neq \mu_R$ and $\sigma_T \neq \sigma_R$. Thus, under the assumption that $\mu_T = \mu_R$ and $\sigma_T = \sigma_R$, the quality range approach for analytical similarity assessment for CQAs from Tier 2 is considered more stringent as compared to equivalence testing for CQAs from Tier 1 (most relevant to clinical outcomes), regardless of the fact that they are mild-to-moderately relevant to clinical outcomes. This is because that equivalence testing allows a possible mean shift of $\sigma_R/8$, while the quality range approach does not. In what follows, several examples for the possible scenarios of (1) $\mu_T \approx \mu_{R,}$ or there is a significant mean shift (either a shift to the right or a shift to the left) and (2) $\sigma_T \approx \sigma_R$, $\sigma_T > \sigma_R$, or $\sigma_T < \sigma_R$.

3.4.2.1 Example 1

First consider the case where $\mu_T \approx \mu_R$ and $\sigma_T \approx \sigma_R$. In this example, if we choose $x = 1.645$, we would expect 90% of the test results from the test lots to lie within the quality range obtained based on the test values of the reference lots. This case is illustrated in Figure 3.2.

3.4.2.2 Example 2

When $\mu_T \approx \mu_R$ but $\sigma_T > \sigma_R$, if we choose $x = 1.645$, we would expect <90% of the test results from test lots to lie within the quality range obtained based on the test values of the reference lots. The percentage of test values from test lots decreases as $C = \sigma_T/\sigma_R > 1$ increases. This case is illustrated in Figure 3.3.

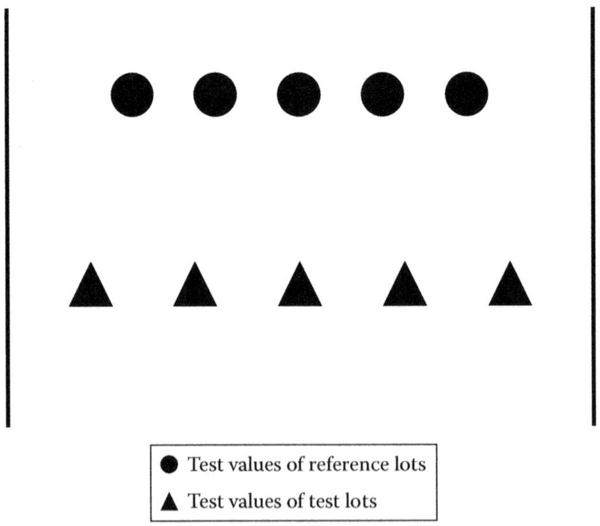

FIGURE 3.2 Quality range approach when $\mu_T \approx \mu_R$ and $\sigma_T \approx \sigma_R$.

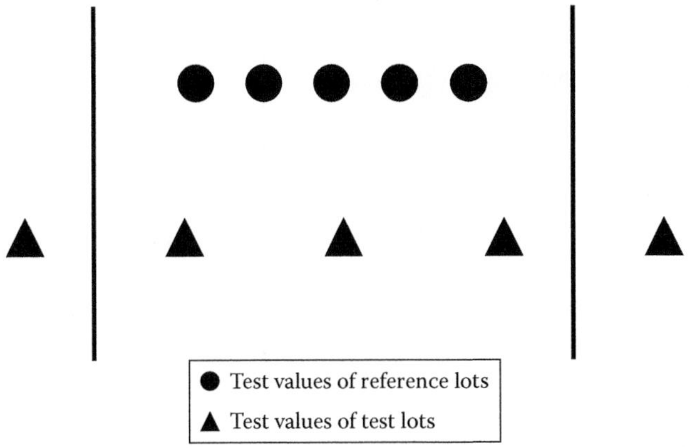

FIGURE 3.3 Quality range approach when $\mu_T \approx \mu_R$ and $\sigma_T > \sigma_R$.

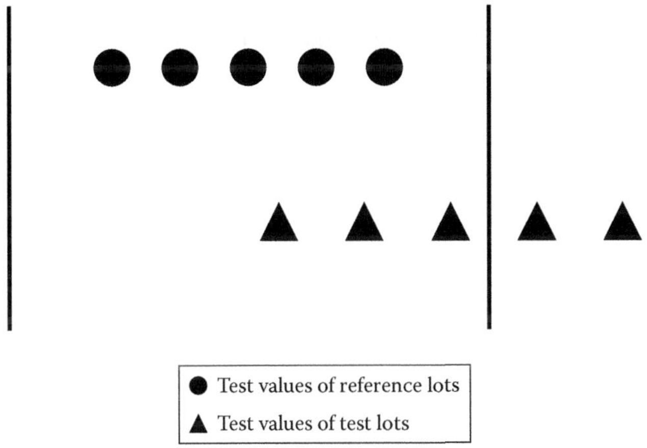

FIGURE 3.4 Quality range approach when $\mu_T > \mu_R$ and $\sigma_T \approx \sigma_R$.

3.4.2.3 Example 3

The case where $\mu_T > \mu_R$ and $\sigma_T \approx \sigma_R$ is illustrated in Figure 3.4. As can be seen from Figure 3.3, if we choose $x = 1.645$, we would expect <90% of the test results from test lots to lie within the quality range obtained based on the test values of the reference lots. The percentage of test values from test lots drops significantly if the difference between $\varepsilon = \mu_T - \mu_R$ increases (i.e., μ_T shifts away from μ_R).

3.4.2.4 Example 4

In practice, it is not uncommon to encounter the case where $\mu_T > \mu_R$ and $\sigma_T > \sigma_R$, which is illustrated in Figure 3.5. As can be seen from Figure 3.4, if we choose $x = 1.645$, we would expect <90% of the test results from test lots to lie within the

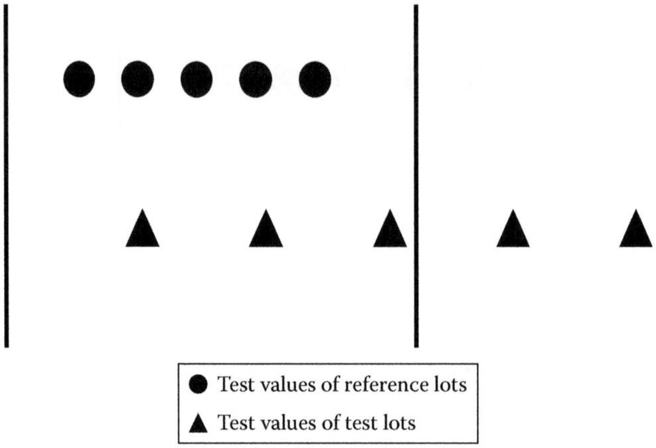

FIGURE 3.5 Quality range approach for the case where $\mu_T > \mu_R$ and $\sigma_T > \sigma_R$.

quality range obtained based on the test values of the reference lots. The percentage of test values from test lots could be very low, especially when both $C = \sigma_T/\sigma_R > 1$ and $\varepsilon = \mu_T - \mu_R$ increases.

3.4.2.5 Remarks

As discussed above, it is suggested that the quality range approach should be modified as follows:

$$\left(\hat{\mu}_R - x\hat{\sigma}_R - |\hat{\varepsilon}|,\ \hat{\mu}_R + x\hat{\sigma}_R + |\hat{\varepsilon}|\right),$$

where $|\hat{\varepsilon}| = |\hat{\mu}_T - \hat{\mu}_R|$ is the mean shift between the proposed biosimilar product and the reference product.

3.4.3 Raw Data and Graphical Comparison for Tier 3

For CQAs in Tier 3 with lowest risk ranking, the FDA recommends an approach that uses raw data/graphical comparisons. The examination of similarity for CQAs in Tier 3 is by no means as stringent, which is acceptable because they have least impact on clinical outcomes in the sense that a notable dissimilarity will not affect clinical outcomes.

The method of raw data and graphical comparison is easy to implement, and yet it is subjective. One of the major criticisms is that it is not clear how the approach can provide totality of the evidence for demonstrating biosimilarity. For CQAs in Tier 1, they are least relevant to clinical outcomes and yet should carry less weight as compared to those CQAs from Tier 1 and Tier 2. There is little or no information regarding what results will be accepted by the method of data and graphical comparison. In practice, if significant differences in graphical comparisons of some CQAs are observed, should this observation raise a concern? In this case, if it is possible, the degree of criticality risk ranking of these CQAs should be assessed whenever possible.

To illustrate the use of the method of raw data and graphical comparison, similarly as above, we consider the following scenarios of (1) $\mu_T \approx \mu_R$ or there is a significant mean shift (either a shift to the right or a shift to the left) and (2) $\sigma_T \approx \sigma_R$, $\sigma_T >> \sigma_R$, or $\sigma_T << \sigma_R$.

3.4.3.1 An Example

Raw data and graphical comparison of test values between test lots and reference lots for CQAs from Tier 3 are rather subjective, and there is a lack of standards for comparison, especially by knowing that (1) CQAs in Tier 3 are least relevant to clinical outcomes and (2) it is expected that $\mu_T \neq \mu_R$ and $\sigma_T \neq \sigma_R$ for biosimilar products. As a result, it is difficult to provide totality of the evidence because it is not clear how much weight Tier 3 will carry. For illustration purposes, Figures 3.6 through 3.8 provide plots for the cases where (1) $\mu_T \approx \mu_R$ and $\sigma_T \neq \sigma_R$, (2) $\mu_T \neq \mu_R$ and $\sigma_T \approx \sigma_R$, and (3) $\mu_T \neq \mu_R$ and $\sigma_T \neq \sigma_R$, respectively. As can be seen from these figures, although graphical comparison may be different, they are least relevant to clinical outcomes,

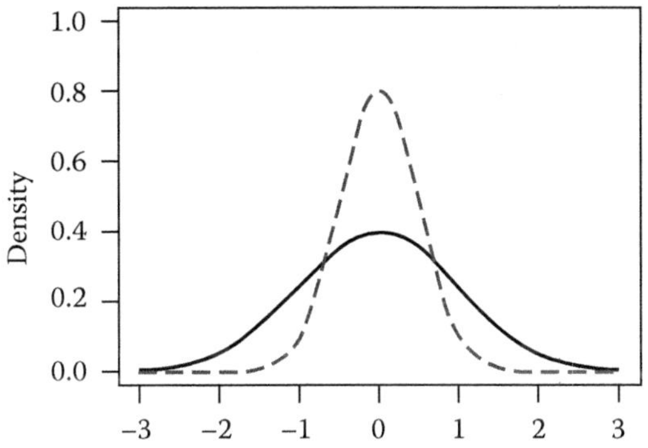

FIGURE 3.6 Graphical comparison for the case where $\mu_T \approx \mu_R$ and $\sigma_T \neq \sigma_R$.

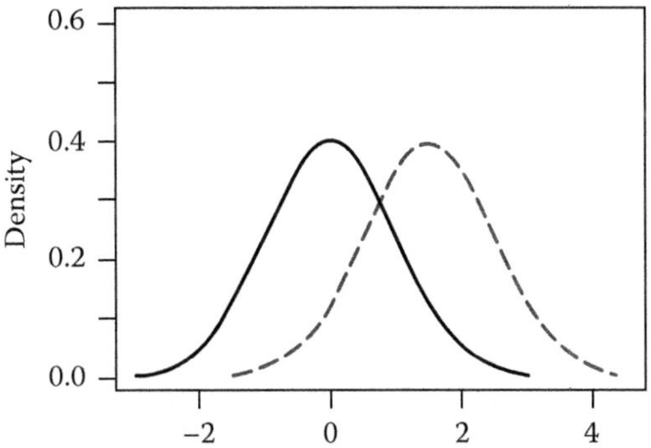

FIGURE 3.7 Graphical comparison for the case where $\mu_T \neq \mu_R$ and $\sigma_T \approx \sigma_R$.

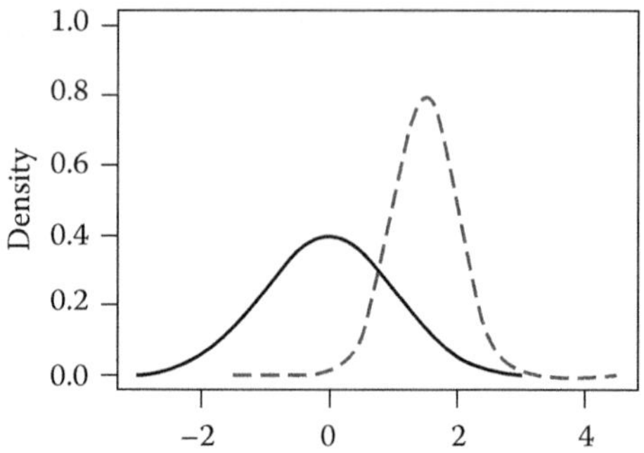

FIGURE 3.8 Graphical comparison for the case where $\mu_T \neq \mu_R$ and $\sigma_T \neq \sigma_R$.

and it is not clear whether a significant difference in distribution of certain CQAs has raised a flag of safety or efficacy concern to demonstrate biosimilarity between the proposed biosimilar product and the reference product (either a US-licensed product or an EU-approved reference product).

3.5 CHALLENGING ISSUES TO FDA'S APPROACHES

The idea of the FDA's proposed equivalence test for Tier 1 CQAs comes from the bioequivalence assessment for generic drugs, which contain the same active ingredient(s) as the reference drug product. It may not be appropriate to apply the idea directly to the assessment of biosimilarity of biosimilar products. The FDA's proposed equivalence test is sensitive to (1) the primary assumptions made, (2) the selection of c, and (3) the estimation of σ_R. Chow (2015) commented on these issues as follows.

3.5.1 Primary Assumptions

Basically, the FDA's proposed equivalence test ignores (1) the lot-to-lot variability of both the reference product and the proposed biosimilar product, (2) the difference between means, and (3) the inflation/deflation in variability between the reference product and the proposed biosimilar product. Suppose that K reference lots will be used to establish EAC for the equivalence test. The FDA suggests that one sample be randomly selected from each lot. The standard deviation of the reference product σ_R can be estimated based on the K test results. Let x_i, $i = 1, 2,\ldots, K$ be the test result of the ith lot. x_i, $i = 1, 2,\ldots, K$ are assumed to be independently and identically distributed with mean μ_R and variance σ_R^2. In other words, we assume that $\mu_{Ri} = \mu_{Rj} = \mu_R$ and $\sigma_{Ri}^2 = \sigma_{Rj}^2 = \sigma_R^2$ for $i \neq j$, $i, j = 1, 2,\ldots, K$. Thus, the expected value of $E(\bar{x}) = \mu_R$ and $\text{var}(\bar{x}) = \sigma_R^2/K$. In practice, it is well recognized that $\mu_{Ri} \neq \mu_{Rj}$ and $\sigma_{Ri}^2 \neq \sigma_{Rj}^2$ for $i \neq j$, where μ_{Ri} and σ_{Ri}^2 are the mean and variance of the ith lot of the reference product. A similar argument applies to the proposed biosimilar (test) product. As a result, the selection of reference lots for the estimation of σ_R is critical for the proposed approach.

In addition, the FDA assumes that the difference in mean responses between the reference product and the proposed biosimilar product is proportional to the variability of the reference product. In other words, $\Delta = \mu_T - \mu_R$ (in log scale) $\propto \sigma_R$. The FDA suggests that the power for detecting a clinically meaningful difference be evaluated at $\sigma_R/8$. Thus, under the assumption, the FDA's proposed equivalence testing is straightforward and easy to implement. However, Chow (2014) indicated that the FDA's proposed testing procedure depends on the selection of the regulatory standard $c = 1.5$, the anticipated difference $\Delta = \mu_T - \mu_R$, and the compromise between the test size (type I error) and statistical power (type II error) for detecting Δ (Chow, 2015).

3.5.2 Justification for the Selection of *c*

The FDA indicates that a potential approach is to assume that the equivalence limit (similarity margin) is proportional to the reference product variability (i.e., $\delta = c \times \sigma_R$). The constant c can be selected as the value that provides adequate power to show equivalence if there is only a small difference in the true mean between the biosimilar and reference products, when a moderate number of reference product and biosimilar lots is available for testing. The FDA's recommended approach for the assessment of analytical similarity for a critical attribute is to choose $\delta = 1.5\sigma_R$ (i.e., $c = 1.5$) and then to select an appropriate sample size for achieving a desired power in order to establish similarity at the $\alpha = 5\%$ level of significance when the true underlying mean difference between the proposed biosimilar and reference product lots is equal to $\sigma_R/8$. The FDA did not provide scientific/statistical justification for the selection of $c = 1.5$ for the EAC. Since the FDA's proposed equivalence test was motivated by the bioequivalence assessment for generic drug products, the selection of $c = 1.5$ can be justified by the following steps:

Step 1. We start with $0.8 = \delta_L \le \mu_T - \mu_R \le \delta_U = 1.25$, where μ_T and μ_R are the reference mean and test mean (in log-scale), respectively.

Step 2. For drug products with large variabilities (i.e., highly variable drug products), the FDA recommends the scaled average bioequivalence (SABE) criterion by adjusting the above bioequivalence limits for variability of the reference product (Haidar et al., 2008; Tothfalusi et al., 2009). This gives

$$0.8\sigma_R = \delta_L \times \sigma_R \le \mu_T - \mu_R \le \delta_U \times \sigma_R = 1.25 \times \sigma_R$$

Step 3. The FDA assumes that the difference between means is proportional to σ_R and allows a mean shift of $\sigma_R/8 = 0.125$, which is the half-width of the margin. The worst possible scenario for the shift is that the true mean difference falls on $1.25 \times \sigma_R$. In this case, the FDA expands the margin by $0.25 \times \sigma_R$. Thus, the upper margin of EAC becomes

$$1.25 \times \sigma_R + 0.25 \times \sigma_R = 1.5 \times \sigma_R.$$

3.5.3 Estimate of Σ_R

The FDA proposed that the equivalence test using available lot values be based mainly on the assumptions that (1) there is no lot-to-lot variability within the reference product and the test product and (2) the difference in mean responses is proportional to the variability of reference product. In practice, however, it is recognized that $\mu_{Ri} \neq \mu_{Rj}$ and $\sigma_{Ri}^2 \neq \sigma_{Rj}^2$ for $i \neq j$. The differences between lots and heterogeneity among lots are major challenges to the *validity* of the FDA's proposed approaches for both equivalence testing for CQAs in Tier 1 and the concept of quality range CQAs from Tier 2. Under the assumptions that $\mu_{Ri} \neq \mu_{Rj}$ and $\sigma_{Ri}^2 \neq \sigma_{Rj}^2$ for $i \neq j$, it is *not* clear what are the statistical properties/finite sample performances and corresponding impact on the assessment of analytical similarity and consequently on providing totality of the evidence to demonstrate similarity.

3.5.4 Heterogeneity within and between Test and Reference Products

Let σ_R^2 and σ_T^2 be the variabilities associated with the reference product and the test product, respectively. Also, let n_R and n_T be the number of lots for analytical similarity assessment for the reference product and the test product, respectively. Thus, we have

$$\sigma_R^2 = \sigma_{WR}^2 + \sigma_{BR}^2 \text{ and } \sigma_T^2 = \sigma_{WT}^2 + \sigma_{BT}^2,$$

where $\sigma_{WR}^2, \sigma_{BR}^2$ and $\sigma_{WT}^2, \sigma_{BT}^2$ are the within-lot variability and between-lot (lot-to-lot) variability for the reference product and the test product, respectively. In practice, it is very likely that $\sigma_R^2 \neq \sigma_T^2$ and often $\sigma_{WR}^2 \neq \sigma_{WT}^2$ and $\sigma_{BR}^2 \neq \sigma_{BT}^2$ even $\sigma_R^2 \approx \sigma_T^2$. This has posed a major challenge to the FDA's proposed approaches for the assessment of analytical similarity for CQAs from both Tier 1 and Tier 2, especially when there is only one test sample from each lot from the reference product and the test product. The FDA's proposal ignores lot-to-lot (between lot) variability, that is, when $\sigma_{BR}^2 = 0$ or $\sigma_{BR}^2 = \sigma_{WR}^2$. In other words, sample variance based on x_i, $i = 1, \ldots, K$ from the reference product may underestimate the true σ_R^2, and consequently may not provide a fair and reliable assessment of analytical similarity for a given quality attribute.

In practice, it is well recognized that $\mu_{Ri} \neq \mu_{Rj}$ and $\sigma_{Ri}^2 \neq \sigma_{Rj}^2$ for $i \neq j$, where μ_{Ri} and σ_{Ri}^2 are the mean and variance of the ith lot of the reference product. A similar argument is applied to the proposed biosimilar (test) product. As a result, the selection of reference lots for the estimation of σ_R is critical for the proposed approach. The selection of reference lots has an impact on the estimation of σ_R and consequently on the EAC. Suppose there are K reference lots available and n lots will be tested for analytical similarity. The FDA suggests using the remaining $K - n$ lots to establish EAC to avoid selection bias. It sounds a reasonable approach if $K >> n$. In practice, however, few lots are available. In this case, the FDA's proposed approach may not be feasible.

3.5.5 Sample Size

In practice, one of the major problems for a biosimilar sponsor is the availability of reference lots for analytical similarity testing. The FDA suggests that an appropriate sample size (the number of lots from the reference product and from the test product) be used to achieve a desired power (say 80%) to establish similarity based on a two-sided test at the 5% level of significance, assuming that the mean response of the test product differs from that of the reference product by $\sigma_R/8$.

Furthermore, since sample size is a function of α (type I error), β (type II error or 1 minus power), δ (treatment effect), and σ^2 (variability), it is a concern that we may have inflated the type I error rate for achieving a desired power to detect a clinically meaningful effect size (adjusted for variability) with a preselected small sample size (i.e., a small number of lots).

3.5.6 Remarks

Different assumptions may lead to different conclusions owing to the difference between mean responses of the various lots and the heterogeneity among lots. It should be noted that the difference between the mean responses of the lots may be offset by the heterogeneity across lots in the FDA's proposed equivalence test. Thus, one of the major criticisms of the FDA's proposed equivalence test procedure is the validity of the primary assumptions, especially the assumption that the difference in the mean responses between the reference product and the proposed biosimilar product is proportional to the variability of the reference product. In addition, for a given CQA, the FDA only requires that a single sample obtained from a lot be tested. In this case, an independent estimate of the variability associated with the test result of the given lot is not available. Similar comments apply to the quality range approach for CQAs from Tier 2.

3.6 RECOMMENDATIONS AND ALTERNATIVE METHODS

3.6.1 Recommendations to Current Approaches for the Assessment of Analytical Similarity

Suppose that there are K reference lots to establish EAC for the equivalence test for Tier 1 CQAs. The FDA suggests that one sample be randomly selected from each lot. The standard deviation of the reference product σ_R can be estimated based on the K test results. Let x_i, $i=1, 2,\ldots, K$ be the test result of the ith lot. x_i, $i = 1, 2,\ldots, K$ are assumed to be independently and identically distributed with mean μ_R and variance σ_R^2. In other words, we assume that $\mu_{Ri} = \mu_{Rj} = \mu_R$ and $\sigma_{Ri}^2 = \sigma_{Rj}^2 = \sigma_R^2$ for $i \neq j$, $i, j = 1, 2,\ldots, K$. Thus, the expected value of $E(\bar{x}) = \mu_R$ and $\text{var}(\bar{x}) = \sigma_R^2/K$. Under the assumption that $\mu_{Ri} \neq \mu_{Rj}$ and $\sigma_{Ri}^2 \neq \sigma_{Rj}^2$ for $i \neq j$, where μ_{Ri} and σ_{Ri}^2 be the mean and variance of the ith lot of the reference product, we have

$$\frac{\sigma_{(1)}^2}{K} \leq \text{var}(\bar{x}) = \frac{\sigma_R^2}{K} \leq \frac{\sigma_{(K)}^2}{K},$$

where $\sigma_{(1)}^2$ and $\sigma_{(K)}^2$ are, respectively, the smallest and largest within-lot variance among the K lots. Thus, it is recommended that the current approach of equivalence test for analytical similarity be modified as follows:

1. Randomly select at least two samples from each lot. The replicates will provide independent estimates of within-lot variability $\left(\sigma_{WR}^2\right)$ and lot-to-lot variability $\left(\sigma_{BR}^2\right)$. σ_R^2 is the sum of σ_{WR}^2 and σ_{RB}^2. In the interest of the same total number of tests, the sponsor can test on two samples from each lot among $K/2$ randomly selected lots.
2. For the establishment of EAC, it is then suggested that $\sigma_{(K)}$ be used to take lot-to-lot and within-lot variabilities into consideration.
3. In the event only one sample from each lot is tested, it is suggested that the upper 95% confidence bound be used as σ_R for establishing EAC for equivalence testing of the identified CQAs in Tier 1. In other words, under the FDA's proposed approach, we will use the following to estimate σ_R:

$$\hat{\sigma}_R = \sqrt{\frac{n-1}{\chi^2_{\alpha/2,n-1}}}\hat{\sigma}_x,$$

where

$\hat{\sigma}_x$ is the sample standard deviation obtained from the n reference lot test values.

$\chi^2_{\alpha/2,n-1}$ is the $(\alpha/2)$th upper quantile of a chi-square distribution with $n-1$ degrees of freedom.

3.6.2 Alternative Approaches

Alternatively, we may consider a Bayesian approach with appropriate choices of priors for the mean and standard deviation of the reference product in order to take into consideration the heterogeneity in mean and variability. The Bayesian approach is to obtain a Bayesian creditable interval which will consider EAC for the assessment of analytical similarity.

3.7 CONCLUDING REMARKS

For purposes of identifying CQAs at various stages of the manufacturing process, most sponsors assign CQAs based on the mechanism of action (MOA) or pharmacokinetics (PK) believed to be relevant to clinical outcomes. It is a reasonable assumption that change in MOA or PK of a given quality attribute is predictive of clinical outcomes. However, the primary assumption that there is a well-established relationship between *in vitro* assays and *in vivo* testing (i.e., *in vitro* assays and *in vivo* testing correlation; IVIVC) needs to be validated. Under the validated IVIVC relationship, the criticality (or risk ranking) can then be assessed based on the degree of the relationship. In practice, however, most sponsors provide clinical rationales for the assignment of the CQAs without using a statistical approach for establishing IVIVC. The assignment of the CQAs without using a statistical approach is considered subjective and hence is somewhat misleading.

For a given quality attribute, the FDA suggests a simple approach by testing one sample (randomly selected) from each of the lots. Basically, the FDA's approach ignores lot-to-lot variability for the reference product. In practice, however, lot-to-lot variability inevitably exists even when the manufacturing process has been validated. In other words, we would expect that there are differences in mean and variability from lot-to-lot, that is, $\mu_{Ri} \neq \mu_{Rj}$ and $\sigma_{Ri}^2 \neq \sigma_{Rj}^2$ for $i \neq j$, $i, j = 1, 2, \ldots, K$. In this case, it is suggested that the FDA's approach be modified (e.g., performing tests on multiple samples from each lot) to account for the within-lot and between-lot (lot-to-lot) variabilities for fair and reliable comparisons.

For the quality range approach for CQAs in Tier 2, the FDA recommends using $x = 3$ by default for 90% of the values of test lots contained in the range. It allows approximately one standard deviation of reference for shifting, which may be adjusted based on biologist reviewers' recommendations. However, some sponsors propose using the concept of tolerance interval to ensure that a high percentage of test values for the lots from the test product fall within the quality range. It should be noted, however, that the percentage decreases when the difference in mean between the reference product and the proposed biosimilar product increases. This is also true when $\sigma_T \ll \sigma_R$. Even the tolerance interval is used as the quality range. This problem is commonly encountered mainly because the quality range approach does not take into consideration (1) the difference in means between the reference product and the proposed biosimilar product and (2) the heterogeneity among lots within and between products. In practice, it is very likely that a biosimilar product with small variability but a mean response that is away from the reference mean (e.g., within the acceptance range of $\sigma_R/8$ per FDA) will fall outside the quality range. In this case, a further evaluation of the data points that fall outside the quality range is necessary to rule out the possibility of chance alone.

The FDA's current thinking for analytical similarity assessment using a three-tier analysis is encouraging. It provides direction for statistical methodology development for a valid and reliable assessment toward providing the totality of the evidence for demonstrating biosimilarity. The three-tier approach is currently under tremendous discussion within the pharmaceutical industry and academia. In addition to the challenging issues discussed above, some issues remain unsolved and require further research. These issues include, but are not limited to, (1) the degree of similarity (i.e., how similar is considered highly similar?), (2) multiplicity (i.e., is there a need to adjust α for controlling the overall type I error at a prespecified level of significance), (3) acceptance criteria (e.g., about what percentage of CQAs in Tier 1 need to pass an equivalence test in order to pass the analytical similarity test for Tier 1?), (4) multiple references (i.e., what if there are two reference products such as US-licensed and EU-approved reference products?), and (5) credibility toward the totality of the evidence.

REFERENCES

Chow SC. (2013) *Biosimilars: Design and Analysis of Follow-on Biologics.* Chapman and Hall/CRC Press, Taylor & Francis, New York.

Chow SC. (2014) On assessment of analytical similarity in biosimilar studies. *Drug Designing* **3**, 119. doi: 10.4172/2169-0138.

Chow SC. (2015) Challenging issues in assessing analytical similarity in biosimilar studies. *Biosimilars* **5**, 33–39.

Chow SC, Liu JP. (2008) *Design and Analysis of Bioavailability and Bioequivalence Studies*, 3rd edition. Chapman Hall/CRC Press, Taylor & Francis, New York.

Christl L. (2015) Overview of the regulatory pathway and FDA's guidance for the development and approval of biosimilar products in the US. Presented at the Oncologic Drugs Advisory Committee Meeting, January 7, 2015, Silver Spring, MD.

FDA. (2003) *Guidance on Bioavailability and Bioequivalence Studies for Orally Administered Drug Products—General Considerations.* Center for Drug Evaluation and Research, Food and Drug Administration, Rockville, MD.

FDA. (2015a) *Guidance for Industry: Scientific Considerations in Demonstrating Biosimilarity to a Reference Product.* Food and Drug Administration, Silver Spring, MD.

FDA. (2015b) *Guidance for Industry: Quality Considerations in Demonstrating Biosimilarity of a Therapeutic Protein Product to Reference Product.* Food and Drug Administration, Silver Spring, MD.

FDA. (2015c) *Guidance for Industry: Questions and Answers Regarding Implementation of the Biologics Price Competition and Innovation Act 2009.* Food and Drug Administration, Silver Spring, MD.

Gutierrez-Lugo MT. (2015) Chemistry, manufacturing, and controls. Presented at ODAC Meeting on BLA 125553 for EP2006, January 7, 2015, Silver Spring, MD.

Haidar SH, Davit B, Chen ML, et al. (2008) Bioequivalence approaches for highly variable drugs and drug products. *Pharmaceutical Research* **25**, 237–241.

Johnson NL, Kotz S. (1970) *Distributions in Statistics—Continuous Univariate Distribution*, Volume 1. Wiley, New York.

Pizzo PA. (2006) *The Dean's Newsletter.* Stanford University School of Medicine, Stanford, CA.

Tothfalusi L, Endrenyi L, Garcia Areta A. (2009) Evaluation of bioequivalence for highly-variable drugs with scaled average bioequivalence. *Clinical Pharmacokinetics* **48**, 725–743.

Tsong Y. (2015) Development of statistical approaches for analytical biosimilarity evaluation. Presented at DIA/FDA Statistics Forum 2015, April 20, 2015, Bethesda, MD.

USP/NF. (2000) *United States Pharmacopeia* **24** and *National Formulary* **19**, United States Pharmacopeial Convention, Inc., Rockville, MD.

Yu LX. (2004) Bioinequivalence: concept and definition. Presented at Advisory Committee for Pharmaceutical Science of the Food and Drug Administration, April 13–14, 2004, Rockville, MD.

4 Characterization of Biosimilar Biologics

The Link between Structure and Functions

Roy Jefferis
University of Birmingham

CONTENTS

4.1 INTRODUCTION

The modern era of biological therapeutics may be identified with the FDA approval of recombinant insulin (Humulin) in 1982, produced in *E. coli*, and recombinant erythropoietin (EPO) in 1989 (Epogen); since glycosylation of EPO is essential to its function, it was necessarily produced in a mammalian cell line [a Chinese hamster ovary (CHO) cell line]. Despite extensive clinical experience, adverse reactions to these recombinant molecules are still encountered: ~2% for insulin (Ghazavi and Johnston, 2011) and rarer, but more devastating, for EPO (Macdougall et al., 2012). These incidences are frequently due to the patient developing antibodies that are specific for the therapeutic and neutralize its activity [antitherapeutic antibodies (ATA): antidrug antibodies (ADA)]. The development of ADA suggests that the therapeutic is being recognized as "foreign" (nonself) by the patient's immune system, owing to the presence of molecules that exhibit structural features different from those of the endogenous protein/glycoprotein (P/GP). Ironically, a high incidence of ADA is encountered for recombinant antibody therapeutics (mAbs), a sort of "good cop/bad cop" situation; the explanation lies in the fact that, by definition, each mAb exhibits a unique specificity due to its unique structure.

Currently, some 90 protein therapeutics have been established in the clinic, and patents for those approved early in the development of biologic drugs have expired or will do so imminently. The high costs of these drugs impose financial pressures on national and private health care bodies, with consequent encouragement for the development and prescribing of copies of innovator biotherapeutics (Biosimilars, EU; Follow-on biologics, FDA). While the availability of biosimilars reduces the "cost of goods" (CoG), advances in formulation and delivery will contribute to further reductions in "cost of treatment" (CoT). Additionally, genetic and glycosylation engineering is being embraced to generate innovator "biobetter" molecules that express "enhanced" attributes, compared to the currently approved biologics. Given the complexity of the human proteome and cellome, there would seem to be virtually unlimited scope for the continued development of recombinant P/GP therapeutics.

The starting point for the generation of recombinant P/GPs is extensive characterization of the structure and function of the endogenous (natural) molecule. This is not a trivial exercise, for while the gene sequence determines the primary amino acid sequence, it does not provide a guide to the precise structure of the active molecule. Additional parameters include the *in vivo*: conformation (secondary, tertiary, and quaternary), chemical properties (charge, hydrophilicity, and hydrophobicity), post-translational (PTM) and chemical modifications (CMs), and the microenvironment in which the molecule is functionally active. Consequently, the endogenous P/GP exhibits structural heterogeneity, and individual isoforms may influence the rate of protein turnover, determining whether or when a given molecule enters a degradative pathway (Welle, 1999). These initial considerations are compounded when one is attempting to establish the structure of a "native" P/GP *ex vivo* because additional heterogeneities may be introduced during its isolation, purification, and characterization.

This results in a "Catch 22" or biological "Heisenberg uncertainty principle" situation; the purer the isolated protein, the lower the yield and the less certain one can be that it is representative of the endogenous molecule active *in vivo*. In practice, a consensus structure for the functional molecule is established as the goal for the development and characterization of a potential recombinant therapeutic. It is mandatory that recombinant P/GP therapeutics exhibit consensus PTMs of the endogenous molecule, with an absence of unnatural PTMs introduced in the production process. If present, the unnatural PTMs may be perceived as nonself by the immune system and result in the generation of ADA. Differences between the recombinant and the "wild-type" molecule are inevitable since recombinant proteins are produced in nonhuman tissues (CHO, NS0, Sp2/0 cells, etc.) exposed to culture medium, the products of intact and effete producer cells, and subjected to rigorous downstream purification, formulation, and storage conditions. Production in a prokaryotic system (e.g., *E. coli*) may result in a protein being recovered as an inclusion body that has to be solubilized and refolded *in vitro* to yield a product that may lack natural PTMs or bear unnatural ones.

In addition to the demonstration of clinical efficacy, an innovator company seeking approval for a potential P/GP therapeutic is required to characterize the drug substance/product structurally and functionally, employing multiple orthogonal technologies. The parameters established define the drug substance/product and, if approved, must be maintained throughout the life cycle of the drug; intentional changes in the production process may be approved by a regulatory authority if it is demonstrated not to compromise efficacy and patient benefit. Consequently, critical quality attributes (CQAs) that contribute to drug efficacy are defined and achieved, employing quality by design (QbD) parameters unique to the production platform, downstream protocols, and formulation employed. The CQA and QbD parameters are the undisclosed intellectual property of an innovator company. Consequently, in principle, it is deemed essentially impossible to produce an identical product employing a similar or alternative platform within another facility, i.e. it is not possible to develop generic biopharmaceuticals. The high cost and large size of the markets for biologics has encouraged "pharma," big and small, to develop biosimilar or follow-on biologic products, and regulatory authorities have developed competence in advising pharma and approving these products (EMA, 2008; FDA, 2014). It is often asserted that for the innovator product "the process defines the product"; for a biosimilar, it may be said that "the product defines the process." As the mechanism(s) of action (MoA) of P/GPs is being elucidated, protein and glycosylation engineering is being employed to develop next-generation molecules exhibiting selected and/or accentuated MoAs. This could be an oversimplistic approach because a mutant molecule is, by definition, an unnatural entity and the totality of its activities *in vivo* may not be the same as the endogenous molecule; it may exhibit enhanced immunogenicity, with consequent induction of ADA that could compromise efficacy and the patient.

It is accepted that "copies" of biologics cannot be structurally identical to an innovator product; however, regulatory authorities demand that they be demonstrated to be "comparable" to the innovator product (EMA, 2008; FDA, 2014). Structural comparability is established by applying multiple orthogonal analytical protocols to characterize the innovator product, sourced from a pharmacy in comparison with the proposed biosimilar. Inevitably, structural differences will be detected that must be demonstrated not to compromise functional efficacy, *ex vivo*, or patient benefit.

The criteria for approval of a biosimilar biologic differ between national regulatory authorities, and biosimilar mAb drugs that have been approved in India (Reditux/Rituxan) and South Korea (Remsima/Remicade) have not been automatically approved by the EMA or the FDA. However, the biosimilar candidate for Remicade (Remsima, Celltrion; Inflectra, Hospira), which was developed under EMA guidelines, received EMA approval in 2013 (Beck and Reichert, 2013; Hospira, 2013). Approval was heralded as a "landmark" event and demonstrates that the EMA has confidence in its ability to evaluate the comparability of biosimilar antibody products; the FDA has recently also scored a "first" in approving Zarxio (filgrastim-sndz), a biosimilar of Filgastrim (FDA, 2015a).

4.2 OVERVIEW OF CO- AND POSTTRANSLATIONAL MODIFICATIONS

Variations in protein structure from that predicted by open reading frame gene sequences are frequently referred to, collectively, as posttranslational modifications. However, heterogeneity may be introduced at an earlier stage [e.g., by misincorporation at the DNA, RNA, amino acid levels and cotranslationally (Harris et al., 2001; NCI, 2016; Zhong and Wright, 2013)]. Commonly encountered co-, post-translational (CTM/PTM), and chemical modifications (CMs) include glycosylation, phosphorylation, sulfation, glycation, deamidation, and deimination (Goetze et al., 2012; Harris et al., 2001; Jefferis, 2012; Khawli et al., 2010; NCI, 2016; Wang et al., 2007; Zhong and Wright, 2013). Additionally, the structural profile may vary with age, sex, health, and disease. The human genome contains ~21,000 protein encoding genes, but it is estimated that the human proteome consists of 1–2 million protein entities, due to the earlier mentioned parameters plus *in vivo* enzymatic and chemical modifications that are essential to systemic physiological function and/or within microenvironments (Harris et al., 2001; NCI, 2016; Zhong and Wright, 2013). Each human individual should, in theory, be immunologically tolerant to all molecules within their proteome, including those exhibiting PTMS and CMs. However, the exquisite sensitivity of current assay systems allows the detection of low-affinity antibody to many self-antigens in healthy individuals. Paradoxically, although healthy individuals exhibit immunological self-recognition antibodies of the same or similar, specificity may be amplified in disease states and be a diagnostic marker for individual disease entities (Burska et al., 2014).

The enumeration of PTMs and CMs generating P/GP heterogeneity has been achieved by "revolutionary" developments in qualitative and quantitative mass spectrometry, with definition of >300 structural CTM/PTMs (Chicooree et al., 2015; Lanucara and Eyers, 2013); analysis of 530,264 sequences in the Swiss-Prot database was shown to yield 87,308 experimentally identified PTMs and 234,938 putative PTMs (Chicooree et al., 2015; Farriol-Mathis et al., 2004; Khoury et al., 2011; Lanucara and Eyers, 2013). The potential for structural and functional complexity can be appreciated from the fact that the human genome encodes 518 protein kinases and 200 phosphatases (Manning et al., 2002; Sacco et al., 2012). The second most frequent CTM/PTM is glycosylation. Oligosaccharides may be attached to asparagine residues to generate *N*-linked glycoproteins or to the hydroxyl groups of serine, threonine, or tyrosine to generate *O*-linked glycoforms. The *N*-linked repertoire contains >500 different oligosaccharide structures that may be differentially attached at multiple glycosylation sites to generate >1000 different types of glycan; a consequence of the activities of >250 glycosyltranferases (Marth and Grewal, 2008; NCI, 2016; Stanley, 2011; Zhong and Wright, 2013).

The potential for complexity/heterogeneity can be illustrated for mAbs. The full-length sequence of an IgG molecule includes ~40 asparagine/glutamine residues; therefore, random deamidation of one of these residues will generate 40 structural variants (isoforms); deamidation of two residues may generate 40 × 40 = 1560 variants; three: 40 × 1560 = 59,280 variants, and so on. When all possible overt PTMs (only) and combinations of PTMs are considered it has been calculated that a full-length IgG molecule may exhibit a heterogeneity embracing 10^8 isoforms (Kozlowski and Swann, 2006). Recombinant mAbs present a unique challenge as they must be evaluated on a case-by-case basis, each being constituted of heavy and light chains having unique variable region sequences. Fortunately, current technologies allow for early screening and selection of clones that do not have amino acid residues susceptible to PTMs within their complementarity determining regions (CDRs); other criteria may be selected to optimize solubility, stability, and so on. The constant region sequence can also be selected (i.e., Ig class and subclass) to define the final drug substance/product developed. A biosimilar candidate must be demonstrated to be structurally and functionally comparable (EMA, 2008; FDA, 2014).

4.3 PHYSICOCHEMICAL CHARACTERIZATION

It has been suggested that comparability between innovator and candidate biosimilar therapeutics may be better established through physicochemical characterization than limited clinical trials. This is posited on the premise that the extensive, and very expensive, clinical trials undertaken to achieve approval of the innovator product is not required for a candidate biosimilar; the added costs would negate the objective of making a less expensive drug available. A limited clinical trial within an outbred human population may yield unreliable results due to concomitant disease, associated and nonassociated differences between patients, and the like. It is essential to select a number of appropriate orthogonal techniques in establishing comparability; the range and sensitivity of methodologies currently available have been introduced in Chapter 2 of this book. The starting point for development of a biosimilar is characterization of the innovator product, employing multiple samples obtained from pharmacies, to inform the preliminary selection of clones and finally to confirm comparability. Unfortunately, the techniques employed and the results obtained by the innovator company are not in the public domain. The biopharmaceutical industry and patient interests could be better served if a consensus was established for the most appropriate techniques and protocols to be applied. The availability of a standard reference material, characterized within state-of-the-art methodologies, and available to any academic or commercial laboratory as an external control for their "in-house" protocols would be a step forward toward achieving these aims.

Such a collaboration revolving around a single humanized IgG1κ (the NISTmAb) involving ten major biopharmaceutical companies, five academic institutions, six instrument manufacturers, and three government agencies is nearing completion. The candidate reference material under development at the National Institute of Standards and Technology (NIST) was distributed to partners for a crowdsourcing characterization approach in which each participant performed a subset of relevant techniques. The project will be reported in a three-volume book set under the collective title: "State of the Art and Emerging Technologies: Therapeutic Monoclonal Antibodies." The first volume has been published, under the title "Monoclonal Antibody Therapeutics: Structure, Function, and Regulatory Space" (Schiel et al., 2014a,b); subsequent volumes utilize the NISTmAb as a case study to present representative materials, methods, data, and detailed discussion on the current state of monoclonal antibody characterization (Volume 2) as well as emerging technologies (Volume 3). The book series and associated reference material (to be released as RM 8671 from NIST) is intended as a fundamental dataset and widely available test molecule to streamline technology development and harmonize characterization approaches throughout industry.

4.4 CHEMICAL, CO-, AND POSTTRANSLATIONAL MODIFICATIONS OF ANTIBODY MOLECULES

The hallmark of a protective polyclonal humoral immune response is specificity; achieved through the generation of a population of antigen-binding sites (paratopes), each specific for a given structure expressed on the surface of the antigen (antigenic determinant, epitope). Additional structural and functional heterogeneity is determined by the relative proportions of the nine antibody isotypes expressed; for human IgG and IgA, an additional heterogeneity may arise due to selection among the multiple polymorphic variants present within an outbred human population (Jefferis, 2012; Jefferis and Lefranc, 2009). A consequence of polyclonal IgG heterogeneity is that multiple amino acids are released at each cycle of the Sanger sequencing technique; this precludes detailed structural analysis but defines the N-terminal region of light and heavy chains as the seat of antibody diversity. This is further confirmed with the sequencing of many homogeneous (monoclonal) IgG molecules isolated from the blood (serum) of patients with multiple myeloma, a cancer of antibody secreting plasma cells. In advanced cases, normal IgG synthesis is suppressed while the paraprotein may reach levels >50 g/L; such proteins have been shown to be representative of normal IgG molecules, although in no case has the antigen been identified. The first full-length sequence of an IgG molecule was determined for a monoclonal human IgG1 protein (paraprotein, Eu) (Edelman et al., 2004), purified from the blood of a patient with multiple myeloma. The numbers assigned to the amino acid residues of this protein defines sequences of the constant regions of the light and heavy chains; thus, the oligosaccharide is consistently reported as being attached to asparagine 297 (Eu), although for any given IgG molecule this asparagine residue may have a different linear heavy chain sequence number, due to differences in the length of the variable region (Jefferis, 2012).

The challenge in generating a biosimilar antibody therapeutic is well illustrated in a report of the analysis of a candidate Herceptin biosimilar (Xie et al., 2010). Values for methionine oxidation, deamidation of asparagine and glutamine were reported; however, a note added in press stated that the values were higher than those obtained by the innovator company. It was therefore posited that additional chemical modifications had been introduced into the protein when reconstituting the innovator Herceptin prior to analysis. This illustrates both the susceptibility of such large molecules to chemical modification *in vitro* and the sensitivity of techniques available for determining comparability. A gross error resulted from generating an IgG1 constant region sequence different from that of the innovator product (Xie et al., 2010). This suggests that rather than sequencing the innovator product prior to embarking on the project, the sequence of Herceptin had been sourced from a sequence database that has been shown to be in error (Jefferis and Lefranc, 2009).

Given the potential structural heterogeneity of endogenous and recombinant biologics, it is outside the scope of this chapter to attempt to summarize all published studies for all currently approved biologics. Therefore, I shall limit this chapter mostly to the consequences of structural heterogeneities for human IgG antibodies, with some appropriate diversions. As previously noted, the heterogeneity of normal polyclonal IgG presents a complexity that has not been amenable to detailed analysis in past decades; however, mass spectrometry is beginning to provide viable approaches (Leblanc et al., 2014; Yang and Zubarev, 2010). The recombinant mAb therapeutics has been a focus of interest for PTM/CM studies due to their impact on structure and function. The usual approach is to determine the PTM/CM profile for an mAb and then to subject it to prolonged exposure to conditions known to induce specific CMs. Subsequent qualitative and quantitative analyses identify the degree of susceptibility of individual amino acid residues to modification (Correia, 2010). An alternative, and possibly more relevant, approach is to recover an mAb therapeutic from patient blood and to determine the PTM/CMs effected *in vivo* after a given time interval (Haberger et al., 2014). Earlier priorities in the development of mAbs centered on antigen specificity and affinity of binding as they transitioned from murine to chimeric, humanized, and fully human antibodies. Residues within light and heavy chain CDRs susceptible to PTM and CM modifications have been shown to compromise binding activity. However, currently available techniques allow for sequence analysis at the clone selection stage, such that molecules with CDR residues susceptible to modification can be discarded (e.g., asparagine, methionine, lysine).

4.5 N- AND C-TERMINAL RESIDUES

Unique N-terminal sequence may be obtained for the heavy and light chains of most monoclonal IgG paraproteins; for others, however, the N-terminal amino acid yield may not be quantitative or appear to be entirely "blocked" (Nakajima et al., 2011). This results when a gene encodes for the incorporation of N-terminal glutamic acid or glutamine residues with subsequent cyclization and the generation of pyroglutamic acid (pGlu), which may occur both *in vivo* and *in vitro* (Kumar and Bachhawat, 2012; Liu et al., 2011; Nakajima et al., 2011; Perez-Garmendia and Gevorkian, 2013;

Yin et al., 2013). The formation of pGlu in antibodies (Liu et al., 2011) and therapeutic proteins is a concern for the biopharmaceutical industry because it introduces charge heterogeneity and variations may be considered to be evidence for lack of process control (Stanley, 2011). Importantly, N-terminal pGlu is also implicated in Alzheimer's disease and dementia because it increases the tendency for proteins to form insoluble fibrils; light chains are particularly prone to such processing with the formation of fibrils (Liu et al., 2011; Merlini et al., 2014). When producing recombinant mAbs in mammalian cells, the balance between production of heavy and light chains is critical and an excess of light chain production is optimal, the excess being secreted into the medium (Li et al., 2007). As there is no evidence of a specific benefit attached to the presence of N-terminal pGlu, to either the heavy or light chain, it may be best avoided during clone selection for a potential mAb therapeutic.

Sequencing studies reported the C-terminal residue of serum-derived IgG heavy chains to be glycine. However, when the genes for the IgG subclasses were sequenced, it was seen that they encoded for a C-terminal lysine residue. It was later shown that the lysine residue is cleaved, *in vivo*, by an endogenous carboxypeptidase B. Recombinant IgG molecules produced in mammalian cells exhibit mixed populations of molecules, with lysine present or absent on each heavy chain. The presence or absence of lysine results in charge heterogeneity, and the proportions of heavy chains bearing C-terminal lysine can vary between clones and the many parameters that define the production platform (Tang et al., 2013). Although the level of C-terminal lysine is not considered to be a CQA, it may be a valuable reporter for production consistency. A concept paper produced by the European Biopharmaceutical Enterprises (EBE), a specialized group of the European Federation of Pharmaceutical Industries and Associations (EFPIA) included the statements: "A number of scientific publications suggest that C-terminal lysine truncation has no impact on biological activity, PK/PD, immunogenicity and safety." And elsewhere in the document: "Lysine truncation does not appear to adversely affect product potency or safety. However, taking a conservative approach potential C-terminal lysine effects on all antibodies cannot be ruled out. Thus, lysine truncation should be characterized, and process consistency should be demonstrated during product development; regulatory agencies suggest that C-terminal lysine content should be reported for both the characterization and development phases" (EBE, 2013). Removal of C-terminal lysine results in the presence of a C-terminal glycine residue that, when produced in CHO cells, may be subject to amidation, introducing further structural and charge heterogeneity (Tsubaki et al., 2013). A recent report demonstrated that this problem has been circumnavigated by genetic engineering CHO cells to "knockdown" expression of the peptidylglycine α-amidating monooxygenase (PAM) enzyme (Skulj et al., 2014).

4.6 CYSTEINE AND DISULFIDE BOND FORMATION

The gene sequence for the human IgG1 subclass protein (Eu) encodes for five light chain and nine heavy chain cysteine residues; 28 for a H_2L_2 protein dimer. The standard structural cartoon for the human IgG1 protein (Eu) exhibits 12 intrachain and 4 interchain disulfide bridges. This general pattern of intrachain disulfide bridge formation is maintained for each of the other IgG subclasses; however, the number of interchain bridges and their architecture varies between and within the IgG subclasses (Liu and May, 2012; Macdougall et al., 2012). Heterogeneity of disulfide bridge formation has been reported for normal serum-derived IgG, myeloma proteins, and recombinant mAbs. Formation of the H_2L_2 dimer occurs following release of heavy and light chains into the endoplasmic reticulum (ER), with evidence that binding of the constant region of the light chain (C_L) to the heavy chain C_H1 domain "catalyses" the generation of a correctly folded H_2L_2 structure (Li et al., 2007). This nascent form explores multiple dynamic structures, with the formation of native and nonnative disulfide bonds that are transiently formed and reduced until a low-energy conformation is achieved (Aricescu and Owens, 2013; Stanley, 2011). It should be noted that little or no processing of the high mannose oligosaccharide will have occurred at this point; therefore, the conformation of the secreted IgG-Fc will not be achieved until oligosaccharide processing is completed.

The IgG1 molecule establishes the "standard" pattern with two interheavy chain disulfide bridges and a single light-heavy chain bridge; IgG2, IgG3, and IgG4 express 3, 11, and 2 interheavy chain bridges, respectively. The cysteine residues that form interchain disulfide bridges are clustered within the hinge region and may be subject to reduction and reformation when present in a reducing environment. Heterogeneity in disulfide bond formation in IgG2 was first reported for recombinant IgG2 proteins but, later, was also observed for normal serum derived IgG2 (Correia, 2010; Dillon et al., 2008; Liu et al., 2011; Wypych et al., 2008). The interconversion of these isoforms is dynamic and promoted by a reducing environment provided by the presence of thioredoxin reductase, released into culture media by effete cells; it can be ameliorated by control of dissolved oxygen levels (Davies et al., 2013; Hutterer et al., 2013; Kao et al., 2013; Koterba et al., 2012; Rispens et al., 2013, 2014). An *in vitro* model revealed that susceptibility to reduction/oxidation differed between IgG subclasses and light chain types, with sensitivity being in the order IgG1λ > IgG1κ > IgG2λ > IgG2κ (Koterba et al., 2012).

A core hinge region sequence of –Cys–Pro–Pro–Cys–, present in IgG1, IgG2, and IgG3, forms a partial helical structure that does not allow for intraheavy chain disulfide bridge formation. However, the homologous sequence in the IgG4 subclass is –Cys–Pro–Ser–Cys–, and this does allow for intraheavy chain disulfide bridge formation. Consequently, natural and recombinant IgG4 antibody populations are a mixture of molecules exhibiting inter- and intrahinge heavy chain disulfide bridge isoforms, generated *in vivo* and *in vitro* (Correia, 2010; Liu et al., 2011; Rispens et al., 2013). The IgG4 form having intrahinge heavy chain bridges is susceptible to dissociation into half-molecules (HLs) that may reassociate randomly to generate bispecific molecules—that is, a molecule that is monovalent for two nonidentical antigens (epitopes). This phenomenon is referred to as Fab arm exchange. The exchange is facilitated by the presence of an arginine residue at position 409 (R409) in the IgG4 heavy chain, in place of the lysine 409 (K409) present in IgG1, IgG2, IgG3 molecules. There exists a polymorphic variant of IgG4 that also has K409 and is not subject to Fab arm exchange (Davies et al., 2013; Rispens et al., 2013, 2014). Lateral noncovalent interactions between the two C_H3 domains of R409 IgG4 are reduced so that, under physiological conditions and in the absence of hinge region intraheavy chain disulfide bridges, they dissociate to form HL heteromonomers.

4.7 OXIDATION OF METHIONINE AND TRYPTOPHAN

Methionine residues exposed, or partially exposed, on the surface of native proteins may be susceptible to oxidation, while mild denaturation may result in exposure of buried methionine residues, which are also susceptible to oxidation. Methionine residues within variable region framework sequences regions have not been reported to be vulnerable to oxidation, owing to the compact domain structure, while residues exposed within CDRs may be (Chumsae et al., 2007; Liu et al., 2011). Consequently, early sequencing is employed to inform clone selection and rejection of clones having methionine within CDRs. For molecules of the IgG1 and IgG2 subclass it has been demonstrated that methionine residues M252 and M428 are prone to oxidation (Bertolotti-Ciarlet et al., 2009; Pan et al., 2009; Wang et al., 2011). Although these residues are distant from each other, in linear sequence, they are proximal at the C_H2/C_H3 interface. The interaction site for the FcRn receptor that regulates IgG catabolism and placental passage is formed at the C_H2/C_H3 interface, and M252, M428 oxidation has been shown to reduce both the affinity of binding to FcRn and half-life *in vivo*, in mice transgenic for human FcRn (Aricescu and Owens, 2013; Wypych et al., 2008). However, it was shown that oxidation of both M252 and M428 on both heavy chains was required to impact IgG1 half-life and was only observed when M252 oxidation was in excess of ~80% (Wang et al., 2011). Both SpA and SpG bind IgG-Fc at the C_H2/C_H3 interface, and methionine oxidation impacts binding affinity for both (Gaza-Bulseco et al., 2008; Pan et al., 2009). Therefore, M252 and M428 oxidation levels may be a useful QbD parameter. The impact of M252 and M428 oxidation on FcγR binding is reported to be minimal, although a "subtle" decrease in binding to the FcγRIIa 131H allele was observed (Bertolotti-Ciarlet et al., 2009). Minimal M252 oxidation levels of 2%–5% are reported for purified IgG antibodies in formulation buffers, while lower levels of oxidation are reported for M428. Oxidation of M252 and M428 increases under conditions of accelerated stability testing and on prolonged storage. Analysis of Herceptin, obtained from a pharmacy, and a potential biosimilar demonstrated that care has to be exercised when resuspending antibody therapeutics since a discrepancy was observed for the level of M252 oxidation between the innovator product (4.39%) and the proposed biosimilar (10.33%). Even so, a note added in proof commented that the value of 4.39% was greater than that determined by the innovator company; no oxidation of M428 was recorded (Xie et al., 2010).

4.8 DEAMIDATION: ASPARAGINE AND GLUTAMINE

Deamidation of asparagine and glutamine residues generates aspartic acid, isoaspartic acid, or glutamic acid, respectively, and is a frequently encountered PTM (Khawli et al., 2010; Wang et al., 2007). Deamidation of asparagine residues is influenced by adjacent amino acid residues, particularly the presence of a glycine residue C-terminal to the asparagine [–N–G–] and the degree of exposure to external environments. Studies of IgG1 and IgG2 proteins, *in vitro* and *in vivo*, have shown that asparagine residues 315 and 384 are susceptible to deamidation with the formation of isoaspartic and aspartic acid residues, respectively (Chelius et al., 2005; Liu et al., 2009; Xie et al., 2010). The relative susceptibility to deamidation at these sites varied between studies; however, the significance may be ameliorated by the finding that ~23% of asparagine 384 residues of normal polyclonal IgG are deaminated to aspartic acid. Thus, it may be assumed that healthy humans are constantly exposed to IgG bearing this PTM and that it might be considered to be a "self" structure. These studies did not identify asparagine deamidation within the constant region of the kappa light chains.

By contrast, deamidation within variable regions, particularly within CDRs, of recombinant antibodies has been shown to compromise antibody specificity and/or binding affinity (Diepold et al., 2012; Huang et al., 2005; Sinha et al., 2009; Vlasak et al., 2009). Interestingly, the approved blockbuster antibody therapeutic Trastuzumab (Herceptin) has asparagine residues in light chain CDR1 (Asn 30) and heavy chain CDR2 (Asn 55) that were shown to be susceptible to deamidation on accelerated degradation studies (Harris et al., 2001). The approved drug substance did not exhibit deamidation of these residues; therefore, their presence or absence could be used as a lot release criterion (Vlasak et al., 2009). As previously discussed, the levels of deamidation of asparagine residues, within both variable and constant regions, of a proposed Herceptin biosimilar were higher than those reported by the innovator molecule (Chelius et al., 2005), underlining the susceptibility to deamidation of these residues and the care that has to be exercised when resuspending this antibody therapeutic.

Glutamine residues are relatively resistant to deamidation, and no glutamine residues were reported to be subject to deamidation under nondenaturing conditions (Harris et al., 2001; Zhong and Wright, 2013). Under conditions of accelerated degradation, six Gln residues of an mAb were shown to be susceptible to deamination, as were four in variable regions and residues 295 and 418 in the IgG-Fc (Harris et al., 2001). The cyclization of N-terminal glutamine to form pyroglutamic acids has already been discussed.

4.9 γ-CARBOXYLATION AND β-HYDROXYLATION

For several proteins of the blood coagulation system, γ-carboxylation and β-hydroxylation are essential to function (Hansson and Stenflo, 2005; Kaufman and Powell, 2013). These PTMs are affected by specific carboxylase and hydroxylase enzymes, with conversion of target glutamate residues to γ-carboxyglutamate (Glu → Gla) and either target aspartate residues to β-hydroxyaspartate (Asp → Hya) or asparagine residues to β-hydroxyasparagine (Asn → Hyn). Both PTMs help mediate the binding of calcium ions, and in some cases they are essential to the functioning of blood factors VII, IX, and X, as well as activated protein C and protein S of the anticoagulant system. All of these proteins are composed of a number of distinct domains, with the so-called N-terminal Gla domain providing the γ-carboxylation sites and the epidermal growth factor (EGF) domains the β-hydroxylation sites. Typically, Gla domains are ~45 amino acids long and contain 9–12 Gla residues. Carboxylation of Gla domain glutamate residues is not dependent on occurrence within a specific consensus sequence, but carboxylase binding is mediated by an immediately adjacent propeptide region, which is subsequently removed by proteolysis (Hansson and Stenflo, 2005; Kaufman and Powell, 2013).

Hydroxylation of target EGF domain aspartate or asparagine residues is catalyzed by a β-hydroxylase located in the ER. EGF domains are ~45 amino acids long and contain one potential hydroxylation site. Hydroxylation is consensus sequence dependent and is usually partial, with only a fraction of target molecules being hydroxylated in practice. Full carboxylation and hydroxylation, on the other hand, is essential to maintaining biological activity of protein C (Grinnell et al., 1991, 2006; Liu et al., 2014; Yan et al., 1990). The native molecule displays nine carboxylation and one hydroxylation sites. Such stringent PTM requirements could not be met by CHO cells, forcing the developers of the recombinant version to develop a modified human cell line (HEK 293) in its manufacture (Liu et al., 2014).

4.10 SULFATION

Sulfation is a PTM required for a limited number of therapeutic proteins. *O*-Sulfation entails the attachment of a sulfate (SO_3^-) group to tyrosine residues by a sulfotransferases-mediated co/posttranslational process within the *trans* Golgi network and is predominantly associated with secretory and membrane proteins (Yang et al., 2015); it is sometimes encountered *O*-linked to serine or threonine (Medzihradszky et al., 2004). In the context of biopharmaceuticals, native hirudin (a leech-derived anticoagulant) and blood factors VIII and IX are usually sulfated. Neither of the approved recombinant forms of hirudin are sulfated, although it has been shown that sulfated hirudin (at Tyr63) displays 10-fold tighter affinity for thrombin than does unsulfated analogues (Costagliola et al., 2002). While over 90% of native factor IX molecules are sulfated, <15% of the approved recombinant form are; with apparently little if any difference in product efficacy (McGrath, 2006). Sulfation of factor VIII is required for optimal binding to its plasma carrier protein (von Willebrand's factor); interestingly, people inheriting a factor VIII Tyr1680 → Phe mutation often display mild hemophilia (Yang et al., 2015). Several hormone cell surface receptors are known to be tyrosine sulfated, and sulfation is required for high-affinity ligand binding and subsequent receptor activation (Choe et al., 2003; Ludeman and Stone, 2014).

4.11 GLYCOSYLATION

As referenced earlier, a recombinant form of the glycoprotein erythropoietin required a mammalian cell production platform since glycosylation and the precise glycoform were shown to be essential to *in vivo* activity (Macdougall et al., 2012). When first produced, in CHO cells, the product was shown to have an enhanced activity *in vitro*, compared to the natural form isolated from urine, but it was inactive *in vivo*. This was due to the absence of terminal sialic residues from the three *N*-linked oligosaccharide moieties; the consequence was exposure of terminal galactose residues, resulting in uptake by the asialo-glycoprotein receptor and rapid clearance through the liver. Process improvements led to a product having ~30% of sialylated EPO, and methods were developed to harvest/purify this fraction, the remainder being discarded. Since expiry of the patent, many biosimilar EPO products have been generated and approved by regulatory authorities; however, increased incidences of the development of ADA, with consequent pure red cell aplasia (PRCA), have been reported

(Macdougall et al., 2012). Prior to patent expiry, the innovator company (Amgen) developed an improved EPO and obtained new patent protection. Improvement was achieved by the introduction of two extra *N*-linked glycosylation sites, resulting in increased sialylation and enhanced biological activity and *in vivo* half-life (Elliott et al., 2004; Sinclair, 2013). Interestingly, while EPO cells secreted from CHO cells is heterogeneously glycosylated when expressed as a transmembrane protein, on CHO cells, it is homogeneously glycosylated and fully sialylated; the membrane bound EPO can be released to yield a fully active EPO product (Singh et al., 2015).

The "hallmark" of an antibody is perceived to be its specificity for a target pathogen (antigen); however, protection against invading microorganisms requires activation of a cascade of downstream biological mechanisms, resulting in the killing and elimination of pathogenic organisms or pathologic targets. These biologic mechanisms are triggered following the binding of immune complexes (ICs) to one or more multiple soluble or cell-bound IgG-Fc effector molecules (Jefferis, 2012; Jefferis and Lefranc, 2009; Vidarsson et al., 2014). It is established that *N*-linked glycosylation of asparagine 297 (N297; Eu numbering) is essential for optimal activation of the effector molecules. Analysis of oligosaccharides released from normal polyclonal human IgG reveals a considerable heterogeneity of diantennary oligosaccharide structures. Similar analysis of monoclonal human IgG proteins, produced by neoplastic plasma cells (multiple myeloma), also reveals a restricted glycoform profile that is characteristic for each protein—that is, each malignant plasma cell population (patient).

The generation of homogeneous antibody glycoforms has allowed each to be evaluated for its functional profile, and these studies have informed the production of selective glycoforms tailored to the disease entity being addressed (Jefferis, 2012; Raju and Jordan, 2012; Raju and Lang, 2014). While the IgG-Fc glycoform can have a profound impact on function, sensitive physical studies have failed to demonstrate a comparable structural difference between different glycoforms. It is evident that each oligosaccharide structure has a subtle but unique impact on the tertiary/quaternary conformation of the IgG-Fc, and hence functions. Before discussing the functional consequences for individual IgG-Fc glycoform(s) of normal, myeloma, and recombinant IgG, an overview of IgG-Fc structure is appropriate.

4.12 QUATERNARY STRUCTURE OF IgG-Fc: THE PROTEIN MOIETY

The first crystal structure of IgG-Fc, resolved at 2.9-Å, was published by Deisenhofer (1981). It was reported that interpretable electron density was not obtained for residues 223–237, which comprise most of the core and lower hinge region, or the C-terminal residues 444–446. It was not known at this time that the C_H3 exon codes for a C-terminal (K, 447) lysine residue is removed by an endogenous serum carboxypeptidase B. Crystal structures have been reported for human, rabbit, and mouse IgG-Fc and chicken IgY-Fc fragments and for human IgG-Fc alone or in complex with Staphylococcal protein A (SpA), Streptococcal protein G (SpG), rheumatoid factor (RF), and soluble (s) recombinant forms of human sFcγRIIa, sFcγRIIIb, and sFcγRIIIa (Corper et al., 1997; Frank et al., 2014; Jefferis, 2012; Padlan, 1990; Radaev et al., 2001; Ramsland et al., 2011; Sauer-Eriksson et al., 1995;

Sondermann et al., 2000). Progressively higher resolution has allowed the generation of more refined models. An α-carbon ribbon structure for a human IgG1 molecule is shown in Figure 4.1.

The observed internal mobility of the lower hinge and hinge proximal regions of the C_H2 domains allows for an equilibrium of high-order conformers to be formed that may differentially bind unique ligands (e.g., the three homologous Fcγ receptor types). Previous proposals that different ligands may bind through "overlapping nonidentical sites" may suggest too rigid a structure and may be modified to propose that each FcγR binds to a unique IgG-Fc conformer present within an equilibrium of transient protein structures. However, amino acid residue side chains and/or main chain atoms may be involved in common (Anumula, 2012; Jefferis, 2012; Jefferis et al., 1998; Ozbabacan et al., 2011). The binding sites for sFcγRIIa, sFcγRIIIa, and sFcγRIIIb are asymmetric, with both heavy chains being engaged such that monomeric IgG is univalent for each Fcγ receptor and the C1 component of complement. This obviates continuous activation of inflammatory cascades by circulating endogenous monomeric IgG *in vivo*; IgG antigen/antibody immune complexes are multivalent and able to cross-link and activate cellular receptors. Residues of the lower hinge region that are disordered in the Fc crystals are ordered in the Fc-sFcγR complexes and are directly involved in binding the receptor (Frank et al., 2014; Radaev

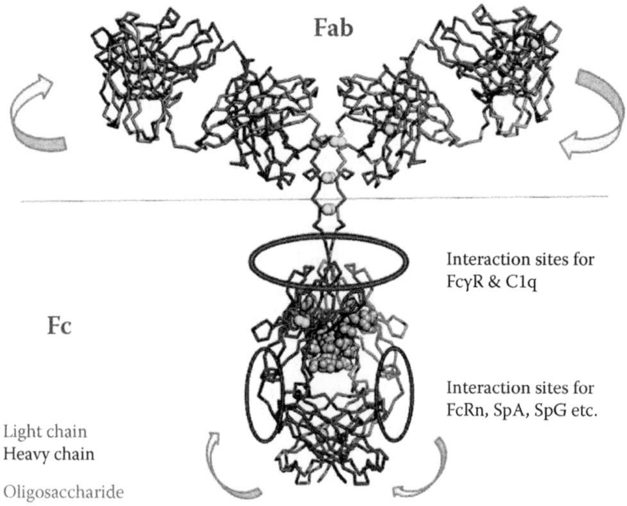

FIGURE 4.1 (See color insert.) An α-carbon cartoon of a human IgG1 molecule indicating binding sites for IgG1-Fc ligand binding sites; arrows indicate independent mobility of Fab & Fc around the hinge region. (1) The C_H3 domains are well defined due to noncovalent pairing, in the $(C_H3)2$ module; (2) There is a significant area of noncovalent contact between the C_H2 and C_H3 domains that stabilizes the C_H2/C_H3 interdomain structure; (3) The C_H2 domains do not pair, and the hydrophobic surface of each C_H2 domain is "overlaid" by the carbohydrate; (4) the structure in the hinge proximal region of the C_H2 domain is, relatively, disordered; (5) The intrinsic stability of the immunoglobulin fold is reflected in higher structural resolution for the β-sheets regions than for β-bends. (Courtesy of Peter Artymiuk, University of Sheffield, using PyMol: http://pymol.sourceforge.net.)

et al., 2001; Sondermann et al., 2000). By contrast, IgG-Fc is functionally divalent for ligands binding at the C_H2–C_H3 interface—for example, the neonatal Fc receptor (FcRn), RF, SpA, SpG. Due to the symmetry of the IgG-Fc, these interaction sites are opposed at ~180°, and each is accessible to bind macromolecular ligands to form multimeric complexes.

4.13 THE IgG-Fc OLIGOSACCHARIDE MOIETY

The oligosaccharide released from normal polyclonal IgG-Fc is heterogeneous and essentially comprises the core heptasaccharide with the variable addition of fucose, galactose, bisecting *N*-acetylglucosamine, and *N*-acetylneuraminic (sialic) acid residues (see Figure 4.2; Anumula, 2012; Reusch et al., 2015a,b; Song et al., 2014). Recent analyses employing high-sensitivity mass spectrometry suggests the possible presence of further minor oligosaccharides (e.g., high mannose and hybrid glycoforms; Goetze et al., 2011a).

Several systems of nomenclature are currently in use to represent oligosaccharide structures (CFG, 2012; NIBRT, 2012), but "antibody practitioners" have adopted a shorthand nomenclature for oligosaccharides released from normal polyclonal IgG. A core heptasaccharide (blue in Figure 4.2) is designated G0 (zero galactose); the core bearing one or two galactose residues are designated G1 and G2, respectively. The core + fucose is designated G0F, the core + fucose + galactose G1F, G2F, and so on; when bisecting *N*-acetylglucosamine is present, a B is added (e.g., G0B, G0BF, G1BF); sialylation at the galactose residues are designated as G1FS, G2FBS, and so on. The approximate composition of neutral oligosaccharides released from normal polyclonal human IgG-Fc is G0, 3%; G1, 3%; G2, 6%; G0F, 23%; G1F, 30%; G2F, 24%; G0BF, 3%; G1BF, 4%; G2BF, 7% (Anumula, 2012; Goetze et al., 2011a; Reusch et al., 2015a,b; Song et al., 2014). It is important to define the glycoform of the intact IgG molecule (e.g., [G0/G1F], [G1F/G2BF], etc.) since it has been shown that individual IgG molecules may be composed of symmetrical or asymmetrical heavy chain glycoform pairs (Matsumiya et al., 2007; Mimura et al., 2007). Consequently, enhanced FcγRIIIa mediated antibody-dependent cellular cytotoxicity (ADCC) may be observed for IgG in which one heavy chain only bears oligosaccharide devoid of fucose (Masuda et al., 2007; Matsumiya et al., 2007; Mimura et al., 2007) and the [G0/G0F] glycoform could be as potent in ADCC as the [G0/G0] glycoform.

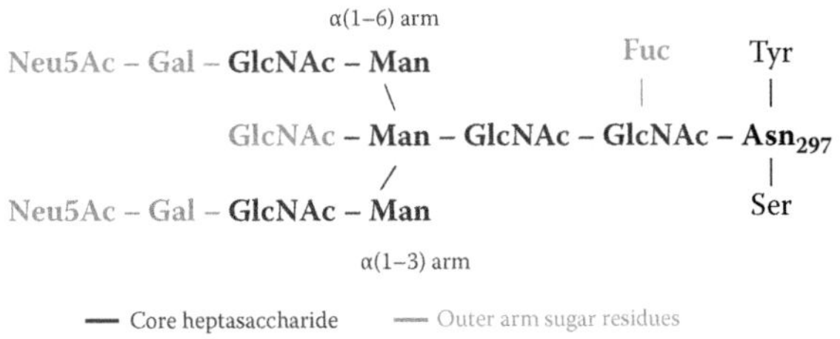

FIGURE 4.2 (See color insert.) The diantennary oligosaccharide showing the heptasaccharide core (dark gray) and possible outer arm sugar residues (light gray).

Minor glycoforms present in polyclonal IgG-Fc may be biologically significant since each could be the predominant glycoform of an individual antibody secreted from a single plasma cell; analysis of monoclonal human IgG, isolated from the sera of a patient with plasma cell cancer (multiple myeloma), has shown that the IgG-Fc glycoform profile of the paraprotein is essentially unique for each protein analyzed (Farooq et al., 1997; Jefferis et al., 1990; Kibe et al., 1996). Subtle differences in oligosaccharide processing is observed, with a preference for addition of galactose to the α(1–6) arm of IgG1-Fc and the α(1–3) arm of IgG2-Fc; the arm preference for IgG3 correlates with an allotypic difference, the presence of tyrosine or phenylalanine at residue 296 (Y296F), immediately adjacent to asparagine 297 (Alavi et al., 2000; Ercan et al., 2010; Jefferis, 2012). These data suggest a critical balance between the conformation of the IgG-Fc and the steric requirements of glycosyltranferases that may be sensitive to niche environments within the Golgi apparatus.

The glycoform profile of serum polyclonal IgG can vary significantly in both health and disease states, particularly in autoimmune and inflammatory diseases (Alavi et al., 2000; Ercan et al., 2010; Farooq et al., 1997; Hayes et al., 2014; Jefferis, 2009a, 2012; Jefferis et al., 1990; Kibe et al., 1996). Methodologies that allow the glycoform profile of antigen-specific serum IgG antibodies to be determined demonstrate significant glycoform profile differences between IgG autoantibodies and the bulk IgG (Holland et al., 2006; Rombouts et al., 2015; Wuhrer et al., 2015). The oligosaccharide profiles of recombinant IgG proteins produced in mammalian cells differ between cell types, the culture method, and precise medium conditions employed. For mAbs the [G0F/G0F] glycoform predominates (see below). Under conditions of stress (e.g., nutrient depletion, acid pH, etc.), deviant glycosylation may be observed (e.g., the presence of high mannose forms and/or incomplete site occupancy) (Bondt et al., 2013).

4.14 IgG-Fc PROTEIN/OLIGOSACCHARIDE INTERACTIONS

The Deisenhofer IgG-Fc structure suggested the potential for 72 protein/oligosaccharide interactions, including six C_H2 protein/oligosaccharide hydrogen bonds and six hydrogen bonds within each oligosaccharide moiety (Corper et al., 1997; Deisenhofer, 1981; Frank et al., 2014; Jefferis, 2012; Padlan, 1990; Radaev et al., 2001; Ramsland et al., 2011; Sauer-Eriksson et al., 1995; Sondermann et al., 2000). These interactions include the sugar residues of the α(1–6) arm, while residues of the α(1–3)-Man-GlcNAc arms are orientated toward the internal space between the C_H2 domains; weak lateral interactions between sugar residues present on opposed heavy chains have been suggested for some structures (Corper et al., 1997; Deisenhofer, 1981; Frank et al., 2014; Jefferis, 2012; Padlan, 1990; Radaev et al., 2001; Ramsland et al., 2011; Sauer-Eriksson et al., 1995; Sondermann et al., 2000).

Structural and functional studies of normal, truncated, and aglycosylated glycoforms of IgG1-Fc, generated *in vitro*, have employed X-ray crystallography (Corper et al., 1997; Deisenhofer, 1981; Frank et al., 2014; Padlan, 1990; Radaev et al., 2001; Ramsland et al., 2011; Sauer-Eriksson et al., 1995; Sondermann et al., 2000), differential scanning microcalorimetry (DSC), and Fcγ receptor binding (isothermal micro-calorimetry, ITC) (Krapp et al., 2003; Mimura et al., 2000, 2001). These studies established that while the galactose residue on the α(1–6) arm has substantial contacts with the protein structure, it does not impact C_H2 domain stability; sequential removal of the terminal GlcNAc and the two-arm mannose residues, generating a (GlcNAc2Man3)2 and (GlcNAc2Man)2 glycoform, resulted in destabilization of the C_H2 domain. DSC has proved to be a valuable tool for probing the contributions of buffers and Fab sequence to stability and solubility of intact IgG and antibody fragments (Fortunato and Colina, 2014; Houde and Engen, 2013; Krapp et al., 2003; Mimura et al., 2000, 2001).

X-ray crystallography of a series of the truncated IgG-Fc glycoforms revealed a progressive increase in temperature factors for the protein moiety of the C_H2 domain, as evidence of increasing structural disorder (destabilization) (Frank et al., 2014; Radaev et al., 2001; Singh et al., 2015; Sondermann et al., 2000). Minimal (weak) FcγRI and C1 binding and activation was observed for the [GlcNAc2Man]2 glycoform that has the potential to form 31 noncovalent contacts with the protein, including at least three hydrogen bonds (Corper et al., 1997; Frank et al., 2014; Padlan, 1990; Radaev et al., 2001; Ramsland et al., 2011; Sauer-Eriksson et al., 1995; Sondermann et al., 2000). Truncation of the sugar residues results in the progressive mutual approach of C_H2 domains with the generation of a "closed" conformation, in contrast to the "open" conformation observed for the fully galactosylated IgG-Fc (Fortunato and Colina, 2014; Houde and Engen, 2013; Krapp et al., 2003; Subedi et al., 2014). The dramatic reduction of FcγR and C1 binding and activation for aglycosylated IgG-Fc contrasts with consistent reports of the minimal structural changes reported.

An extensive NMR study of a series of truncated glycoforms showed that trimming of the oligosaccharide was accompanied by concomitant increase in the number of amino acid residues perturbed within the C_H2 domains (Yamaguchi et al., 2006). Cleavage between the primary and secondary GlcNAc sugar residues induced conformational changes within the lower hinge region, at sites having no direct contact with the carbohydrate moieties but formed the major FcγR-binding site. Conformation at the C_H2/C_H3 interface, which forms the FcRn and SpA binding sites, was minimally perturbed. A dynamic model was also proposed from an NMR study of differentially galactosylated and sialylated IgG-Fc glycoforms (Yamaguchi et al., 2006). It was proposed that interactions of sugar residues of the α(1–6) arm and the protein surface may be a dynamic equilibrium between the bound and unbound state; the latter state may allow for increased accessibility to glycosyltransferases.

Although the oligosaccharide is integral to the IgG-Fc structure and appears to be sequestered within the space between the C_H2 domains, some microorganisms produce endoglycosidases that cleave the oligosaccharides from native IgG-Fc. Thus, *Streptococcus pyogenes* produces endoglycosidase S (EndoS) that cleaves between the primary and secondary *N*-acetlyglucosamine residues (Allhorn et al., 2010; Zheng et al., 2011). Peptide: *N*-Glycosidase F (PNGase F), isolated from culture filtrate of *Flavobacterium meningoscepticum*, cleaves the peptide/oligosaccharide bond to generate deglycosylated IgG (Flynn et al., 2010; Jefferis, 2009c). Since the consequence of oligosaccharide cleavage is loss of effector functions and the ability to kill and remove target bacteria, it is tempting to conclude that this is evidence of coevolution.

4.15 IgG-Fc GLYCOFORM PROFILES OF RECOMBINANT IgG ANTIBODY THERAPEUTICS

Glycosylation of IgG-Fc has a profound influence on the range and magnitude of effector functions activated; however, residual effector activity has been observed for large immune complexes formed with nonglycosylated IgG antibodies, owing to the increased avidity afforded by multiple IgG-Fc/ligand interactions (Crispin, 2013; Hristodorov et al., 2013; Jefferis, 2009b,c, 2012; Ju and Jung, 2014; Nesspor et al., 2012). Downregulation (elimination!) of effector functions has also been achieved through protein engineering, with retention of glycosylation (Vafa et al., 2014), while reverse engineering has been employed to restore effector functions to aglycosylated antibody formats. This could open up the potential for commercial production in a simpler system (e.g., *E. coli*) and a consequent reduction in CoG; the significant protein and glycosylation engineering required may be expected to result in increased immunogenicity (Borrok et al., 2012). When the MoA includes the activation of antibody effector functions, 100% oligosaccharide occupancy at asparagine 297 is a CQA. By the same token, if an aglycosylated IgG is to be employed, 0% occupancy is a CQA (Borrok et al., 2012; Jefferis, 2009b; Ju and Jung, 2014; Vafa et al., 2014).

Initially, CHO, NSO, and Sp2/0 cells were used for the production of mAbs (Burnouf, 2011; Estes and Melville, 2014; Lai et al., 2013; Li et al., 2012) that add predominantly G0F oligosaccharides with relatively low levels of G1F and G2F, relative to normal polyclonal IgG-Fc. CHO cells may also add *N*-acetylneuraminic acid residues but in α(2–3) linkage rather than the α(2–6) linkage present in humans. In addition, these cell lines may add sugars that are not present in normal serum-derived IgG that can be immunogenic in humans. A particular concern is the addition, by NSO and Sp2/0 cells, of galactose in α(1–3) linkage to galactose linked β(1–4) to the *N*-acetylglucosamine residues (Bosques et al., 2010; Estes and Melville, 2014; Galili, 2013; Ghaderi et al., 2012; Jefferis, 2012; Lai et al., 2013). Humans and higher primates do not have a functional gene encoding the transferase that adds galactose in α(1–3) linkage. However, due to continual environmental exposure to the gal α(1–3)gal epitope (e.g., in red meats), humans develop IgG antibody specific for this antigen (Estes and Melville, 2014; Galili, 2013; Ghaderi et al., 2012). The (gal α(1–3)gal) epitope is widely expressed on hamster cells, and it has recently been reported that some CHO cell lines are capable of (gal α(1–3)gal) addition. Similarly, CHO, NSO, and Sp2/0 cells may add *N*-glycolylneuraminic acid, in α(2–3) and/or α(2–6) linkage, that is not present in human oligosaccharides and is reported to be immunogenic (Borrok et al., 2012; Burnouf, 2011; Estes and Melville, 2014; Ghaderi et al., 2012; Lai et al., 2013; Li et al., 2012; Vafa et al., 2014). A significant population of normal human IgG-Fc bears a bisecting *N*-acetylglucosamine residue that is absent from IgG-Fc produced in CHO, NSO, or Sp2/0 cells.

The generation of homogeneous IgG-Fc glycoforms *in vitro* has shown that the effector functions activated, both qualitatively and quantitatively, differ between IgG subclasses and antibody glycoforms. It has not proved possible to manipulate culture medium conditions to generate predetermined mAb glycoform profiles. However, significant "tweaking" of the profile can be achieved during a production run (Maeda et al., 2012; Pacis et al., 2011; Yu et al., 2011b) to maintain the product within specification; cellular engineering has been employed to enhance production of particular human IgG-Fc glycoforms—vide infra.

Two apparently unnatural glycoforms that are invariably present in recombinant antibody preparations are a hexasaccharide core bearing a fucose residue but lacking an *N*-acetylglucosamine residue on either the α(1–3) or α(1–6) arm. The levels present are low (1–5%), and while the impact of this glycoform on mAb performance has not been reported, it does not appear to have generated concern. Of particular moment is the level of fucosylation, both fucosylated and nonfucosylated being natural glycoforms, and oligomannose glycoforms not reported for serum IgG. These issues will be explored in turn.

4.16 CONTROL OF THE ADDITION OF FUCOSE AND ITS IMPACT ON FUNCTIONAL ACTIVITIES

The influence of recombinant protein glycoforms on biological activity has been explored through their production in mutant CHO cells lacking the ability to add one or more sugar residue (Marth and Grewal, 2008; Stanley, 2011). The cell line Lec 13 lacks the ability to add fucose to the primary *N*-acetylglucosamine residue, and antibodies of the IgG1 subclass produced in this cell line were shown to exhibit enhanced abilities to kill cancer cells by the mechanism referred to as antibody dependent cellular cytotoxicity (ADCC) (Ripka et al., 1986). This finding was confirmed and extended to all IgG subclasses when antibodies were produced in an α(1,6)–fucosyltransferase “knock-out” CHO cell line (Niwa et al., 2005; Shields et al., 2002). This cell line is available commercially from the company Biowa and provides access to the “Potelligent” platform technology (Shibata-Koyama et al., 2009a; Yamane-Ohnuki and Satoh, 2009). A nonfucosylated anti-CCR4 antibody (Mogamulizumab) produced by this cell line has been approved in Japan for the treatment of patients with relapsed or refractory CCR4-positive Adult T-Cell Leukaemia-Lymphoma (ATL) (Subramaniam et al., 2012) and is in Phase III trials in Europe and the US. A similar improvement in ADCC was reported for IgG1 antibody produced in a “knock-in” CHO cell line transfected with the human β 1,4-*N*-acetylglucosaminyltransferase III (GnTIII) gene, resulting in the addition of bisecting *N*-acetylglucosamine residues (Davies et al., 2001; Ferrara et al., 2006; Mossner et al., 2010; Umana et al., 1999). The early addition of bisecting *N*-acetylglucosamine during passage through the Golgi was shown to inhibit the addition of fucose by the endogenous α(1–6)—fucosyltransferase (Mossner et al., 2010). It is posited, therefore, that the absence of fucose is the main factor determining increased ADCC for these glycoforms. This platform has been consolidated, by Glycart-Roche, and the nonfucosylated anti-CD20 antibody Obinutuzumab was approved by the FDA in 2013, for previously untreated chronic lymphocytic leukemia (CLL) (FDA, 2015b; Mossner et al., 2010; Ogorek et al., 2012); an anti-EGFR (Imgatuzumab) produced employing this platform is in phase II clinical trials (Delord et al., 2014). Multiple technologies are being developed to generate antibodies homogeneous for a single naturally occurring glycoform; customized to deliver MoAs deemed appropriate for treatment of specific disease indications. These IgG glycoforms may be minor components of the oligosaccharides released from normal polyclonal human IgG-Fc; however, being normal components, they do not present immunogenicity issues (Farooq et al., 1997; Jefferis, 2012, 2014; Jefferis et al., 1990; Kobata, 2008). By contrast, nonhuman (mammalian) cell lines may add sugars in a nonhuman linkage or sugars that are not present in the human glycome and may be immunogenic (Bosques et al., 2010; Galili, 2013; Ghaderi et al., 2012; Jefferis, 2012; Maeda et al., 2012; Stanley, 2011).

The foregoing discussion was essentially confined to ADCC mediated by peripheral blood mononuclear leucocytes; however, the impact of fucosylation is reported to differ for polymorphonuclear cells (e.g., neutrophils). A study employing two batches of a monoclonal IgG antibody, one of high fucose content and one of low fucose content, reported that higher fucose content resulted in more active neutrophil-mediated ADCC (Peipp et al., 2008). Contrary outcomes are reported for neutrophil-mediated phagocytosis and apoptosis (Derer et al., 2014; Mizushima et al., 2011; Nakagawa et al., 2010; Shibata-Koyama et al., 2009a). It should be noted that results obtained by *in vitro* ADCC studies employing cell lines may vary since the glycoform of the receptor is also a critical parameter and will differ between cell lines (Hayes et al., 2014). I am not aware of publications reporting a positive or negative impact of fucosylation on complement dependent cytotoxicity (CDC); however, an IgG1/IgG3 hybrid molecule has been shown to exhibit enhanced CDC for both fucosylated and nonfucosylated IgG-Fc glycoforms (Natsume et al., 2008).

The perceived advantage of nonfucosylated antibodies has led other groups and companies to explore alternative routes for the generation of nonfucosylated glycoproteins. Cell lines have been engineered to produce homogeneous Man5 glycoforms that lack the addition of fucose (Gloster and Vocadlo, 2012; Haryadi et al., 2013; Ogorek et al., 2012; Reeves et al., 2002; Sealover et al., 2013; Yu et al., 2011a; Zhang et al., 2013; Zhou et al., 2008). Inhibitor targeting of enzymes within the Golgi apparatus provides multiple opportunities for producing nonfucosylated molecules, and the action of kifunensine has been employed by several groups for the generation of nonfucosylated high mannose (Man6–Man9) glycoforms (Yu et al., 2011a; Zhang et al., 2013; Zhou et al., 2008). Other platforms generated include GlymaxX (Ogorek et al., 2012) which engineers mammalian cells to express a bacterial enzyme that inhibits the pathway leading to the addition of fucose. A further platform employs sugar analogues that, when added to the culture medium, inhibit incorporation of the natural sugar (Gloster and Vocadlo, 2012).

The influence of fucose on FcγRIIIa-mediated ADCC has been shown to be dependent also on the glycoforms of the receptor. The natural FcγRIIIa receptor expresses five *N*-linked glycosylation sites, with one, at N162, being at the interface of the receptor/IgG-Fc interaction. Interactions of this oligosaccharide with the IgG-Fc are favored for afucosylated IgG-Fc, accounting for enhanced ADCC; aglycosylated FcγRIIIa has the same binding affinity for fucosylated and afucosylated IgG-Fc (Ferrara et al., 2011; Zeck et al., 2011). The presence of a further *N*-linked oligosaccharide, at N45, has been shown to have a negative impact on FcγRIIIa binding (Shibata-Koyama et al., 2009b).

4.17 THE INFLUENCE OF GALACTOSYLATION ON IgG-Fc ACTIVITIES

The extent of galactose addition to IgG-Fc oligosaccharides is a major source of glycoform heterogeneity in both health and disease states. Accepting the levels of galactosylation observed for young adults as the norm, one finds a decline on ageing and a small but significant gender difference (Ruhaak et al., 2011; Yamada et al., 1997). An increase in IgG-Fc galactosylation occurs over the course of normal pregnancy, with levels returning to the adult norm following parturition (Alavi et al., 2000; Einarsdottir et al., 2013; Kibe et al., 1996). Hypogalactosylation of IgG-Fc is reported for a number of inflammatory states associated with autoimmune diseases (Bondt et al., 2013; Holland et al., 2006; Jefferis, 2009a; Rombouts et al., 2015; Wuhrer et al., 2015), The extent of IgG-Fc galactosylation observed for monoclonal human myeloma IgG proteins is highly variable, indicating that the level of IgG-Fc galactosylation is a clonal property (Farooq et al., 1997; Jefferis et al., 1990; Kobata, 2008). The antibody products of CHO, Sp2/0, and NS0 cell lines used in commercial production of mAbs are generally highly fucosylated but hypogalactosylated, relative to normal polyclonal human IgG (Boyd et al., 1995; Pascoe et al., 2007; Wacker et al., 2011). Therefore, it is necessary to consider the possible impact of differential IgG-Fc galactosylation on functional activity.

The variations in galactosylation observed in health and disease suggest that it is either of functional significance or an epiphenomenon. The increase in galactosylation in pregnancy is particularly intriguing as it coincides with FcRn-mediated transcytosis of IgG from mother to fetus in the third trimester (Alavi et al., 2000; Bondt et al., 2013; Einarsdottir et al., 2013; Jefferis, 2012; Kibe et al., 1996; Kobata, 2008). Although studies, *in vitro*, have not revealed a relationship between the natural glycoform of IgG-Fc and FcRn binding, the glycoform at the single glycosylation site in human FcRn does show this relationship (Natsume et al., 2008). The possible consequences for hypogalactosylated recombinant antibodies on *in vivo* activity have been extrapolated from *in vitro* cell-based assays and animal experiments. Removal of terminal galactose residues from Campath-1H was shown to reduce classical complement activation but to be without effect on FcγR-mediated functions (Boyd et al., 1995). Similarly, the ability of rituximab to kill tumor cells by the classical complement route has been shown to be maximal for the [G2F]2 glycoform, in comparison to the [G0F]2 glycoform (Raju and Jordan, 2012). The product that gained licensing approval was composed of ~25% galactosylated oligosaccharides; therefore, the level of galactosylation is a CQA for some mAbs and must be maintained, within limits, over the lifetime of the drug. It may also be an indicator and measure of control over the production process. In the absence of galactose, the terminal sugar residue is *N*-acetylglucosamine, which may be accessible to bind the mannose receptor, expressed on many cell types, including antigen presenting dendritic cells. Similarly, immune complexes formed with agalactosylated IgG may bind the mannan binding lectin (MBL) and activate the lectin complement pathway (Arnold et al., 2006; Jefferis, 2012; Malhotra et al., 1995).

As previously stated, CHO, NS0, and Sp2/0 cells have the potential to add galactose in an α(1–3) linkage that is not present in the human glycome and is immunogenic.

While the addition of galactose in α(1–3) linkage to IgG-Fc is minimal it is readily added to oligosaccharides when present in antibody V regions. This presents a particular challenge for the generation of a cetuximab biosimilar, as the innovator product is known to bear α(1–3)galactose residues that have proven to be immunogenic in patients and, in a minority of cases, the production of IgE ADA that carries with it the risk of immediate hypersensitivity reactions (Bosques et al., 2010; Estes and Melville, 2014; Galili, 2013; Ghaderi et al., 2012; Jefferis, 2012; Lai et al., 2013).

4.18 SIALYLATION OF IgG-Fc OLIGOSACCHARIDES

Although ~70% of oligosaccharides released from polyclonal IgG-Fc bear galactose residues, only a minority are sialylated, bearing terminal α(2–6) *N*-acetylneuraminic acid residues (NANA) (Anumula, 2012; Reusch et al., 2015a,b; Song et al., 2014). The paucity of sialylation has been presumed to reflect restricted access of terminal galactose residues to the α(2–6) *N*-acetylneuraminic transferase enzyme, rather than being due to any deficit in the sialylation machinery. This conclusion is supported by the finding that when oligosaccharides are present in both IgG-Fc and IgG-Fab, the latter bears highly galactosylated and sialylated structures, demonstrating that the glycosylation machinery is fully functional (Jefferis, 2012; Mimura et al., 2007). The presence or absence of terminal galactose and/or sialic acid residues does not influence IgG half-life since it is not catabolized in the liver, via the asialo-glycoprotein receptor (ASGPR), but by multiple cell types that express the FcRn receptor.

The balance between structure and accessibility is illustrated by a panel of IgGs in which individual amino acid residues making contact with the oligosaccharide were replaced by alanine. The phenylanaline/alanine 241 and 243 mutants (F241A; F243A) resulted in the generation of hypergalactosylated and highly sialylated glycoforms, suggesting some relaxation of structure allowing access to glycosyl transferases (Lund et al., 1996). This explanation is supported by X-ray crystallographic and *in silico* molecular dynamic studies that support a model in which the oligosaccharides of the α(1–6) arm alternate between a bound and free state with the C_H2 domain protein surface (Ahmed et al., 2014; Barb and Prestegard, 2011; Barb et al., 2012). Interactions between the different oligosaccharide structures and the C_H2 domain protein impact C_H2/C_H3 interdomain dynamics that, in turn, might influence conformational dynamics at the N-terminal region of the C_H2 domain and interaction sites for Fcγ receptors and the C1 component of complement.

Currently, there are contradictory reports in the literature relating to the functional activities of IgG and IgG-Fc glycoforms bearing terminal NANA residues. This relates particularly to attempts to elucidate the mechanism(s) by which intravenous human IgG (IVIG) mediates an anti-inflammatory activity in a range of human inflammatory diseases. Structural and functional studies have employed α(2–6) sialylated IgG purified from IVIG in *in vivo* mouse models of immune thrombocytopenic purpura (ITP) and rheumatoid arthritis (RA). One series of publications has consistently reported that the presence of α(2–6) sialylation, and not α(2–3) sialylation, is essential for anti-inflammatory outcomes and that sialylated human IgG acts by binding to the lectin receptor SIGN-R1 in mouse models, or DC-SIGN in humans; a "knock-on" effect is upregulated expression of the inhibitory FγRIIb receptor, with consequent attenuation of inflammation mediated by autoantibody/antigen immune complexes (Ahmed et al., 2014; Ballow, 2014; Oaks et al., 2013; Schwab et al., 2012; Sondermann et al., 2013; Washburn et al., 2015). It has been observed that caution should be exercised when extrapolating from mouse models to humans since the tissue distribution of SIGN-1, in the mouse, and DC-SIGN, in man, differ. Other studies have demonstrated anti-inflammatory activity for IVIg to be independent of IgG-Fc sialylation and/or may be associated with F(ab')2 sialylation (Campbell et al., 2014; Guhr et al., 2011; Leontyev et al., 2012; Othy et al., 2014; Yu et al., 2013). As usual for apparently contradictory scientific reports, it now appears that both outcomes may be valid but critically dependent on precise experimental protocols and animal models both *in vitro* and *in vivo* (Schwab and Nimmerjahn, 2014).

Clearly, these findings have potential relevance for the generation of mAbs, biosimilars, and the development of biobetters; however, as previously stated, the current production platforms employ CHO cells that add NANA in α(2–3) linkage only, if at all. The murine NS0 and Sp2/0 cells may add *N*-glycolylneuraminic acid (Neu5Gc) residues in α(2–3) or α(2–6) that are absent from human glycoproteins and consequently are immunogenic (Raju and Lang, 2014).

4.19 RECOMBINANT GLYCOPROTEINS BEARING HIGH MANNOSE (MAN5–MAN9) OLIGOSACCHARIDES

I am aware of only one publication in which high mannose (Man5–Man9) glycoforms of normal serum IgG were reported, at a level of <1% (Flynn et al., 2010); however, it is difficult to exclude contamination by other serum proteins. In contrast, high mannose glycoforms are frequently present in CHO, NS0, and Sp2/0 produced recombinant mAbs. There has been a concern that this glycoform may compromise the efficacy of an mAb therapeutic through enhanced clearance (Goetze et al., 2011b; Shi and Goudar, 2014; Yu et al., 2012; Zhong et al., 2012) and/or enhanced immunogenicity, due to uptake of immune complexes via the mannose receptor present on antigen presenting cells, for example, dendritic cells (Adams et al., 2008; Dong et al., 1999; Royer et al., 2010) and/or activation of the lectin pathway of complement via the MBL (Arnold et al., 2006; Banda et al., 2008). Alternatively, purposeful generation of Man5–Man9 glycoforms has been achieved by inhibition of the glycosylation processing pathway at a point prior to the addition of fucose, conferring high ADCC activity on the product.

The CHO-lec 3.2.8.1 and HEK293S cell lines have been generated that lack expression of the *N*-acetylglucosamine transferase 1 enzyme (GnT1) activity, limiting processing to the (GlcNAc2Man5)2 glycoform (Goetze et al., 2011b; Shi and Goudar, 2014; Stanley, 2011; Yu et al., 2012; Zhong et al., 2012). Ideally, control of the IgG glycoform profile might be achieved by manipulation of culture conditions and, while multiple parameters impact Golgi-mediated glycoprotein processing, some control of Man5 levels by manipulation of cell culture conditions has been reported (Doores et al., 2010; Franze et al., 2012; Hossler, 2012; Pacis et al., 2011;

Yu et al., 2011b; Zhong et al., 2012). Inhibition of enzymes within the Golgi apparatus provides another avenue for the production of high mannose glycoforms. Thus, kifunensine inhibits mannosidases with the production Man6–Man9 glycoforms (Ferrara et al., 2011; Reeves et al., 2002; Sealover et al., 2013; Shibata-Koyama et al., 2009b; Zeck et al., 2011). It has recently been demonstrated that incomplete processing *in vivo*, with consequent generation of truncated mannose oligosaccharides, can result from restricted access for mannose transferases. Thus, while the surface of recombinant HIV GP120 glycoprotein is almost entirely covered by *N*-linked high mannose oligosaccharide structures, it has been shown that native GP120, expressed on HIV virus isolates, bears a number of truncated oligomannose structures. It appears that the density of the early oligomannose structures limits accessibility to mannose transferases (Doores et al., 2010).

4.20 IgG-Fab GLYCOSYLATION

It is established that ~30% of polyclonal human IgG molecules bear *N*-linked oligosaccharides within the IgG-Fab region, in addition to the conserved glycosylation site at Asn 297 in the IgG-Fc (Anumula, 2012; Borrok et al., 2012; Coloma et al., 1999; Ercan et al., 2010; Girardi et al., 2009; Holland et al., 2006; Jacquemin, 2010; Mimura et al., 2007; Spencer et al., 2012; Stanfield and Wilson, 2014; Youings et al., 1996). When present, they are attached within the variable regions of the kappa (V_κ), lambda (V_λ), or heavy (V_H) chains and sometimes both. In the immunoglobulin sequence database, ~20% of expressed IgG variable regions have *N*-linked glycosylation consensus sequences (Asn-X-Thr/Ser; where *X* can be any amino acid except proline). Interestingly, these consensus sequences are mostly not germline encoded but result from somatic hypermutation—suggestive of positive selection for improved antigen binding. Analysis of polyclonal human IgG-Fab reveals the presence of diantennary oligosaccharides that are extensively galactosylated and substantially sialylated, in contrast to the oligosaccharides released from IgG-Fc (Anumula, 2012; Borrok et al., 2012; Coloma et al., 1999; Ercan et al., 2010; Girardi et al., 2009; Holland et al., 2006; Jacquemin, 2010; Mimura et al., 2007; Spencer et al., 2012; Stanfield and Wilson, 2014; Youings et al., 1996). This pattern was maintained for IgG-Fab prepared from hypogalactosylated IgG isolated from the sera of patients with Wegner's granulomatosis or microscopic polyangiitis (Youings et al., 1996). Thus, the *in vivo* environment of IgG producing plasma cells influences the efficacy of glyco-processing of IgG-Fc but not IgG-Fab during passage through the Golgi apparatus. The functional significance for IgG-Fab glycosylation of polyclonal IgG has not been fully determined, but data emerging for monoclonal antibodies suggest that V_κ, V_λ or V_H glycosylation can have a neutral, positive, or negative influence on antigen binding (Borrok et al., 2012; Coloma et al., 1999; Holland et al., 2006; Jacquemin, 2010). The differences observed for polyclonal IgG-Fc and IgG-Fab glycoforms has been maintained for recombinant antibodies produced in CHO cells and myeloma IgG proteins (Anumula, 2012; Coloma et al., 1999; Girardi et al., 2009; Jacquemin, 2010; Mimura et al., 2000; Yamaguchi et al., 2006). Since it is generally observed that the oligosaccharide present in glycoproteins contributes to solubility and stability, it is possible that IgG-Fab glycosylation may similarly be beneficial, particularly when formulating IgG therapeutics at concentrations of 100–150 mg/mL (Spencer et al., 2012). Such high concentration formulations allow the development of self-administration protocols and can reduce dosing intervals, resulting in reduced CoT. Controlling glycoform fidelity at two sites offers a further challenge to the biopharmaceutical industry.

The licensed antibody therapeutic Erbitux (cetuximab), bears an *N*-linked oligosaccharide at Asn 88 of the V_H region; interestingly, there is also a glycosylation consensus sequence at Asn 41 of the V_L but it is not occupied (Qian et al., 2007; Wiegandt and Meyer, 2014). Analysis of the IgG-Fc and IgG-Fab oligosaccharides of Erbitux, produced from Sp2/0 cells, reveal highly significant differences in composition. While the IgG-Fc oligosaccharides are typical, that is, comprised predominantly of diantennary G0F oligosaccharides, the IgG-Fab oligosaccharides are extremely heterogeneous and include complex diantennary, triantennary, and hybrid oligosaccharides; nonhuman oligosaccharides were also present (e.g., galactose in α(1–3) linkage to galactose and *N*-glycylneuraminic acid residues).

Severe adverse reaction to cetuximab therapy have been reported, and in a study of 76 patients treated with Erbitux, 25 experienced hypersensitivity reactions; this was shown to be due to the presence of IgE anti-gal α(1,3)gal antibodies. Interestingly, environmental factors appeared to influence the development of IgE anti-gal α(1,3)gal responses, and IgE antibodies were detected in pretreatment samples from 17 of the patients (Chung et al., 2008; Daguet and Watier, 2011; Lammerts van Bueren et al., 2011). The incidence varied significantly between treatment centers and may be linked to differences in predominant infectious agents present in local environments. Subsequently, it has been demonstrated that many individuals who consume meat (beef, lamb, pork, etc.) have IgG anti-Gal α(1–3)Gal antibodies and a minority IgE anti-Gal α(1–3)Gal antibodies. It is becoming routine, therefore, to monitor patients for the presence of IgE anti-Gal α(1–3)Gal antibodies prior to exposure to Erbitux (Berg et al., 2014; Daguet and Watier, 2011; Mullins et al., 2012; Pointreau et al., 2012).

A detailed analysis of the glycoforms of a humanized IgG rMAb bearing oligosaccharides at Asn 56 of the V_H and Asn 297, also produced in Sp2/0 cells, reveals the expected IgG-Fc glycoform profile of predominantly G0F oligosaccharides. However, eleven oligosaccharides were released from the IgG-Fab, including diantennary and triantennary oligosaccharides bearing gal α(1,3)gal, *N*-glycylneuraminic acid, and *N*-acetylgalactosamine residues (Huang et al., 2006). The consistent observation of higher levels of galactosylation and sialylation for IgG-Fab *N*-linked oligosaccharides, in comparison to IgG-Fc, is thought to reflect increased exposure and/or accessibility. In view of these experiences, it would seem that the perceived virtues of the NS0 and Sp2/0 cells might best be pursued by engineering to inactivate the gal α(1–3) and *N*-glycylneuraminic acid transferases.

The double challenge to produce rMAbs having appropriately glycosylated IgG-Fc and IgG-Fab has led some companies to engineer out V_H or V_L glycosylation motifs when present in candidate rMAbs (Carter et al., 1992). However, present reports suggest that CHO cells can glycosylate V_H and/or V_L motifs in a similar manner to that observed for normal polyclonal IgG (Lim et al., 2008). Since oligosaccharides are hydrophilic, the addition of glycans within V_H and/or V_L regions may impact the physicochemical properties of an antibody molecule and consequently pharmacokinetics (Lim et al., 2008; Millward et al., 2008), solubility (Wu et al., 2010), aggregation, and so on. While V_H glycosylation of a human IgG antibody was shown to have the same pharmacokinetics as the V_H deglycosylated molecule in a mouse model (Lim et al., 2008), introduction of a glycosylation site within bispecific single-chain diabodies resulted in a significant increase in serum half-lives (Stork et al., 2008). An antibody with specificity for IL-13 was generated that included a glycosylation sequon (53NSS55) within the heavy chain CDR2 (Wu et al., 2010). Initially, this site was engineered out by replacing N53 with an aspartic acid residue; however, the product exhibited very limited solubility (~13 mg/mL) and consequent aggregation. By contrast, reintroduction of N53, together with engineering within V_L, resulted in a V_H glycosylated antibody product with a solubility >110 mg/mL (Wu et al., 2010).

4.21 CAVEAT: ONE CANNOT EXTRAPOLATE FROM *IN VITRO/EX VIVO* TO *IN VIVO* BIOLOGICAL ACTIVITIES

The impact of the expression/production platform on product characteristics has been emphasized in the foregoing; however, it should equally be emphasized that when reporting functional activities, the parameters of the assay system employed determine the results obtained. A clear example is the determination/reporting of complement mediated lysis and binding. Early studies employed hamster or rabbit serum as source, each being more stable than human complement. However, there are significant structural and functional nuances between these species' proteins. Another parameter is the epitope density expressed by the target cell (Rojko et al., 2014; Voice and Lachmann, 1997; Zhang et al., 1995). An important parameter in contemporary effector function studies is the source and/or expression of Fcγ receptors. Binding studies conducted, *in vitro*, have employed the soluble external domain of FcγR (sFcγR) produced in a variety of cell types, mostly sourced from commercial companies. The external domain of sFcγRIIIa bears five *N*-linked glycosylation sites, all occupied, one of which makes direct contacts with the IgG-Fc and determines binding affinity; therefore, the glycoform of the receptor is of equal importance. The impact of glycoform on binding to the other sFγR has not been reported. Similarly, when conducting ADCC studies, the FcγR expression level on the effector cell and the epitope density on the target cell and the sensitivity of the readout impact the apparent outcome. In the context of this chapter, it is relevant to include the following section emphasizing FcγR heterogeneities.

4.22 FcγR RECEPTORS MEDIATING ADCC AND/OR THE REMOVAL AND DESTRUCTION OF IgG/ANTIGEN IMMUNE COMPLEXES

Three types or classes of membrane bound human FcγR have been defined by immunochemical, biochemical, and gene sequencing studies: FcγRI [CD64], FcγRII [CD32], FcγRIII [CD16], with an additional six subtypes: FcγRIIA, FcγRIIB1, FcγRIIB2, FcγRIIC, FcγRIIIA, FcγRIIIB (Anthony et al., 2012; Cartron et al., 2002;

Lux et al., 2013; Sanz et al., 2007; van de Winkel, 2010). These receptors are constitutively, but differentially, expressed on a wide range of leucocytes, and expression may be upregulated and/or induced by cytokines generated and released within an inflammatory response. The effector mechanisms activated are diverse and include "killing" and removal (e.g., phagocytosis, respiratory burst, and cytolysis), accessory functions such as the enhancement of antigen presentation by dendritic cells, and the downregulation of growth and differentiation of lymphocytes. It is evident, therefore, that FcγRs play an essential role in the induction, establishment, and resolution of protective immune responses.

Multiple parameters determine the structure and biological activities of immune complexes formed within a polyclonal antibody response: (1) valency, (2) average affinity/avidity of the antibody population, (3) isotype profile, (4) IgG-Fc glycoforms, (5) valency or epitope density of the antigen, (6) density of cell surface effector ligands (e.g., FcγR), (7) cumulative valency when multiple ligands are engaged (e.g., FcγR and complement receptors), and (8) proportions of each antibody isotype within a polyclonal response (Anthony et al., 2012; Borrok et al., 2012; Cartron et al., 2002; Crispin, 2013; Gillis et al., 2014; Hristodorov et al., 2013; Jefferis, 2009c, 2011, 2012; Louis et al., 2004; Lux et al., 2013; Miescher et al., 2004; Mimura et al., 2000, 2001; Nesspor et al., 2012; Pascoe et al., 2007; Sanz et al., 2007; Sjogren et al., 2013; van de Winkel, 2010).

The FcγRI receptor is constitutively expressed on mononuclear phagocytes and dendritic cells; however, expression can be upregulated and/or induced by the action of cytokines. FcγRIIa is the most widely expressed FcγR and is found on most hemopoietic cells. Polymorphic variants of FcγIIa are identified by the presence of histidine (FcγRIIa-131H) or arginine (FcγRIIa-131R) at amino acid residues 131. The higher affinity of the FcγRIIa-131H form for IgG2 results in differing cellular responses to engagement by IgG2 immune complexes (Anthony et al., 2012; Borrok et al., 2012; Cartron et al., 2002; Gillis et al., 2014; Jefferis, 2011, 2012; Louis et al., 2004; Lux et al., 2013; Miescher et al., 2004; Pascoe et al., 2007; Sanz et al., 2007; Sjogren et al., 2013; van de Winkel, 2010); higher phagocytic capacity for *Streptococcus pneumoniae* opsonized with IgG2 antibody was observed for neutrophils of donors homozygous for FcγRIIa-131H than for FcγIIa-131R. The FcγRIIb receptor is expressed on B-lymphocytes and monocytes, and ligation of this receptor results in growth and differentiation inhibition (Anthony et al., 2012; Borrok et al., 2012; Cartron et al., 2002; Gillis et al., 2014; Jefferis, 2011, 2012; Louis et al., 2004; Lux et al., 2013; Miescher et al., 2004; Pascoe et al., 2007; Sanz et al., 2007; Sjogren et al., 2013; van de Winkel, 2010).

Initially, the FcγRIIIa receptor was reported to bind and be activated by IgG1 and IgG3 only; however, recognition of a polymorphism in the receptor and the differential influence of IgG glycoforms have radically changed our understanding, with important clinical consequences. The avidity of binding of IgG differs between the FcγRIIIa-158V and FcγRIIIa-158F polymorphic variants (Cartron et al., 2002). It was demonstrated, *in vitro*, that IgG1 antibody is more efficient at mediating ADCC through homozygous FcγRIIIa-158V bearing cells than homozygous FcγRIIIa-158F or heterozygous FcγRIIIa-158V/FcγRIIIa-158F cells (Cartron et al., 2002;

Pascoe et al., 2007). Similar differences in ADCC efficacy might pertain *in vivo* since more favorable responses were reported for patients, diagnosed with systemic lupus erythematosus or leukemia and homozygous for FcγRIIIa-158V when exposed to Rituxan than for homozygous FcγRIIIa-158F patients (Cartron et al., 2002; Pascoe et al., 2007). Similarly, FcγRIIIa polymorphisms were shown to influence the response of Crohn's disease patients to infliximab (99) and red blood cell clearance by anti-D antibody (Miescher et al., 2004).

All FcγR are transmembrane molecules, with the exception of FcγRIIIb which is glycosylphosphatidylinositol (GPI)-anchored within the membrane of neutrophils. FcγRI and FcγRIIIa are members of the multichain immune recognition receptor (MIRR) family and are present in the membrane as heterooligomeric complexes comprised of an α and a γ chain: an IgG/antigen complex binds the α chain to initiate signaling through the γ chain; the FcγRIIIa α chain of NK cells is also associated with a signaling ζ chain. FcγRIIa and FcγRIIb molecules are composed of an α chain only (Gillis et al., 2014; Lux et al., 2013; van de Winkel, 2010). The FcγR α chains show a high degree of sequence homology in their extracellular domains (70%–98%) but differ significantly in their cytoplasmic domains. The cytoplasmic domains of γ chains and the FcγRIIa α chain express the immunoreceptor tyrosine-based activation motif (ITAM) that is involved in the early stages of intracellular signal generation. By contrast, the FcγRIIb receptor α chain expresses an immunoreceptor tyrosine-based inhibition motif (ITIM) (Deisenhofer, 1981; Gillis et al., 2014; Jefferis, 2012; Lux et al., 2013; Stanfield and Wilson, 2014). Cellular activation may be dependent on the balance between the relative levels of expression of these two isoforms and hence the balance of signals generated through the ITAM and ITIM motifs (Gillis et al., 2014; Lux et al., 2013).

4.23 CONCLUDING COMMENTS

The challenge to develop biosimilar therapeutics is being met. There seems to be no issues relating to antigen-binding specificity and affinity; given that the sequence is determined by that of the innovator molecule. However, the unique structure of the paratope is principally responsible for immunogenicity and the development of ADA regardless of whether the antibody is a chimeric, humanized, or fully human molecule. A further immunogenicity issue is evident for Erbitux, produced in Sp2/0 cells, since the glycosylation sequon present at asparagine 88 of the heavy chain results in the addition of a complex array of oligosaccharides, including the allergenic gal α(1–3)gal residue. Consequently, biosimilars of this antibody are being developed with production in CHO cells that add familiar diantennary oligosaccharide structures. The focus of interest rests with the structure/conformation of the IgG-Fc region that is a composite of the protein and oligosaccharide moieties.

There remains unresolved understanding of the finesse of structural changes induced by the presence or absence of a fucose and/or bisecting *N*-acetylglucosamine sugar residue, the functional activity of the human IgG-Fc and consequently, presumably, on the MoA. While the glycoform is identified as a CQA, it is in fact the conformation of the IgG-Fc that determines interactions with effector ligands. It is essential, therefore; that multiple orthogonal techniques should be applied to define structural parameters. Both industry and academia would be best served by having access to a reference standard that has been characterized by all relevant techniques (Jefferis, 2014; Schiel et al., 2014b) and that can serve to standardize instrumentation performance. It is interesting to note that disparate ligands may bind to the Fc through common amino acid residues, within the hinge proximal region, for FcγR and C1q, and at the C_H2/C_H3 interface for FcRn, SpA, SpG, RFs, and IgG-Fc-like receptors encoded within the genomes of some viruses. The presence of sialic acid might further influence Fc/ligand interactions at this interface.

A rationalization for the topography of ligand binding sites may be the functional necessity for circulating IgG to be monovalent for FcγRs and C1q, to prevent continuous cellular activation while providing opportunity for divalency at the C_H2/C_H3 interface. The influence of the IgG-Fc glycoform on functional activity is being exploited to generate glycoforms having a preselected MoA considered to be optimal for a given disease indication. It is important to note that it is natural glycoforms that are being generated; therefore, they are not immunogenic. By contrast, innovative studies have explored engineering of the protein moiety to selectively enhance biologic activities; however, these are mutant forms of IgG, that is, nonself, that may enhance immunogenicity. This may not be an issue when treating patients for cancer, since they may be immune-suppressed; however, it is a concern for long-term treatment of chronic diseases. I would caution that our current understanding has mostly been accrued employing reductionist approaches, that is, measuring antibody/ligand interaction employing monomeric antibody and ligand species *in vitro* while interactions *in vivo* are necessarily initiated by multivalent immune complexes.

REFERENCES

Adams EW, Ratner DM, Seeberger PH, Hacohen N. (2008) Carbohydrate-mediated targeting of antigen to dendritic cells leads to enhanced presentation of antigen to T cells. *Chembiochem* **9(2)**, 294–303.

Ahmed AA, Giddens J, Pincetic A, et al. (2014) Structural characterization of anti-inflammatory immunoglobulin G Fc proteins. *Journal of Molecular Biology* **426(18)**, 3166–3179.

Alavi A, Arden N, Spector TD, Axford JS. (2000) Immunoglobulin G glycosylation and clinical outcome in rheumatoid arthritis during pregnancy. *Journal of Rheumatology* **27(6)**, 1379–1385.

Allhorn M, Briceno JG, Baudino L, et al. (2010) The IgG-specific endoglycosidase EndoS inhibits both cellular and complement-mediated autoimmune hemolysis. *Blood* **115(24)**, 5080–5088.

Anthony RM, Wermeling F, Ravetch JV. (2012) Novel roles for the IgG Fc glycan. *Annals of New York Academy of Sciences* **1253**, 170–180.

Anumula KR. (2012) Quantitative glycan profiling of normal human plasma derived immunoglobulin and its fragments Fab and Fc. *Journal of Immunology Methods* **382(1–2)**, 167–176.

Aricescu AR, Owens RJ. (2013) Expression of recombinant glycoproteins in mammalian cells: towards an integrative approach to structural biology. *Current Opinion in Structural Biology* **23(3)**, 345–356.

Arnold JN, Dwek RA, Rudd PM, Sim RB. (2006) Mannan binding lectin and its interaction with immunoglobulins in health and in disease. *Immunology Letters* **106(2)**, 103–110.

Ballow M. (2014) Mechanisms of immune regulation by IVIG. *Current Opinion in Allergy and Clinical Immunology* **14(6)**, 509–515.

Banda NK, Wood AK, Takahashi K, et al. (2008) Initiation of the alternative pathway of murine complement by immune complexes is dependent on N-glycans in IgG antibodies. *Arthritis & Rheumatology* **58(10)**, 3081–3089.

Barb AW, Meng L, Gao Z, et al. (2012) NMR characterization of immunoglobulin G Fc glycan motion on enzymatic sialylation. *Biochemistry* **51(22)**, 4618–4626.

Barb AW, Prestegard JH. (2011) NMR analysis demonstrates immunoglobulin G N-glycans are accessible and dynamic. *Nature Chemistry and Biology* **7(3)**, 147–153.

Beck A, Reichert JM. (2013) Approval of the first biosimilar antibodies in Europe: a major landmark for the biopharmaceutical industry. *MAbs* **5(5)**, 621–623.

Berg EA, Platts-Mills TA, Commins SP. (2014) Drug allergens and food—the cetuximab and galactose-α-1,3-galactose story. *Annals of Allergy, Asthma and Immunology* **112(2)**, 97–101.

Bertolotti-Ciarlet A, Wang W, Lownes R, et al. (2009) Impact of methionine oxidation on the binding of human IgG1 to Fc Rn and Fcγ receptors. *Molecular Immunology* **46(8–9)**, 1878–1882.

Bondt A, Selman MH, Deelder AM, et al. (2013) Association between galactosylation of immunoglobulin G and improvement of rheumatoid arthritis during pregnancy is independent of sialylation. *Journal of Proteome Research* **12(10)**, 4522–4531.

Borrok MJ, Jung ST, Kang TH, et al. (2012) Revisiting the role of glycosylation in the structure of human IgG Fc. *ACS Chemistry and Biology* **7(9)**, 1596–1602.

Bosques CJ, Collins BE, Meador JW, 3rd, et al. (2010) Chinese hamster ovary cells can produce galactose-α-1,3-galactose antigens on proteins. *Nature Biotechnology* **28(11)**, 1153–1156.

Boyd PN, Lines AC, Patel AK. (1995) The effect of the removal of sialic acid, galactose and total carbohydrate on the functional activity of Campath-1H. *Molecular Immunology* **32(17–18)**, 1311–1318.

Burnouf T. (2011) Recombinant plasma proteins. *Vox Sang* **100(1)**, 68–83.

Burska AN, Hunt L, Boissinot M, et al. (2014) Autoantibodies to posttranslational modifications in rheumatoid arthritis. *Mediators of Inflammation* **2014**, 492873.

Campbell IK, Miescher S, Branch DR, et al. (2014) Therapeutic effect of IVIG on inflammatory arthritis in mice is dependent on the Fc portion and independent of sialylation or basophils. *Journal of Immunology* **192(11)**, 5031–5038.

Carter P, Presta L, Gorman CM, et al. (1992) Humanization of an anti-p185HER2 antibody for human cancer therapy. *Proceedings of National Academy of Science USA* **89(10)**, 4285–4289.

Cartron G, Dacheux L, Salles G, et al. (2002) Therapeutic activity of humanized anti-CD20 monoclonal antibody and polymorphism in IgG Fc receptor FcγRIIIa gene. *Blood* **99(3)**, 754–758.

CFG. (2012) Symbol and text nomenclature for representation of glycan structure. June 15, 2015. Available from: http://glycomics.scripps.edu/CFGnomenclature.pdf.

Chelius D, Rehder DS, Bondarenko PV. (2005) Identification and characterization of deamidation sites in the conserved regions of human immunoglobulin gamma antibodies. *Analytical Chemistry* **77(18)**, 6004–6011.

Chicooree N, Unwin RD, Griffiths JR. (2015) The application of targeted mass spectrometry-based strategies to the detection and localization of post-translational modifications. *Mass Spectrometry Reviews* **34(6)**, 595–626.

Choe H, Li W, Wright PL, et al. (2003) Tyrosine sulfation of human antibodies contributes to recognition of the CCR5 binding region of HIV-1 gp120. *Cell* **114(2)**, 161–170.

Chumsae C, Gaza-Bulseco G, Sun J, Liu H. (2007) Comparison of methionine oxidation in thermal stability and chemically stressed samples of a fully human monoclonal antibody. *Journal of Chromatography B Analytical Technology Biomedical Life Sciences* **850(1–2)**, 285–294.

Chung CH, Mirakhur B, Chan E, et al. (2008) Cetuximab-induced anaphylaxis and IgE specific for galactose-α-1,3-galactose. *New England Journal of Medicine* **358(11)**, 1109–1117.

Coloma MJ, Trinh RK, Martinez AR, Morrison SL. (1999) Position effects of variable region carbohydrate on the affinity and in vivo behavior of an anti-(1→6) dextran antibody. *Journal of Immunology* **162(4)**, 2162–2170.

Corper AL, Sohi MK, Bonagura VR, et al. (1997) Structure of human IgM rheumatoid factor Fab bound to its autoantigen IgG Fc reveals a novel topology of antibody-antigen interaction. *Nature Structural Biology* **4(5)**, 374–381.

Correia IR. (2010) Stability of IgG isotypes in serum. *MAbs* **2(3)**, 221–232.

Costagliola S, Panneels V, Bonomi M, et al. (2002) Tyrosine sulfation is required for agonist recognition by glycoprotein hormone receptors. *EMBO Journal* **21(4)**, 504–513.

Crispin M. (2013) Therapeutic potential of deglycosylated antibodies. *Proceedings of National Academy of Science USA* **110(25)**, 10059–10060.

Daguet A, Watier H. (2011) 2nd Charles Richet et Jules Hericourt workshop: therapeutic antibodies and anaphylaxis; May 31–June 1, 2011, Tours, France. *MAbs* **3(5)**, 417–421.

Davies AM, Rispens T, den Bleker TH, et al. (2013) Crystal structure of the human IgG4 C(H)3 dimer reveals the role of Arg409 in the mechanism of Fab-arm exchange. *Molecular Immunology* **54(1)**, 1–7.

Davies J, Jiang L, Pan LZ, et al. (2001) Expression of GnTIII in a recombinant anti-CD20 CHO production cell line: expression of antibodies with altered glycoforms leads to an increase in ADCC through higher affinity for FCγRIII. *Biotechnology and Bioengineering* **74(4)**, 288–294.

Deisenhofer J. (1981) Crystallographic refinement and atomic models of a human Fc fragment and its complex with fragment B of protein A from *Staphylococcus aureus* at 2.9- and 2.8-A resolution. *Biochemistry* **20(9)**, 2361–2370.

Delord JP, Tabernero J, Garcia-Carbonero R, et al. (2014) Open-label, multicentre expansion cohort to evaluate imgatuzumab in pre-treated patients with KRAS-mutant advanced colorectal carcinoma. *European Journal of Cancer* **50(3)**, 496–505.

Derer S, Kellner C, Rosner T, et al. (2014) Fc engineering of antibodies and antibody derivatives by primary sequence alteration and their functional characterization. *Methods in Molecular Biology* **1131**, 525–540.

Diepold K, Bomans K, Wiedmann M, et al. (2012) Simultaneous assessment of Asp isomerization and Asn deamidation in recombinant antibodies by LC-MS following incubation at elevated temperatures. *PLoS One* **7(1)**, e30295.

Dillon TM, Ricci MS, Vezina C, et al. (2008) Structural and functional characterization of disulfide isoforms of the human IgG2 subclass. *Journal of Biological Chemistry* **283(23)**, 16206–16215.

Dong X, Storkus WJ, Salter RD. (1999) Binding and uptake of agalactosyl IgG by mannose receptor on macrophages and dendritic cells. *Journal of Immunology* **163(10)**, 5427–5434.

Doores KJ, Bonomelli C, Harvey DJ, et al. (2010) Envelope glycans of immunodeficiency virions are almost entirely oligomannose antigens. *Proceedings of National Academy of Science USA* **107(31)**, 13800–13805.

EBE. (2013) Concept paper—considerations in setting specifications. April 12, 2016. Available from: http://www.ebe-biopharma.eu/uploads/Modules/Documents/ebe-concept-paper-%E2%80%93-considerations-in-setting-specifications.pdf.

Edelman GM, Cunningham BA, Gall WE, et al. (2004) The covalent structure of an entire γG immunoglobulin molecule. *Journal of Immunology* **173(9)**, 5335–5342.

Einarsdottir HK, Selman MH, Kapur R, et al. (2013) Comparison of the Fc glycosylation of fetal and maternal immunoglobulin G. *Glycoconjugation Journal* **30(2)**, 147–157.

Elliott S, Egrie J, Browne J, et al. (2004) Control of rHuEPO biological activity: the role of carbohydrate. *Experimental Hematology* **32(12)**, 1146–1155.

EMA. (2008) Guideline on development, production, characterisation and specifications for monoclonal antibodies and related products. April 1, 2014. Available from: http://www.ema.europa.eu/docs/en_GB/document_library/Scientific_guideline/2009/09/WC500003074.pdf.

Ercan A, Cui J, Chatterton DE, et al. (2010) Aberrant IgG galactosylation precedes disease onset, correlates with disease activity, and is prevalent in autoantibodies in rheumatoid arthritis. *Arthritis and Rheumatism* **62(8)**, 2239–2248.

Estes S, Melville M. (2014) Mammalian cell line developments in speed and efficiency. *Advances in Biochemical Engineering and Biotechnology* **139**, 11–33.

Farooq M, Takahashi N, Arrol H, et al. (1997) Glycosylation of polyclonal and paraprotein IgG in multiple myeloma. *Glycoconjugation Journal* **14(4)**, 489–492.

Farriol-Mathis N, Garavelli JS, Boeckmann B, et al. (2004) Annotation of post-translational modifications in the Swiss-Prot knowledge base. *Proteomics* **4(6)**, 1537–1550.

FDA. (2014) Guidance for industry: clinical pharmacology data to support a demonstration of biosimilarity to a reference product. May 20, 2014. Available from: http://www.fda.gov/downloads/Drugs/GuidanceComplianceRegulatoryInformation/Guidances/UCM397017.pdf.

FDA. (2015a) News release: FDA approves first biosimilar product Zarxio. April 12, 2016. Available from: http://www.fda.gov/NewsEvents/Newsroom/PressAnnouncements/ucm436648.htm.

FDA. (2015b) Gazyva (obinutuzumab). April 12, 2016. Available from: http://www.fda.gov/Drugs/InformationOnDrugs/ApprovedDrugs/ucm373263.htm.

Ferrara C, Brunker P, Suter T, et al. (2006) Modulation of therapeutic antibody effector functions by glycosylation engineering: influence of Golgi enzyme localization domain and co-expression of heterologous β1, 4-*N*-acetylglucosaminyltransferase III and Golgi α-mannosidase II. *Biotechnology and Bioengineering* **93(5)**, 851–861.

Ferrara C, Grau S, Jager C, et al. (2011) Unique carbohydrate-carbohydrate interactions are required for high affinity binding between FcγRIII and antibodies lacking core fucose. *Proceedings of National Academy of Science USA* **108(31)**, 12669–12674.

Flynn GC, Chen X, Liu YD, et al. (2010) Naturally occurring glycan forms of human immunoglobulins G1 and G2. *Molecular Immunology* **47(11–12)**, 2074–2082.

Fortunato ME, Colina CM. (2014) Effects of galactosylation in immunoglobulin G from all-atom molecular dynamics simulations. *Journal of Physical Chemistry B* **118(33)**, 9844–9851.

Frank M, Walker RC, Lanzilotta WN, et al. (2014) Immunoglobulin G1 Fc domain motions: implications for Fc engineering. *Journal of Molecular Biology* **426(8)**, 1799–1811.

Franze R, Hirashima C, Link T, et al. (2012) Method for the production of a glycosylated immunoglobulin. EP2493922 A1.

Galili U. (2013) Discovery of the natural anti-Gal antibody and its past and future relevance to medicine. *Xenotransplantation* **20(3)**, 138–147.

Gaza-Bulseco G, Faldu S, Hurkmans K, et al. (2008) Effect of methionine oxidation of a recombinant monoclonal antibody on the binding affinity to protein A and protein G. *Journal of Chromatography B Analytical Technology Biomedical Life Science* **870(1)**, 55–62.

Ghaderi D, Zhang M, Hurtado-Ziola N, Varki A. (2012) Production platforms for biotherapeutic glycoproteins. Occurrence, impact, and challenges of non-human sialylation. *Biotechnology and Genetic Engineering Reviews* **28**, 147–175.

Ghazavi MK, Johnston GA. (2011) Insulin allergy. *Clinical Dermatology* **29(3)**, 300–305.
Gillis C, Gouel-Cheron A, Jonsson F, Bruhns P. (2014) Contribution of human FcγRs to disease with evidence from human polymorphisms and transgenic animal studies. *Frontiers of Immunology* **5**, 254.
Girardi E, Holdom MD, Davies AM, et al. (2009) The crystal structure of rabbit IgG-Fc. *Biochemical Journal* **417(1)**, 77–83.
Gloster TM, Vocadlo DJ. (2012) Developing inhibitors of glycan processing enzymes as tools for enabling glycobiology. *Nature Chemistry and Biology* **8(8)**, 683–694.
Goetze AM, Liu YD, Arroll T, et al. (2012) Rates and impact of human antibody glycation in vivo. *Glycobiology* **22(2)**, 221–234.
Goetze AM, Liu YD, Zhang Z, et al. (2011b) High-mannose glycans on the Fc region of therapeutic IgG antibodies increase serum clearance in humans. *Glycobiology* **21(7)**, 949–959.
Goetze AM, Zhang Z, Liu L, et al. (2011a) Rapid LC-MS screening for IgG Fc modifications and allelic variants in blood. *Molecular Immunology* **49(1–2)**, 338–352.
Grinnell BW, Walls JD, Gerlitz B. (1991) Glycosylation of human protein C affects its secretion, processing, functional activities, and activation by thrombin. *Journal of Biological Chemistry* **266(15)**, 9778–9785.
Grinnell BW, Yan SB, Macias WL. (2006) Activated protein C. In: *Directory of Therapeutic Enzymes*. McGrath BM and Walsh G, eds., 69–95. CRC Press, Boca Raton, FL.
Guhr T, Bloem J, Derksen NI, et al. (2011) Enrichment of sialylated IgG by lectin fractionation does not enhance the efficacy of immunoglobulin G in a murine model of immune thrombocytopenia. *PLoS One* **6(6)**, e21246.
Haberger M, Bomans K, Diepold K, et al. (2014) Assessment of chemical modifications of sites in the CDRs of recombinant antibodies: susceptibility vs. functionality of critical quality attributes. *MAbs* **6(2)**, 327–339.
Hansson K, Stenflo J. (2005) Post-translational modifications in proteins involved in blood coagulation. *Journal of Thrombosis and Haemostasis* **3(12)**, 2633–2648.
Harris RJ, Kabakoff B, Macchi FD, et al. (2001) Identification of multiple sources of charge heterogeneity in a recombinant antibody. *Journal of Chromatography B Biomedical Science Applications* **752(2)**, 233–245.
Haryadi R, Zhang PQ, Chan KF, Song ZW. (2013) CHO-gmt5, a novel CHO glycosylation mutant for producing afucosylated and asialylated recombinant antibodies. *Bioengineered* **4(2)**, 90–94.
Hayes JM, Cosgrave EF, Struwe WB, et al. (2014) Glycosylation and Fc receptors. *Current Topics in Microbiology and Immunology* **382**, 165–199.
Holland M, Yagi H, Takahashi N, et al. (2006) Differential glycosylation of polyclonal IgG, IgG-Fc and IgG-Fab isolated from the sera of patients with ANCA-associated systemic vasculitis. *Biochimica Biophysica Acta* **1760(4)**, 669–677.
Hospira. (2013) Press release: Hospira's Inflectra (infliximab) the first biosimilar monoclonal antibody to be approved in Europe. April 12, 2016. Available from: http://phx.corporate-ir.net/phoenix.zhtml?c=175550&p=irol-newsArticle&ID=1853480&highlight.
Hossler P. (2012) Protein glycosylation control in mammalian cell culture: past precedents and contemporary prospects. *Advances in Biochemical Engineering and Biotechnology* **127**, 187–219.
Houde D, Engen JR. (2013) Conformational analysis of recombinant monoclonal antibodies with hydrogen/deuterium exchange mass spectrometry. *Methods in Molecular Biology* **988**, 269–289.
Hristodorov D, Fischer R, Linden L. (2013) With or without sugar? (A)glycosylation of therapeutic antibodies. *Molecular Biotechnology* **54(3)**, 1056–1068.
Huang L, Biolsi S, Bales KR, Kuchibhotla U. (2006) Impact of variable domain glycosylation on antibody clearance: an LC/MS characterization. *Analytical Biochemistry* **349(2)**, 197–207.

Huang L, Lu J, Wroblewski VJ, et al. (2005) In vivo deamidation characterization of monoclonal antibody by LC/MS/MS. *Analytical Chemistry* **77(5)**, 1432–1439.

Hutterer KM, Hong RW, Lull J, et al. (2013) Monoclonal antibody disulfide reduction during manufacturing: untangling process effects from product effects. *MAbs* **5(4)**, 608–613.

Jacquemin M. (2010) Variable region heavy chain glycosylation determines the anticoagulant activity of a factor VIII antibody. *Haemophilia* **16(102)**, 16–19.

Jefferis R. (2009a) Glycoforms of human IgG in health and disease. *Trends in Glycoscience and Glycotechnology* **21(2009)**, 105–117.

Jefferis R. (2009b) Aglycosylated antibodies and the methods of making and using them: WO2008030564. *Expert Opinion Therapy and Pathology* **19(1)**, 101–105.

Jefferis R. (2009c) Recombinant antibody therapeutics: the impact of glycosylation on mechanisms of action. *Trends in Pharmacological Science* **30(7)**, 356–362.

Jefferis R. (2011) Aggregation, immune complexes and immunogenicity. *MAbs* **3(6)**, 503–504.

Jefferis R. (2012) Isotype and glycoform selection for antibody therapeutics. *Archives of Biochemistry and Biophysics* **526(2)**, 159–166.

Jefferis R. (2014) Monoclonal antibodies: mechanisms of action. In: *State-of-the-Art and Emerging Technologies for Therapeutic Monoclonal Antibody Characterization Volume 1. Monoclonal Antibody Therapeutics: Structure, Function, and Regulatory Space*. Schiel JE, Davis DL, Borisov OV, eds., 35–68. American Chemical Society, Washington, DC.

Jefferis R, Lefranc MP. (2009) Human immunoglobulin allotypes: possible implications for immunogenicity. *MAbs* **1(4)**, 332–338.

Jefferis R, Lund J, Mizutani H, et al. (1990) A comparative study of the *N*-linked oligosaccharide structures of human IgG subclass proteins. *Biochemical Journal* **268(3)**, 529–537.

Jefferis R, Lund J, Pound JD. (1998) IgG-Fc-mediated effector functions: molecular definition of interaction sites for effector ligands and the role of glycosylation. *Immunological Reviews* **163**, 59–76.

Ju MS, Jung ST. (2014) Aglycosylated full-length IgG antibodies: steps toward next-generation immunotherapeutics. *Current Opinion in Biotechnology* **30**, 128–139.

Kao YH, Laird MW, Schmidt MT, et al. (2013) Prevention of disulfide bond reduction during recombinant production of polypeptides. US8574869 B2.

Kaufman RJ, Powell JS. (2013) Molecular approaches for improved clotting factors for hemophilia. *Blood* **122(22)**, 3568–3574.

Khawli LA, Goswami S, Hutchinson R, et al. (2010) Charge variants in IgG1: isolation, characterization, *in vitro* binding properties and pharmacokinetics in rats. *MAbs* **2(6)**, 613–624.

Khoury GA, Baliban RC, Floudas CA. (2011) Proteome-wide post-translational modification statistics: frequency analysis and curation of the swiss-prot database. *Science Reports* **1**, Article number 90.

Kibe T, Fujimoto S, Ishida C, et al. (1996) Glycosylation and placental transport of immunoglobulin G. *Journal of Clinical Biochemistry and Nutrition* **21(1)**, 57–63.

Kobata A. (2008) The N-linked sugar chains of human immunoglobulin G: their unique pattern, and their functional roles. *Biochimica Biophysica Acta* **1780(3)**, 472–478.

Koterba KL, Borgschulte T, Laird MW. (2012) Thioredoxin 1 is responsible for antibody disulfide reduction in CHO cell culture. *Journal of Biotechnology* **157(1)**, 261–267.

Kozlowski S, Swann P. (2006) Current and future issues in the manufacturing and development of monoclonal antibodies. *Advanced Drug Delivery Reviews* **58(5–6)**, 707–722.

Krapp S, Mimura Y, Jefferis R, et al. (2003) Structural analysis of human IgG-Fc glycoforms reveals a correlation between glycosylation and structural integrity. *Journal of Molecular Biology* **325(5)**, 979–989.

Kumar A, Bachhawat AK. (2012) Pyroglutamic acid: throwing light on a lightly studied metabolite. *Current Science* **102**, 288–297.

Lai T, Yang Y, Ng SK. (2013) Advances in Mammalian cell line development technologies for recombinant protein production. *Pharmaceuticals (Basel)* **6(5)**, 579–603.

Lammerts van Bueren JJ, Rispens T, Verploegen S, et al. (2011) Anti-galactose-α-1,3-galactose IgE from allergic patients does not bind α-galactosylated glycans on intact therapeutic antibody Fc domains. *Nature Biotechnology* **29(7)**, 574–576.

Lanucara F, Eyers CE. (2013) Top-down mass spectrometry for the analysis of combinatorial post-translational modifications. *Mass Spectrometry Reviews* **32(1)**, 27–42.

Leblanc Y, Romanin M, Bihoreau N, Chevreux G. (2014) LC-MS analysis of polyclonal IgGs using IdeS enzymatic proteolysis for oxidation monitoring. *Journal of Chromatography B Analytical Technology Biomedical Life Sciences* **961**, 1–4.

Leontyev D, Katsman Y, Ma XZ, et al. (2012) Sialylation-independent mechanism involved in the amelioration of murine immune thrombocytopenia using intravenous gamma-globulin. *Transfusion* **52(8)**, 1799–1805.

Li J, Menzel C, Meier D, Zhang C, Dubel S, Jostock T. (2007) A comparative study of different vector designs for the mammalian expression of recombinant IgG antibodies. *Journal of Immunological Methods* **318(1–2)**, 113–124.

Li J, Wong CL, Vijayasankaran N, et al. (2012) Feeding lactate for CHO cell culture processes: impact on culture metabolism and performance. *Biotechnology and Bioengineering* **109(5)**, 1173–1186.

Lim A, Reed-Bogan A, Harmon BJ. (2008) Glycosylation profiling of a therapeutic recombinant monoclonal antibody with two N-linked glycosylation sites using liquid chromatography coupled to a hybrid quadrupole time-of-flight mass spectrometer. *Analytical Biochemistry* **375(2)**, 163–172.

Liu H, May K. (2012) Disulfide bond structures of IgG molecules: structural variations, chemical modifications and possible impacts to stability and biological function. *MAbs* **4(1)**, 17–23.

Liu J, Jonebring A, Hagstrom J, Nystrom AC, Lovgren A. (2014) Improved expression of recombinant human factor IX by co-expression of GGCX, VKOR and furin. *Protein Journal* **33(2)**, 174–183.

Liu YD, Goetze AM, Bass RB, Flynn GC. (2011) N-terminal glutamate to pyroglutamate conversion *in vivo* for human IgG2 antibodies. *Journal of Biological Chemistry* **286(13)**, 11211–11217.

Liu YD, van Enk JZ, Flynn GC. (2009) Human antibody Fc deamidation in vivo. *Biologicals* **37(5)**, 313–322.

Louis E, El Ghoul Z, Vermeire S, et al. (2004) Association between polymorphism in IgG Fc receptor IIIa coding gene and biological response to infliximab in Crohn's disease. *Alimentary Pharmacology Therapy* **19(5)**, 511–519.

Ludeman JP, Stone MJ. (2014) The structural role of receptor tyrosine sulfation in chemokine recognition. *British Journal of Pharmacology* **171(5)**, 1167–1179.

Lund J, Takahashi N, Pound JD, et al. (1996) Multiple interactions of IgG with its core oligosaccharide can modulate recognition by complement and human Fc gamma receptor I and influence the synthesis of its oligosaccharide chains. *Journal of Immunology* **157(11)**, 4963–4969.

Lux A, Yu X, Scanlan CN, Nimmerjahn F. (2013) Impact of immune complex size and glycosylation on IgG binding to human FcγRs. *Journal of Immunology* **190(8)**, 4315–4323.

Macdougall IC, Roger SD, de Francisco A, et al. (2012) Antibody-mediated pure red cell aplasia in chronic kidney disease patients receiving erythropoiesis-stimulating agents: new insights. *Kidney International* **81(8)**, 727–732.

Maeda E, Kita S, Kinoshita M, et al. (2012) Analysis of nonhuman *N*-glycans as the minor constituents in recombinant monoclonal antibody pharmaceuticals. *Analytical Chemistry* **84(5)**, 2373–2379.

Malhotra R, Wormald MR, Rudd PM, et al. (1995) Glycosylation changes of IgG associated with rheumatoid arthritis can activate complement via the mannose-binding protein. *Nature Medicine* **1(3)**, 237–243.

Manning G, Whyte DB, Martinez R, et al. (2002) The protein kinase complement of the human genome. *Science* **298(5600)**, 1912–1934.

Marth JD, Grewal PK. (2008) Mammalian glycosylation in immunity. *Nature Reviews in Immunology* **8(11)**, 874–887.

Masuda K, Kubota T, Kaneko E, et al. (2007) Enhanced binding affinity for FcγRIIIa of fucose-negative antibody is sufficient to induce maximal antibody-dependent cellular cytotoxicity. *Molecular Immunology* **44(12)**, 3122–3131.

Matsumiya S, Yamaguchi Y, Saito J, et al. (2007) Structural comparison of fucosylated and nonfucosylated Fc fragments of human immunoglobulin G1. *Journal of Molecular Biology* **368(3)**, 767–779.

McGrath B. (2006) Factor IX (Protease Zymogen). In: *Directory of Therapeutic Enzymes*. McGrath BM and Walsh G, eds., 209–238. CRC Press, Boca Raton, FL.

Medzihradszky KF, Darula Z, Perlson E, et al. (2004) *O*-sulfonation of serine and threonine: mass spectrometric detection and characterization of a new posttranslational modification in diverse proteins throughout the eukaryotes. *Molecular Cell Proteomics* **3(5)**, 429–440.

Merlini G, Comenzo RL, Seldin DC, et al. (2014) Immunoglobulin light chain amyloidosis. *Expert Reviews in Hematology* **7(1)**, 143–156.

Miescher S, Spycher MO, Amstutz H, et al. (2004) A single recombinant anti-RhD IgG prevents RhD immunization: association of RhD-positive red blood cell clearance rate with polymorphisms in the FcγRIIA and FcγIIIA genes. *Blood* **103(11)**, 4028–4035.

Millward TA, Heitzmann M, Bill K, et al. (2008) Effect of constant and variable domain glycosylation on pharmacokinetics of therapeutic antibodies in mice. *Biologicals* **36(1)**, 41–47.

Mimura Y, Ashton PR, Takahashi N, et al. (2007) Contrasting glycosylation profiles between Fab and Fc of a human IgG protein studied by electrospray ionization mass spectrometry. *Journal of Immunological Methods* **326(1–2)**, 116–126.

Mimura Y, Church S, Ghirlando R, et al. (2000) The influence of glycosylation on the thermal stability and effector function expression of human IgG1-Fc: properties of a series of truncated glycoforms. *Molecular Immunology* **37(12–13)**, 697–706.

Mimura Y, Sondermann P, Ghirlando R, et al. (2001) Role of oligosaccharide residues of IgG1-Fc in FcγRIIb binding. *Journal of Biological Chemistry* **276(49)**, 45539–45547.

Mizushima T, Yagi H, Takemoto E, et al. (2011) Structural basis for improved efficacy of therapeutic antibodies on defucosylation of their Fc glycans. *Genes to Cells* **16(11)**, 1071–1080.

Mossner E, Brunker P, Moser S, et al. (2010) Increasing the efficacy of CD20 antibody therapy through the engineering of a new type II anti-CD20 antibody with enhanced direct and immune effector cell-mediated B-cell cytotoxicity. *Blood* **115(22)**, 4393–4402.

Mullins RJ, James H, Platts-Mills TA, Commins S. (2012) Relationship between red meat allergy and sensitization to gelatin and galactose-α-1,3-galactose. *Journal of Allergy and Clinical Immunology* **129(5)**, 1334–1342.

Nakagawa T, Natsume A, Satoh M, Niwa R. (2010) Nonfucosylated anti-CD20 antibody potentially induces apoptosis in lymphoma cells through enhanced interaction with FcγRIIIb on neutrophils. *Leukemia Research* **34(5)**, 666–671.

Nakajima C, Kuyama H, Nakazawa T, et al. (2011) A method for N-terminal *de novo* sequencing of Nα-blocked proteins by mass spectrometry. *Analyst* **136(1)**, 113–119.

Natsume A, In M, Takamura H, et al. (2008) Engineered antibodies of IgG1/IgG3 mixed isotype with enhanced cytotoxic activities. *Cancer Research* **68(10)**, 3863–3872.

NCI. (2016) What is cancer proteomics? April 12, 2016. Available from: http://proteomics.cancer.gov/whatisproteomics.

Nesspor TC, Raju TS, Chin CN, et al. (2012) Avidity confers FcγR binding and immune effector function to aglycosylated immunoglobulin G1. *Journal of Molecular Recognition* **25(3)**, 147–154.

NIBRT. (2012) Glycobase. June 15, 2015. Available from: http://glycobase.nibrt.ie/glycobase/about.action.

Niwa R, Natsume A, Uehara A, et al. (2005) IgG subclass-independent improvement of antibody-dependent cellular cytotoxicity by fucose removal from Asn297-linked oligosaccharides. *Journal of Immunological Methods* **306(1–2)**, 151–160.

Oaks M, Taylor S, Shaffer J. (2013) Autoantibodies targeting tumor-associated antigens in metastatic cancer: sialylated IgGs as candidate anti-inflammatory antibodies. *Oncoimmunology* **2(6)**, e24841.

Ogorek C, Jordan I, Sandig V, von Horsten HH. (2012) Fucose-targeted glycoengineering of pharmaceutical cell lines. *Methods of Molecular Biology* **907**, 507–517.

Othy S, Topcu S, Saha C, et al. (2014) Sialylation may be dispensable for reciprocal modulation of helper T cells by intravenous immunoglobulin. *European Journal of Immunology* **44(7)**, 2059–2063.

Ozbabacan SEA, Engin HB, Gursoy A, Keskin O. (2011) Transient protein-protein interactions. *Protein Engineering Design & Selection* **24(9)**, 635–648.

Pacis E, Yu M, Autsen J, et al. (2011) Effects of cell culture conditions on antibody *N*-linked glycosylation—what affects high mannose 5 glycoform. *Biotechnology and Bioengineering* **108(10)**, 2348–2358.

Padlan EA. (1990) X-ray diffraction studies of antibody constant regions. In: *Fc Receptors and the Action of Antibodies*. Metzger H, ed., 12–30. American Society for Microbiology, Washington, DC.

Pan H, Chen K, Chu L, et al. (2009) Methionine oxidation in human IgG2 Fc decreases binding affinities to protein A and FcRn. *Protein Science* **18(2)**, 424–433.

Pascoe DE, Arnott D, Papoutsakis ET, et al. (2007) Proteome analysis of antibody-producing CHO cell lines with different metabolic profiles. *Biotechnology and Bioengineering* **98(2)**, 391–410.

Peipp M, Lammerts van Bueren JJ, Schneider-Merck T, et al. (2008) Antibody fucosylation differentially impacts cytotoxicity mediated by NK and PMN effector cells. *Blood* **112(6)**, 2390–2399.

Perez-Garmendia R, Gevorkian G. (2013) Pyroglutamate-modified amyloid beta peptides: emerging targets for Alzheimer's disease immunotherapy. *Current Neuropharmacology* **11(5)**, 491–498.

Pointreau Y, Commins SP, Calais G, et al. (2012) Fatal infusion reactions to cetuximab: role of immunoglobulin e-mediated anaphylaxis. *Journal of Clinical Oncology* **30(3)**, 334–335.

Qian J, Liu T, Yang L, et al. (2007) Structural characterization of N-linked oligosaccharides on monoclonal antibody cetuximab by the combination of orthogonal matrix-assisted laser desorption/ionization hybrid quadrupole-quadrupole time-of-flight tandem mass spectrometry and sequential enzymatic digestion. *Analytical Biochemistry* **364(1)**, 8–18.

Radaev S, Motyka S, Fridman WH, et al. (2001) The structure of a human type III Fcγ receptor in complex with Fc. *Journal of Biological Chemistry* **276(19)**, 16469–16477.

Raju TS, Jordan RE. (2012) Galactosylation variations in marketed therapeutic antibodies. *MAbs* **4(3)**, 385–391.

Raju TS, Lang SE. (2014) Diversity in structure and functions of antibody sialylation in the Fc. *Current Opinions in Biotechnology* **30**, 147–152.

Ramsland PA, Farrugia W, Bradford TM, et al. (2011) Structural basis for FcγRIIa recognition of human IgG and formation of inflammatory signaling complexes. *Journal of Immunology* **187(6)**, 3208–3217.

Reeves PJ, Callewaert N, Contreras R, Khorana HG. (2002) Structure and function in rhodopsin: high-level expression of rhodopsin with restricted and homogeneous *N*-glycosylation by a tetracycline-inducible *N*-acetylglucosaminyltransferase I-negative HEK293S stable mammalian cell line. *Proceedings of National Academy of Science USA* **99(21)**, 13419–13424.

Reusch D, Haberger M, Falck D, et al. (2015a) Comparison of methods for the analysis of therapeutic immunoglobulin G Fc-glycosylation profiles—Part 2: mass spectrometric methods. *MAbs* **7(4)**, 732–742.

Reusch D, Haberger M, Maier B, et al. (2015b) Comparison of methods for the analysis of therapeutic immunoglobulin G Fc-glycosylation profiles—Part 1: separation-based methods. *MAbs* **7(1)**, 167–179.

Ripka J, Adamany A, Stanley P. (1986) Two Chinese hamster ovary glycosylation mutants affected in the conversion of GDP-mannose to GDP-fucose. *Archives of Biochemistry and Biophysics* **249(2)**, 533–545.

Rispens T, Davies AM, Ooijevaar-de Heer P, et al. (2014) Dynamics of inter-heavy chain interactions in human immunoglobulin G (IgG) subclasses studied by kinetic Fab arm exchange. *Journal of Biological Chemistry* **289(9)**, 6098–6109.

Rispens T, Meesters J, den Bleker TH, et al. (2013) Fc-Fc interactions of human IgG4 require dissociation of heavy chains and are formed predominantly by the intra-chain hinge isomer. *Molecular Immunology* **53(1–2)**, 35–42.

Rojko JL, Evans MG, Price SA, et al. (2014) Formation, clearance, deposition, pathogenicity, and identification of biopharmaceutical-related immune complexes: review and case studies. *Toxicology and Pathology* **42(4)**, 725–764.

Rombouts Y, Ewing E, van de Stadt LA, et al. (2015) Anti-citrullinated protein antibodies acquire a pro-inflammatory Fc glycosylation phenotype prior to the onset of rheumatoid arthritis. *Annals of Rheumatic Diseases* **74(1)**, 234–241.

Royer PJ, Emara M, Yang C, et al. (2010) The mannose receptor mediates the uptake of diverse native allergens by dendritic cells and determines allergen-induced T cell polarization through modulation of IDO activity. *Journal of Immunology* **185(3)**, 1522–1531.

Ruhaak LR, Uh HW, Beekman M, et al. (2011) Plasma protein N-glycan profiles are associated with calendar age, familial longevity and health. *Journal of Proteome Research* **10(4)**, 1667–1674.

Sacco F, Perfetto L, Castagnoli L, Cesareni G. (2012) The human phosphatase interactome: an intricate family portrait. *FEBS Letters* **586(17)**, 2732–2739.

Sanz I, Anolik JH, Looney RJ. (2007) B cell depletion therapy in autoimmune diseases. *Frontiers of Bioscience* **12**, 2546–2567.

Sauer-Eriksson AE, Kleywegt GJ, Uhlen M, Jones TA. (1995) Crystal structure of the C2 fragment of streptococcal protein G in complex with the Fc domain of human IgG. *Structure* **3(3)**, 265–278.

Schiel JE, Davis DL, Borisov OV. (2014a) *State-of-the-Art and Emerging Technologies for Therapeutic Monoclonal Antibody Characterization Volume 1. Monoclonal Antibody Therapeutics: Structure, Function, and Regulatory Space.* American Chemical Society, Washington, DC.

Schiel JE, Mire-Sluis A, Davis D. (2014b) Monoclonal antibody therapeutics: the need for biopharmaceutical reference materials. In: *State-of-the-Art and Emerging Technologies for Therapeutic Monoclonal Antibody Characterization Volume 1. Monoclonal Antibody Therapeutics: Structure, Function, and Regulatory Space.* Schiel JE, Davis DL, Borisov OV, eds., 1–34. American Chemical Society, Washington, DC.

Schwab I, Biburger M, Kronke G, et al. (2012) IVIg-mediated amelioration of ITP in mice is dependent on sialic acid and SIGNR1. *European Journal of Immunology* **42(4)**, 826–830.

Schwab I, Nimmerjahn F. (2014) Role of sialylation in the anti-inflammatory activity of intravenous immunoglobulin—F(ab')(2) versus Fc sialylation. *Clinical and Experimental Immunology* **178(Suppl. 1)**, 197–199.

Sealover NR, Davis AM, Brooks JK, et al. (2013) Engineering Chinese hamster ovary (CHO) cells for producing recombinant proteins with simple glycoforms by zinc-finger nuclease (ZFN)-mediated gene knockout of mannosyl (alpha-1,3-)-glycoprotein beta-1,2-N-acetylglucosaminyltransferase (Mgat1). *Journal of Biotechnology* **167(1)**, 24–32.

Shi HH, Goudar CT. (2014) Recent advances in the understanding of biological implications and modulation methodologies of monoclonal antibody N-linked high mannose glycans. *Biotechnology and Bioengineering* **111(10)**, 1907–1919.

Shibata-Koyama M, Iida S, Misaka H, et al. (2009a) Nonfucosylated rituximab potentiates human neutrophil phagocytosis through its high binding for FcγRIIIb and MHC class II expression on the phagocytotic neutrophils. *Experimental Hematology* **37(3)**, 309–321.

Shibata-Koyama M, Iida S, Okazaki A, et al. (2009b) The *N*-linked oligosaccharide at FcγRIIIa Asn-45: an inhibitory element for high FcγRIIIa binding affinity to IgG glycoforms lacking core fucosylation. *Glycobiology* **19(2)**, 126–134.

Shields RL, Lai J, Keck R, et al. (2002) Lack of fucose on human IgG1 *N*-linked oligosaccharide improves binding to human FcγRIII and antibody-dependent cellular toxicity. *Journal of Biological Chemistry* **277(30)**, 26733–26740.

Sinclair AM. (2013) Erythropoiesis stimulating agents: approaches to modulate activity. *Biologics* **7**, 161–174.

Singh PK, Devasahayam M, Devi S. (2015) Expression of GPI anchored human recombinant erythropoietin in CHO cells is devoid of glycosylation heterogeneity. *Indian Journal of Experimental Biology* **53(4)**, 195–201.

Sinha S, Zhang L, Duan S, et al. (2009) Effect of protein structure on deamidation rate in the Fc fragment of an IgG1 monoclonal antibody. *Protein Science* **18(8)**, 1573–1584.

Sjogren J, Struwe WB, Cosgrave EF, et al. (2013) EndoS2 is a unique and conserved enzyme of serotype M49 group A *Streptococcus* that hydrolyses N-linked glycans on IgG and α1-acid glycoprotein. *Biochemical Journal* **455(1)**, 107–118.

Skulj M, Pezdirec D, Gaser D, et al. (2014) Reduction in C-terminal amidated species of recombinant monoclonal antibodies by genetic modification of CHO cells. *BMC Biotechnology* **14**, 76.

Sondermann P, Huber R, Oosthuizen V, Jacob U. (2000) The 3.2-Å crystal structure of the human IgG1 Fc fragment-FcγRIII complex. *Nature* **406(6793)**, 267–273.

Sondermann P, Pincetic A, Maamary J, et al. (2013) General mechanism for modulating immunoglobulin effector function. *Proceedings of the National Academy of Science USA* **110(24)**, 9868–9872.

Song T, Ozcan S, Becker A, Lebrilla CB. (2014) In-depth method for the characterization of glycosylation in manufactured recombinant monoclonal antibody drugs. *Analytical Chemistry* **86(12)**, 5661–5666.

Spencer S, Bethea D, Raju TS, et al. (2012) Solubility evaluation of murine hybridoma antibodies. *MAbs* **4(3)**, 319–325.

Stanfield RL, Wilson IA. (2014) Antibody structure. *Microbiology Spectra* **2(2)**, doi:10.1128/microbiolspec.AID-0012-2013.

Stanley P. (2011) Golgi glycosylation. *Cold Spring Harbour Perspectives in Biology* **3(4)**, a005199.

Stork R, Zettlitz KA, Muller D, et al. (2008) *N*-glycosylation as novel strategy to improve pharmacokinetic properties of bispecific single-chain diabodies. *Journal of Biological Chemistry* **283(12)**, 7804–7812.

Subedi GP, Hanson QM, Barb AW. (2014) Restricted motion of the conserved immunoglobulin G1 N-glycan is essential for efficient FcγRIIIa binding. *Structure* **22(10)**, 1478–1488.

Subramaniam JM, Whiteside G, McKeage K, Croxtall JC. (2012) Mogamulizumab: first global approval. *Drugs* **72(9)**, 1293–1298.

Tang L, Sundaram S, Zhang J, et al. (2013) Conformational characterization of the charge variants of a human IgG1 monoclonal antibody using H/D exchange mass spectrometry. *MAbs* **5(1)**, 114–125.

Tsubaki M, Terashima I, Kamata K, Koga A. (2013) C-terminal modification of monoclonal antibody drugs: amidated species as a general product-related substance. *International Journal of Biological Macromolecules* **52**, 139–147.

Umana P, Jean-Mairet J, Moudry R, et al. (1999) Engineered glycoforms of an antineuroblastoma IgG1 with optimized antibody-dependent cellular cytotoxic activity. *Nature Biotechnology* **17(2)**, 176–180.

Vafa O, Gilliland GL, Brezski RJ, et al. (2014) An engineered Fc variant of an IgG eliminates all immune effector functions via structural perturbations. *Methods* **65(1)**, 114–126.

van de Winkel JG. (2010) Fc receptors: role in biology and antibody therapy. *Immunology Letters* **128(1)**, 4–5.

Vidarsson G, Dekkers G, Rispens T. (2014) IgG subclasses and allotypes: from structure to effector functions. *Frontiers of Immunology* **5**, 520.

Vlasak J, Bussat MC, Wang S, et al. (2009) Identification and characterization of asparagine deamidation in the light chain CDR1 of a humanized IgG1 antibody. *Analytical Biochemistry* **392(2)**, 145–154.

Voice JK, Lachmann PJ. (1997) Neutrophil Fcγ and complement receptors involved in binding soluble IgG immune complexes and in specific granule release induced by soluble IgG immune complexes. *European Journal of Immunology* **27(10)**, 2514–2523.

Wacker C, Berger CN, Girard P, Meier R. (2011) Glycosylation profiles of therapeutic antibody pharmaceuticals. *European Journal of Pharmaceutics and Biopharmaceutics* **79(3)**, 503–507.

Wang W, Singh S, Zeng DL, et al. (2007) Antibody structure, instability, and formulation. *Journal of Pharmaceutical Sciences* **96(1)**, 1–26.

Wang W, Vlasak J, Li Y, et al. (2011) Impact of methionine oxidation in human IgG1 Fc on serum half-life of monoclonal antibodies. *Molecular Immunology* **48(6–7)**, 860–866.

Washburn N, Schwab I, Ortiz D, et al. (2015) Controlled tetra-Fc sialylation of IVIg results in a drug candidate with consistent enhanced anti-inflammatory activity. *Proceedings of National Academy of Science USA* **112(11)**, E1297–1306.

Welle S. (1999) *Human Protein Metabolism*. Springer-Verlag New York Inc., New York.

Wiegandt A, Meyer B. (2014) Unambiguous characterization of *N*-glycans of monoclonal antibody cetuximab by integration of LC-MS/MS and (1)H NMR spectroscopy. *Analytical Chemistry* **86(10)**, 4807–4814.

Wu SJ, Luo J, O'Neil KT, et al. (2010) Structure-based engineering of a monoclonal antibody for improved solubility. *Protein Engineering Design & Selection* **23(8)**, 643–651.

Wuhrer M, Stavenhagen K, Koeleman CA, et al. (2015) Skewed Fc glycosylation profiles of anti-proteinase 3 immunoglobulin G1 autoantibodies from granulomatosis with polyangiitis patients show low levels of bisection, galactosylation, and sialylation. *Journal of Proteome Research* **14(4)**, 1657–1665.

Wypych J, Li M, Guo A, et al. (2008) Human IgG2 antibodies display disulfide-mediated structural isoforms. *Journal of Biological Chemistry* **283(23)**, 16194–16205.

Xie H, Chakraborty A, Ahn J, et al. (2010) Rapid comparison of a candidate biosimilar to an innovator monoclonal antibody with advanced liquid chromatography and mass spectrometry technologies. *MAbs* **2(4)**, 379–394.

Yamada E, Tsukamoto Y, Sasaki R, et al. (1997) Structural changes of immunoglobulin G oligosaccharides with age in healthy human serum. *Glycoconjugation Journal* **14(3)**, 401–405.

Yamaguchi Y, Nishimura M, Nagano M, et al. (2006) Glycoform-dependent conformational alteration of the Fc region of human immunoglobulin G1 as revealed by NMR spectroscopy. *Biochimica Biophysica Acta* **1760(4)**, 693–700.

Yamane-Ohnuki N, Satoh M. (2009) Production of therapeutic antibodies with controlled fucosylation. *MAbs* **1(3)**, 230–236.

Yan SCB, Razzano P, Chao YB, et al. (1990) Characterization and novel purification of recombinant human protein C from three mammalian cell lines. *Bio-Technology* **8(7)**, 655–661.

Yang H, Zubarev RA. (2010) Mass spectrometric analysis of asparagine deamidation and aspartate isomerization in polypeptides. *Electrophoresis* **31(11)**, 1764–1772.

Yang YS, Wang CC, Chen BH, et al. (2015) Tyrosine sulfation as a protein post-translational modification. *Molecules* **20(2)**, 2138–2164.

Yin S, Pastuskovas CV, Khawli LA, Stults JT. (2013) Characterization of therapeutic monoclonal antibodies reveals differences between *in vitro* and *in vivo* time-course studies. *Pharmaceutical Research* **30(1)**, 167–178.

Youings A, Chang SC, Dwek RA, Scragg IG. (1996) Site-specific glycosylation of human immunoglobulin G is altered in four rheumatoid arthritis patients. *Biochemical Journal* **314(Pt. 2)**, 621–630.

Yu C, Crispin M, Sonnen AF, et al. (2011a) Use of the α-mannosidase I inhibitor kifunensine allows the crystallization of apo CTLA-4 homodimer produced in long-term cultures of Chinese hamster ovary cells. *Acta Crystallographica, Section F, Structural Biology and Crystallization Communications* **67(Pt. 7)**, 785–789.

Yu M, Brown D, Reed C, et al. (2012) Production, characterization, and pharmacokinetic properties of antibodies with N-linked mannose-5 glycans. *MAbs* **4(4)**, 475–487.

Yu M, Hu Z, Pacis E, et al. (2011b) Understanding the intracellular effect of enhanced nutrient feeding toward high titer antibody production process. *Biotechnology and Bioengineering* **108(5)**, 1078–1088.

Yu X, Vasiljevic S, Mitchell DA, Crispin M, Scanlan CN. (2013) Dissecting the molecular mechanism of IVIg therapy: the interaction between serum IgG and DC-SIGN is independent of antibody glycoform or Fc domain. *Journal of Molecular Biology* **425(8)**, 1253–1258.

Zeck A, Pohlentz G, Schlothauer T, et al. (2011) Cell type-specific and site directed *N*-glycosylation pattern of FcγRIIIa. *Journal of Proteome Research* **10(7)**, 3031–3039.

Zhang P, Chan KF, Haryadi R, et al. (2013) CHO glycosylation mutants as potential host cells to produce therapeutic proteins with enhanced efficacy. *Advances in Biochemical Engineering and Biotechnology* **131**, 63–87.

Zhang W, Voice J, Lachmann PJ. (1995) A systematic study of neutrophil degranulation and respiratory burst *in vitro* by defined immune complexes. *Clinical and Experimental Immunology* **101(3)**, 507–514.

Zheng K, Bantog C, Bayer R. (2011) The impact of glycosylation on monoclonal antibody conformation and stability. *MAbs* **3(6)**, 568–576.

Zhong X, Cooley C, Seth N, et al. (2012) Engineering novel Lec1 glycosylation mutants in CHO-DUKX cells: molecular insights and effector modulation of *N*-acetylglucosaminyltransferase I. *Biotechnology and Bioengineering* **109(7)**, 1723–1734.

Zhong X, Wright JF. (2013) Biological insights into therapeutic protein modifications throughout trafficking and their biopharmaceutical applications. *International Journal of Cell Biology* **2013**, 273086.

Zhou Q, Shankara S, Roy A, et al. (2008) Development of a simple and rapid method for producing non-fucosylated oligomannose containing antibodies with increased effector function. *Biotechnology and Bioengineering* **99(3)**, 652–665.

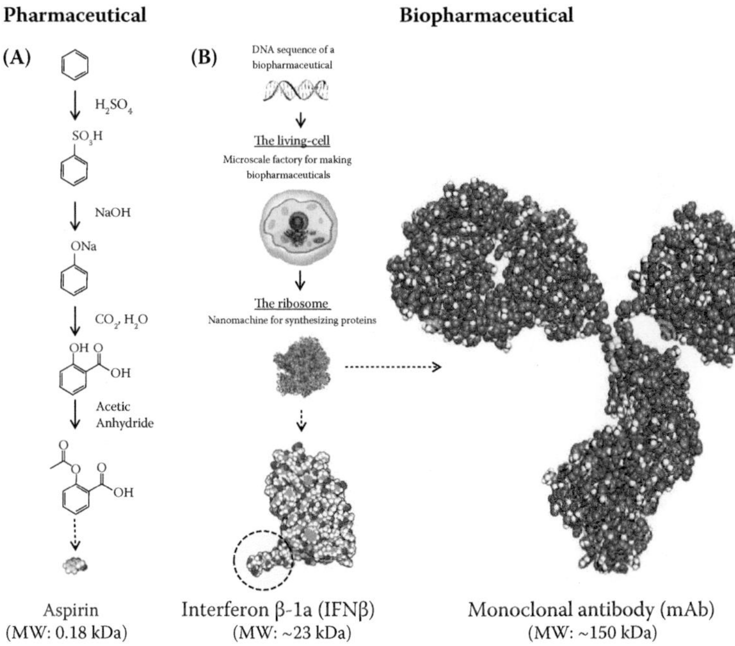

FIGURE 2.2 A simple comparison illustrating differences in the process of making a pharmaceutical versus making a biopharmaceutical.

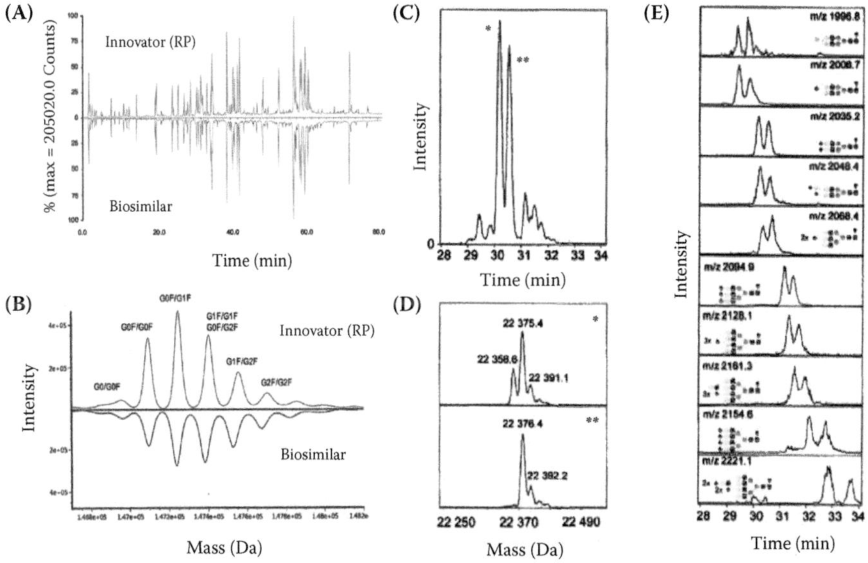

FIGURE 2.9 Some examples of MS being used to characterize biopharmaceuticals.

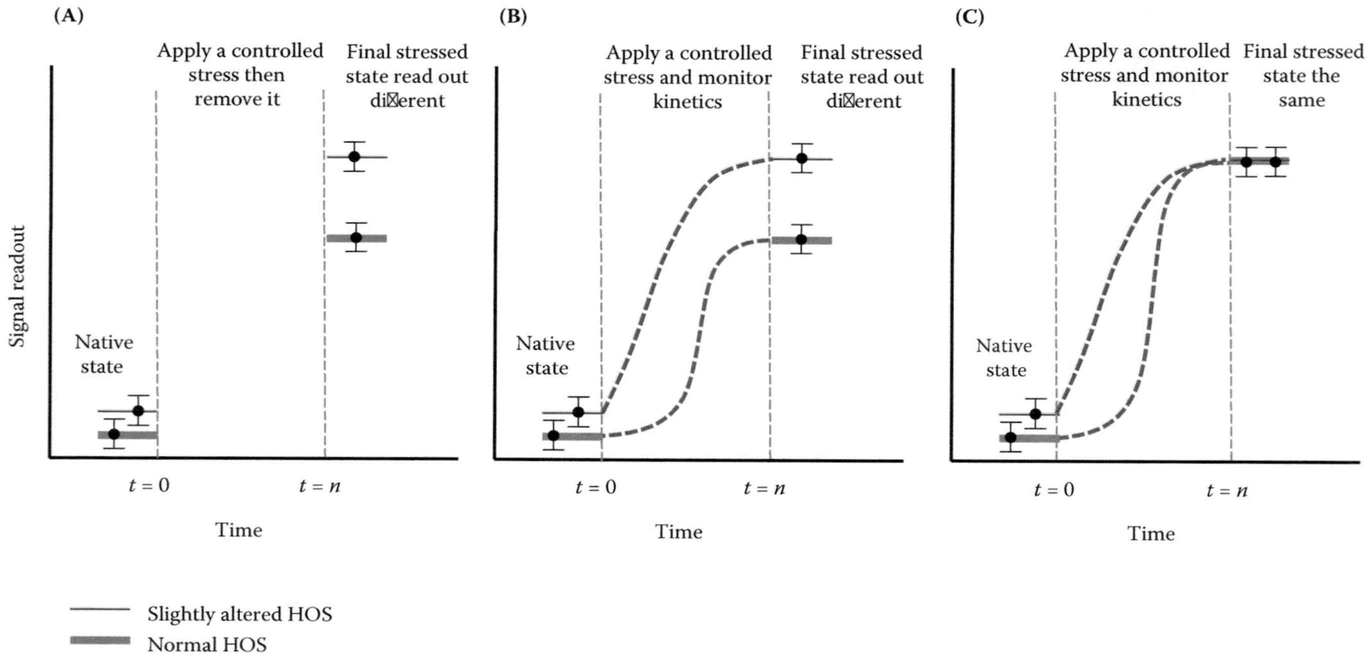

FIGURE 2.11 The utility of stress studies to facilitate the detection of differences in the HOS between two biopharmaceuticals using low-resolution biophysical methods.

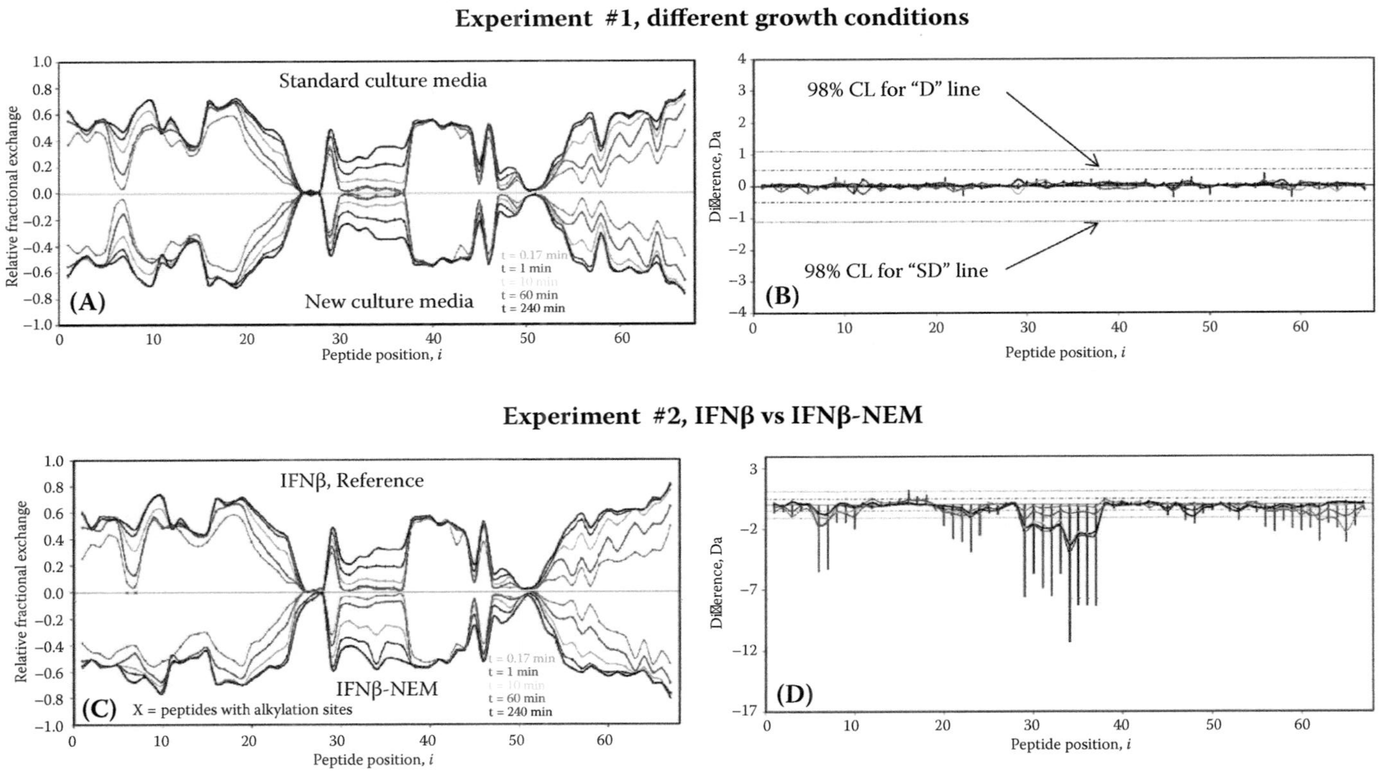

FIGURE 2.12 Two different local H/DX-MS comparison experiments conducted on a biopharmaceutical (IFNβ).

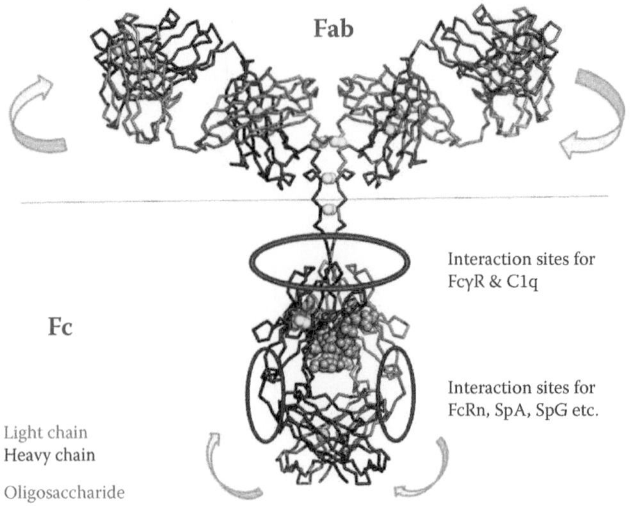

FIGURE 4.1 An α carbon cartoon of a human IgG1 molecule indicating binding sites for IgG1-Fc ligand binding sites; arrows indicate independent mobility of Fab & Fc around the hinge region.

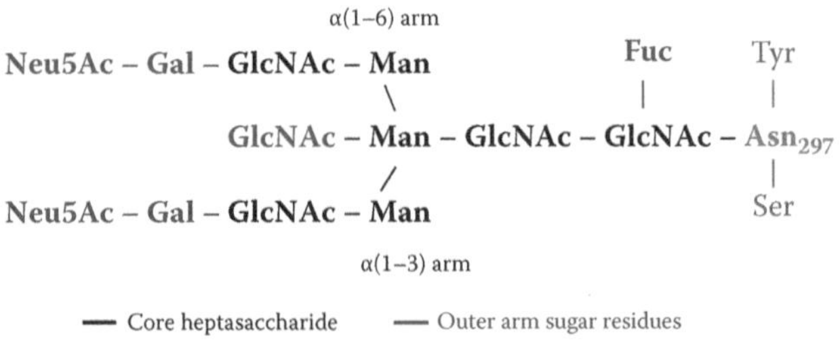

FIGURE 4.2 The diantennary oligosaccharide showing the heptasaccharide core (blue) and possible outer arm sugar residues (red).

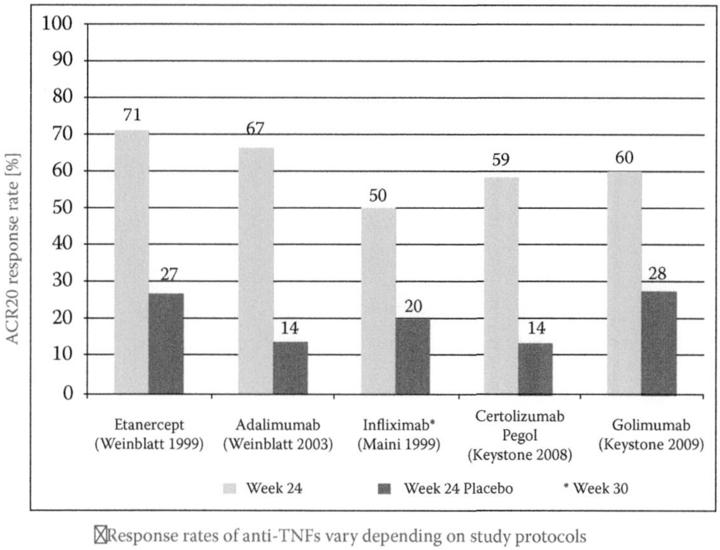

FIGURE 7.2 The clinical ACR 20 responses to Etanercept, Adalimumab, Infliximab, Certolizumab, and Golimumab at 24 weeks are clinically comparable.

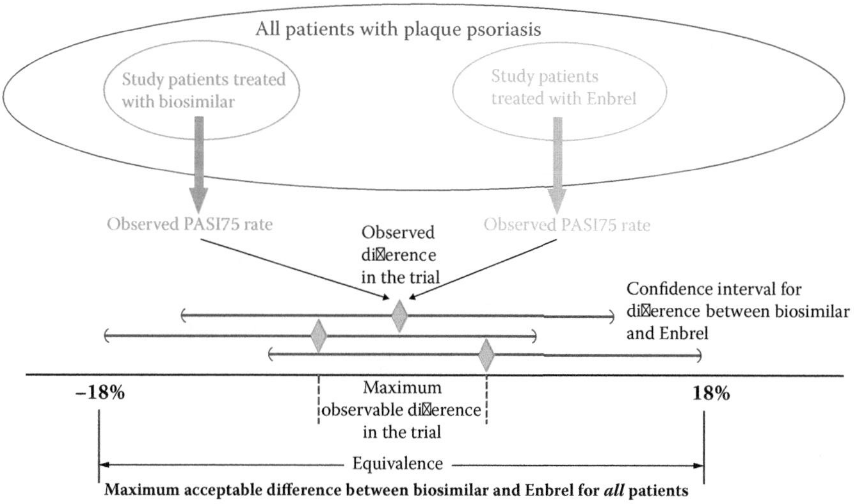

FIGURE 7.3 For all patient responses to be within the equivalence margin, the actual point estimates or average responses for the patients treated with either drug must be fairly close to each other.

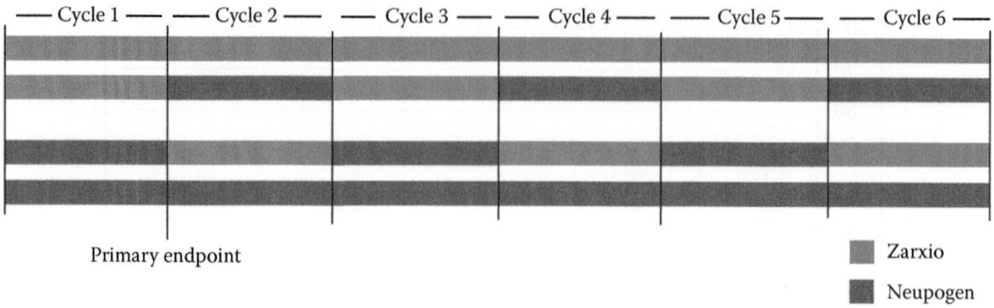

FIGURE 7.4 Clinical trial design of the Zarxio biosimilar confirmatory clinical trial presented at the January 7, 2015, Oncologic Drug Advisory Board (Sandoz, 2015c).

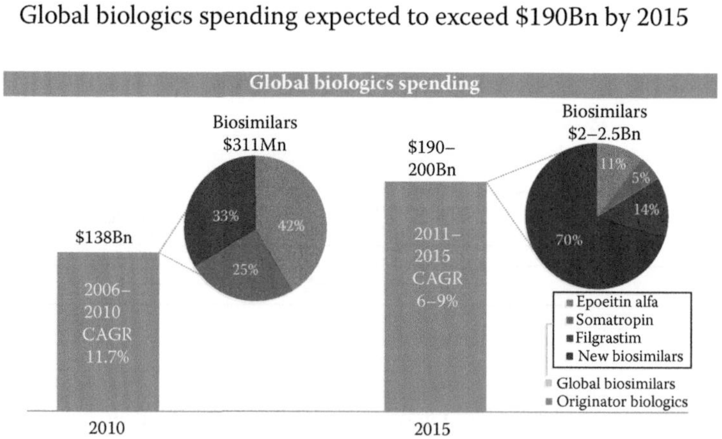

FIGURE 8.1 Global biologics spending (in billion dollars). (IMS, Multinational integrated data analysis system (MIDAS). IMS Institute for Healthcare and Informatics, Danbury, CT, 2010.)

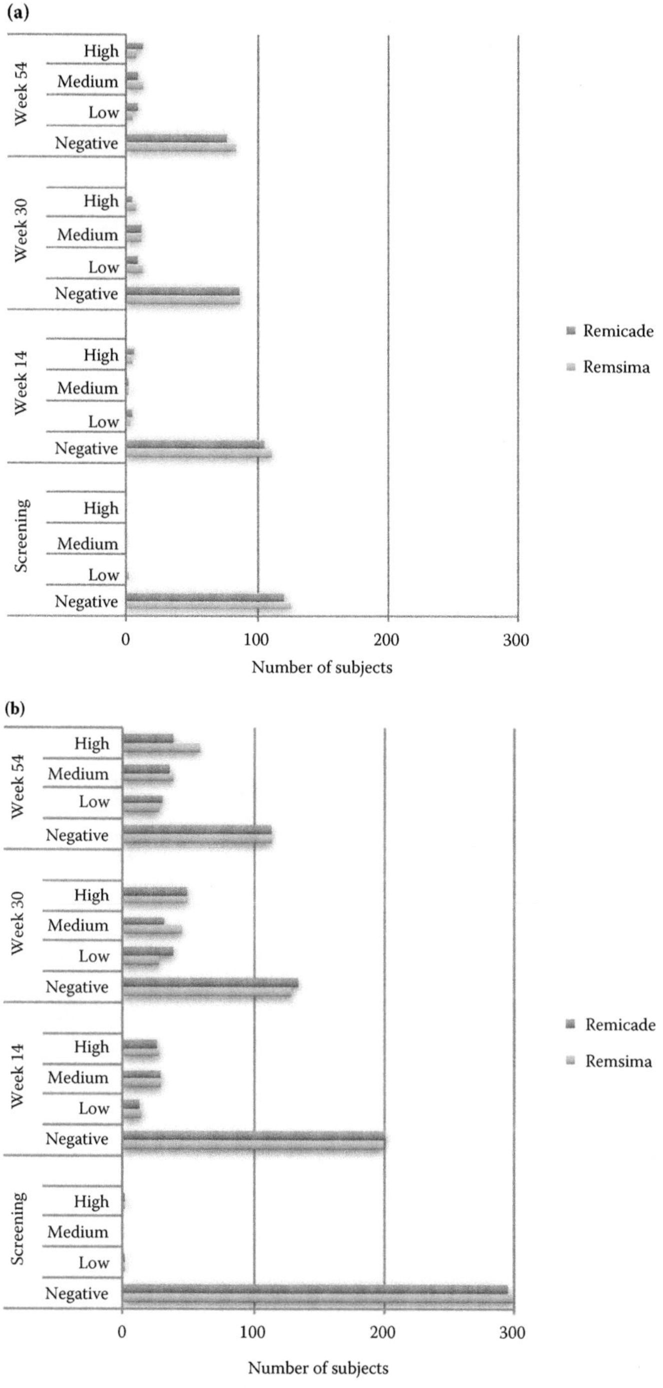

FIGURE 12.3 ADA response; number of subjects (x-axis) vs. treatment time/titer category from screening visit to Week 54 of treatment period (y-axis) (based on EPAR for Remsima).

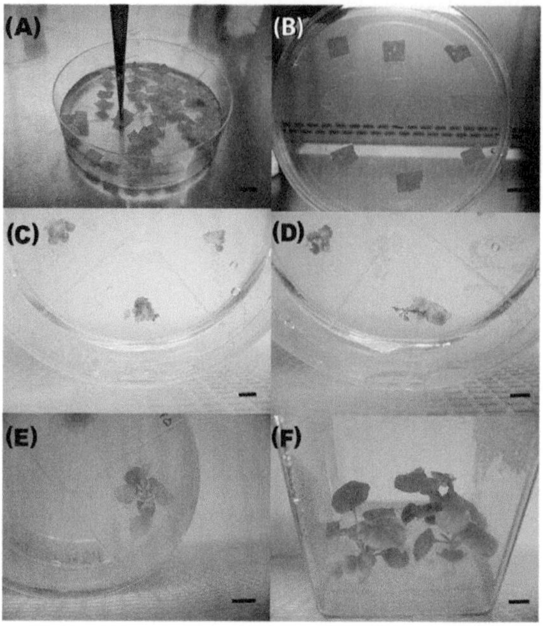

FIGURE 17.1 Agrobacterium-mediated transformation of *Nicotiana benthamiana.*

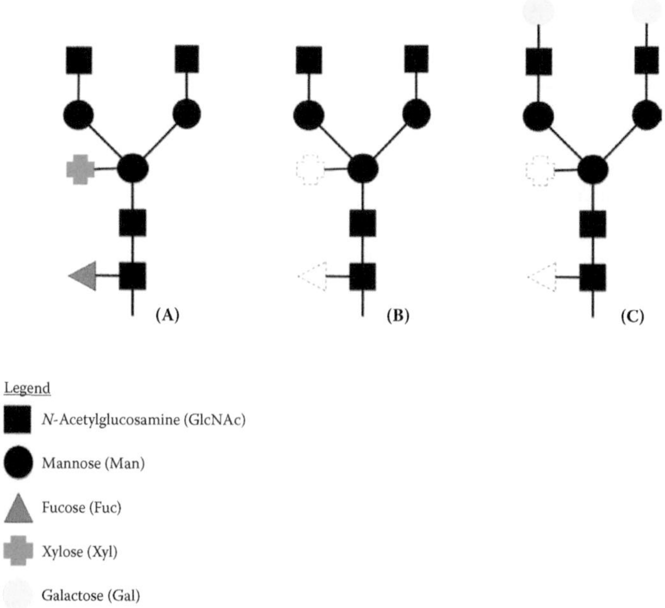

FIGURE 17.2 Schematic representations of native plant *N*-glycans and their humanized glycoforms. The black core composed of *N*-acetylglucosamine and mannose is conserved across all eukaryotes.

5 Manufacturing and Process Control Issues

Quality Development of Biosimilar Medicinal Products

Alan Fauconnier
Federal Agency for Medicines and Health Products
Culture in vivo ASBL

Lyudmil Antonov
Bulgarian Drug Agency

CONTENTS

5.1 INTRODUCTION

The concept of biosimilarity emerged in the early 2000s, when the patents of innovative biotechnological medicinal products expired and/or when they started losing data protection, opening the door to the application for the marketing authorization of putative biological generics. Regulators promptly realized that the abbreviated procedures at that time might not be appropriate for licensing biological medicinal products, or "biologics."

Indeed, these procedures, namely the abbreviated new drug application (ANDA) in the United States and the derogation granted by the European Union (EU) to essentially similar medicinal products, usually referred to as generics, are intended for the marketing authorization application (MAA) of products that are "essentially comparable" to a reference, originator, or innovator medicinal product. The generic and its reference must have the same qualitative and quantitative composition in active substance and the same dosage form and route of administration. On this basis, full nonclinical and clinical development is not required, and clinical studies are usually reduced to bioavailability studies aimed at demonstrating bioequivalence. In essence, the generics approach relies on the availability of chemical pharmaceutical and biological documentation compiled and presented in the Quality/CMC part of the application. The quality development is expected to bring the information needed for establishing the comparison between the generic and its reference. Since chemical entities can be subjected to in-depth and precise characterization, this allows drawing a reliable picture of the drug and determining its quality, facilitating the comparability exercise between the generics and its reference.

In contrast, biological medicinal products have high-molecular-weight active substance(s) with high structural complexity, usually displaying a high degree of inherent structural heterogeneity. Instead of being expressed in content (i.e., mg, µg, mL), their strength is often based on potency testing, expressed in units as determined by a bioassay. In most cases, the validation of these assays is challenging. Indeed, international standards are rarely available, thereby complicating the determination of accuracy. Precision and linearity are significantly less performing than those of physicochemical assays. Most biological substances are labile and sensitive to heat, oxidation, and sometimes light. Subtle differences in their manufacturing process may affect their quality attributes with a possible impact on the safety and efficacy profile of the medicinal product. Moreover, a number of raw and starting materials used in their manufacturing process are of biological origin. Added to the fact that their manufacturing process is conducted under mild physiological conditions of temperature, pH, ionic strength, and so on and that their chemical nature (i.e., proteins, glycoproteins, polysaccharides) may support microbial growth, biologics thus pose a higher risk of contamination by microorganisms, viruses, and nonconventional infectious agents as compared to chemical drugs. This risk cannot be mitigated by a terminal sterilization step, which is not applicable to heat labile substances. Instead, an aseptic process has to be set up and validated.

Finally, most biological substances cannot be administered orally without being digested. As a consequence, biological medicinal products are mainly parenteral forms. Parenterally administered medicinal products need to be sterile and free of pyrogens. Another particular feature of biologics comes from their capacity to raise an immune response. Setting aside vaccines, such a response is not desirable since it can elicit the production of antibodies that may neutralize the biological active substance. This risk is enhanced for medicinal products intended for repeated or chronic administration.

The quality development of biologics thus appears significantly more complex than that of chemical drugs. Accordingly, establishing the sameness of a biological generic, assuming that such a product exists, with a reference medicinal product would be particularly challenging.

The EU pioneered in the field by adapting the Community's legislation governing medicinal products in 2003. Directives 2003/63/EC and 2004/27/EC, which are amendments to the basic Directive 2001/83/EC on the EU code relating to medicinal products for human use introduced the Similar Biological Medicinal Products through a new provision, that is, Article 10(4), stating that where "the information required in the case of essentially similar products (generics) does not permit the demonstration of the similar nature of two biological medicinal products, additional data, in particular, the toxicological and clinical profile shall be provided" (EC, 2003, 2004, 2012). This approach further spread to other regions of the world, including emerging markets (Mintz, 2013; GaBI, 2014a). In the United States, the Biologics Price Competition and Innovation Act of 2009 (BPCI Act) was enacted as part of the Affordable Care Act, the health reform introduced in 2010. The BPCI Act creates a new pathway for approval of biosimilar biological products or follow-on biologics (FOBs) (USC, 2010).

In contrast to the EU and US biosimilar regulatory pathways, which were enabled through legislation, in Canada, the subsequent entry biologics (SEBs) are regulated through guidance documents (Health Canada, 2010; Pandya, 2014). In Japan, the overarching *Guideline for the Quality, Safety and Effectiveness of Biosimilar Products* paved the way for the follow-on biologic products regulatory pathway (Derbyshire, 2013; GaBI, 2014b). As can be seen, different terms such as similar biological medicinal products, biosimilar biological products, follow-on biologics, follow-on protein products, similar biotherapeutic products, and subsequent entry biologics were coined to capture the concept of biosimilar, with more or less pronounced conceptual and/or regulatory differences. The corresponding regulatory pathways have in common that biosimilars are biological medicinal products approved under an abbreviated procedure. A stepwise approach is recommended, which includes a thorough quality development aimed in particular at comparing the quality attributes of the biosimilar and its reference product, a biological product already licensed, usually within the same national territory. Depending on the degree of similarity that can be achieved at the quality level, the nonclinical and clinical data package will be reduced since a more or less important part of it is covered by the prior nonclinical and clinical data of the corresponding reference medicinal product (EMA, 2015a; FDA, 2015a). Quality development thus appears as the cornerstone of the stepwise biosimilar approach.

Biosimilars are not generics. Whereas a generic is essentially similar and contains the same active substance as the original medicinal product, a biosimilar is merely similar to a reference biological product and its active substance is most likely not identical. Substitution is ordinarily allowed for the generics but excluded for biosimilars. However, it was anticipated that advances in analytical technologies might lead to comprehensive characterization of recombinant proteins, opening the perspective of well-characterized biologicals (Chirino and Mire-Sluis, 2004). It is indeed not excluded that in the future, at least as regards recombinant low-molecular-weight nonglycosylated proteins such as insulin, biologics could be completely defined and therefore subject to generic application. In this regard, it should be emphasized that the provision in Article 10(4) establishing biosimilars in European Directive 2001/83/EC does not exclude the submission of an application for a biological generic

(EC, 2012). However, even if feasible in theory, in the practical order this possibility is currently not open, even for the simplest recombinant proteins.

Another difference between theory and practice can be seen in the scope of biosimilar regulation. Legal provisions are generally applicable to all biological medicinal products whether or not they are derived from biotechnologies. However, with the notable exception of a European product-specific guideline on low-molecular-weight heparins (EMA, 2013a), virtually all guidance documents are focusing on recombinant proteins. The following sections of this review thus address the biosimilars that contain a recombinant protein as an active substance, focusing on their quality development.

5.2 QUALITY DEVELOPMENT

Several guidelines issued from different regulatory bodies address the quality development of biosimilars. The issues discussed below are based on the recommendations provided in the guidelines listed in Table 5.1. This list is nonexhaustive since other regulatory authorities issued their own, such as Japan's Ministry of Health (GaBI, 2014b) or the South Korean Ministry of Food and Drug Safety (GaBI, 2015), which, because of linguistic constraints, are less easily accessible.

5.2.1 Origin of the Reference Medicinal Product

Depending on the region, the reference does or does not need to be country specific. Country-specific approval and, sometimes, marketing, is usually requested in important regulatory areas such as the US, EU, Canada, and Japan, sometimes with some

TABLE 5.1
Guidelines Addressing the Quality Development of Biosimilars

Institution	Title	Scope
WHO	*Guidelines on Evaluation of Similar Biotherapeutic Products* (SBPs) (WHO, 2013)	Well-established and well-characterized biotherapeutic products such as recombinant DNA-derived therapeutic proteins
Health Canada/ Minister of Health	*Guidance for Sponsors: Information and Submission Requirements for Subsequent Entry Biologics* (SEBs) (Health Canada, 2010)	All biologics provided they can be well characterized by a set of modern analytical methods
CDER and CBER/FDA	*Quality Considerations in Demonstrating Biosimilarity of a Therapeutic Protein Product to a Reference Product* (FDA, 2015b)	Therapeutic protein products
CHMP/EMA	*Guideline on Similar Biological Medicinal Products Containing Biotechnology-Derived Protein as Active Substance: Quality Issues* (revision 1) (EMA, 2014a)	Recombinant DNA-derived proteins

flexibility. For instance, the Canadian regulation allows use of a non-Canadian reference product, provided that it is shown to be a suitable proxy for the version actually licensed in Canada. However, this requirement cannot be met in countries that do not have a nationally licensed reference. The WHO guidance considers this possibility and, accordingly, makes some recommendations for the choice of the reference product: the reference product should be marketed for a suitable duration in a jurisdiction that has a well-established regulatory framework, and its original licensing should be based on the submission of full quality, nonclinical and clinical data, and so on.

5.2.2 Quality Package

Just as for any normal MAA, a complete quality/CMC section (CTD Module 3) is to be provided in a biosimilar application, addressing the manufacture, characterization, specifications, adventitious agent safety, stability, and the like, of the drug substance and drug product. These data need to be supplemented with a more specific package aimed at demonstrating comparability with the reference medicinal product. The assessment of similarity is viewed as a distinct collection of data supplementing the normal requirements. Accordingly, it may be recommended that this additional information be presented in a separate section such as Section 3.2.R in the EU application.

It is noteworthy that the reduction of the nonclinical and clinical data packages needs to be counterbalanced by a significant extension of the quality part.

5.2.2.1 The Manufacturing Process

The adage “the process is the product” is often quoted to indicate that the quality attributes of biological products are the results of the manufacturing process by which they are produced. Differences in the manufacturing process may impact the quality profile of the product, affecting its safety and/or efficacy. Manufacturers of biosimilars have a very limited view of the manufacturing process of the reference product, only based on the information publicly available in the Scientific Discussion of the European Public Assessment Report (published by the EMA) and/or in the Chemistry Review of the Drug Approval Package (published by the FDA), for instance. The manufacturing processes of the biosimilar and its reference are thus not expected to be the same. This is particularly the case inasmuch a decade or more may separate the application for a biosimilar from its reference. The manufacturing process technology will most likely have evolved during this period of time.

The biosimilar is thus developed on its own, produced by a different manufacturing process (e.g., different starting materials, raw materials, equipment, process, in-process control, acceptance criteria). On some occasions, even the host cell type may be different. For example, in 2006, Valtropin was granted a marketing authorization in the EU under the legal base of similar biological medicinal product (EMA, 2012). The reference medicinal product for this application was Humatrope (EMA, 2014b). Both products contain somatropin (recombinant human growth hormone) as an active substance. Remarkably, whereas the active substance of Humatrope is produced in *Escherichia coli*, that of Valtropine is produced in *Saccharomyces cerevisiae*. This is quite an unusual situation, however. Guidelines generally recommend expressing the active ingredient of biosimilars in the same or in a comparable host cell type as the reference in order to both minimize the differences in quality profile of the products and avoid affecting the type of product-related substances as well as the product- and process-related impurities.

In spite of the expected differences between the manufacturing processes of the biosimilar and its reference, the biosimilar quality profile must converge toward that of the reference. This can possibly be achieved by applying the enhanced approach principles developed in the International Council for Harmonization (ICH) guideline Q8 (R2) (ICH, 2009). In this respect, the quality target product profile (QTPP) of a biosimilar should be set up to match the reference product, and its manufacturing process should be designed to meet the QTPP. In addition, it should be demonstrated that the manufacturing process is robust and shows consistency as normally required. Should a manufacturing change being introduced during the development phase of the biosimilar but after completing the initial analytical similarity assessment, a comparability between pre- and postchange products should then be demonstrated.

5.2.2.2 Characterization

The characterization of biosimilars does not differ from that of biological/biotechnological products and includes the determination of physicochemical properties, biological activity, immunochemical properties, and purity and impurities as defined in ICH Q6B (ICH, 1999). As is typically the case for biologics, the emphasis is primarily on the characterization of the drug substance. Essentially all tests and methods implemented for characterizing the biological drug substance will be used during the comparability exercise in the side-by-side analysis with the reference medicinal product. The characterization thus serves a dual purpose in the biosimilar pathway. It is aimed not only at establishing the quality profile of the biosimilar, but also at detecting the differences (or their absence) in its quality attributes and the reference product.

5.2.2.2.1 Physicochemical Properties

The physicochemical characterization comprises the determination of the primary, secondary, tertiary, and, where relevant, quaternary structure, using appropriate analytical methods such as amino acid sequencing, peptide mapping, circular dichroism, and mass spectrometry. Special attention should be paid to posttranslational modifications (e.g., oxidation, deamidation, truncation) and, where relevant, to glycosylation. In contrast to the primary structure, which can be unequivocally determined by the gene-coding sequence, posttranslational modifications depend on the "environmental conditions" in which the protein expression is taking place. For instance, the host cell and expression system will drive the glycosylation profile of the protein. Whereas no glycosylation occurs in prokaryotic systems, lower eukaryotes such as the yeast *S. cerevisiae* are synthesizing glycoproteins but with a different glycosylation profile than human proteins. Thus, host expression systems more closely related to human cells such as Chinese hamster ovary (CHO) cells will be preferred for the production of therapeutic glycoproteins.

Even with cells issued from more closely related species, however, differences still persist. For instance, the end of the sugar chains are usually capped by a sialic acid of the type *N*-acetylneuraminic acid (NANA) in human glycoproteins, whereas *N*-glycosylneuraminic acid (NGNA) is capping glycoproteins of other mammalians, including the great apes (Chou et al., 1998). Posttranslational modifications also depend on the manufacturing process. A same host cell expression system may give rise to proteins displaying different glycosylation profiles depending on cell culture conditions. This protein population as it stands after harvest could be further widened by the environment met by the protein during the production process (pH, ionic strength, temperature), including its purification and its formulation in the final dosage/galenic form. All these modifications show that, within the drug product, a protein will not be present as a single molecular form but instead, as a mix of molecules sharing the same primary structure but displaying different chemical modifications and branched-chain sugar moieties, each individual molecule having its own physicochemical and biological properties.

Analysis of such a mix will reveal not a single form but instead a population of isoforms that, to some extent, can be fractionated and characterized qualitatively and quantitatively by appropriate analytical methods. The emergence of these different isoforms and their relative proportion introduces the notion of microheterogeneity of proteins (Trouvin, 2011). Glycosylation, reported as the "most common, complex and heterogeneic modifications that may occur to therapeutic proteins" (quoted from Vugmeister et al., 2012), and microheterogeneity can thus significantly modify the quality attributes of the proteins by changing the stability, solubility, immunogenicity, aggregation, and affinity to receptor, putatively impacting pharmacokinetic and pharmacodynamic behaviors as well as product efficacy (Solá and Griebenow, 2010; Li and d'Anjou, 2009). Therefore, carbohydrate structures should be investigated in depth, including the overall glycan profile, site-specific glycosylation patterns, and site occupancy.

Determination of physicochemical properties for biosimilars is discussed in greater detail in Chapter 2 of this book.

5.2.2.2.2 Biological Activity

Biological activity is defined as the specific ability or capacity of the product to achieve a defined biological effect. As mentioned in ICH Q5E, it serves different purposes in the assessment of quality attributes that are useful for the characterization and for the batch analysis of the product (ICH, 2004). Ideally, the biological assay should reflect the protein's mechanism of action. In this best case scenario, potency, defined as the quantitative measure of biological activity, could serve as a link to clinical activity. Where relevant, cross reference to the clinical and/or nonclinical sections should be made. Biological activity analysis also allows discrimination between protein product variants that display an appropriate level of activity, referred to as product-related substance, from those being inactive, therefore considered as a product-related impurity. Particularly, the biological assay complements the physicochemical analyses by confirming the correct higher-order structure, integrity, and function of the protein. This is of particular importance for complex proteins, where physicochemical analyses alone may be unable to confirm the integrity of these structures. For proteins consisting of multiple functional domains, it makes sense to develop several biological assays addressing the different functional activities of these proteins.

Different types of biological assays can be used such as enzyme activity, ligand or receptor binding assays, and cell-based assays. However, these assays have their limitations. For instance, *in vitro* biological assays are generally not predictive of the bioavailability of the product, which can affect pharmacodynamics and clinical performance.

5.2.2.2.3 Immunochemical Properties

Immunological functions of antibodies and antibody-based products (e.g., fusion proteins based on IgG Fc) should be investigated in terms of specificity, affinity, binding kinetics, and Fc functional activity (e.g., binding to FcγR, C1q, FcRn).

5.2.2.2.4 Purity and Impurities

Product- and process-related impurities of the biosimilar and its reference should be characterized and compared both qualitatively and quantitatively. As mentioned in ICH Q1A(R2) (ICH, 2003), the degradation pathway of the protein as well as the shelf life of the reference at the time of testing need to be considered. Forced degradation studies and/or stability under stress conditions (e.g., exposure to heat, light, agitation, freeze–thaw, base hydrolysis, acid hydrolysis, and oxidation) intended to identify the degradation products that can provide insight into the similarity of the degradation pathway of the reference product and the biosimilar. By nature, process-related impurities depend on the manufacturing process. As indicated above, the manufacturing process of a biosimilar is not expected to be the same, not even similar to the reference product manufacturing process. Accordingly, comparing process-related impurities (e.g., host cell DNA, host cell proteins, cell culture components, downstream process impurities) could prove to be difficult and of limited meaning. The process-related impurity profile of the biosimilar thus needs to be addressed on its own. Generally, process-related impurities are not expected to be comparable, but need to be characterized and clinically qualified on their own for each specific medicinal product. Thus, depending on the type and amount of impurity introduced by the manufacturing process, additional toxicological studies could be required to rule out the risk of adverse impact on the safety of the product.

5.2.2.3 Analytical Procedures

Unlike routine release testing as per specification, the assessment of the quality attributes performed in the characterization does not necessarily require the use of fully validated assays according to ICH Q2(R1) (ICH, 2005a). However, the methods used should be scientifically sound and qualified, meaning fit for their intended use for the purpose of finding comparability and providing reproducible and reliable results. The discrimination ability of the tests is of concern since high variability and/or low sensitivity may prevent detection of differences between the biosimilar and its reference. This holds true especially for biological assays usually displaying a high level of variation. Assays that are less variable and are sufficiently sensitive to detect small but possibly significant differences between the biosimilar and the reference product should be developed. Finally, using orthogonal methods is of particular importance. Indeed, assessing a same-quality attribute with independent methods based on different physicochemical or biological principles enlarge the possibility of detecting variants with one method, whereas the others do not.

Analytical characterization for biosimilars is discussed in greater detail in Chapter 2. A typical list of tests and methods is provided in Table 5.2.

5.2.2.4 The Comparability Exercise

The comparability exercise is defined in the WHO guideline as a "head-to-head comparison of a biotherapeutic product with a licensed originator product with the goal of establishing similarity in quality, safety and efficacy. Products should be compared in the same study using the same procedures" (WHO, 2013). The quality part is of decisive importance. Indeed, demonstration of a high degree of similarity as determined by robust quality data allows reducing the nonclinical and clinical requirements for licensing. It should be emphasized that the similarity is to be shown primarily at the level of the finished product. However, in guidelines, it is clearly stipulated or implicitly recognized that the active substances must also be similar, especially as the head-to-head comparison of formulated drug products is not always feasible, either because of the low concentration of the active substance or because of the presence of interfering excipients. Thus, the drug substance must be isolated from the reference medicinal product in order to be analyzed on its own. In this case, the isolation process should be outlined and carried out in such a way that the relevant quality attributes of the originator drug substance are not significantly altered, for example, by comparing the active substance before and after formulation/deformulation preparation.

The guideline ICH Q5E is aimed at providing principles for the comparability of biotechnological/biological products subject to changes in their manufacturing process (ICH, 2004). Although the scope of this guideline does not cover the comparison of two different products, a number of principles discussed in ICH Q5E may be applicable to the comparability exercise aimed at demonstrating similarity at the quality level between the biosimilar and its reference. However, it should be stressed that developing a biosimilar is different from introducing a change in the manufacturing process of a same product. In the latter case, the manufacturer has acquired solid knowledge of the product and of its manufacturing process and control. Drug substance as well as a range of values coming from historical data are available to the manufacturer. In contrast, the biosimilar manufacturer has access neither to the originator drug substance nor to the product information which remains proprietary, a hindrance referred to as the "knowledge gap" (Declerck et al., 2016). It has to rely on the originator drug product that is available on the market. The development of a biosimilar thus starts with the finished reference product and proceeds backwards.

Data collected during the comparability exercise from both the biosimilar and the reference product will be used to establish the range of variability of quantitative quality attributes. Apart from general statements on the need to follow an appropriate statistical approach, the guidelines provide little guidance on the number of lots to be analyzed and the statistical approach to be used in order to establish equivalence between the biosimilar and its reference for a given quality attribute and to set limits to its ranges. The answer to this question is not straightforward. The need for more detailed guidance was acknowledged by the authorities and should materialize through publication of a Reflection Paper on statistical methodology for the

TABLE 5.2
Typical Analytical Testing for Characterization and Release Specification of Drug Substances

Quality Attribute	Analytical Method	Charact.	Spec.
	Identity/Biophysical Characteristics		
Primary structure	Peptide mapping (LC-MS)	x	
	N-terminal sequencing	x	
	C-terminal sequencing	x	
	Amino acid composition	x	
Size	Molecular weight by MALDI-TOF MS	x	
Higher-order structure	Far UV circular dichroism (FAR UV-CD)	x	
	Differential scanning fluorimetry (DSF)	x	
	Hydrodynamic radius by dynamic light scattering (DLS)	x	
	FT-IR	x	
Thermal stability	Differential scanning calorimetry (DSC) thermal analysis	x	
	Identity/Immunological Properties		
Immunological identification	Western blot	x	x
	ELISA	x	x
	Posttranslational Modifications		
Posttranslational modifications	Disulfide bonds by peptide mapping	x	
	Free thiols	x	
	Glycosylation		
Intact glycoprotein mass	LC-MS		
Glycopeptides (glycosylation site profiling)	ETD-CID	x	
Released *N*-linked oligosaccharide analysis	PNGAse F/MS or HPAE-PAD	x	
Released *O*-linked oligosaccharide analysis	Alkaline β-elimination/MS or HPAE-PAD	x	

(Continued)

TABLE 5.2 ***(Continued)***
Typical Analytical Testing for Characterization and Release Specification of Drug Substances

Quality Attribute	Analytical Method	Charact.	Spec.
Monosaccharide analysis on released glycans	Neutral monosaccharide analysis HPAE-PAD	x	
	Charged monosaccharide analysis with HPAE-PAD	x	
	Sialic acids analysis (NeuGc, *O*-acetylated sialic acids)/HPAE-PAD	x	
	Purity/Product-Related Species/Impurities		
Purity	RP-HPLC	x	x
	SEC-HPLC (% monomer)	x	x
	SDS-PAGE (reduced and nonreduced silver stained)	x	x
Aggregates	SEC-HPLC (% HMWS)	x	x
	Analytical ultracentrifugation	x	
Isoforms/degradation product	Isoforms IEF	x	
	Isoforms 2-D gel	x	
	Oxidized species by RP-HPLC/MS	x	
	Deamidated forms by RP-HPLC/MS	x	
	C-terminal lysine variability by IEC-HPLC and/or C-terminal sequencing	x	
	SDS-PAGE (reduced and non-reduced silver stained)	x	
	Process-Related Impurities		
Host cell proteins	ELISA	x	x
Residual DNA	qPCR	x	x
Protein A	ELISA	x	
Other agents (Triton X100, residual solvents, antifoam, etc.)	RP-HPLC	x	x
	GC	x	x
	ICP-OES/MS	x	x
Microbiology	Bioburden	x	x
	Sterility	x	x
Endotoxin	LAL (kinetics, endpoint)	x	x

(Continued)

TABLE 5.2 (*Continued*)

Typical Analytical Testing for Characterization and Release Specification of Drug Substances

Quality Attribute	Analytical Method	Charact.	Spec.
	Quantity		
Protein concentration	UV-A280	x	x
Protein quantity	BCA, Lowry, Bradford	x	
Potency	Immunoassays	x	x
	Cell-based bioassays	x	x
	Enzymatic activity	x	x
	Biological activity *in vivo*	x	
	Functional Activity of Monoclonal Antibodies		
	Binding affinity to antigen	x	
	Binding to Fcγ receptors	x	
	Binding to FcRn receptors	x	
	C1q binding	x	
	ADCC	x	
	CDC	x	
	Cytokines release	x	

comparative assessment of quality attributes in drug development (EMA, 2013b). Without being exhaustive, some considerations are presented here.

The use of an inappropriate statistical approach is not uncommon. A common intuitive but inappropriate statistical method consists of establishing equivalence by setting tolerance intervals (TIs) for the reference product. Under this approach, a TI calculated on historical data for the reference product is compared with a TI calculated on the test product data. If the test product TI fits inside the reference product TI, the test product is considered similar to the reference product. From a statistical point of view, the main objection against this approach is that it is based on an erroneous hypothesis. Actually, the question of significance is not even addressed in the TI approach since it is not based on probabilistic inference. Because the TI approach is not a hypothesis testing method, its hypothesis is implicit: if the test product is not significantly different from the reference product, then it is significantly similar. This is a wrong assumption. Various seemingly paradoxical situations arise in connection with the wrong hypothesis. For example, the reference TI is wider with smaller reference sample size, higher reference variability, and inconsistency of the production process over time. All these undesirable characteristics of the reference product, in a rather contradictory manner, make it easier to establish the similarity with the test product.

The correct approach to be used in establishing biosimilarity is equivalence testing, which is based on a correct hypothesis. In this approach, a 90% confidence interval (CI) is established about the difference between reference and test sample means (mean difference). If the mean difference and the CI fit inside some equivalence limits (ELs) established prior to the test, the reference and test samples are significantly equivalent (similar) (Figure 5.1). Otherwise, they are not significantly similar.

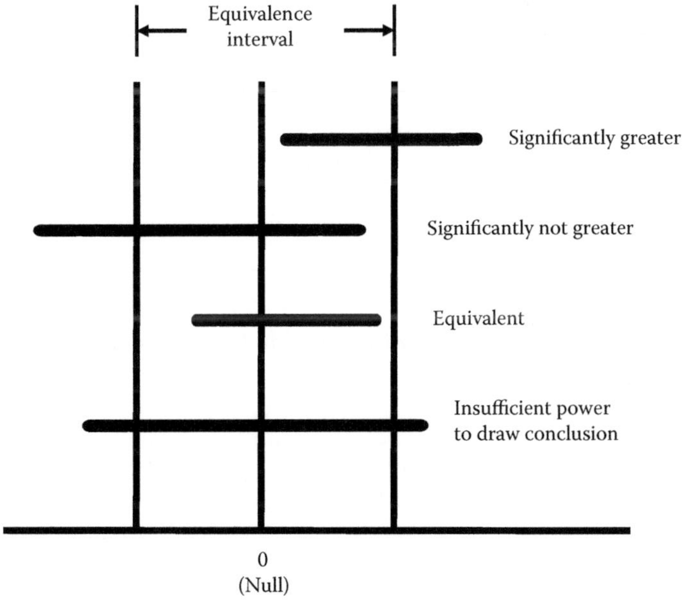

FIGURE 5.1 Equivalence tests, equivalent limits, and types of significance.

This is the two-sided equivalence hypothesis. There are two one-sided variants of this hypothesis. In the noninferiority testing, one attempts to establish that mean difference and CI do not fall below the lower EL, while in the nonsuperiority testing, those should not be above the higher EL. It makes sense, for example, to test impurities with a one-sided nonsuperiority test rather than with a two-sided equivalence test. Like other tests, the equivalence testing starts with a null hypothesis (which for the two-sided test is that reference and test samples are inequivalent) and then asks if one has enough evidence (data) to reject that null hypothesis, and, in effect, to prove the alternative hypothesis that the samples are equal for the particular quality attribute (QA) tested. It is said that one makes a Type I error when one incorrectly rejects a valid null hypothesis/accepts a false alternative hypothesis. The probability for a Type I error is the significance level of the test (α-level, p-value). The significance level of the equivalence test is usually .10 (90% CI), while for the one-sided tests (noninferiority and nonsuperiority) the significance level is .05 (95% CI). Type II error is an incorrect acceptance of a false null hypothesis/rejection of a valid alternative hypothesis. With regard to equivalence testing, α is the probability for concluding equivalence when reference and test are inequivalent and β is the probability for concluding inequivalence when reference and test are equivalent. From the aspect of risk, α is the consumer's risk and β is the producer's risk. Power is the reverse of β (power = $1 - \beta$) and corresponds to the probability for correctly concluding equivalence when reference and test are equivalent (Table 5.3).

Several equivalence tests can be used presenting their own advantages and weaknesses. Among these, the two one-sided t-tests (TOST) (Schuirmann, 1981) is the best known and most widely used equivalence test to be applied to normally distributed interval-ratio data. As its name shows, TOST consists of two t-tests that test the two one-sided hypotheses (one t-test for noninferiority and the other t-test for nonsuperiority). Each of the two tests has a significance level .05, and their combined use results in a significance level .10. Because they are based on the correct hypotheses, equivalence tests have expected outcomes. With a fixed α (.05 or .10), the test should have enough power ($1 - \beta$) in order to pass. To achieve this power, one should have a sufficient sample size (number of batches, repetitions within a batch) and high precision (that is, low variance) data. The chance to pass (the power) increases with widening of the ELs, so the usual question that arises when testing with equivalence

TABLE 5.3
Possible Outcomes with Regard to the Alternative Hypothesis (H_1)

	Type I and Type II Errors	
	True	**False**
Accept	PASS	α Type I error
Reject	β Type II error	FAIL

tests is not about the sample size (as large as required to pass the test) but about the width of the equivalence interval (the ELs). When dealing with bioequivalence studies, the ELs often chosen are ±20% of the reference, which is acceptable within the context of the clinical testing variability. However, setting a universal limit is not possible for quality attributes because they are based on a great variety of analytical methods and parameters.

The most rigorous statistical approach to set ELs, as well as the other interrelated parameters used in inferential statistical tests such as Type I and Type II errors (α and β), is to rely on decision theory in combination with inference. Decision theory serves to determine these parameters in a way that corresponds to the problem at hand. The derivation of these parameters is based on an objective function (minimize a loss function or maximize its negative, the reward function) and in the hazard rate, which is the expectation of the objective function. The decision theory relies on the quantitative concept of utility, which, in many applications and particularly in the field of health and medical sciences, is hard to quantify. In the absence of universal limits and decision theory-derived parameters, a third approach consists of setting limits on the basis of experimental data variability and employs the reference interval (RI). RI depends only on the standard deviation of the reference samples (= square root of the reference variability). Common applications of tolerance intervals for comparisons also use RI as a criterion.

Therefore, it makes sense to use RI as EL, with the justification that RI fixes the EL on the basis of the variability, sometimes expressed as a coefficient of variation (CV), which is an intrinsic summary characteristic of the concrete system.

Since power is not a consumer's risk, when assessing a formal test of equivalence, regulators are *a priori* less interested in the test's power. As for the biosimilar manufacturer, its best interest is to increase the power of the TOST since it will increase the probability of demonstrating the equivalence of the products. The most important and feasible way to increase power is to increase sample size. There is no consensus about what should be the sample size unit: several batches or repetitions (more than one sample of the same batch), respectively, reflecting the "between-batch variability" (σ^2_b) and the "within-batch variability" (σ^2_w). As the total variability σ^2 shall be the sum of σ^2_b and σ^2_w, both several batches and/or several repetitions per batch could be used with a repetition as a sample size unit. However, considering the significance of the "between-batch variability" (σ^2_b) for assessing similarity, it can be stated with some certainty that several batches of each reference and test product should be tested. Based on the suggestion from the stability guidelines (ICH Q1A(R2), Stability Testing of New Drug Substances and Products), it is assumed that three batches per product would be the minimum requirement to assess the between-batch variability. Power is also highly dependent on the variability of the methods and drops sharply when CV is close to the EL. Thus, for the biosimilar manufacturer, there are two not mutually exclusive options for increasing the probability of demonstrating equivalence: either by increasing the sample size and preferably the number of different batches or by reducing the variability of the methods used to assess the QA. The equivalence tests (e.g., TOST) thus appear to be the appropriate approach for demonstrating similarity between the biosimilar and its reference and does not suffer the deficiencies that were identified in the TI methodology. Indeed, in equivalence tests, a smaller sample size and/or more variance in the collected data will lessen the chance of demonstrating similarity.

5.2.2.5 Pharmaceutical Development

As set out above, the concepts of critical quality attributes (CQAs) or QTPP exposed in ICH Q8(R2) *Pharmaceutical Development* (ICH, 2009) appear to be valuable tools for identifying the meaningful target acceptance criteria for biosimilars. Likewise, the risk management principles described in ICH Q9 *Quality Risk Management* may help define comparability criteria and assist in the evaluation and management of inevitable differences between the biosimilar and its reference (ICH, 2005b; Schiestl, 2009). However, it is generally recognized that the design space largely depends on a particular manufacturing process. ICH Q8 and ICH Q9 are applicable to biosimilars but primarily for their own development in the same way as for the originators (Kresse, 2009).

5.2.2.6 Specifications

As mentioned in guidelines (ICH, 1999; WHO, 2013), specifications are not aimed at fully characterizing the drug substance and the drug product but, instead at verifying their routine quality and consistency. Biosimilars do not differ from other medicinal products in this regard. Where applicable, their specification and limits are required to comply with pharmacopoeial prescriptions. Having said that, we find that compliance with compendial standards is generally regarded as a minimum set of requirements, additional test parameters being usually required.

As for any biotechnological and biological products, the establishment of specifications for biosimilars is guided by the principles laid down in ICH Q6B (ICH, 1999). In line with these principles, the establishment of a specification is a global process and is not simply based on the range of the quality attributes of the reference but instead on the manufacturer's experience on a sufficient number of lots and upon a full data package, encompassing manufacturing process, analytical procedure, stability, characterization, and preclinical and clinical studies.

As mentioned earlier, the manufacturing processes of a biosimilar and its reference are different. Accordingly, a number of tests and acceptance criteria will be customized for a given biosimilar and established on their own, without referring to the originator. This is typically the case for specifications addressing the manufacturing process consistency and its control (e.g., uniformity of dosage units, residual humidity, visible and subvisible particles) or the process-related impurities (e.g., Host Cell Proteins, residual DNA, media component).

Analytical procedures are also developed and validated by the manufacturer of the biosimilar, regardless of the methods used by the reference manufacturer.

In contrast, the stability studies needed to set up the specification are to be conducted on the biosimilar but not independently from the reference. For instance, it makes sense to conduct accelerated and stress stability studies in parallel on both the biosimilar and its reference in order to establish their degradation profiles which, in turn, will reverberate on the specifications.

Similarly, the characterization of the biosimilar, since it is closely related to the comparability exercise, is not an exercise conducted independently from the reference either. The establishment of specifications based on the characterization is inevitably influenced by the comparability exercise. Indeed, testing of the reference product during the comparability exercise and the data thereby obtained are needed for determining the quality attributes of the product, which, in turn, will assist in setting the biosimilar product specification. Generally speaking, the limits set for the specifications of a biosimilar should not be wider than the range of variability of the reference product.

Finally, preclinical and clinical testing is of critical importance in establishing the specification and the acceptance criteria. This is particularly true for setting the potency limits which establish the range within which the expected clinical outcome occurs. The setting of specifications is a multistage process of adjustment of clinical batches' manufacturing specifications (±3 SD tolerance interval, the so-called natural tolerance interval) to the clinical limits determined by clinical trials. The lower clinical limit (the minimal effective dose) is determined in efficacy studies, and the upper clinical limit (the maximal tolerable dose) is determined in safety studies. Clinical limits encompass a range called the therapeutic window, which is the basis for calculating the therapeutic index. In the various phases of the preclinical and clinical studies, both the minimal effective dose and the maximal tolerable dose are refined, taking into account populations, age group, sex, drug interactions, and other factors. Often, clinical limits do not overlap the manufacturing limits, and the latter are narrowed on one or both sides, taking into account the process capacity index, which measures the overlap of clinical limits with the manufacturing ones, and employing various calibration techniques such as precision-to-tolerance ratio and Gage R&R, which are taking into account the variation due to the measurement error. This is where the biosimilars mostly display their peculiarity with regard to the setting of specification. Indeed, preclinical and clinical development is reduced for this type of product. A high level of similarity demonstrated at the quality level could possibly lead to skipping the efficacy clinical study, as already suggested for biosimilar products containing recombinant human insulin and insulin analogs (EMA, 2015b), thereby leaving open the question of the potency limits and their clinical validation.

The final question is thus whether the potency limits of a biosimilar could be deduced from the potency range of its reference, without further clinical validation. Assuming that the range of the potency variability of the biosimilar is not wider than the reference, the answer to this question is probably positive, notwithstanding the EMA guideline wording which warns that *acceptable ranges used for the biosimilar comparability exercise versus the reference medicinal product should be handled separately from release specifications* (EMA, 2014a). In the authors' opinion, the biosimilar potency limits could be based mainly on the ±3 SD tolerance interval, the so-called natural tolerance interval, determined on a sufficient number of batches—usually 30 or more depending on the variability. The biosimilar manufacturer could initially rely on the natural tolerance interval of the originator and then add its own data as they are gathered and recalculate the acceptance limits with each new biosimilar lot. A system of statistical monitoring aimed at detecting any putative trend and out-of-specification results should be set up. Any systematic deviation from these specifications or high number of out-of-specification results indicate a need for adjustment after conducting additional clinical studies sponsored by the biosimilar manufacturer.

5.2.2.7 Stability

Since minor changes in a manufacturing process may affect the stability profile of a product, the stability claims of biosimilars cannot be extrapolated from their respective reference products. Instead, as is true of any medicinal product, the shelf life should be supported by real-time/real-conditions stability data, in compliance with the principles outlined in Q5C Quality of Biotechnological Products: Stability Testing of Biotechnological/Biological Products and Q1A(R2) Stability Testing of New Drug Substances and Products (ICH, 1995, 2003). As a consequence, the shelf life of a biosimilar is not necessarily the same as that for the reference product. Whereas some guidelines prescribe undertaking comparative real-time stability studies on both the biosimilar and the reference, with the aim of detecting possible subtle differences that went unnoticed during the characterization studies (Health Canada, 2010), others don't (EMA, 2014a).

Yet all guidelines recognize the need to conduct accelerated stability and forced degradation studies not only on the biosimilar but also on the reference. In particular, forced degradation or stress conditions studies are standardly conducted on the medicinal products and their drug substance in order to reveal the pattern of degradation, to identify the likely degradation products, and, as a corollary, to gain insight into the analytical procedures that offer the best stability indicating potential (ICH, 1995, 2003). In the particular case of biosimilars, conducting comparative accelerated stability and stress conditions studies could reveal otherwise hidden properties of a product that warrant additional evaluation and/or controls.

5.3 APPARENT INCONSISTENCIES

The quality matter of biosimilars shows some singular features that merit emphasis. For instance, the complexity of biological medicinal products derives primarily from their drug substance. Accordingly, the emphasis of the quality comparability exercise is predominantly placed on the drug substance. However, it should be kept in mind that biosimilars are similar medicinal products, and not similar drug substances. Therefore, at the end of the day, biosimilarity is to be demonstrated at the level of the drug product. Guidelines are unambiguous in this regard. Considering the medicinal product as a whole in the comparability exercise makes sense at all levels, and the emphasis given to the drug substance may hide the consideration that the drug product deserves. For instance, safety issues related to inappropriate formulation were reported. The increased incidence of pure red cell aplasia linked to a change in the formulation of a recombinant human erythropoietin (EPO) is a notorious example. To comply with a new European regulation, human serum albumin used as an excipient was replaced by polysorbate 80. Through the action of polysorbate 80, organic compounds were leached from rubber stoppers, which in turn were eliciting a deleterious immune response to EPO (Sharma et al., 2004; Boven et al., 2005).

In principle, the biosimilar concept applies to all biological drug submissions. In practice, similarity is demonstrable only on products that contain well-characterized drug substances, typically recombinant proteins. For this reason, drug substances that are more difficult to characterize, such as those sourced from organs, tissues, or fluids are less prone to, not to say excluded from, biosimilar applications. The same perception prevails for vaccines that are usually excluded from the scope of biosimilar guidelines. Whereas complex vaccines such as those including conjugate polysaccharides would indeed be difficult to characterize, recombinant protein vaccines could be easier to copy. By the way, the Japanese guideline differs from other guidelines since well-characterized recombinant vaccines can possibly be developed as follow-on biologics (Yamaguchi and Arato, 2011). As for the live attenuated vaccine, although this view is not universally shared, the authors have no objection in principle to their possible development under the biosimilar approach. Such applications would obviously require clinical studies, but these might be significantly lighter than the full clinical development of the vaccines and could possibly be based on surrogate markers, making this an attractive approach.

More than a decade may elapse between the application for an originator and for its copy. In the meantime, the technology will have changed with the emergence of possible new production processes and more powerful analytical methods. This gives rise to biosimilar products that are more effectively controlled, display a higher purity and show a better lot-to-lot consistency. In this context, paradoxically, a lesser similarity of biosimilars to their reference would merely be the result of a better quality, in stark contrast to their negative image of wrongly alleged second-grade products (Schneider, 2013).

The improvement of physicochemical and biological analytical methods is also pushing forward the biosimilar paradigm. State-of-the-art, more powerful methods allow detection of subtle differences in the quality profiles of the biosimilar and the reference, and the greater the resolution of the analytical methods used, the more heterogeneity will be apparent (Kozlowski and Swann, 2006). The former question, "*are the methods sensitive enough?*" is now changing to "*what do differences mean?*" (quoted from Schneider, 2011). The most straightforward way to answer this latter question would be to conduct additional animal and/or human testing, while the biosimilar pathway is aimed primarily at reducing the burden of the nonclinical and clinical development. In addition, the clinical measurements have their own relatively large variability. As a consequence, it is not unlikely that the power of standard scale clinical trials will not be sufficient for revealing the clinical impact (if any) of slight quality differences. Finally, assuming that there is a minor, hardly detectable but established impact of quality differences on the clinical performance of a biosimilar, do these subtle differences really matter? Where knowledge acquisition through clinical studies is neither technically, ethically, nor economically feasible, instead of relying on the premarketing clinical development, experience could be gleaned from the postmarketing pharmacovigilance which thus appears of particular significance for biosimilars.

Finally, and it is not the least odd feature of biosimilars, demonstration of biosimilarity against the reference product is a "one-shot" exercise to be achieved prior to submitting a marketing authorization application. There is no regulatory requirement to repeat this demonstration once the marketing authorization has been granted. However, both the biosimilar and the reference could evolve during their commercial life, and over time their respective quality profiles could diverge from each other. The Boy Scouts' motto "Once a scout, always a scout," is thus not applicable to the biosimilar. "Once a biosimilar—not necessarily always a biosimilar." This may have far-reaching implications for the highly topical questions of substitution and interchangeability.

5.4 CONCLUSION

As presented in this chapter, the quality development, particularly the characterization and quality comparability exercise, appear to be the starting point for the stepwise development of biosimilars. Quality considerations and, in particular, the extent of observed differences between biosimilars and their reference product will be decisive in determining the nonclinical and clinical program.

The complexity of biologics ranges over a broad spectrum with, at one end, medicinal products containing recombinant proteins of little complexity such as insulin and insulin analogs. As the analytical technologies are improving, these products are getting closer to the small-molecule drug standards (Chirino and Mire-Sluis, 2004). Accordingly, these well-characterized biologicals are becoming more prone to a generic-like development. Thus, there is no anticipated need for efficacy studies for biosimilars including these active ingredients (EMA, 2015b). At the other end of the spectrum lie extremely complex medicinal products such as the advanced therapy medicinal products, which are currently far beyond the reach of biosimilar development. Between these two extremes is a range of biopharmaceuticals for which the biosimilar pathway, initially introduced in 2003 (EC, 2003), has been specially tailored. Thus, this approach is gradually coming into maturity. Not all related concepts are fully mature, and a number of acknowledged or alleged inconsistencies still remain. In spite of these inconsistencies, the procedure has on several occasions proved successful in making top-quality medicinal products available to clinicians and their patients.

REFERENCES

Boven K, Knight J, Bader F, et al. (2005) Epoetin-associated pure red cell aplasia in patients with chronic kidney disease: solving the mystery. *Nephrology Dialysis Transplantation* **20(Suppl. 3)**, iii33–iii40. Available from: http://ndt.oxfordjournals.org/content/20/suppl_3/iii33.full.pdf (Accessed June 14, 2015).

Chirino AJ, Mire-Sluis A. (2004) Characterizing biological products and assessing comparability following manufacturing changes. *Nature Biotechnology* **22(11)**, 1383–1391.

Chou HH, Takematsu H, Diaz S, et al. (1998) A mutation in human CMP-sialic acid hydroxylase occurred after the *Homo-Pan* divergence. *Proceedings of the National Academy of Sciences of the USA* **95**, 11751–11756. Available from: http://www.ncbi.nlm.nih.gov/pmc/articles/PMC21712/pdf/pq011751.pdf (Accessed June 14, 2015).

Declerck P, Farouk-Rezk M, Rudd PM. (2016) Biosimilarity versus manufacturing change: two distinct concepts. *Pharmaceutical Research* **33(2)**, 261–268.

Derbyshire M. (2013) Biosimilar development and regulation in Japan. *Generics and Biosimilars Initiative Journal* **2(4)**, 207–208. Available from: http://gabi-journal.net/biosimilar-development-and-regulation-in-japan.html (Accessed June 14, 2015).

EC. (2003) Directive 2003/63/EC of the Commission of 25 June 2003 amending Directive 2001/83/EC of the European Parliament and of the Council on the Community code relating to medicinal products for human use. *Official Journal of the European Union 2003 L* **159**, 46–94. Available from: http://eur-lex.europa.eu/legal-content/EN/TXT/PDF/?uri=CELEX:32003L0063&rid=3 (Accessed June 13, 2015).

EC. (2004) Directive 2004/27/EC of the European Parliament and of the Council of 31 March 2004 amending Directive 2001/83/EC on the Community code relating to medicinal products for human use. *Official Journal of the European Union 2004 L* **136**, 34–57. Available from: http://eur-lex.europa.eu/legal-content/EN/TXT/PDF/?uri=CELEX:32004L0027&rid=3 (Accessed June 13, 2015).

EC. (2012) Directive 2001/83/EC of the European Parliament and of the Council of 6 November 2001 on the Community code relating to medicinal products for human use (consolidated text). Available from: http://eur-lex.europa.eu/legal-content/EN/TXT/PDF/?uri=CELEX:02001L0083–20121116&rid=1 (Accessed June 15, 2015).

EMA. (2012) European public assessment reports (EPAR)—scientific discussion for Valtropin (no longer authorized). Available from: http://www.ema.europa.eu/docs/en_GB/document_library/EPAR_Scientific_Discussion/human/000602/WC500047158.pdf (Accessed June 14, 2015).

EMA. (2013a) EMEA/CHMP/BMWP/118264/2007. Guideline on non-clinical and clinical development of similar biological medicinal products containing low molecular-weight-heparins (draft). Available from: http://www.ema.europa.eu/docs/en_GB/document_library/Scientific_guideline/2013/01/WC500138309.pdf (Accessed June 14, 2015).

EMA. (2013b) EMA/297149/2013. Concept paper on the need for a reflection paper on statistical methodology for the comparative assessment of quality attributes in drug development (revision 1). Available from: http://www.ema.europa.eu/docs/en_GB/document_library/Scientific_guideline/2013/06/WC500144945.pdf (Accessed June 14, 2015).

EMA. (2014a) EMA/CHMP/BWP/247713/2012. Guideline on similar biological medicinal products containing biotechnology-derived proteins as active substance: quality issues (revision 1). Available from: http://www.ema.europa.eu/docs/en_GB/document_library/Scientific_guideline/2014/06/WC500167838.pdf (Accessed May 19, 2015).

EMA. (2014b) electronic Medicines Compendium (eMC). Summary of product characteristics (SmPC) of Humatrope. Available from: https://www.medicines.org.uk/emc/medicine/601 (Accessed June 14, 2015).

EMA. (2015a) CHMP/437/04 Rev 1. Guideline on similar biological medicinal products. Available from: http://www.ema.europa.eu/docs/en_GB/document_library/Scientific_guideline/2014/10/WC500176768.pdf (Accessed June 3, 2015).

EMA. (2015b) EMEA/CHMP/BMWP/32775/2005_Rev. 1. Guideline on non-clinical and clinical development of similar biological medicinal products containing recombinant human insulin and insulin analogues (revision 1). Available from: http://www.ema.europa.eu/docs/en_GB/document_library/Scientific_guideline/2015/03/WC500184161.pdf (Accessed June 10, 2015).

FDA. (2015a) Guidance for industry: scientific considerations in demonstrating biosimilarity to a reference product. Available from: http://www.fda.gov/downloads/Drugs/GuidanceComplianceRegulatoryInformation/Guidances/UCM291128.pdf (Accessed May 19, 2015).

FDA. (2015b) Guidance for industry: quality considerations in demonstrating biosimilarity of a therapeutic protein product to a reference product. Available from: http://www.fda.gov/downloads/Drugs/GuidanceComplianceRegulatoryInformation/Guidances/UCM291134.pdf (Accessed May 21, 2015).

GaBI. (2014a) Biosimilars in emerging markets. *Generics and Biosimilars Initiative Journal.* Available from: http://www.gabionline.net/Reports/Biosimilars-in-emerging-markets (Accessed June 13, 2015).

GaBI. (2014b) Japanese guidelines for biosimilars. *Generics and Biosimilars Initiative Journal.* Available from: http://www.gabionline.net/Guidelines/Japanese-guidelines-for-biosimilars (Accessed June 14, 2015).

GaBI. (2015) South Korean guidelines for biosimilars. *Generics and Biosimilars Initiative Journal.* Available from: http://www.gabionline.net/Guidelines/South-Korean-guidelines-for-biosimilars (Accessed June 14, 2015).

Health Canada. (2010) Guidance for sponsors: information and submission requirements for subsequent entry biologics (SEBs). Available from: http://www.hc-sc.gc.ca/dhp-mps/alt_formats/pdf/brgtherap/applic-demande/guides/seb-pbu/seb-pbu-2010-eng.pdf (Accessed May 18, 2015).

ICH. (1995) ICH harmonised tripartite guideline. Quality of biotechnological products: stability testing of biotechnological/biological products Q5C. Available from: http://www.ich.org/fileadmin/Public_Web_Site/ICH_Products/Guidelines/Quality/Q5C/Step4/Q5C_Guideline.pdf (Accessed June 14, 2015).

ICH. (1999) ICH harmonised tripartite guideline. Specifications: test procedures and acceptance criteria for biotechnological/biological products Q6B. Available from: http://www.ich.org/fileadmin/Public_Web_Site/ICH_Products/Guidelines/Quality/Q6B/Step4/Q6B_Guideline.pdf (Accessed June 14, 2015).

ICH. (2003) ICH harmonised tripartite guideline. Stability testing of new drug substances and product Q1A(R2). Available from: http://www.ich.org/fileadmin/Public_Web_Site/ICH_Products/Guidelines/Quality/Q1A_R2/Step4/Q1A_R2_Guideline.pdf (Accessed June 14, 2015).

ICH. (2004) ICH harmonised tripartite guideline. Comparability of biotechnological/biological products subject to changes in their manufacturing process Q5E. Available from: http://www.ich.org/fileadmin/Public_Web_Site/ICH_Products/Guidelines/Quality/Q5E/Step4/Q5E_Guideline.pdf (Accessed June 14, 2015).

ICH. (2005a) ICH harmonised tripartite guideline. Validation of analytical procedures: text and methodology Q2(R1). Available from: http://www.ich.org/fileadmin/Public_Web_Site/ICH_Products/Guidelines/Quality/Q2_R1/Step4/Q2_R1_Guideline.pdf (Accessed June 14, 2015).

ICH. (2005b) ICH harmonised tripartite guideline. Quality risk management Q9. Available from: http://www.ich.org/fileadmin/Public_Web_Site/ICH_Products/Guidelines/Quality/Q9/Step4/Q9_Guideline.pdf (Accessed June 14, 2015).

ICH. (2009) ICH harmonised tripartite guideline. Pharmaceutical development Q8(R2). Available from: http://www.ich.org/fileadmin/Public_Web_Site/ICH_Products/Guidelines/Quality/Q8_R1/Step4/Q8_R2_Guideline.pdf (Accessed June 14, 2015).

Kozlowski S, Swann P. (2006) Current and future issues in the manufacturing and development of monoclonal antibodies. *Advanced Drug Delivery Reviews* **58**, 707–722.

Kresse GB. (2009) Workshop on biosimilar monoclonal antibodies. Session 1 CMC Innovator Industry Presentation. London, July 2, 2009. Available from: http://www.ema.europa.eu/docs/en_GB/document_library/Presentation/2009/11/WC500008474.pdf (Accessed June 14, 2015).

Li H, d'Anjou M. (2009) Pharmacological significance of glycosylation in therapeutic proteins. *Current Opinion in Biotechnology* **20(6)**, 678–684.

Mintz C. (2013) Biosimilars in emerging markets. *BioProcess Online* Available from: http://www.bioprocessonline.com/doc/biosimilars-in-emerging-markets-0001 (Accessed June 13, 2015).

Pandya N. (2014) Regulation of subsequent entry biologics. In: *Fundamentals of Canadian Regulatory Affairs Fourth Edition (e-Book)* RAPS, ed., 147–158. Available from: http://www.raps.org/uploadedFiles/PDF_Assets/Canadian%20Fundamentals,%20Ch.%2019.pdf (Accessed May 18, 2015).

Schiestl M. (2009) EMEA Workshop on biosimilar monoclonal antibodies. EGA-EBG's quality perspective on biosimilar mAbs. London, July 2, 2009. Available from: http://www.ema.europa.eu/docs/en_GB/document_library/Presentation/2009/11/WC500008475.pdf (Accessed June 4, 2015).

Schneider CK. (2011) Perspectives: challenges with biosimilars. In: *Les biosimilaires*, Prugnaud JL and Trouvin JH, eds., 115. Springer-Verlag, France, Paris.

Schneider CK. (2013) Biosimilars in rheumatology: the wind of change. *Annals of the Rheumatic Diseases* **72**, 315–318. Available from: http://ard.bmj.com/content/72/3/315.full (Accessed June 14, 2015).

Schuirmann DJ. (1981) On hypothesis testing to determine if the mean of a normal distribution is continued in a known interval. *Biometrics* **37**, 617.

Sharma B, Bader F, Templeman T, et al. (2004) Technical investigations into the cause of the increased incidence of antibody-mediated pure red cell aplasia associated with Eprex. *European Journal of Hospital Pharmacy* **5**, 86–91.

Solá RJ, Griebenow K. (2010) Glycosylation of therapeutic proteins: an effective strategy to optimize efficacy. *BioDrugs* **24**, 9–21. Available from: http://www.ncbi.nlm.nih.gov/pmc/articles/PMC2805475/ (Accessed June 14, 2015).

Trouvin JH. (2011) Caractéristiques des biosimilaires. In: *Les biosimilaires*, Prugnaud JL and Trouvin JH, eds., 1–26. Springer-Verlag France, Paris.

USC. (2010) Biologics Price Competition and Innovation Act of 2009. Section 7001 *ff.* Available from: http://www.gpo.gov/fdsys/pkg/BILLS-111hr3590enr/pdf/BILLS-111hr3590enr.pdf (Accessed June 13, 2015).

Vugmeyster Y, Xu X, Theil FP, et al. (2012) Pharmacokinetics and toxicology of therapeutic proteins: Advances and challenges. *World Journal of Biological Chemistry* **3(4)**, 73–92. Available from: http://www.ncbi.nlm.nih.gov/pmc/articles/PMC3342576/ (Accessed June 15, 2015).

WHO. (2013) Guidelines on evaluation of similar biotherapeutic products (SBPs), Annex 2, Technical Report Series No. 977. Available from: http://www.who.int/biologicals/publications/trs/areas/biological_therapeutics/TRS_977_Annex_2.pdf?ua=1 (Accessed May 18, 2015).

Yamaguchi T, Arato T. (2011) Quality, safety and efficacy of follow-on biologics in Japan. *Biologicals* **39**, 328–332. Available from: http://www.sciencedirect.com/science/article/pii/S1045105611000807 (Accessed June 16, 2015).

6 Nonclinical Studies for Biosimilars

Karen De Smet
Federal Agency for Medicines and Health Products

Leon AGJM van Aerts
Medicines Evaluation Board

CONTENTS

6.1 INTRODUCTION

The current EMA, FDA, and WHO guidelines for biosimilar development recommend a scientifically rigorous stepwise process that is different from that for generic small-molecule drugs.

This chapter highlights the key recommendations included in the guidances for biosimilar development, focusing on those related to nonclinical development.

Contrary to applications for generic medicinal products, animal studies have traditionally been requested for biosimilars from early on.

While reviewing historical and current regulatory thinking on nonclinical testing of biosimilars and focusing on the evolution in the recommendations in the European Union (EU) guidelines, we aim to clarify the shift in paradigm on the need for *in vivo* testing.

6.2 EUROPE HAS BEEN SETTING THE SCENE

In 2005, the EU was the first region to set up a legal framework and regulatory path for biosimilar development (EC, 2001). The most critical feature of biosimilar development, as explained in the very first European guidance documents (EMA, 2005, 2006), is that it should be demonstrated that the product is (highly) similar to the reference product in terms of quality, safety, and efficacy. It is not to demonstrate *de novo* nonclinical or clinical safety or efficacy because there is already established knowledge from the approved product. Rather, it must be demonstrated that potential subtle differences in quality attributes between the products do not result in clinically relevant differences in safety or efficacy. The importance of the comparability at the level of the pharmaceutical quality, together with the limitations to the extent to which biosimilars could be characterized using physicochemical methods, was already emphasized in the early guidance documents. The final product is often a complex mixture of closely related large-sized molecules, which makes it difficult to provide a complete characterization. The complexity of the final product can be influenced by various factors such as protein aggregation and glycosylation. Today, the methods used to detect differences in quality attributes between products have advanced to a considerable extent. New analytical methods are suitable to detect subtle structural and compositional differences between the biosimilar and the reference product. For a more detailed discussion, see the chapters in this book related to the analytical assessment of biosimilars.

Still, the clinical relevance of the observed subtle quality differences (i.e., the effect on efficacy and safety) is often not clear based on the analytical data alone and further nonclinical and clinical studies may be needed, as reflected in the overarching biosimilar guidelines (EMA, 2014a,b, 2015a). It can be anticipated that in the future control of the biotechnological processes and analytical abilities will have advanced to such a level that the need for additional nonclinical and clinical studies will be further diminished.

6.3 A HISTORICAL REQUEST FOR STANDARD *IN VIVO* STUDIES

Decisions on the marketing authorization of medicinal products, and to an even larger extent approval of clinical trials, are partly based on nonclinical data, including animal studies. Before initiating clinical development, nonclinical studies should be conducted. Nonclinical development of new biologicals requires *in vivo* toxicity studies in relevant animal species, as described in ICH guideline S6 (R1) (ICH, 2011). Typically, this consists of repeated dose toxicity, as well as studies regarding safety pharmacology and reproduction toxicology.

TABLE 6.1
Updates of Product-Specific EMA Biosimilar Guidelines

Product-Specific EMA Biosimilar Guidelines	Date of First Publication	Date of Revision
Recombinant human insulin	June 2006	September 2015, includes stepwise approach
Somatropin	June 2006	Revision is needed to include the stepwise approach
Recombinant granulocyte-colony stimulating factor	June 2006	Ongoing Concept May 2015, includes stepwise approach
Recombinant erythropoietins	July 2006	Revised in September 2010, but further revision is needed to include the stepwise approach
Recombinant interferon alpha	April 2009	Revision is planned to include the stepwise approach
Low molecular weight heparins	October 2009	Ongoing Draft January 2013, includes stepwise approach
Monoclonal antibodies	December 2012, includes stepwise approach	
Interferon beta	September 2013, includes stepwise approach	
Recombinant follicle stimulating hormone	September 2013, includes stepwise approach	

In the initial European overarching and product-specific guidance documents on biosimilars published between June 2006 and October 2009 [including recombinant human insulin, somatropin, recombinant granulocyte-colony stimulating factor, recombinant erythropoietins, recombinant interferon alpha, low-molecular-weight heparins (Table 6.1)], a nonclinical package for a biosimilar development was expected to consist of comparative *in vitro* bioassays and *in vivo* studies (when a relevant animal model is available), including a pharmacodynamic study and/or at least a repeated dose toxicology study, and in many cases local tolerance studies. However, as more knowledge and experience in the development of biosimilars were gained, and more specifically, when the field was widened to biosimilar monoclonal antibodies (mAbs), it became clear that this standard approach was not appropriate.

6.4 THE THREE Rs PRINCIPLE

Since the introduction of the regulatory need to study medicinal products in animals, there has been a continuous debate on the scientific basis of the predictability of animal studies—in other words its translational value for the pharmacological and toxicological effects of medicinal products in humans and how and when animal studies should be conducted (van Meer et al., 2015b).

In EU legislation, a new directive on the protection of animals used for scientific purposes was issued in 2010 (EU, 2010), which updates and replaces the 1986 Directive 86/609/EEC (EEC, 1986). This new directive took full effect on January 1, 2013, and aims to anchor the principle of the "Three Rs" to Replace, Reduce, and Refine the use of animals.

In the EU, the use of animals for scientific or educational purposes should only be considered when a nonanimal alternative is unavailable (preamble 12). Member states need to ensure that, wherever possible, a scientifically satisfactory method or testing strategy, not entailing the use of live animals, is used instead (Article 4.1). Moreover, according to the directive, nonhuman primates (NHPs) are exempted from use in animal studies whenever possible. This is reflected in Article 8.1(b) as there should be scientific justification that the purpose of the procedure (animal study) cannot be achieved by the use of species other than NHPs.

6.5 THE EU BIOSIMILAR mAbs GUIDELINE

At the same time that the EU legislation on the protection of animals used for scientific purposes was being finalized, the EMA Working Party on Similar Biological Medicinal Products (BMWP) was drafting a guideline on the nonclinical and clinical issues of the development of biosimilar mAbs (EMA, 2012). The EMA has organized several workshops to discuss the feasibility of the development of biosimilar mAbs with the scientific community and pharmaceutical companies. With respect to nonclinical development, the debate was focusing on the need of animal studies in the comparability exercise and more specifically, on the justification of using nonhuman primates as laboratory animals.

As specified in ICH guideline S6 (R1), safety evaluation programs should include the use of relevant species (ICH, 2011). In the case of monoclonal antibodies, a relevant species is one in which the test material is pharmacologically active due to the expression of an epitope. Monoclonal antibodies are biologicals showing high species and target specificity, and mostly only animal species closely related to humans are pharmacologically responsive to the investigational medicinal product. For many mAbs, the nonhuman primate is often the only relevant species, and therefore, pharmacology and toxicity studies are often performed in the cynomolgus macaque (*Macaca fascicularis*) (Chapman et al., 2012). When comparative nonclinical studies have to be powered to detect small differences in efficacy and safety, these studies must be of considerable size to reach this goal. Studies in NHPs (and nonrodents in general) have small group sizes, and interindividual variability further reduces the sensitivity of these studies to detect differences in pharmacological response.

The usually small group size in animal studies (especially when nonrodents are being used) limits the sensitivity of these studies to detect relevant differences in safety and efficacy.

Another approach to provide safety data for some mAbs may be the use of animal models of disease, in which the pharmacological activity of a pharmaceutical can be shown and which can be used to demonstrate pharmacodynamics activity. Examples are the use of SCID mice with xenotransplants of tumors for oncology products, or transgenic mouse models, such as Tg197 carrying a modified human

TNF gene construct, for products used in rheumatoid arthritis (Keffer et al., 1991). The outcome of a comparability exercise comparing the pharmacological activity of a biosimilar and a reference product in such models depends on the robustness of the model. Variable growth of xenotransplants and semiquantitative scoring of pathological features would, however, decrease the usefulness of these models to detect differences in biological activity (van Aerts et al., 2014).

It is always important to understand the limitations of animal studies when interpreting the results. In the case of animal studies for biosimilar mAbs development, these limitations include, next to target specificity, small sample size and intraspecies variations. If the proposed biosimilar has already shown a high degree of *in vitro* similarity, it is recognized that animal studies in standard species or in animal models of disease rarely provide additional information needed for a decision. This limitation has led to the strategy proposed in the guideline on the nonclinical and clinical issues of the development of biosimilar mAbs: the major decisions on similarity should be based on similarity of quality attributes and comparable performance in *in vitro* studies. This is considered the foundation of biosimilar mAbs development.

Where in previous guideline recommendations on biosimilars a nonclinical package was expected to consist of comparative studies, including a pharmacodynamic study (bioassay) and a repeated dose toxicology study, a new paradigm emerged for biosimilar mAbs in which the use of animals was obviated in most cases by a thorough stepwise approach of testing. Moreover, the conduct of large comparative toxicological studies in NHPs is unacceptable from both a scientific and an ethical point of view.

The EU biosimilar mAbs guideline (EMA, 2012) indicates that relevant assays on the binding of the complementarity determining region (CDR) to its primary target and the binding to representative isoforms of the relevant three Fc gamma receptors (FcγRI, FcγRII, and FcγRIII), FcRn and complement (C1q) need to be performed. Since both the Fab as the Fc portion of the molecule may elicit several effector functions, both need to be evaluated, even though some may not be considered essential for the therapeutic mode of action. Notably, Fab-associated functions like neutralization of a soluble ligand, receptor activation or blockade, and Fc-associated functions such as antibody-dependent cell-mediated cytotoxicity (ADCC), complement-dependent cytotoxicity (CDC), and complement activation need to be evaluated. As these assays are considered paramount in the nonclinical comparability exercise, the qualification of these assays is of utmost importance. Variability due to assay format and reagents utilized needs to be minimized.

6.6 A PARADIGM SHIFT IN EUROPE

The discussions on the relevance of *in vivo* animal testing for biosimilar mAbs were soon broadened to other biologicals, since target specificity is not a property that is specific for mAbs only. Target binding, such as to a receptor, ligand, or substrate, and the resulting functional effects can be evaluated *in vitro* for most biological medicinal products. If there is a high degree of confidence that the *in vitro* functional assays are reflective of the mode of action of the biological, these assays could be of greater value in demonstrating similarity. This may be the case for biosimilars, since there is experience in correlation between *in vitro* and *in vivo* data from the reference medicinal product. Using cellular systems, we can make more extensive and precise comparisons in *in vitro* assays, and thus these could have greater sensitivity than comparative animal *in vivo* studies.

Potential issues may arise when differences are detected between the biosimilar and the reference product in *in vitro* assays, of which the impact of those differences on *in vivo* safety and efficacy in humans is uncertain. In case of subtle differences, additional *in vitro* assays or an assessment of human clinical data may provide an answer to the question on the clinical relevance of the observed differences, while data from nonclinical *in vivo* studies are unlikely to provide a definitive answer. It seems that the contribution of animal studies to the totality of evidence for establishing biosimilarity in such cases is rather limited. Only when the scientific question cannot be answered by *in vitro* studies should *in vivo* studies be considered.

Also, in case of substantial differences in physicochemical and biological characteristics, it is not acceptable to overrule these data by less robust data from *in vivo* studies, either from animal studies or from clinical trials.

Some of the initial product-specific biosimilar guidelines have been revised since the publication of the biosimilar mAbs guideline (EMA, 2012), and recently the overarching guideline on nonclinical and clinical issues of biosimilars containing biotechnology-derived proteins as active substances has been revised giving similar recommendations (EMA, 2015a). The basic concept is a three-step tiered approach to regulatory authorization and approval. However, before considering nonclinical development, the biosimilar comparability exercise for the physicochemical and biological characteristics needs to be considered satisfactory.

Step one of the nonclinical development process includes *in vitro* methods such as binding studies and signal transduction and functional activity/viability of cells known to be of relevance for the pharmaco-toxicological effects of the reference product.

For some of these assays, there is overlap with quality-related assays to identify the biological characteristics. The limitations, sensitivity, and specificity of the functional assays need to be considered when assessing the extent of additional animal and clinical data required to demonstrate biosimilarity. An important point that is often neglected during the early development phases is that the assays should compare an appropriate number of batches of the reference product procured over a period of time and of the biosimilar representative of the material intended for clinical use. A factor to be taken into account is the available quantity of test material, and the number of available batches of adequate quality may be limited.

The guideline indicates that these *in vitro* assays can be considered paramount for the nonclinical biosimilar comparability exercise, since *in vitro* assays may often be more specific and sensitive to detect differences between the biosimilar and the reference product than studies in animals (or humans).

Step two is the evaluation of the need for *in vivo* nonclinical studies, as indicated in Table 6.2. If no specific concerns are raised and if the quality and nonclinical *in vitro* studies (see Step 1) are considered satisfactory, an *in vivo* animal study is usually not considered necessary. For those biosimilars, approval for clinical testing and marketing authorization could occur with no *in vivo* animal data.

TABLE 6.2
Evaluation of the Need for *In Vivo* Nonclinical Studies (EMA, 2014b, 2015a)

Factors To Be Considered When the Need for *In Vivo* Nonclinical Studies Is Evaluated:

- Presence of potentially relevant quality attributes that have not been detected in the reference product
  - New posttranslational modification structures—atypical glycosylation structures or variants not observed in the reference medicinal product, with particular attention to nonhuman structures (nonhuman linkages, sequences, or sugars)[a]
- Presence of potentially relevant quantitative differences in quality attributes between the biosimilar and the reference product
  - Quantitative differences in posttranslational modification structures (e.g., glycosylation, oxidation, deamidation, truncation)
- Relevant differences in formulation
  - Use of excipients not widely used for biotechnology-derived proteins

[a] Introduction of novel structures, especially when nonhuman, may preclude biosimilarity due to unresolved residual uncertainties on safety.

Step three describes the 3Rs approach to *in vivo* nonclinical studies, reminding us that the type of studies to be performed needs to be critically weighed against the type of additional information needed (PK and/or PD and/or safety issues, with appropriate justification of study design). Testing at one dose level, in one sex, and the omission of the recovery group should be considered wherever possible. The conduct of repeat-dose studies in NHPs is usually not recommended, and studies regarding safety pharmacology, reproduction toxicology, and carcinogenicity are not required for nonclinical testing of biosimilars.

van Meer et al. (2015a) nicely summarized the new European approach to *in vivo* animal studies for biosimilars as “animal studies in the new European Union biosimilar guidance: No longer ‘yes, but’ but ‘no, unless.’”

6.7 WORLDWIDE INTERPRETATION OF EU GUIDELINES/GLOBAL DEVELOPMENT

Currently, there are guidance documents on the development of biosimilars from 26 countries or regions (Krishnan et al., 2015). Following the 2006 EMA guidelines, in 2009 the World Health Organization (WHO) issued a biosimilar guideline and includes a similar approach as initially was done in the EU, that is, requesting animal testing for the nonclinical development step (WHO, 2009). The WHO guidance was accepted by many countries and has been translated to national guidelines over the world, and *in vivo* animal studies are considered mandatory (e.g., Brazil, India, South Korea).

The need to revise the WHO guideline on biosimilars as a consequence of the new European paradigm on requirements has been highlighted by European regulators. Discussions on an international level as to whether a revision of the WHO guideline on biosimilars is feasible are currently ongoing.

In some countries, even more extensive animal studies are demanded. The Chinese Food and Drug Administration published a guideline requesting manufacturers to perform a toxicity comparability study next to *in vivo* comparability testing for pharmacodynamics, pharmacokinetics, and immunogenicity, if the results indicated differences that might be related to safety (Chinese FDA, 2014).

The Japanese guideline for follow-on biologics requires extensive animal testing, such as that expected for an new biological, and encourages manufacturers to follow the ICH S6 guideline, albeit including the statement that “where the similarity of bioactivity between a follow-on biologic and the original biologic is fully evaluated by in vitro comparability studies, in vivo comparative studies of pharmacodynamics may not be necessary. However, useful information may often be obtained through in vivo pharmacological studies conducted at the stage prior to clinical study” (Japanese PMDA, 2009). The default request for *in vivo* toxicity studies has been recently revised in a new Q&A on the Japanese guideline on biosimilars published by PMDA (Japanese PMDA, 2015). Q&A 6 indicates that in cases where there is no concern about nonclinical safety based on characterization studies and comparative comparison of the physicochemical and pharmacological properties, *in vivo* toxicity studies may not be required.

The FDA guidance on biosimilars was only recently finalized (April 2015) and is consistent with the revised EMA guidelines (FDA, 2015). It advocates a flexible approach rather than defaulting to 4-week comparative toxicology studies that have historically been conducted. A stepwise totality of evidence approach should be used in developing a biosimilar, and the first step should be extensive structural and functional testing. *In vivo* studies in animals and clinical studies should be undertaken to address any residual uncertainties. *In vivo* animal comparability studies may consist of PK/PD studies and/or toxicity studies, preferably in pharmacologically relevant species.

Based on the results of *in vitro* structural and functional assessments, before initiation of clinical studies in humans, the FDA requires animal toxicity data when unexplained analytical differences exist. However, animal toxicity data are only required when a nonclinical species is available showing pharmacological activity and if the results from these animal studies can meaningfully address the remaining uncertainties.

According to the FDA, comparative PK and systemic tolerability studies from a pharmacologically nonresponsive species (including rodents) may be useful in some instances. In the US guidance, it is not specifically stated that approval can be granted without *in vivo* studies. Nevertheless, in both the final FDA guidelines and the revised EMA guidelines, the emphasis is on structural and functional analytical data, and the *in vivo* nonclinical studies should only be used to address remaining concerns.

Although, in the past 10 years, a lot of effort has enabled the introduction of biosimilars globally, there exist major differences in regulation of biosimilars in different parts of the world. Companies developing products intended for global marketing may conduct *in vivo* animal studies to meet the requirements of the region that requires the most animal testing.

No specific ICH guideline covers biosimilars. However, a more global consideration of the 3Rs principles where the nonclinical development of biosimilars is concerned would be welcomed. However, there is currently no plan for development of an ICH guideline on biosimilars.

6.8 FREQUENT CONCERNS WITH RESPECT TO OMITTING *IN VIVO* STUDIES

In the publication by van Aerts et al. (2014), the regulatory strategy that helped to shape the new guidance to favor a risk-based approach that could lead to biosimilars entering the clinic without conducting any animal study is explained in detail. We refer to this publication for an explanation of the strategy, substantiated with numerous case examples. The following paragraphs review the most essential elements with respect to the frequent concerns to omitting *in vivo* studies.

6.8.1 Need to Establish Biological Activity in IU in Pharmacopoeial *In Vivo* Bioassays

Pharmacopoeial bioassays with potency determination in animals exist for some biologicals. Such assays are employed to express the biological activity of the product in international units (IUs). However, there may be considerable variability in these *in vivo* assays, which limits their use in a comparability exercise (Mulders et al., 1999; Zimmermann et al., 2011). Therefore, the results are not expected to contribute significantly to establishing biosimilarity.

For poorly characterized products where the active substance is extracted from a biological matrix, this approach to express biological activity in IU based on *in vivo* potency appears sensible, but these products may not be developed as biosimilars.

Whereas, for well-characterized products such as recombinant proteins, *in vitro* alternatives to the pharmacopoeial *in vivo* assays may be available, and the biological activity expressed in μg comparing the biosimilar and the reference product based on *in vitro* assays could be employed.

6.8.2 Establish Pharmacokinetics before Administration to Humans

Quality attributes, such as the presence of the binding target, whether this target is soluble or not, and the mechanism of clearance (e.g., receptor-complex mediated or not) may affect the PK behavior of a biological. In the case of mAbs, PK properties are also mediated by Fc-dependent mechanisms, and differences in glycosylation patterns of an mAb may influence Fc receptor binding. Although several functional aspects of the impact of a difference in glycosylation can be evaluated *in vitro* (as discussed in the paragraph in this chapter on product-related differences), the PK behavior cannot be covered by *in vitro* methods. However, comparative PK assessment in animals would require large numbers of animals, and both immunogenicity issues and species differences may limit the relevance of animal data with regard to PK in humans.

Although it is acknowledged that it may be possible to compare the PK properties of a biosimilar and a reference product in an animal study to some extent, these studies will always have their limitations. Many factors are important in determining pharmacokinetic bioequivalence in clinical trials, and even differences between healthy volunteers and patients may be relevant. Therefore, the PK properties of a biosimilar and the reference product should be established in plasma samples obtained from human volunteers or patients when there are no safety concerns.

Only when it is ethically difficult to study the compound in volunteers, which could result in the need to study suboptimal doses that are equally unacceptable in patients, there could be a reason to compare the PK behavior of biosimilar and reference product in animals first. This could be the case when a product with a poor safety profile shows a steep dose response curve. Since for mAbs it is known that rodents recognize the Fc fragment of human mAbs, within its limitations PK studies in a nontarget binding species could provide some information (Ober et al., 2001).

Also, when target tissue PK data are needed, nonclinical PK data could be useful, since this is difficult to obtain from humans. An example of a difference in target tissue distribution could be different levels of mannosylation of recombinant proteins used for replacement therapy where the differences in glycosylation may affect the uptake in target tissues (van Aerts et al., 2014). On the other hand, not all cases where traditional systemic PK studies in humans would not yield comparative exposure data (e.g., the intravitreal administration of ranibizumab) need a default *in vivo* PK assessment in animals. This need is dependent on the available data package and on whether there are remaining concerns with respect to differences in PK after extensive structural and functional assessments.

The FDA guideline indicates that a PK study in animals could be useful even in a nonpharmacologically relevant nonclinical species to compare the kinetics of non-target-mediated clearance of the biosimilar with the reference biological product (FDA, 2015).

6.8.3 Differences in Formulation

Although EMA guidance states that a biosimilar must have the same posology and route of administration as the reference biological product, some formulation changes (strength, pharmaceutical form, formulation, excipients, or presentation) might be permitted, provided the manufacturer of the biosimilar submits evidence demonstrating that the differences are without consequences for efficacy and safety.

Since a biosimilar product may contain excipients that are different from the reference product, it is conceivable that the *in vivo* PK behavior of the active substance is affected. For known excipients, *in vivo* studies are not generally necessary when they have already been used in humans at the same concentrations and via the same administration route and maintain the right osmolality and pH.

For less well-known substances, there may be a concern with respect to differences in PK behavior, but this can in principle be evaluated in humans, negating the need for animal studies in these cases. See the discussion in the paragraph above on the pharmacokinetics for cases where a nonclinical *in vivo* study would be justified to compare the PK behavior of biosimilar and reference product in animals first. Also, if excipients are introduced for which there is no or little experience with the intended clinical route of administration, local tolerance may need to be evaluated. If other *in vivo* studies are performed, evaluation of local tolerance may be part of the design of that study instead of the performance of separate local tolerance studies (EMA, 2015a).

When the formulation includes a completely new excipient, or when the intended route of administration is new for a known excipient, additional nonclinical data are needed in accordance with relevant guidance. In case insufficient information is available, toxicology studies may also be required (van Aerts et al., 2014).

6.8.4 Unexpected Toxicity

The value of extensive use of nonhuman primates in routine safety studies of mAbs has been questioned, as the toxicity of these products is characterized as usually related to on-target effects, so-called exaggerated pharmacology (van Meer et al., 2013). Most adverse events related to biologicals in general are linked to exaggerated pharmacology. For biosimilars, exaggerated pharmacology is a predictable type of toxicity that the innovator has established for the reference product. When the *in vitro* pharmacological activity of the biosimilar has been shown to be comparable, there is no need to confirm these properties in a less sensitive animal model. Moreover, given the clinical experience with the reference product, the clinical relevance of exaggerated pharmacology observed in animals is already known.

Another type of adverse event is one that is not related pharmacologically and in which variable terminology is used, such as unexpected toxicity, off-target toxicity, and nonspecific toxicity. As specified in ICH S6, this type of toxicity needs to be investigated for new biologicals. In rare cases, unexpected toxicity may be encountered in animal studies during the development of new biologicals. For biosimilars, however, off-target toxicity may be regarded as a predictable type of toxicity, since this has already been established for the reference product.

Nevertheless, some concern has been voiced about whether new off-target toxicities that have not been observed with the reference product could occur with a biosimilar. In the EMA workshop discussions on biosimilar development, no examples of off-target toxicity of biosimilars were presented in the EMA mAbs workshop (EMA, 2011). In addition, unexpected toxicity by postchange products was discussed, and it was stated that following changes to the production process of an already marketed biological, unexpected toxicity was never encountered in animal studies. The examples of unexpected toxicity presented were scarce and concerned rather new biologicals that were still in development (EMA, 2011). The well-known case of pure red cell aplasia (PRCA) following the use of inferior erythropoietin is often mentioned in connection with the biosimilar discussion. As explained later in this chapter, this example is not a reason to perform additional *in vivo* studies.

6.8.5 Process-Related Impurities

Safety concerns have also been expressed regarding different levels and types of process-related impurities occurring from different production processes used by biosimilar and reference product manufacturers.

Qualitative and quantitative differences of process-related impurities can occur from cell substrates (e.g., host cell DNA and host cell proteins), cell culture components (e.g., antibiotics and media components), and downstream processing steps (e.g., reagents, residual solvents, leachables, and endotoxins).

As for any biological, analytical procedures are implemented to detect, identify, and accurately quantify biologically significant levels of impurities, in accordance with ICH regulations for testing impurities. The impurity profile of the proposed biosimilar should be well understood prior to dosing humans and should be controlled at acceptable levels.

The biggest issue of concern remains the occurrence of host cell proteins that could trigger an immune response in patients or might cause a hypersensitivity reaction in sensitized individuals. Upon applying good manufacturing techniques, host cell proteins will be reduced as much as possible in the final product. As discussed in further detail below, *in vivo* animal studies do not add value in assessing immunogenicity and hypersensitivity reactions; thus, it is not meaningful to conduct animal studies to evaluate or qualify host cell proteins (van Aerts et al., 2014).

When there is a concern related to the presence of process-related impurities acting through TOLL-like receptors (TLRs), *in vitro* TLR assays may be able to detect and identify these impurities, while in animals the TLR-mediated immune response shows high species variability, thereby making animal models poor predictors of these infusion reactions in humans (Huang et al., 2009).

The biosimilar guidelines recommend that process-related impurities should be kept to a minimum, which is the best strategy to minimize any associated risk. It may be necessary to reevaluate the manufacturing processes and introduce additional purification steps. This strategy is also highlighted in the ICH S6 (R1) guideline of new biologicals, where it is preferable to rely on purification processes to remove impurities and contaminants rather than to establish a nonclinical testing program for their qualification (ICH, 2011).

6.8.6 Product-Related Substances

When a biosimilar is produced, the final composition of the product is likely to be slightly different from the reference product regarding the relative contribution of all the product-related substances. As part of the analytical characterization of a biosimilar, the presence and extent of posttranslational modifications should be appropriately characterized and compared to those of the reference product. Deamidated and oxidated forms may be present, and terminal amino acid truncations or modifications may occur. If present, carbohydrate structures should be thoroughly compared; including the overall glycan profile, site-specific glycosylation patterns, as well as site occupancy (EMA, 2014b).

Posttranslational changes in a biological molecule may affect its biological function. A change in a relevant part of the structure of the molecule is likely to lead to a change of biological activity, whereas minor variations in other parts are most likely not going to affect the functional properties of the molecule.

Qualitative or quantitative differences of product-related substances may affect the biological functions of the proposed biosimilar. It is therefore important to determine that the differences in composition of biosimilar and reference product are only minor and to establish that these minor differences do not affect safety and efficacy.

In the case of quantitative differences in protein variants that may have pharmacologic activity, an assessment of this activity is expected to be evaluated by appropriate sensitive binding affinity and functional *in vitro* assays rather than *in vivo* studies.

In IgG molecules, parts of the sugar moiety in the Fc region are known to contribute to the binding to Fc gamma receptors (FcγR) (Radaev and Sun, 2002).

A well-known example is that differences in the fucosylation of an mAb will impact the FcγRIIIa receptor binding. A higher degree of afucosylation leads to an increased binding affinity for the receptor, which may be reflected by an increased ADCC activity (Shields et al., 2002; Houde et al., 2010). However, depending on the mechanism of action, this difference may or may not be clinically meaningful. If the biosimilar is rituximab, where the mechanism of action of the drug is highly dependent on ADCC activity, this change is expected to have clinical consequences, while for mAbs where ADCC activity is not the central mechanism of action (e.g., etanacerpt), slight differences in ADCC activity may not be clinically relevant. For infliximab, the first mAb granted marketing authorization, differences in FcγRIIIa binding and ADCC in an NK cell assay were shown. Such differences were not considered relevant for the rheumatoid indications, but the relevance for inflammatory bowel disease (IBD) was discussed extensively. The company provided additional *in vitro* data to demonstrate that the observed differences were not measurable under more physiological conditions (e.g., in the presence of serum). In addition, it was demonstrated that induction of regulatory macrophages did not differ between biosimilar and reference products. These additional *in vitro* data convinced the European authorities that the observed differences in afucosylation were not relevant for all indications applied for and the IBD indication was accepted by extrapolation (Weise et al., 2014).

Qualitative differences of product-related substances may raise more concerns and will require appropriate justification. This may be the case for the presence of glycosylation structures or variants not observed in the reference medicinal product, especially nonhuman structures (nonhuman linkages, sequences, or sugars). These differences may have an effect on immunogenic potential and the potential to cause hypersensitivity. While glycosylation does not appear to play a major role, nonhuman or nonmammalian glycosylations within the product due to the expression system used can induce immune responses (Wadhwa et al., 2015).

A well-known example of qualitative differences in product-related substances is the presence of a glycan with a terminal galactose-α-1-3-galactose configuration in cetuximab because of the expression system used, causing hypersensitivity reactions in individuals with preexisting IgE antibodies against this structure. This immune response was not induced in naïve individuals (Chung et al., 2008). It is recognized that for structures for which it is not known if hypersensitivity is an issue, animal studies do not provide a predictive model to find out. Therefore, the best precaution to prevent safety issues in patients is to avoid the presence of glycoforms not observed in the reference medicinal product (van Aerts et al., 2014).

6.8.7 Immunogenicity

A systematic evaluation of formation of antidrug antibodies (ADAs) in patients is necessary for approval of all biological and biosimilar medicinal products.

Binding ADAs may affect the PK, and neutralizing antibodies can lead to diminished pharmacodynamic effects, compromising clinical efficacy.

Clinical safety may also be impacted in those situations when ADAs are cross-reactive with endogenous proteins. This may even constitute a serious safety issue, as is the case for pure red cell aplasia in chronic renal disease patients treated with erythropoietin, following induction of neutralizing antibodies that cross-reacted with the functionally nonredundant endogenous erythropoietin (Casadevall et al., 2002).

Nevertheless, in humans, there are also examples in which neutralizing antibodies appear to have no impact on clinical efficacy, pharmacodynamics, or adverse events (Wadhwa et al., 2015).

As animals will recognize human proteins or humanized biologicals as foreign proteins, they will produce ADAs, which may or may not be cross-reactive with the animal's endogenous molecule. When a biosimilar and a reference product induce different levels of ADA in an animal experiment, this experiment cannot be used as a reliable predictor of immunogenicity in humans. Therefore, the value of nonclinical studies for evaluation of immunogenicity in humans is considered low. Nonclinical studies aiming at predicting immunogenicity in humans are normally not required (EMA, 2015b).

Immunogenic potential can only be reliably assessed in clinical trials and should be further evaluated through postmarketing monitoring and considered in risk management and risk mitigation plans (Wadhwa et al., 2015).

Although immunogenicity assessment in animals is generally not predictive for immunogenicity in humans, it may be needed for interpretation of *in vivo* studies in animals. In those cases where animal studies are needed, blood samples should be taken and stored for future evaluations of pharmacokinetic/toxicokinetic data if needed (EMA, 2015a).

Due to the low predictivity for immunogenicty in humans, comparison of the antidrug antibody response to the biosimilar and the reference product in an animal model is not recommended as part of the biosimilar comparability exercise. Small differences in frequency and titers of antidrug antibodies likely reflect only interindividual variability in response, whereas large differences are likely associated with notable differences in quality characteristics that can be detected otherwise (EMA, 2015b).

The immunogenicity of biosimilars is discussed in greater detail in Chapter 12 of this book.

6.8.8 Lack of Full Analytical Characterization at the Start of the First in Human Clinical Trial

In vivo animal studies are sometimes performed to cover safety issues due to lack of appropriate analytical comparability data. Since fast development is key in many pharmaceutical companies, animal studies may be performed in order not to delay the start of clinical development. *In vivo* animal studies are sometimes used to reassure clinical investigators of the safety of the product. This may occur when the analytical comparability has not yet been performed, as ideally should be done to adequately characterize possible differences between both molecules and with a sufficient number of batches to enable a conclusion on safety from *in vitro* data alone. This "comfort factor" from first exposing animals before dosing humans is not based on a scientifically sound rationale and may even be misleading.

Even when the provided *in vitro* package of comparability data is considered appropriate by some regulators, it may be challenging for others or for clinical investigators or ethics committees to approve the start of biosimilar clinical trials in the absence of *in vivo* data. This could occur when dose escalation in the clinic is not possible (e.g., for biosimilars used in oncology where studies in healthy volunteers are not possible). In such cases, a direct dosing at the therapeutic dose in patients is needed, since an inefficient dose would ethically not be acceptable. To avoid program delays in these cases, companies may perform *in vivo* studies before submitting the clinical trial application to the different national competent authorities. However, when analytical and *in vitro* data do not raise concerns, additional *in vivo* animal studies are not considered needed.

6.9 CURRENT PRACTICE

The provided *in vitro* package in a nonclinical dossier for submission of a marketing authorization can be variable, depending on the complexity of the molecule. This information is publicly available in the European Public Assessment Reports (EPARs) published on the EMA website (EPARs of EMA). In this section, we describe the nonclinical study programs of the approved marketing authorization applications of all biosimilar products registered in the EU until October 1, 2015, which includes the nonclinical study programs of twelve different biosimilar molecules. The *in vitro* results are summarized in Table 6.3.

The nonclinical study programs of the approved, rejected, and withdrawn EU biosimilars up to October 1, 2013, have also been recently reviewed by van Meer et al. (2015a).

For the very first biosimilars that were approved in Europe, the two somatropin biosimilars, *in vitro* comparability assays were not performed. However, these products were developed before the legal framework for biosimilars existed in Europe. A year later two different erythropoietin biosimilar molecules were approved in the EU; here, *in vitro* comparability assays were included, consisting of receptor binding and biological activity assessment. Also, the *in vitro* studies to support filgrastim developments included a bioassay and a receptor binding assay for the four different filgrastim biosimilar developments.

TABLE 6.3
Results from the *In Vitro* Nonclinical Studies for the Approved Biosimilar Applications in the EU up to October 1, 2015

Name	Active Substance	Authorization Date	*In Vitro* Results
Omnitrope	Somatropin	April 12, 2006	No *in vitro* potency assays were available, only *in vivo* bioassays
Valtropin	Somatropin	April 24, 2006 Withdrawn after approval	No *in vitro* potency assays were available, only *in vivo* bioassays
Abseamed/Binocrit/Epoetin Alfa Hexal	Epoetin alfa	August 28, 2007	Characterization of receptor binding and signal transmission; dose response curves of reference and biosimilar were similar
Retacrit/Silapo	Epoetin zeta	December 18, 2007	Receptor binding, proliferation, and second messenger activation were evaluated; similarity at the level of receptor binding as well as in functional assays was demonstrated
Biograstim/Ratiograstim/Tevagrastim	Filgrastim	September 15, 2008	Relative receptor binding and biological activity showed similar binding affinities of the biosimilar and reference product and were equally effective in inducing cellular proliferation
Zarzio/Filgrastim Hexal	Filgrastim	February 6, 2009	*In vitro* potency of the product samples was comparable to that of the reference product; comparative receptor binding studies showed similar affinity to the receptor
Nivestim	Filgrastim	June 8, 2010	A bioassay and receptor binding assay was performed to evaluate PD; receptor-binding affinity and effects of the biosimilar and the reference product were similar
Grastofil/Accofil	Filgrastim	October 18, 2013 September 18, 2014	Receptor binding data for the Apo-Filgrastim DP batches from Process VII and IX demonstrated comparability

(Continued)

TABLE 6.3 (*Continued*)
Results from the *In Vitro* Nonclinical Studies for the Approved Biosimilar Applications in the EU up to October 1, 2015

Name	Active Substance	Authorization Date	*In Vitro* Results
Remsima/Inflectra	Infliximab	September 10, 2013	Thirty-three *in vitro* studies were done and included comparative tissue and species cross-reactivity studies, receptor-binding affinity assays, TNFα-neutralizing capacity assays, and assays to investigate FcgRI, FcgRII, and C1q complement binding; CDC, ADCC, and apoptosis activation were also investigated. In all studies the biosimilar and its reference product were comparable. Binding affinity of the biosimilar to FcgRIIIa and ADCC in NK cell assay was not fully comparable to its reference product but was not considered to have clinical consequences
Ovaleap	Follitropin alfa	September 27, 2013	Receptor activity and receptor affinity were similar between biosimilar and reference product
Bemfola	Follitropin alfa	March 27, 2014	Receptor binding and functional activity (in human FSH-receptor bearing cells) were similar between biosimilar and reference product
Abasaglar	Insulin glargin	September 9, 2014	Receptor binding affinity, functional activity, potential to stimulate lipogenic activity and mitogenicity were similar between biosimilar and reference product

Source: Adapted from van Meer et al., *Drug Discov. Today*, 120, 483, 2015a.
Note: Different marketing authorizations, but based on the same molecular development are grouped.

For the development of the biosimilar infliximab, extensive *in vitro* testing to establish PD biosimilarity was done exclusively *in vitro*. The Remsima EPAR shows that primary PD consisted of 33 *in vitro* studies, including the binding affinity of CT-P13 and Remicade to soluble (from different species) and transmembrane form of human TNFα, TNFβ, FcγRI, FcγRIIa, FcγRIIIa, FcRn, and C1q; the TNFα neutralization activity; the CDC, ADCC, apoptotic effects; and the cross-reactivity with various human tissues.

The *in vitro* comparability studies performed for the two biosimilar follicle stimulating hormones were less extensive as compared to the first biosimilar mAb, but included comparative studies for receptor activity, receptor affinity, and functional activity.

The most recent *in vitro* package was for the biosimilar insulin glargine, which included a series of comparative *in vitro* pharmacology studies to evaluate the binding affinity and functional activity at the relevant insulin receptors. In addition, the potential to stimulate lipogenic activity in adipocytes and mitogenicity in IR and IGF-1 receptor dominant cells has been evaluated. The carcinogenic risk of the insulin glargine in a proposed biosimilar was considered similar to that of the reference product based on additional *in vitro* data that was provided during the evaluation procedure.

With respect to the *in vivo* studies performed, repeated dose toxicity studies were included for all 12 biosimilar nonclinical development programs for the approved biosimilar applications in Europe. For 10 molecules, the package also included investigation of *in vivo* pharmacodynamics in animals. There was no *in vivo* PD study performed for the insulin glargine and for the infliximab biosimilar. A separate standalone local tolerance study in animals was performed for nine molecules. There was no local tolerance study performed for the insulin glargine, the infliximab, and the Bemfola FSH biosimilar.

When comparing the number of *in vivo* studies performed for filgrastim biosimilar products, a clear evolution over time can be observed.

For Tevagrastim, one of the first filgrastim biosimilars that was approved, multiple *in vivo* nonclinical studies were conducted in nonhuman primates, dogs, rats, rabbits, and mouse neutropenia models. Some of these studies were chronic toxicology studies that were up to 26 weeks in length.

For Nivestim, only a single pharmacodynamic study in a neutropenic rat model and a repeat-dose toxicity study in healthy rats were conducted.

Finally, for Grastofil, the most recently approved filgrastim, no *in vivo* studies have been performed. However, this is based on bridging to another nonclinical development program where several *in vivo* studies were conducted. Grastofil and Accofil filgrastim are based on the same molecular development (i.e., Apo-Filgrastim process IX), but most nonclinical *in vivo* comparability studies with the reference product have been performed with the process V drug substance (i.e., Neukine). This approach has been considered acceptable given the fact that receptor binding data for the Apo-Filgrastim DP batches from Process VII and IX demonstrated comparability. In addition, a recent comparative clinical study that investigated the PK/PD profile of Apo-Filgrastim versus Neupogen used Process IX drug substance material.

There was regulatory acceptance not to request any *in vivo* studies with the product using the intended manufacturing process.

Although mAbs are more complex recombinant proteins, for infliximab, it is considered feasible to characterize the biological properties *in vitro*. Therefore, reestablishing the pharmacodynamic response in an *in vivo* model could be considered redundant. For Remsima, comparability has been extensively evaluated *in vitro*. Nevertheless, two 2-weeks repeat-dose toxicity studies in rats were performed to compare the off-target toxicity profiles of CT-P13 and Remicade, and the observed effects were similar for both test articles.

van Meer et al. (2015a) critically reviewed animal studies submitted for the approved biosimilar applications in Europe up to October 1, 2013, and concluded that none of these showed relevant differences to the reference product. Moreover, it was questioned whether the data submitted for EU biosimilar marketing authorizations from animal studies were suited to detect subtle differences between biosimilars and reference products or to translate them into measurable endpoints.

6.10 CONCLUSION

The rapid scientific progress in biotechnological capabilities, analytical tools, and *in vitro* techniques over the last decade has changed the paradigm in assessment of biosimilars. The physicochemical and functional properties of a biosimilar can be extensively compared to those of its reference medicinal product. It is recognized that *in vitro* assays are generally more sensitive than animal studies or studies in humans. Therefore, a combination of both the physicochemical and *in vitro* functional properties of a biosimilar forms a solid basis for establishing biosimilarity.

It may be too soon to try to evaluate how the new approach in the EU biosimilar guidance with respect to recommendations for *in vivo* animal studies has influenced nonclinical studies.

The NC3Rs, a UK-based scientific organization dedicated to a 3Rs approach to the use of animals in research, together with the UK medicines agency [Medicines and Health Care Products Regulatory Agency (MHRA)], set up a biosimilar working group in 2014 (Chapman et al., 2016). This working group is composed of contract research organizations (CROs), regulatory bodies, pharma, biotech, and global biosimilar companies to address the issue of animal use and the feasibility of developing a biosimilar product using *in vitro* data alone. The group is sharing experiences and reviewing current practices. While the EU is actively promoting clinical trials with *in vitro* data only (if appropriate), the members of the NC3Rs working group on animal use for biosimilars indicate that *in vivo* studies are still performed.

For a well-characterized biological for which close similarity with a well-known reference product has been demonstrated, based on extensive analytical and *in vitro* data, occurrence of a safety concern different from the reference product is highly unlikely.

The exception could be issues related to immunogenicity, but these issues cannot be adequately addressed in *in vivo* animal studies. Only in rare cases animal *in vivo* data could be relevant and considered needed in the biosimilar comparability exercise.

Through this chapter, the authors hope to contribute to a more global discussion of these aspects in order to change practice and some regulatory guidances and reach a more harmonized approach in animal use in the development of biosimilars.

DISCLAIMER

The views expressed in this chapter are the personal views of the authors and may not be understood or quoted as being made on behalf of or reflecting the position of any agency or one of its committees or working parties.

REFERENCES

Casadevall N, Nataf J, Viron B, et al. (2002) Pure red-cell aplasia and antierythropoietin antibodies in patients treated with recombinant erythropoietin. *New England Journal of Medicine* **346**, 469–475.

Chapman K, Adjei A, Baldrick P, et al. (2016) Waiving in vivo studies for monoclonal antibody biosimilar development: national and global challenges. *MAbs* **8(3)**, 427–435.

Chapman K, Andrews L, Bajramovic JJ, et al. (2012) The design of chronic toxicology studies of monoclonal antibodies: implications for the reduction in the use of non-human primates. *Regulatory Toxicology and Pharmacology* **62(2)**, 347–354.

Chinese FDA. (2014) Draft guidelines for R&D and evaluation techniques of biosimilars. Available from: http://www.cde.org.cn (Accessed September 21, 2015).

Chung CH, Mirakhur B, Chan E, et al. (2008) Cetuximab-induced anaphylaxis and IgE specific for galactose-α-1,3-galactose. *New England Journal of Medicine* **358**, 1109–1117.

EC. (2001) Directive 2001/83/EC of the European Parliament and of the Council of 6 November 2001 on the community code relating to medicinal products for human use. *Official Journal of the European Union L 311*, 67–128. November 28, 2004. Consolidated version: November 16, 2012. Available from: http://ec.europa.eu/health/files/eudralex/vol-1/dir_2001_83_consol_2012/dir_2001_83_cons_2012_en.pdf (Accessed September 21, 2015).

EEC. (1986) Directive 86/609/EEC Council Directive: the approximation of laws, regulations and administrative provisions of the Member States regarding the protection of animals used for experimental and other scientific purposes. November 24, 1986. Available from: http://ec.europa.eu/food/fs/aw/aw_legislation/scientific/86–609-eec_en.pdf (Accessed September 21, 2015).

EMA. (2005) Guideline on similar biological medicinal products. CHMP/437/04. Available from: http://www.ema.europa.eu/docs/en_GB/document_library/Scientific_guideline/2009/09/WC500003517.pdf (Accessed September 21, 2015).

EMA. (2006) Guideline on similar biological medicinal products containing biotechnology-derived proteins as active substance: non-clinical and clinical issues. EMEA/CHMP/BMWP/42832/2005. Available from: http://www.ema.europa.eu/docs/en_GB/document_library/Scientific_guideline/2009/09/WC500003920.pdf (Accessed September 21, 2015).

EMA. (2011) Closed workshop on biosimilar monoclonal antibodies and immunogenicity of monoclonal antibodies. EMA/97951/2012. October 24, 2011, London. Available from: http://www.ema.europa.eu/docs/en_GB/document_library/Report/2012/06/WC500128739.pdf (Accessed September 21, 2015).

EMA. (2012) Guideline on similar biological medicinal products containing monoclonal antibodies—non-clinical and clinical issues. EMA/CHMP/BMWP/403543/2010. Available from: http://www.ema.europa.eu/docs/en_GB/document_library/Scientific_guideline/2012/06/WC500128686.pdf (Accessed September 21, 2015).

EMA. (2014a) Guideline on similar biological medicinal products. CHMP/437/04 Rev 1. Available from: http://www.ema.europa.eu/docs/en_GB/document_library/Scientific_guideline/2014/10/WC500176768.pdf (Accessed September 21, 2015).

EMA. (2014b) Guideline on similar biological medicinal products containing biotechnology-derived proteins as active substance: quality issues. Revision 1 EMEA/CHMP/BWP/247713/2012. Available from: http://www.ema.europa.eu/docs/en_GB/document_library/Scientific_guideline/2014/06/WC500167838.pdf (Accessed September 21, 2015).

EMA. (2015a) Guideline on similar biological medicinal products containing biotechnology-derived proteins as active substance: non-clinical and clinical issues. EMEA/CHMP/BMWP/42832/2005 Rev. 1. Available from: http://www.ema.europa.eu/docs/en_GB/document_library/Scientific_guideline/2015/01/WC500180219.pdf (Accessed September 21, 2015).

EMA. (2015b) Draft Guideline on immunogenicity assessment of biotechnology-derived therapeutic proteins. EMEA/CHMP/BMWP/14327/2006 Rev. 1. Available from: http://www.ema.europa.eu/ema/doc_index.jsp?curl=pages/includes/document/document_detail.jsp?webContentId=WC500194507&murl=menus/document_library/document_library.jsp&mid=0b01ac058009a3dc (Accessed October 1, 2015).

EPARs of EMA. European Public Assessment Reports. Available from: http://www.ema.europa.eu/ema/index.jsp?curl=pages/medicines/landing/epar_search.jsp&mid=WC0b01ac058001d124 (Accessed September 21, 2015).

EU. (2010) Directive 2010/63/EU of the European Parliament and of the Council of 22 September 2010 on the protection of animals used for scientific purposes. *Official Journal of the European Union L 276*, 33–79. October 20, 2010. Available from: http://eurlex.europa.eu/LexUriServ/LexUriServ.do?uri=OJ:L:2010:276:0033:0079:EN:PDF (Accessed September 21, 2015).

FDA. (2015) Scientific considerations in demonstrating biosimilarity to a reference product. Guidance for Industry. April 2015. Available from: http://www.fda.gov/downloads/DrugsGuidanceComplianceRegulatoryInformation/Guidances/UCM291128.pdf (Accessed September 21, 2015).

Houde D, Peng Y, Berkowitz SA, Engen JR. (2010) Post-translation modifications differentially affect IgG1 conformation and receptor binding. *Molecular and Cellular Proteomics* **9**, 1716–1728.

Huang L-Y, DuMontelle JL, Zolodz M, et al. (2009) Use of toll-like receptor assays to detect and identify microbial contaminants in biological products. *Journal of Clinical Microbiology* **47**, 3427–3434.

ICH. (2011) ICH guideline S6 (R1): preclinical safety evaluation of biotechnology-derived pharmaceuticals. EMA/CHMP/ICH/731268/1998. Available from: http://www.ema.europa.eu/docs/en_GB/document_library/Scientific_guideline/2009/09/WC500002828.pdf (Accessed September 21, 2015).

Japanese PMDA. (2009) Guideline for the quality, safety and effectiveness of follow-on biological medicinal products (interim translation from Japanese by the Japanese generics association). Available from: http://www.jga.gr.jp/english/wp-content/uploads/sites/4/2011/03/Interim_Translation_of_Notification_0304007Follow-on_Biologics1.pdf (Accessed September 21, 2015).

Japanese PMDA. (2015) Q&A on the guideline for the quality, safety and effectiveness of follow-on biological medicinal products. PSEHB/ELD Administrative Notice. Available from: http://www.pmda.go.jp (published in Japanese on December 15, 2015).

Keffer J, Probert L, Cazlaris H, et al. (1991) Transgenic mice expressing human tumour necrosis factor: a predictive genetic model of arthritis. *EMBO Journal* **10**, 4025–4031.

Krishnan A, Mody R, Malhotra H. (2015) Global regulatory landscape of biosimilars: emerging and established market perspectives. *Biosimilars* **5**, 19–32.

Mulders JWM, Wijn H, Theunissen F, et al. (1999) Prediction of the in-vivo biological activity of human recombinant follicle-stimulating hormone using quantitative isoelectric focusing, optimization of the model (Conference Paper). *Pharmacy and Pharmacology Communications* **5**, 51–55.

Ober RJ, Radu CG, Ghetie V, Ward ES. (2001) Differences in promiscuity for antibody-FcRn interactions across species: implications for therapeutic antibodies. *International Immunology* **13(2)**, 1551–1559.

Radaev S, Sun P. (2002) Recognition of immunoglobulins by Fcγ receptors. *Molecular Immunology* **38**, 1073–1083.

Shields RL, Lai J, Keck R, et al. (2002) Lack of fucose on human IgG1 *N*-linked oligosaccharide improves binding to human FcγRIII and antibody-dependent cellular toxicity. *Journal of Biological Chemistry* **277**, 26733–26740.

van Aerts LAGJM, Smet K, Reichmann G, et al. (2014) Biosimilars entering the clinic without animal studies: a paradigm shift in the European Union. *mAbs* **6(5)**, 1155–1162.

van Meer PJK, Ebbers HC, Kooijman M, et al. (2015a) Contribution of animal studies to evaluate the similarity of biosimilars to reference products. *Drug Discovery Today* **120(4)**, 483–490.

van Meer PJK, Graham ML, Schuurman H-J. (2015b) The safety, efficacy and regulatory triangle in drug development: impact for animal models and the use of animals. *European Journal of Pharmacology* **759**, 3–13.

wvan Meer PJK, Kooijman M, van der Laan JW, et al. (2013) The value of non-human primates in the development of monoclonal antibodies. *Nature Biotechnology* **31**, 882–883.

Wadhwa M, Knezevic I, Kang H-N, Thorpe R. (2015) Immunogenicity assessment of biotherapeutic products: an overview of assays and their utility. *Biologicals* **43(5)**, 298–306.

Weise M, Kurki P, Wolff-Holz E, et al. (2014) Biosimilars: the science of extrapolation. *Blood* **124(22)**, 3191–3196.

WHO. (2009) Guidelines on evaluation of similar biotherapeutic products (SBPs). World Health Organization. Available from: http://www.who.int/biologicals/areas/biological_therapeutics/BIOTHERAPEUTICS_FOR_WEB_22APRIL2010.pdf (Accessed September 21, 2015).

Zimmermann H, Gerhard D, Hothorn LA, et al. (2011) An alternative to animal testing in the quality control of erythropoietin. *Pharmeuropa Bio & Scientific Notes* **1**, 66–80.

7 The Clinical Development of Biosimilar Drugs

Mark McCamish
Forty Seven, Inc.

Gillian Woollett
Avalere Health

Sigrid Balser
Hexal AG

CONTENTS

7.1 TOTALITY OF EVIDENCE

As documented in other chapters, the development of biosimilars in "highly regulated" regions of the world [i.e., those countries that subscribe to the standards of the International Committee for Harmonization (ICH)—including Europe, the US, and Japan, as the formal members of ICH, and also Canada and Australia] follows a "totality of the evidence" model, with the development proceeding in a "stepwise" fashion. The overall focus of the biosimilar development program is to prove that the biosimilar contains "essentially the same biological substance" as the reference product (EMA, 2012). It is important for the practicing physician to have confidence when using a biosimilar that it is just like changing the batch of the reference product, and this is indeed the case considering the natural variation existing in all biological products (Schiestl et al., 2011). Consequently, the clinical trials in support of regulatory approval of a biosimilar are designed to "confirm similarity" established analytically and not to demonstrate the safety and efficacy of the biosimilar in all indications for which the reference product is already approved. Because of this unique role of clinical trials to confirm similarity, there are many design features, outlined below, that fundamentally differ from traditional drug development clinical trials used for initial approval of a novel biological drug or for the addition of an indication.

Before discussing the design characteristics of biosimilar clinical trials, it will be helpful to understand how the approach to clinical development of biosimilars has evolved beyond the studies applied in traditional regulatory science for originator products. This evolution began with the introduction of recombinant biological products in the 1980s. Adoption of these new drugs into clinical practice required sponsors to make more material than could be originally anticipated, and this scale-up plus the rapid developments in biotechnology belied the old "product is the process" approach to biologics. The production of chemically synthesized small-molecule drugs involves the creation of the identical active pharmaceutical ingredient with every batch. However, manufacturing recombinant biological drugs does not usually produce a single identical biological product. Just like in living organisms, the host cells involved in synthesizing the biological drug, coded by the recombinant DNA program, modify the protein after synthesis of the basic amino acid and protein structure. This is called posttranslational modification. These modifications, such as glycosylation or terminal amino acid modification, produce a mixture of biological product in an equilibrium within the recombinant cell system. The final product, purified from these cells, is nonetheless designed to have the same clinical effect over multiple batches.

The pharmaceutical sponsors must provide evidence that they have good control of their manufacturing process so that batch-to-batch variability is within specific acceptance criteria. This applies to all biologics irrespective of whether they are recombinant or naturally sourced, but recombinant technology has provided new opportunities for scale-up and control. Increasing acceptance and expanded use of recombinant biologics led to the need to manufacture more drugs than the maximal capacity of the original manufacturing process. To expand the manufacturing capacity, the sponsors had to "scale-up" the manufacturing process or transfer the manufacturing process to a larger facility. As pointed out by Schiestl et al., such a scale-up or transfer induces a "manufacturing change" that can be measurable in the final product. However, such changes in the manufacturing process cannot be allowed to create clinically meaningful differences, and approval is required for each such change by each regulatory authority for every jurisdiction in which the product is made available (Schiestl et al., 2011). This regulatory approval is required prior to the manufacturing change being made. Consequently, the "new" biological product is introduced to the marketplace after the regulatory authority has agreed to the manufacturing change.

The goal of the regulatory approval is to ensure that all prescribers and patients have confidence that the product produced with the newly amended, likely expanded, manufacturing process is "highly similar" to the original biologic and produces the same clinical response as the original product (ICH, 2004). The regulatory process involved with evaluating a manufacturing change focuses on confirming that the postmanufacturing change product is "highly similar" to the premanufacturing change product. The "highly similar" terminology was established as it is impossible to produce an "identical" biological product even from batch to batch, much less with a manufacturing change, such as scale-up or transfer to a different site, or even change in cell line (ICH, 2004). Regulators' review and approval of these technical transfers or manufacturing changes are not revealed to the public in the United States, as the product label does not change and there is no requirement for publication of a summary basis of approval (SBA) for these manufacturing changes.

In Europe, however, the regulatory process involved in the review and approval of manufacturing changes is often published in European Public Assessment Reports (EPARs). In reviewing these EPARs, it is clear that clinical trials are rarely used to justify that these manufacturing changes do not confer a clinically meaningful difference (Vezer, 2016). This is reasonable given the extensive knowledge accumulated over the past three decades on the clinical impact of specific glycosylation patterns and other posttranslational modifications. It is only when glycosylation or posttranslational modifications are introduced by the new process that have not been seen before, or where insufficient data exist to conclude that no clinically meaningful difference is expected, that clinical trials may be an appropriate requirement for regulatory approval (Woodcock, 2007). Reviewing an example can help illustrate the evolution of the clinical trial requirements for manufacturing changes; this is important as the same principles apply to the development of a biosimilar.

7.1.1 Case Study

Amgen changed its manufacturing process for Aranesp (darbepoetin alfa) from a roller bottle process to a more modern high-throughput process to enable an increase in manufacturing capacity. This change required the establishment of a new master cell bank and the use of different media components (EMA, 2008). The glycosylation changes seen from different batches of Aranesp purchased from the market are documented by Schiestl et al. (2011). In order for the EMA to conclude that this manufacturing change resulted in a product that would not produce a clinically meaningful difference (i.e., that the pre- and postmanufacturing process change products were "highly similar"), multiple human clinical trials were required. These trials were not designed to reestablish the safety and efficacy of Aranesp, but to provide confirmatory evidence to the regulating authority that the old and new products produced the same clinical effect. This approach of comparability in support of manufacturing changes is analogous to the principles used in the development of biosimilars and the clinical trials required to confirm analytical similarity when developing a biosimilar to an existing reference product.

7.2 CLINICAL TRIALS ARE LESS SENSITIVE THAN ANALYTICS TO DOCUMENT DIFFERENCES

Development of a biosimilar, just like that of an originator biologic, proceeds in a stepwise manner with the clinical trial(s) being the last step. However, for a biosimilar, those studies do not establish safety and efficacy; rather, they confirm the "sameness" of the biosimilar to the reference product, the "sameness" having been established analytically before any clinical studies are undertaken. This stepwise process provides a methodology for confirming that the active biological substance of the biosimilar is essentially the same as the reference product, and through having addressed any "residual uncertainty" (FDA, 2015a) prior to studies in humans. The extensive state-of-the-art analytical data are able to document that the amino acid sequence and protein structure of the biosimilar are identical to the reference product and that the biosimilar has indistinguishable binding characteristics and biological function compared to the reference product. Such data provide reassurance that the biosimilar will produce the same clinical effects as the reference product. Hence, at initiation of the clinical trial, limited residual uncertainty remains. Therefore, the clinical trial is designed to confirm the findings of analytical "sameness" generated in prior steps in developing the biosimilar. Clinical trials are substantially less sensitive in picking differences between the two products (proposed biosimilar and reference product) than analytical methods. In fact, analytical methodologies are able to document many differences between batches of reference product or between products undergoing manufacturing changes that have no clinical relevance (Schiestl et al., 2011). This is an essential point for all biologics. It is not that biologics don't have differences, whether different batches of the reference product or a biosimilar of that reference product. Indeed they do. However, those differences cannot have clinical relevance.

7.2.1 Case Study

An example is the difference in glycosylation patterns from Enbrel's pre- and post-manufacturing change (Schiestl et al., 2011). Schiestl et al. documented a 20% change in enrichment of the G2F glycosylation species after the manufacturing change of this anti-tumor necrosis factor biologic, decreasing from over 50% enrichment to 30% enrichment (see Figure 7.1).

This G2F glycosylation does not have an impact on pharmacokinetics or pharmacodynamics (Goetze et al., 2011), does not impact binding, and is not associated with increased immunogenicity. Therefore, this analytical difference, though statistically different, would not be expected to have any meaningful clinical effect. Indeed,

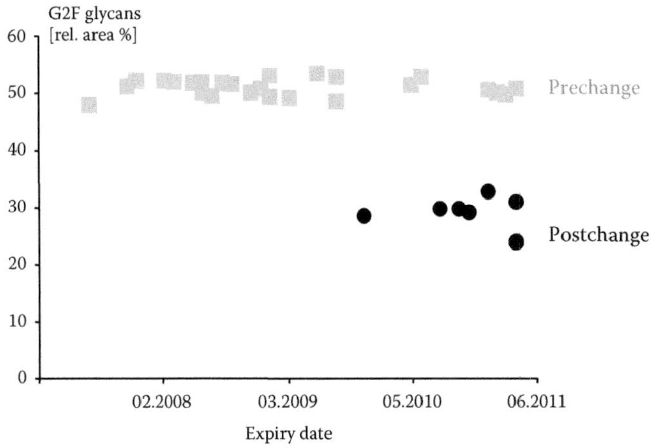

FIGURE 7.1 Comparisons of different pre- and postchange batches of Enbrel. (From Schiestl, M. et al., *Nature Biotechnology.*, 29, 310–312, 2011. With permission.)

when using the traditional clinical trial design for anti-TNF biologics in rheumatoid arthritis patients using ACR20 (American College of Rheumatology: 20% improvement) as the primary endpoint, such a difference in this particular glycosylation pattern would not impact efficacy. This absence of clinically meaningful differences, and hence the lack of sensitivity of clinical studies, is all the more apparent when this same established clinical metric is used to compare the clinical outcomes of various anti-TNF agents, each containing totally different active ingredient molecules. For example, Enbrel (etanercept), Humira (adalimumab), Remicade (infliximab), Cimzia (certolizumab pegol), and Simponi (golimumab) all produce similar clinical outcomes but have different active ingredients. These similar efficacy results with quite different molecules provide convincing evidence that the clinical trial is far less sensitive in picking up differences between biologics whether originator or biosimilar than analytical studies (Figure 7.2). One has to conclude that different molecules can produce the same clinical outcomes, but the same molecules cannot produce different clinical outcomes.

7.2.2 The Contribution of Clinical Trials—A Paradigm Shift

Historically, clinicians are trained to look at clinical data when making choices about which drug to use; therefore, they look predominantly for clinical data. However, for biosimilars the clinical trials are not the primary data on which regulatory approval is based. For biosimilars the clinical trials are not designed in the same way as those trials used for approval of novel drugs. The evaluation of clinical trials for biosimilars by both regulators and clinicians requires a paradigm shift, as stated by Janet Woodcock, director of the Center for Drug Evaluation and Research (CDER) at the US Food and Drug Administration (Woodcock, 2012). This paradigm shift requires physicians to understand that the clinical trial for approval of a biosimilar drug is

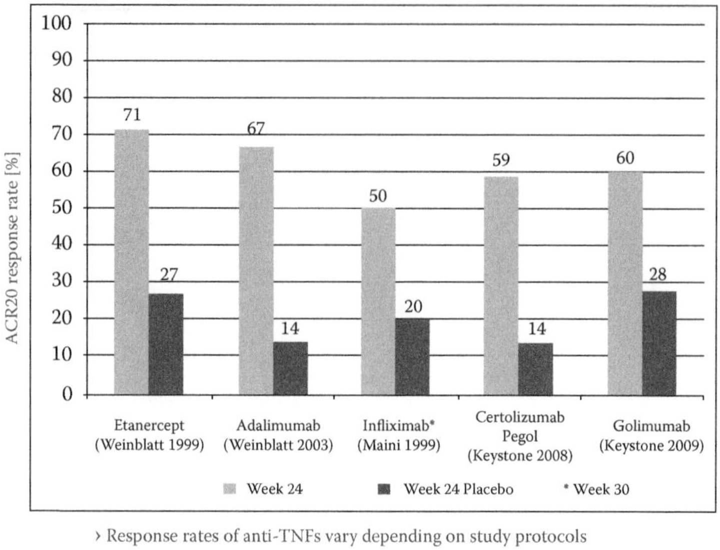

FIGURE 7.2 **(See color insert.)** The clinical ACR 20 responses to Etanercept, Adalimumab, Infliximab, Certolizumab, and Golimumab at 24 weeks are clinically comparable.

designed to complement all other analytical data and to provide confirmation that the biosimilar is essentially the same as the reference product. Without this paradigm shift of understanding, physicians will often consider biosimilar clinical trials to be inadequate. This issue will be addressed more specifically in Section 7.4.

7.3 IMMUNOGENICITY

As with the development of any biologic, including the original reference product used to create the biosimilar, it is difficult to predict human immunogenicity using preclinical models. Therefore, although the analytical data on the molecular attributes of the biosimilar can be evaluated for similarity to its reference product, it is challenging to predict overall immunogenicity. A clinical trial can evaluate if there are major signals of differential immune responses. Since the reference product has been used for many years (sometimes multiple decades), there is a substantial database regarding the types of immunogenicity that would be expected from these products. Therefore, the biosimilar sponsor can look specifically for these types of immune responses. It is less complicated to evaluate immunogenicity for a biosimilar than to evaluate a completely new biologic that has not been exposed to humans previously.

Unfortunately, some have erroneously used the experiences of pure red cell aplasia (PRCA) caused by erythropoietic agents in Europe to suggest that there is an increased risk of immunogenicity for biosimilars. The Eprex/Erypo (epoetin alfa) immunogenicity problem was due to a formulation change in the originator product (Casadevall, 2002; Macdougall, 2005), well before the existence of any biosimilars and prior to Eprex/Erypo subsequently being used as a reference product for a number of biosimilars. In this historical case of immunogenicity with a branded biologic, Eprex underwent a formulation change that precipitated aggregation of the erythropoietin in the prefilled syringe (PFS); this in turn stimulated a neutralizing antibody response that cross-reacted to endogenous erythropoietin, causing PRCA (Casadevall, 2002). Extensive root cause analyses were performed to address this issue and highlight the importance of maintaining control and the consistency of manufacturing controls for all biologics whether branded biologics or biosimilars. This PRCA episode in the early 2000s does not implicate safety risks with biosimilars, but it does highlight that manufacturing of biologics of all types requires a quality mindset with consistent and appropriate manufacturing controls, and that all manufacturing changes must be undertaken with care.

Each biologic has a unique immunogenicity profile, with some agents frequently stimulating development of neutralizing antibodies (such as with Humira) (Bartelds et al., 2011), whereas others rarely cause immunological responses at all (such as with Neupogen or Zarxio) (Holzmann et al., 2015). Knowledge and a sound understanding of immunological experience with the chosen reference product is critical to evaluating the probability for demonstrating differences in immunogenicity between the biosimilar and the reference product. The goal in such a comparison is to demonstrate that there are no worrisome signals of increased immunogenicity for the biosimilar. This would be unexpected based on the high degree of similarity established in the biosimilarity evaluation. In unique situations, lower immunogenicity may be acceptable for a biosimilar such as in using an alternate expression system. This is the case with centuximab (Erbitux) which is manufactured using an SP2/0 murine expression system that incorporates an immunogenic glycosylation involving alpha galactosylation, Gal α1–3Gal (Galili, 2005). Biosimilar sponsors can use a CHO (Chinese hamster ovary) expression system to produce a product with less of this same immunogenic glycosylation pattern (Bosques et al., 2011). In this case, the biosimilar would be produced with less of the Gal α1–3Gal glycosylation providing less of an immunogenic stimulus, and this would be acceptable provided the other quality aspects or physicochemical properties resulted in the same clinical performance.

Therefore, overall, the biosimilar is designed to be highly similar or essentially the same as the reference product and is expected to have the same immunogenic profile. However, as with all biologics, stability of the final product or formulation is critical and may impact aggregation, thus requiring diligent manufacturing quality control.

7.4 BIOSIMILAR CLINICAL TRIAL DESIGNS

The rationale for clinical trials in the development of biosimilars is to provide confirmation of the "sameness" of the biosimilar to the reference product. This has already been established by the analytical and preclinical comparisons that build the foundation for the "totality of evidence" that the biosimilar can be expected to have the same clinical effect in patients. With this consideration in mind, the clinical trial is specifically designed to demonstrate differences between the biosimilar and reference product should any such differences exist. Hence, the question is how to optimize the sensitivity of the clinical trial to show those differences, even while knowing that it is so much less sensitive than analytical methodologies previously employed. Since substantially greater variability is expected in any clinical trial, especially considering patient factors, it is critical to try to limit the variability in the conduct of the clinical trial to enable a signal from differences between the biosimilar and the reference product to be seen. This means that the patient population should be selected to minimize variability of clinical response. By way of contrast, during the development of a novel biological drug the sponsor must demonstrate the clinical effect of the product in a variety of populations that vary by gender, race, age, and weight. However, for a biosimilar it minimizes these sources of variability allowing for focus on the key question as to whether the biosimilar produces the same biological response as the reference product in a head-to-head study. Once again, this requires a paradigm shift in thought and strategy for clinical trials for biosimilar development.

7.4.1 Case Study

One example of patient factors contributing to variability is with pegylated proteins (proteins wherein a single long chain of polyethylene glycol is added covalently to the protein to eliminate renal clearance). Pegfilgrastim (Neulasta which is a longer acting neutrophil stimulator for patients given chemotherapy) was developed to produce a longer residence time of the product in the blood so that a single dose could be used with each course of chemotherapy (Amgen, 2015d). The product was initially dosed in a weight-based approach with pharmacokinetic studies, but the phase III investigation was undertaken using a fixed dose of 6 mg. The variability of the initial studies produced a coefficient of variation of approximately 35%; however, when larger fixed-dose pharmacokinetic studies were performed the variation was approximately 85%, complicating the demonstration of pharmacokinetic bioequivalence (Yang et al., 2015). With such high intersubject variability, "noise" may mask the differences between a proposed biosimilar and the reference product. To improve the sensitivity of a pharmacokinetic study of pegylated protein products, it is necessary to limit subject factors such as gender and weight. With pegfilgrastim it is also critical to limit the variation of baseline neutrophil counts because the clearance of this drug is determined by receptor-mediated uptake (receptors on neutrophils bind and initiate clearance). These factors are far more variable than any analytical difference between pegfilgrastim products.

7.4.2 Pharmacokinetics

Pharmacokinetic studies are often called "phase I" studies in the development of a novel compound. This terminology is not appropriate for the development of a biosimilar because the pharmacokinetic study is pivotal to confirming the similarity of the biosimilar to the originator. Therefore, biosimilar sponsors also refer to the pharmacokinetic study as a pivotal study for demonstrating biosimilarity. Indeed, only a pharmacokinetic/pharmacodynamic study may be required to confirm biosimilarity to the reference product, and no additional efficacy study may be needed. This possible approach is allowed by several regulators (EMA, 2014; FDA, 2014) when justified.

Because of the pivotal nature of the pharmacokinetic study, the study is usually performed with clinical material made at the final manufacturing scale so that regulatory authorities can be assured that additional manufacturing changes will not be needed that could alter product characteristics or clinical responses. When performing pharmacokinetic (PK) studies to determine if a biosimilar is bioequivalent to the reference product, the least variability is encountered with normal healthy volunteers. However, with some biological drugs with a relatively long half-life or prolonged pharmacodynamic effects, it is not possible to use normal healthy volunteers. For example, normal volunteers have been used in PK studies of filgrastim, erythropoietin, anti-TNF biologics, and even bevacizumab (Sorgel et al., 2009; Elmeliegy et al., 2015; Sorgel et al., 2015). However, it is not possible to perform PK studies in normal volunteers with rituximab (Maloney et al., 1997) due to the prolonged depletion of B lymphocytes that would make these studies unethical. Therefore, the initial PK study of biosimilarity may be fairly straightforward with some biologics (except for pegfilgrastim as mentioned above), but for others like rituximab, PK studies have to be performed in patients. This adds variability and difficulty standardizing the multiple blood sample draws required for dense PK profiling in routine clinical centers. Limiting patient variability to have a completely homogeneous population is antithetical in the development of novel biologics. For novel biologics the evaluation of PK while assessing the impact of patient factors on the exposure and use of the product is actively sought.

Traditional approaches to statistical justification of bioequivalence margins hold true for biosimilar PK studies just as with all other molecules (FDA, 2013). Regulatory authorities generally agree that using a 90% confidence interval is acceptable. Thus, it is possible to reasonably harmonize biosimilar development between regulatory regions. This is not the case for efficacy equivalence designs discussed below. Some regulators consider it more sensitive to pick up differences between a reference product and a proposed biosimilar when lower doses than those used in routine clinical treatment are tested. However, when conducting PK/PD studies in patients, it is difficult to use lower doses than are indicated for that disease state, as this would be ethically inappropriate. When using normal healthy volunteers, it is possible to use lower doses than specified on the label for therapeutic use. As mentioned above, dosing on the steep part of the dose response curve may be more sensitive to pick up differences between the biosimilar and the reference product if differences exist. Once analytical or PK differences are detected, whether statistically significant or not, these differences have to be evaluated to determine if these differences would be clinically meaningful in practice.

7.4.3 Confirmatory Efficacy Study

The typical nomenclature for a pivotal trial for a novel drug uses the "phase III" terminology. However, in the development of biosimilars, the PK/PD study is also pivotal (as mentioned above), while the confirmatory efficacy study addresses residual uncertainty about the biological function of the proposed biosimilar compared to the reference product. Therefore, biosimilar sponsors often simply refer to these studies as confirmatory safety and efficacy studies and avoid the phase III moniker. Even simple terminology such as "phase III" can wrongly communicate to physicians and clinical reviewers that the purpose of the study is to point to the safety and efficacy of the biosimilar when the goal is to confirm similarity and address any residual uncertainty based on the totality of evidence available at the time the clinical program starts.

When designing the confirmatory study evaluating the efficacy and safety of the biosimilar, it is obviously necessary to compare the response of the biosimilar to the reference product in a head-to-head study. Comparisons to historical data are not useful in picking up differences between the biosimilar and the reference product if differences exist. Comparing responses to the reference product is a major challenge for biosimilar sponsors because they must purchase reference products from the market in order to facilitate the best scientific comparison. This is a significant financial undertaking, and it is an operational challenge in many ways. The biosimilar sponsor must acquire and relabel the reference product to enable blinded clinical trials, repackage and ship to all clinical sites, and analyze every batch of reference product acquired and used in clinical trials so as to prove the reference product actually used is analytically "highly similar" to the specific biosimilar lots used in the same trial. Such analytical characterization often requires purchasing 100 additional vials or prefilled syringes of each batch of the reference product used. Lastly, some originator companies create barriers to biosimilar sponsors being able to acquire consistent lots for use in biosimilar trials.

7.4.4 Trial Designs

Most regulatory authorities prefer equivalence trial designs that test to exclude a biosimilar that is inferior or superior to the reference product (two-sided testing) versus noninferiority designs that focus on excluding only an inferior response (EMA, 2007; Health Canada, 2010). The rationale for this approach is that a biosimilar with superior efficacy should not be considered a biosimilar. Thus, the burden on the biosimilar sponsor is to prove that the proposed biosimilar is neither inferior nor superior to the reference product. This is in contrast to originator product development wherein noninferiority studies are often used when developing novel drugs using the same receptor or mechanism of action. Regulatory authorities approve such a drug when it is shown to have at least the same clinical effect as the comparator drug.

7.4.4.1 Case Study

With the development of Neulasta and Aranesp, which were designed to have the same clinical effect as their forerunners with less frequent injections, noninferiority studies were used (Amgen, 2015a,d).

To always require an equivalence trial design for development of biosimilars is inappropriate as it is not possible for a biosimilar with highly similar product attributes to produce statistically superior clinical responses when compared to the reference product. When small differences in product attributes, clearly determinable by analytics, are able to produce a superior clinical response, the originator company will have discovered and patented this minor change as part of the lifecycle development of its drug.

7.4.4.2 Case Study

This is the case, for example, with Genentech/Roche's approval of Gazyva, which is a follow-on product or lifecycle management strategy for Rituxan, wherein increased antibody-dependent cellular cytotoxicity (ADCC) was achieved by altering the glycosylation structure of the molecule. Such modifications are easily determined and are known to have an impact on ADCC. Therefore, it is unrealistic to consider that a biosimilar sponsor creating essentially the same active substance as an originator would produce a product with superior clinical efficacy without evidence of analytical differences. This is an issue because the design and conduct of an equivalence trial generally requires more subjects than a noninferiority design, which can complicate the operational and financial aspects of developing biosimilars. This is an example of biosimilars seemingly being held to a higher standard than development of novel drugs.

Equivalence trial designs require biosimilar sponsors to justify the statistical assumptions for the comparability margin (noninferiority or equivalence margin). To do so there has to be sufficient literature using the reference product to describe the clinical effect of the dose and other comedications used to produce a given effect size or difference between the group treated with placebo or standard of care. The determination of the comparability margin is critical as both sample size and interpretation of the study depend on this margin. This margin is essentially the largest difference between the use of the two products that can be judged as clinically acceptable or the "same." This margin must clearly be smaller than differences observed in superiority trials of the reference product. Regulatory guidance suggests that this margin should allow retention of at least 50% of the effect size of the reference product (EMA, 2005; FDA, 2010). In general, the larger the effect-size of the reference product, the smaller the sample size of the biosimilarity equivalence trial. Those products with smaller effect sizes such as with Avastin are challenging as they therefore require quite large clinical trials. Most sponsors are conducting clinical trials in non-small-cell lung cancer (NSCLC) with Avastin biosimilars with patient sample sizes over 700 (Amgen, 2015b). Considering that acquiring the reference product from the US market is very expensive, this can add $100,000 to each patient enrolled in the trial, depending on how long the trial lasts (Jirillo et al., 2008).

Physicians are not as familiar with equivalence trial designs and become concerned when the protocol or discussion mentions retaining only 50% of the effect size of the reference product as demonstrated in comparison to a placebo or standard of care active control. Let us attempt to put this in the context of clinical relevance by using an example.

7.4.4.3 Case Study

To establish the justification for an equivalence trial design for the use of an anti-TNF biologic in psoriasis, in this case etanercept, one uses existing literature to establish a justified overall response rate as the starting point. In this case, Leonardi et al. (2003) have published a pivotal trial used for approval of etanercept in the psoriasis indication. In this study, they demonstrated a 55% effect size, that is, the difference between using etanercept (59%) versus the placebo (4%) based on those subjects achieving at least a 75% improvement in their PASI (Psoriasis Area Severity Index) score (Fredriksson and Pettersson, 1978). In other words, 55% more patients achieved a 75% improvement of their psoriasis lesions compared to placebo. To set up an equivalence trial design that retains at least 60% of this effect size, we could propose a margin of ±18% covering 37%–77% of patients having a 75% PASI improvement.

When looking at this range, physicians may say that a product that only produces a 37% PASI 75 response is not equivalent to a product that produces a 59% PASI 75 response; however, this does not reflect the clinical reality. For this, we have to look at how equivalence is assessed in such a trial.

The ultimate goal is to get an understanding of how different the biosimilar and Enbrel could potentially be in the whole psoriasis population based on the data generated in the clinical study—that is, based on a sample of all psoriasis patients eligible for etanercept treatment. This potential difference is estimated by a confidence interval, which will contain the true difference between the two products assessed at a certain probability, usually 90% (e.g., required by FDA) or 95% (e.g., required by EMA). In order for the whole confidence interval for the difference in PASI responses to fall within the justified equivalence margins, the actual difference between the two products must be far less than the 18% margin. In other words, the "point estimate" of the difference between the two products may only be a few percentage points different to meet this criterion. The actual clinical difference between the two products is best represented by the point estimate difference. It is clearly not clinically relevant if the point estimate difference is only a few percentage points between the two products. Clinicians would not consider it problematic or clinically important if 55% of patients achieved a PASI 75 response in one group of patients, while 58% of patients in a second group achieved a PASI 75 response. This is well within the variability of using PASI as a clinical tool. Therefore, to actually meet clinical equivalence statistically retaining 50% or 60% of the effect size requires the point estimates of the two treatments to be fairly similar (Figure 7.3).

One unique aspect that impacts a sponsor's ability to harmonize the clinical trial design assumptions across regions is the confidence interval used for justification of the statistical approach as outlined above. There is a longstanding disagreement in the design of equivalence studies between FDA and EMA regarding the confidence intervals needed to demonstrate equivalence. The FDA along with other regulatory authorities is comfortable with a 90% confidence interval (CI), whereas the EMA routinely insists that sponsors use a 95% CI for equivalence testing in pharmacodynamic and efficacy studies. There is no scientifically valid reason why PD or efficacy should be subject to a higher level of scrutiny (significance level of 2.5%) as compared to showing bioequivalence in PK studies where the use of 90% confidence intervals (significance level of 5%) is a globally accepted norm. Using a 95%, instead of a 90%, confidence interval as described above has a dramatic impact on the sample size in that it adds approximately 20% to the total number of test subjects. If the sponsor is willing to take a risk in harmonizing between regulatory authorities, it may justify a 90% confidence interval outside EMA with comfortable power to achieve success. However, using the same number of subjects while increasing the confidence interval for EMA to 95% reduces the statistical power by approximately 10%. If complying with EMA requirements and retaining reasonable power (≥80%),

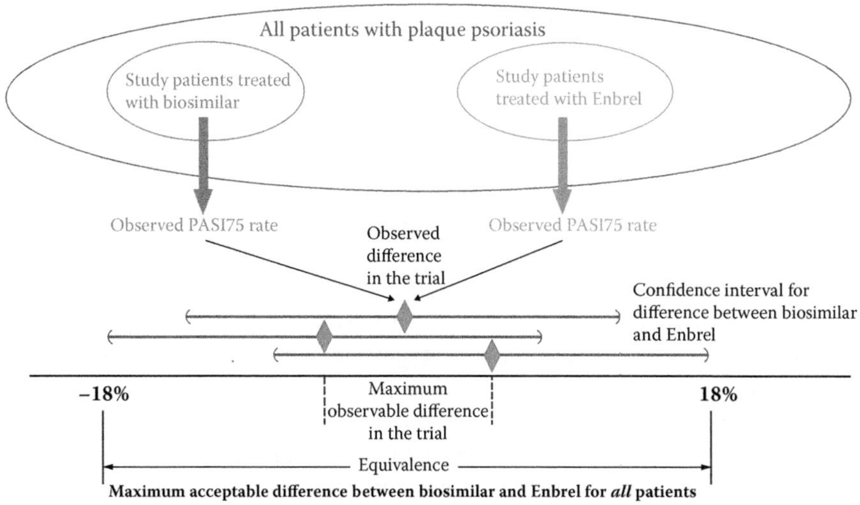

FIGURE 7.3 **(See color insert.)** For all patient responses to be within the equivalence margin, the actual point estimates or average responses for the patients treated with either drug must be fairly close to each other. Clinicians would be reassured about average responses of the patients treated with the biosimilar and the reference product being relatively close. Focus on the maximum margin justified does not represent the clinically relevant representation of the data.

the increased sample size induces clinical trial operational challenges and increases costs commensurately. Finally, a sponsor may have to have two different statistical analysis plans (SAPs), one for a 90% confidence interval outside of Europe and one for a 95% confidence interval for EMA.

The clinical effect size of the reference product (the difference in clinical response between the reference product and placebo) is important not only for sizing the study but also in context of sensitivity to pick up differences. The clinical indication with the largest effect size would be able to pick up smaller differences between the reference product and the proposed biosimilar. Therefore, regulatory authorities are also interested in using the "most sensitive" patient population when conducting the final efficacy study to confirm similarity (McCamish and Woollett, 2012). This can be illustrated in the choice of patient populations for biosimilar studies with anti-TNF biosimilar candidates. Many sponsors have or are using rheumatoid arthritis patients in the confirmatory study for their biosimilar (Yoo et al., 2013; Amgen, 2015c). For this class of drugs, psoriasis is actually a more sensitive population to use since the effect size is generally much larger. In addition, cotreatment with other immunosuppressive drugs is not as common in treating psoriasis as it is in treating rheumatoid arthritis (where methotrexate as a comedication is almost universal). Such immunosuppressants can impact the evaluation of immunogenicity, which is usually a key point of interest in the final confirmatory clinical trial for biosimilars. Therefore, use of a psoriasis population would be a more sensitive "model" for testing biosimilarity and, as with rheumatoid arthritis, published data allows for a robust statistical justification of the equivalence margin.

7.4.5 Primary Endpoints

Traditional primary endpoints for pivotal efficacy trials may or may not be used in the confirmatory biosimilar efficacy trial. Since the goal of the biosimilar trial is to confirm "sameness" or similarity to the reference product, and clinical testing is far less sensitive in discovering differences than analytical methods, the primary endpoint of the clinical trial has to focus on the biological effect of the molecule where there is greater sensitivity to pick up a difference.

7.4.5.1 Case Study

As mentioned above and outlined in Figure 7.2, ACR 20 in the evaluation of anti-TNF biologics is not sufficiently sensitive. Regulatory authorities' views on this topic have evolved, and they have now accepted the use of endpoints that can be evaluated sequentially and are not dichotomous. The routine biosimilar trials in rheumatoid arthritis often used ACR 20 as a primary endpoint measured after 6 months of treatment. As the regulatory mindset has evolved over time, it is acceptable now to use DAS28 (disease activity score in 28 joints) for the assessment of efficacy in rheumatoid arthritis patients (Hoffmann-La Roche, 2010; NRAS, 2016). Importantly, DAS28 can be measured repetitively and earlier on in the clinical trial, and some regulators understand that differences between biologics could be better documented on the upswing of drug exposure to these biologics which happens earlier in the clinical trial.

Transitioning to oncology studies, primary endpoints in trials for regulatory approval are overwhelmingly focused on the overall survival of patients after treatment with a new agent. This is appropriate as one wants to understand whether the new treatment actually has an impact on the survival of patients. Measuring other parameters such as size of the tumor after treatment or overall response could reveal a difference in tumor mass that does not benefit the patient's survival. Since trials looking at overall survival may take substantial time, some regulatory authorities, such as the FDA, came up with approaches for accelerated approval using progression free survival (PFS). Drugs approved using a primary endpoint of progression free survival are usually required to ultimately demonstrate a positive impact on overall survival.

However, for confirming "sameness" for a biosimilar after all of the analytics have demonstrated similarity of the molecules, conducting an oncology study using a primary endpoint of overall survival is not appropriate. Such an approach does not provide sensitivity to pick up differences between the molecules due to many confounding factors and would require huge and unfeasible studies. Biosimilar confirmatory clinical trials often focus on an aspect related to the biology of the biosimilar that is more sensitive to pick up differences between the molecules.

7.4.5.2 Case Study

For the biosimilar confirmatory clinical trial for trastuzumab, which is an anti-HER2 mAb, authorities and sponsors initially proposed clinical trials in metastatic breast cancer patients using a primary endpoint of PFS. As the understanding of biosimilar clinical trial design matured, most sponsors then focused on early breast cancer specifically using neoadjuvant treatment. In such patients, it is possible to evaluate response rate or tumor size prior to surgery when treated with trastuzumab or a proposed trastuzumab biosimilar. The primary endpoint in this setting can be pathological complete response (PCR) during surgery as this more sensitively measures the actual biologic function of the agent being used.

Validated biomarkers have also been used in the approval of novel drugs. Such biomarkers have been validated to predict the ultimate clinical outcome.

7.4.5.3 Case Study

Validated biomarkers include hemoglobin A1c (HbA1c) in approval of antidiabetic drugs, duration of severe neutropenia (DSN) for hematologic growth factors that prevent infection in chemotherapy patients, as well as LDL cholesterol in patients at risk for myocardial infarctions.

Unfortunately, there are very few validated biomarkers. Hence, only rarely is it possible to use one of these biomarkers in the development of biosimilars. However, it is possible to provide information to regulatory agencies about potential biomarkers in biosimilar clinical trials. This is particularly the case when a sponsor provides extended characterization data that provide assurance that there is "fingerprint-like" (FDA, 2015b) similarity of the proposed biosimilar to the reference product.

The medications used for cotreatments are also an important factor not only in considerations of immunogenicity, but also as part of the statistical justification of the equivalence margin. As mentioned above for statistical justification, published studies are required to define the clinical effect size, and this may depend on the comedication utilized. Since the reference biologic was often originally approved well over a decade ago, if not two, the comedications may have evolved. Often there are far less clinical data produced with newer comedications making it more difficult to justify the statistical margins and sample size needed to establish equivalence.

7.4.5.4 Case Study

The largest published database for efficacy of rituximab in treatment of follicular lymphoma includes cotreatment with CVP (Cyclophosphamide, Vincristine, and Prednisone) (Sandoz, 2015a); however, other chemotherapy treatments may be preferred in current clinical practice such as bortezomib (Coiffier et al., 2011). Therefore, although the sponsor can provide statistical justification for effect size, equivalence margin, and number of patients required; enrollment of the clinical trial may be challenging if the cotreatment is no longer the standard of care in major countries. Physicians prefer to use a more modern comedication either because of improved efficacy or better tolerance or safety and, therefore, they prefer to not be involved in a study that uses an older treatment.

This is the same situation as with filgrastim and pegfilgrastim where the most robust published information utilized breast cancer patients treated with TAC (docetaxel, doxorubicin, and cyclophosphamide) (Blackwell et al., 2015) chemotherapy, and current regimens utilize various dose-dense regimens with less risk of neutropenia, thus limiting interest in such a biosimilar clinical trial using TAC chemotherapy. Therefore, a sponsor can design a statistically justified biosimilar clinical trial that is accepted by multiple regulatory agencies but has challenges with recruitment and conduct of the trial.

7.4.6 Extrapolation

A unique aspect of biosimilar development and the resulting approval is the concept of extrapolation. As presented throughout this chapter, the goal of biosimilar development is to provide a stepwise set of data that can be interpreted within a "totality of the evidence" approach to enable a regulatory determination that the proposed biosimilar is "highly similar" or "essentially the same" as the reference product (EMA, 2012).

The clinical trial is the final stage of this stepwise approach, and it is designed to confirm the similarity of the biosimilar to the reference product. The trials are not designed to establish safety and efficacy, neither a priori nor in every indication of the reference product. Physicians are often unfamiliar with this unique biosimilar regulatory pathway where it is not required to conduct multiple clinical trials when the reference product is approved in multiple indications. The anti-TNF biologics are approved in multiple indications, with some having a far greater number of indications (such as infliximab) than others (such as etanercept). Since many physicians have direct experience with drugs that may work in one indication and not another, they struggle without seeing data in the specific indication for which they would use the biosimilar. However, once they understand the unique scientific approach used for development and approval of a biosimilar, and its one-to-one relationship with its originator reference product in addition to the fact that the data generated by the sponsor provides evidence that the biosimilar is essentially the same biological substance as the reference product, extrapolation becomes not only feasible but inevitable. If the biological substances are essentially the same (the biosimilar is essentially the same as the reference product), then they must have the same clinical effect in the same indications as the reference product. To the health care provider and patient, using a biosimilar is clinically just like using another batch of the reference product. It is just that in the case of the biosimilar the sponsor is different. If such a complete dataset is acceptable by regulators as proving the sameness of the biosimilar, then it is inappropriate, and indeed unethical, to perform clinical trials of the biosimilar in all additional indications of the reference product.

7.4.7 Interchangeability

There is one final unique aspect of biosimilar clinical trial design if the sponsor is intending to obtain an interchangeability designation in the US. The approach currently envisaged includes switching patients from one treatment (either the biosimilar or the reference product) to the opposite product multiple times. Such switching is usually performed after reaching the primary endpoint for the efficacy analysis. This can be after 12 weeks of treatment for psoriasis (Sandoz, 2015b), 24 weeks of treatment for rheumatoid arthritis (Amgen, 2015c), or after the first course of chemotherapy in breast cancer patients treated with TAC chemotherapy (Blackwell et al., 2015). Such a switching protocol must allow for statistical evaluation of equivalent efficacy of the switching group with a nonswitched group. In addition, immunogenicity is evaluated between the switched group versus the nonswitched group. This makes the clinical trial design somewhat complicated as patients treated with the biosimilar must initially be compared with patients treated with the reference

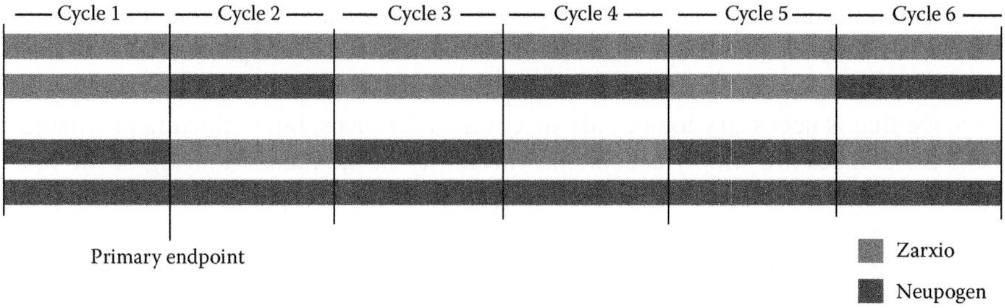

FIGURE 7.4 **(See color insert.)** Clinical trial design of the Zarxio biosimilar confirmatory clinical trial presented at the January 7, 2015, Oncologic Drug Advisory Board (Sandoz, 2015c). The upper and lower groups were treated continuously with either the biosimilar (light gray) or the reference product (dark gray). The inner groups were switched back and forth five times between the initial product assigned and the other product (biosimilar or reference product). Initial statistical comparison at the primary endpoint time point was between the two groups treated with the biosimilar compared with the two groups treated with the reference product. Thereafter, efficacy or immunogenicity could be compared between the groups treated continuously versus the groups switched back and forth between products.

product at the time of the primary endpoint evaluation. After the primary endpoint time point is reached, subjects must be randomized thereafter into a switched treatment group or a nonswitched group (that is, a portion of the patients continue consistent treatment throughout the entire protocol).

Sandoz's approach to accomplish this design is to start with four groups upon initiation to include two groups being treated with the reference product and two groups being treated with the biosimilar, all in a blinded fashion. Upon reaching the primary endpoint, one group being treated with the biosimilar and one group being treated with the reference product enters a switching period where all subjects assigned to that group are switched back and forth multiple times between the reference product and the biosimilar (Blackwell et al., 2015; Figure 7.4).

7.5 SUMMARY

Clinical trial designs to support the approval of biosimilars in the highly regulated markets are quite different from those used for the approval of novel drugs. There is a paradigm shift required in approaching biosimilar clinical trials wherein the focus is on confirming the similarity of the structure and function of the biosimilar molecule to that of the reference product (Woodcock, 2012).

This is performed in a stepwise fashion beginning with well-established and proven analytical and functional comparison methods that can sensitively pick up very minor differences between the molecules, some of which may well have no clinical relevance. Since the focus is on comparing the molecules and not challenging the safety and efficacy of the reference molecule, the clinical design uses different approaches to minimize clinical variability and using endpoints that take into consideration the biologic function of the molecule. Equivalence designs are often required, although noninferiority designs are justified. The designs and methodologies used are selected because they are the most powerful and sensitive to enable discovery of differences between the proposed biosimilar and the reference product should they exist. Indeed, it is the combination of the analytical and clinical data or totality of evidence that is necessary to not only discover differences, but to document whether those differences are simply acceptable variability in the manufacturing of the biologics or clinically relevant differences that would be unacceptable and may have an impact on patients.

Clinicians are often less familiar with the principles of this unique approach, but when information is shared about biosimilarity, and it is explained that the goal is to prove that the proposed biosimilar product is essentially the same as the reference product by relying heavily on many very sensitive analytical tools that probe every aspect of the molecules, then they become quite intrigued. Clinicians are aware of the access limitations involved in prescribing biological drugs to patients due to the expense of these agents and see the opportunities created for their patients by biosimilars. There is intense passion and interest in addressing this access issue, and clinicians become very interested in being involved with this process, thereby enabling more patients to benefit and to be treated earlier in the development of their disease.

REFERENCES

Amgen. (2015a) ARANESP (darbepoetin alfa) injection product insert. July 23, 2015. Available from: http://www.accessdata.fda.gov/drugsatfda_docs/label/2015/103951s5363lbl.pdf.

Amgen. (2015b) Efficacy and safety study of ABP 215 compared with bevacizumab in subjects with advanced non-small cell lung cancer. Clinicaltrials.gov. NCT01966003. November 2015. Available from: https://clinicaltrials.gov/ct2/show/NCT01966003?term=ABP+215&rank=1.

Amgen. (2015c) Efficacy and safety study of ABP 501 compared to adalimumab in subjects with moderate to severe rheumatoid arthritis (RA). Clinicaltrials.gov. NCT01970475. May 2015. Available from: https://clinicaltrials.gov/ct2/show/NCT01970475?term=amgen+ABP501&rank=3.

Amgen. (2015d) NEULASTA (pegfilgrastim) injection product insert. November 13, 2015. Available from: http://www.accessdata.fda.gov/drugsatfda_docs/label/2015/125031s180lbl.pdf.

Bartelds GM, Krieckaert CLM, Nurmohamed MT, et al. (2011) Development of antidrug antibodies against adalimumab and association with disease activity and treatment failure during long-term follow-up. *JAMA* **305(14)**, 1460–1468.

Blackwell K, Semiglazov V, Krasnozhon D, et al. (2015) Comparison of EP2006, a filgrastim biosimilar, to the reference: a phase III randomized, double-blind clinical study in the prevention of severe neutropenia in patients with breast cancer receiving myelosuppressive chemotherapy. *Annals of Oncology* **26(9)**, 1948–1953.

Bosques CJ, Collins BE, Meador JW, et al. (2011) Chinese hamster ovary cells can produce galactose-α-1,3-galactose antigens on proteins. *Nature Biotechnology* **29(459)**, doi:10.1038/nbt0511-459e.

Casadevall N. (2002) Antibodies against rHuEPO: native and recombinant. *Nephrology Dialysis Transplantation* **17(Suppl. 5)**, 42–47.

Coiffier B, Osmanov EA, Hong Z, et al. (2011) LYM-3001 study investigators. Bortzomib plus rituximab versus rituximab alone in patients with relapsed, rituximab-naïve or rituximab-sensitive, follicular lymphoma: a randomized phase 3 trial. *The Lancet Oncology* **12(8)**, 773–784. Available from: http://www.sciencedirect.com/science/article/pii/S1470204511701504.

Elmeliegy M, Potocka E, Renard D, et al. (2015) Population pharmacokinetic and pharmacodynamic modeling of bevacizumab and a proposed biosimilar (GP2019) in healthy male volunteers. *European Society for Medical Oncology*. European Cancer Congress 2015, Poster #307. Available from: http://www.europeancancercongress.org/Scientific-Programme/Abstract-search#.

EMA. (2005) Guideline on the choice of the non-inferiority margin, EMEA/CPMP/EWP/2158/99, July 27, 2005, European Medicines Agency, London, UK. Available from: http://www.ema.europa.eu/docs/en_GB/document_library/Scientific_guideline/2009/09/WC500003636.pdf.

EMA. (2007) Committee for medicinal products for human use guideline: the clinical investigation of the pharmacokinetics of therapeutic proteins, CHMP/EWP/89249/2004. January 24, 2007, European Medicines Agency, London, UK. Available from: http://www.ema.europa.eu/docs/en_GB/document_library/Scientific_guideline/2009/09/WC500003029.pdf.

EMA. (2008) Assessment report for Aranesp Procedure No. EMA//C/332/X/0042, EMEA/478499/2008. July 3, 2008, European Medicines Agency, London, UK. Available from: http://www.ema.europa.eu/docs/en_GB/document_library/EPAR_-_Assessment_Report_ _Variation/human/000332/WC500026148.pdf.

EMA. (2012) Questions and answers on biosimilar medicines (similar biological medicinal products). EMA/837805/2011. September 27, 2012, European Medicines Agency, London, UK. Available from: http://www.ema.europa.eu/ema/index.jsp?curl=pages/special_topics/document_listing/document_listing_000318.jsp&murl=menus/special_topics/special_topics.jsp&mid=WC0b01ac0580281bf0.

EMA. (2014) Guideline on similar biological medicinal products containing biotechnology-derived proteins as active substance: non-clinical and clinical issues, EMEA/CHMP/BMWP/42832/2005 Rev. 1. December 18, 2014, European Medicines Agency, London, UK. Available from: http://www.ema.europa.eu/docs/en_GB/document_library/Scientific_guideline/2015/01/WC500180219.pdf.

FDA. (2010) Draft guidance for industry: non-inferiority clinical trials. March 2010, U.S. Food and Drug Administration. Available from: http://www.fda.gov/downloads/Drugs/GuidanceComplianceRegulatoryInformation/Guidances/UCM202140.pdf.

FDA. (2013) Draft guidance for industry: bioequivalence studies with pharmacokinetic endpoints for drugs submitted under an ANDA. December 2013, U.S. Food and Drug Administration. Available from: http://www.fda.gov/downloads/Drugs/GuidanceComplianceRegulatoryInformation/Guidances/UCM377465.pdf.

FDA. (2014) Draft guidance for industry: clinical pharmacology data to support a demonstration of biosimilarity to a reference product. May 2014, U.S. Food and Drug Administration. Available from: http://www.fda.gov/downloads/Drugs/GuidanceComplianceRegulatory Information/Guidances/UCM397017.pdf.

FDA. (2015a) Guidance for industry: scientific considerations in demonstrating biosimilarity to a reference product. April 2015, U.S. Food and Drug Administration. Available from: http://www.fda.gov/downloads/Drugs/GuidanceComplianceRegulatoryInformation/Guidances/UCM291128.pdf.

FDA. (2015b) Guidance for industry: quality considerations in demonstrating biosimilarity of a therapeutic protein product to a reference product. April 2015, U.S. Food and Drug Administration. Available from: http://www.fda.gov/downloads/Drugs/GuidanceComplianceRegulatoryInformation/Guidances/UCM291134.pdf.

Fredriksson T, Pettersson U. (1978) Severe psoriasis—oral therapy with a new retinoid. *Dermatologica* **157(4)**, 238–244.

Galili U. (2005) The α-gal epitope and the anti-Gal antibody in xenotransplantation and in cancer immunotherapy. *Immunology and Cell Biology* **83**, 674–686. Available from: http://www.nature.com/icb/journal/v83/n6/full/icb200593a.html.

Goetze AM, Liu YD, Shang Z, et al. (2011) High-mannose glycans on the Fc region of therapeutic IgG antibodies increase serum clearance in humans. *Glycobiology* **21**, 949–959.

Health Canada. (2010) Guidance for sponsors: information and submission requirements for subsequent entry biologics (SEBs). March 5, 2010, Health Canada, Ottawa, ON. Available from: http://hc-sc.gc.ca/dhp-mps/alt_formats/pdf/brgtherap/applic-demande/guides/seb-pbu/seb-pbu-2010-eng.pdf.

Hoffmann-La Roche. (2010) SMART study: a study of re-treatment with MabThera (Rituximab) in patients with rheumatoid arthritis who have failed on anti-TNF alfa therapy. Clinicaltrials.gov NCT01126541. September 2010. Available from: https://clinicaltrials.gov/ct2/show/NCT01126541?no_unk=Y&outc=DAS28&rank=2.

Holzmann, J, Balser S, Windisch J. (2015) Totality of the evidence at work: the first U.S. biosimilar. *Expert Opinion on Biological Therapy* **16**, 137–142, doi:10.1517/14712598.2016.1128410.

ICH. (2004) Technical requirements for registration of pharmaceuticals for human use. 2004. International conference on harmonisation: harmonised tripartite guideline, comparability of biotechnological/biological products subject to changes in their manufacturing process Q5E. November 18, 2004. Available from: http://www.ich.org/fileadmin/Public_Web_Site/ICH_Products/Guidelines/Quality/Q5E/Step4/Q5E_Guideline.pdf.

Jirillo A, Vascon F, Giacobbo M. (2008) Bevacizumab in advanced cancer, too much or too little? *Annals of Oncology* **19(10)**, 1817–1818. Available from: http://annonc.oxfordjournals.org/content/19/10/1817.full.

Keystone E, Genovese M, Klareskog L, et al. (2009) Golimumab, a human antibody to tumour necrosis factor α given by monthly subcutaneous injections, in active rheumatoid arthritis despite methotrexate therapy: the GO-FORWARD Study. *Annals of Rheumatic Disease* **68**, 789–796.

Keystone E, van der Heijde D, Mason, Jr. D, et al. (2008) Certolizumab Pegol plus methotrexate is significantly more effective than placebo plus methotrexate in active rheumatoid arthritis. *Arthritis and Rheumatism* **58(11)**, 3319–3329.

Leonardi CL, Powers JL, Matheson RT, et al. (2003) Etanercept Psoriasis Study Group. Etanercept as monotherapy in patients with psoriasis. *New England Journal of Medicine* **349**, 2014–2022. Available from: http://www.nejm.org/doi/pdf/10.1056/NEJMoa030409.

Macdougall IC. (2005) Antibody-mediated pure red cell aplasia (PRCA): epidemiology, immunogenicity and risks. *Nephrology Dialysis Transplantation* **20(Suppl. 4)**, iv9–iv15.

Maini R, St Clair E, Breedveld F, et al. (1999) Infliximab (chimeric anti-tumour necrosis factor α monoclonal antibody) versus placebo in rheumatoid arthritis patients receiving concomitant methotrexate: a randomised phase III trial. *The Lancet* **354**, 1932–1939.

Maloney DG, Grillo-Lopez AJ, Bodkin DJ, et al. (1997) IDEC-C2B8: results of a phase I multiple-dose trial in patients with relapsed non-Hodgkin's lymphoma. *Journal of Clinical Oncology* **15(10)**, 3266–3274.

McCamish M, Woollett G. (2012) The state of the art in the development of biosimilars. *Clinical Pharmacology and Therapy* **91(3)**, 405–417.

NRAS. (2016) The DAS28 score. National Rheumatoid Arthritis Society. Available from: http://www.nras.org.uk/the-das28-score.

Sandoz. (2015a) GP2013 in the treatment of patients with previously untreated, advanced stage follicular lymphoma (ASSIST_FL). Clinicaltrials.gov. NCT01419665. March 2015. Available from: https://clinicaltrials.gov/ct2/show/NCT01419665?term=GP2013+sandoz&rank=4.

Sandoz. (2015b) Study to demonstrate equivalent efficacy and to compare safety of biosimilar Etanercept (GP2015) and Enbrel (EGALITY). Clinicaltrials.gov. NCT01891864. October 2015. Available from: https://clinicaltrials.gov/ct2/show/NCT01891864.

Sandoz. (2015c) Presentation to the Oncologic Drug Advisory Board of FDA. Available from: http://www.fda.gov/downloads/AdvisoryCommittees/CommitteesMeetingMaterials/Drugs/OncologicDrugsAdvisoryCommittee/UCM431119.pdf.

Schiestl M, Stangler T, Torella C, et al. (2011) Acceptable changes in quality attributes in glycosylated biopharmaceuticals. *Nature Biotechnology* **29(4)**, 310–312.

Sorgel F, Schwebig A, Holzmann J, et al. (2015) Comparability of biosimilar filgrastim with originator filgrastim: protein characterization, pharmacodynamics, and pharmacokinetics. *BioDrugs* **29**, 123–131.

Sorgel F, Thyroff-Friesinger U, Vetter A, et al. (2009) Bioequivalence of HX575 (recombinant human epoetin alfa) and a comparator epoetin alfa after multiple subcutaneous administrations. *Pharmacology* **83**, 122–130.

Vezér B, Buzás Z, Sebeszta M, Zrubka Z. (2016) Authorized manufacturing changes for therapeutic monoclonal antibodies (mAbs) in European Public Assessment Report (EPAR) documents. *Current Medical Research and Opinion*, doi:10.1185/03007995.2016.1145579.

Weinblatt M, Keystone E, Furst D, et al. (2003) Adalimumab, a fully human anti-tumor necrosis factor α monoclonal antibody, for the treatment of rheumatoid arthritis in patients taking concomitant methotrexate. *Arthritis and Rheumatism* **48(1)**, 35–45.

Weinblatt M, Kremer J, Bankhurst A, et al. (1999) A trial of etanercept, a recombinant tumor necrosis factor receptor:Fc fusion protein, in patients with rheumatoid arthritis receiving methotrexate. *New England Journal of Medicine* **340(4)**, 253–259.

Woodcock J. (2007) Assessing the impact of a safe and equitable biosimilar policy in the United States. Statement to the House Committee on Energy and Commerce, Subcommittee on Health. Available from: http://www.fda.gov/NewsEvents/Testimony/ucm154017.htm.

Woodcock J. (2012) Keynote address to the Drug Information Association (DIA)/FDA Biosimilars Conference, September 2012, Washington, DC.

Yang BB, Morrow PK, Wu X, et al. (2015) Comparison of pharmacokinetics and safety of pegfilgrastim administered by two delivery methods: on-body injector and manual injection with a prefilled syringe. *Cancer Chemotherapy and Pharmacology* **75(6)**, 1199–1206.

Yoo DH, Hrycai P, Miranda P, et al. (2013) A randomised, double-blind, parallel-group study to demonstrate equivalence in efficacy and safety of CT-P13 compared with innovator infliximab when coadministered with methotrexate in patients with active rheumatoid arthritis: the PLANETRA study. *Annals of Rheumatic Disease* **72(10)**, 1613–1620.

8 Statistical Methods for Assessing Biosimilarity

Shein-Chung Chow
Duke University School of Medicine

Fuyu Song
China Food and Drug Administration

CONTENTS

8.1 INTRODUCTION

Biological drugs make up one of the fastest-growing sectors of the pharmaceutical and biotechnology industry (IMS, 2010). For example, in 2010, spending on global biologics totaled about $138 billion. The spending of biosimilars has grown rapidly in the past few years. It is expected that spending on biologics will roar to over $200 billion beyond 2015 (see, e.g., Figure 8.1). This has given the pharmaceutical and biotechnology industry the opportunity to develop biosimilars when the innovative biologics go off patent protection.

Because of the high costs involved in the production and consumption of many drug products, regulatory regimes have been created to balance the intellectual property interests (patent protection) and investment made by innovative (originator) companies, with the need for wider patient access through generic forms of the drugs. In the United States, for the traditional chemical (small-molecule) drug market, such a regime was created by the Hatch–Waxman Act. This Act added

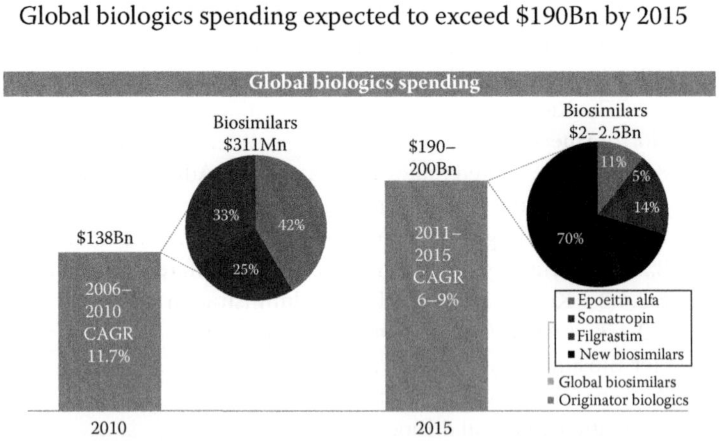

FIGURE 8.1 **(See color insert.)** Global biologics spending (in billion dollars). (IMS, Multinational integrated data analysis system (MIDAS). IMS Institute for Healthcare and Informatics, Danbury, CT, 2010.)

Section 505 (j) to the US Federal Food, Drug, and Cosmetic Act. This section and its accompanying regulations created the abbreviated new drug application (ANDA) process which was designed to provide independent generic firms with a strong incentive to develop and introduce lower-cost generic drugs to consumers. By virtually all accounts, the Hatch–Waxman Act has been extremely successful in bringing cheaper generic products to the market while maintaining incentives for the development and discovery of new drugs. Because of the effectiveness of this law's approach in the context of traditional small-molecule drugs, some have called for applying a similar regime for biologics, given the similarly high costs in this class of drugs. In particular, Representative Henry Waxman (D-CA), one of the original authors of the Hatch–Waxman Act, has sponsored the Access to Life-Saving Medicine Act to determine whether such a legal infrastructure was appropriate for regulating the follow-on biologics, and a thorough policy assessment was necessary. This effort has led to passage of the Biologics Price Competition and Innovation (BPCI) Act.

Following passage of the BPCI Act, the FDA hosted a public hearing on November 2–3, 2010 to obtain public input regarding scientific factors for assessing biosimilarity and drug interchangeability of biosimilar products. After extensive discussions, the FDA developed and circulated three draft guidances on biosimilars on February 9, 2012, and hosted another public hearing to obtain public input and comments on the draft guidances on May 11, 2012. These guidances were subsequently finalized in 2015 (FDA, 2015a,b,c). At the May 11 public hearing, special attention was directed to discussion of drug interchangeability in terms of the concepts of switching and alternating as described in the BPCI Act. On March 23, 2010, the BPCI Act (as part of the Affordable Care Act) was written into law, giving the FDA the authority to approve similar biological drug products.

As indicated in the BPCI Act, a biosimilar product is defined as a product that is *highly similar* to the reference product notwithstanding minor differences in clinically inactive components, and there are no clinically meaningful differences in terms of safety, purity, and potency. However, the BPCI Act provides little or no discussion regarding how similar is considered highly similar (see also Chow, 2013a,b). On the other hand, the European Union (EU) has a well-established and well-documented legal and regulatory pathway for the review and approval of biosimilars. Other countries and organizations around the world, including Australia, Canada, China, Japan, Switzerland, and the World Health Organization (WHO), are also following the same scientific principles for an abbreviated approval pathway for biosimilars. In contrast, the United States is at the very beginning of the process, with a legal pathway being discussed by the United States Congress for a number of years. The FDA's views on biosimilars can be examined from a publication (Woodcock et al., 2007) and from communications between the FDA and Congress. These documents seem to indicate that the FDA is contemplating similar scientific principles to those established by the EU's EMA.

This chapter provides a comprehensive review of regulatory requirements, criteria for biosimilarity, and statistical methods, including power calculations for determining the sample size needed for the assessment of biosimilarity between a proposed biosimilar product and an innovative biologic product. In addition, the chapter introduces a unified approach for demonstrating biosimilarity using the concept of reproducibility.

8.2 REGULATORY REQUIREMENTS

For the assessment of similar biological products, regulatory requirements from the WHO and different regions such as the EU, the US, and the Asian Pacific Region such as Japan, Taiwan, South Korea, and China are similar and yet slightly different (Wang and Chow, 2012; Chow et al., 2013, 2015), which are briefly described below. To provide a better understanding of regulatory requirements from different regions, Table 8.1 presents a comparison of requirements for the evaluation of biosimilars.

8.2.1 The World Health Organization

As an increasingly wide range of similar biotherapeutic products (SBPs) was under development or already licensed in many countries, WHO formally recognized the need for guidance for their evaluation and overall regulation in 2007. The *Guidelines on Evaluation of Similar Biotherapeutic Products (SBPs)* were developed and adopted by the 60th meeting of the WHO Expert Committee on Biological Standardization in 2009. The intention of the guidelines was to provide globally acceptable principles for licensing biotherapeutic products claimed to be similar to the reference products that had been licensed based on a full licensing dossier (WHO, 2009). The scope of the guidelines includes well-established and well-characterized biotherapeutic products that have been marketed for a suitable period of time with proven quality, efficacy, and safety, such as recombinant DNA-derived therapeutic proteins.

8.2.1.1 Key Principles and Basic Concept

As indicated in the WHO guidelines, one of the most important principles of developing an SBP is the stepwise approach, starting with the characterization of quality attributes of the product and followed by nonclinical and clinical evaluations. Manufacturers should submit a full quality dossier that includes a complete characterization of the product, a demonstration of the consistent and robust manufacture of their product, and the comparability exercise between the SBP and the reference biotherapeutic product (RBP) in the quality part, which together serve as the basis for the possible reduction in data requirements in the nonclinical and clinical development. This principle indicates that data reduction is only possible for the nonclinical and clinical parts of the development program, and significant differences between the SBP and the chosen RBP detected during the comparability exercise would result in the requirement for more extensive nonclinical and clinical data. In addition, the amount of nonclinical and clinical data considered necessary also depends on the class of products, which calls for a case-by-case approach for different classes of products.

8.2.1.2 Reference Biotherapeutic Product

The choice of the reference biotherapeutic product is another important issue covered in the WHO guidelines. Traditionally, national regulatory authorities (NRAs) have required the use of a nationally-licensed reference product for registering generic medicines, but this may not be feasible in countries lacking nationally-licensed RBPs. Thus, additional criteria may be needed to guide the acceptability of using an RBP licensed in other jurisdictions. Considering the choice of the RBP, WHO

TABLE 8.1
Comparison of Requirements for the Evaluation of SBPs between Different Regions

	WHO	Canada	Korea	EU	Japan
Term	SBPs	SEBs	Biosimilars	Biosimilars	Follow-on Biologics
Scope	Recombinant protein drugs			Mainly recombinant protein drugs	Recombinant protein drugs
Efficacy	Double-blind or observer-blind: Equivalence or noninferiority design		Equivalence design	Comparability margins should be prespecified and justified	
Reference product	Authorized in a jurisdiction with well-established regulatory framework			Authorized in EU	Authorized in Japan
Stability	• Accelerated degradation studies • Studies under various stress conditions				Not necessary
Purity	Process-related and product-related impurities				
Manufacture	• Same standards required by the NRA for originator products • Full chemistry and manufacture data package				
Physicochemical	• Primary and higher-order structure • Posttranslational modifications				
Biological activity	• Qualitative measure of the function • Quantitative measure (e.g., enzyme assays or binding assays)				
Nonclinical studies	• *In vitro* (e.g., receptor-binding, cell-based assays) • *In vivo* (pharmacodynamics activity, at least one repeat dose toxicity study, antibody measurements, local tolerance)				

(*Continued*)

TABLE 8.1 (*Continued*)
Comparison of Requirements for the Evaluation of SBPs between Different Regions

	WHO	Canada	Korea	EU	Japan
PK study design and criteria	• Single-dose, steady-state studies, or repeated determination of PK • Crossover or parallel • Include absorption and elimination characteristics • Traditional 80%–125% equivalence range is used				
PD	Pharmacodynamics markers should be selected, and comparative PK/PD studies may be appropriate				
Safety	Prelicensing safety data and risk management plan				
Principles	• Generic approach is not appropriate for follow-on biologic • Follow-on biologic should be similar to the reference in terms of quality, safety, efficacy • Stepwise comparability approach: similarity of the SBP to RBP in terms of quality is a prerequisite for reduction of nonclinical and clinical data required for approval • Case-by-case approach for different classes of products • Pharmacovigilance is stressed				

Source: Wang, J and Chow, SC, *Pharmaceuticals*, 5, 353–368, 2012.

requires that it should have been marketed for a *suitable* duration and have a volume of marketed use, and also that it should be licensed based on full quality, safety, and efficacy data. Besides, the same RBP should be used throughout the development of SBP, and the drug substance, dosage form, and route of administration of the SBP should be the same as that of the RBP.

8.2.1.3 Quality

As mentioned in the preceding section, a comprehensive comparison showing quality similarity between SBP and RBP is a prerequisite for applying the clinical safety and efficacy profile of RBP to SBP. Thus, a full quality dossier for both drug substance and drug product is always required. To evaluate comparability, WHO recommends that the manufacturer conduct a comprehensive physicochemical and biological characterization of the SBP in head-to-head comparisons with RBP. The following aspects of product quality and heterogeneity should be assessed.

8.2.1.4 Manufacturing Process

The manufacturing process should meet the same standards as required by the NRAs for the originator products and should implement good manufacturing practices, modern quality control and assurance procedures, in-process controls, and process validation. The SBP manufacturer should assemble all available knowledge of the RBP with regard to the type of host cell, formulation, and container closure system, and submit a complete description and data package delineating the whole manufacturing process, including obtaining an expression of target genes, the optimization and fermentation of gene engineering cells, the clarification and purification of the products, formulation and testing, aseptic filling, and packaging.

8.2.1.5 Characterization

Thorough characterization and comparability exercise are required, and details should be provided on the primary and higher-order structure, posttranslational modifications, biological activity, process- and product-related impurities, relevant immunochemical properties, and results from accelerated degradation studies and studies under various stress conditions.

8.2.1.6 Nonclinical and Clinical Studies

After demonstrating the similarity of SBP and RBP in quality, proving the safety and efficacy of an SBP usually requires further nonclinical and clinical data. Nonclinical evaluations should be undertaken both *in vitro* (e.g., receptor-binding studies, cell-proliferation, cytotoxicity assays) and *in vivo* (e.g., biological/pharmacodynamics activity, repeat dose toxicity study, toxicokinetic measurements, antiproduct antibody titers, cross reactivity with homologous endogenous proteins, product neutralizing capacity).

In terms of the clinical evaluation, the comparability exercise should begin with pharmacokinetic (PK) and pharmacodynamic (PD) studies, followed by the pivotal clinical trials. PK studies should be designed to enable detection of potential differences between SBP and RBP. Single-dose crossover PK studies in homogeneous populations are recommended by WHO. The manufacturer should justify the choice of single-dose studies, steady-state studies, or repeated determination of PK parameters, and the study population. Owing to the lack of established acceptance criteria for the demonstration of similar PK between SBP and RBP, the traditional 80%–125% equivalence range is often used. Besides, PD studies and confirmatory PK/PD studies may be appropriate if there are clinically relevant PD markers. In addition, the similar efficacy of SBP and RBP has to be demonstrated in randomized and well-controlled clinical trials, which should preferably be double-blind or at least observer-blind. In principle, equivalence designs (requiring lower and upper comparability margins) are clearly preferred for the comparison of the efficacy and safety of SBP with RBP. Noninferiority designs (requiring only one margin) may be considered if appropriately justified. The WHO also suggests that prelicensing safety and immunogenicity data should be obtained from the comparative efficacy trials.

In addition to the nonclinical and clinical data, applicants also need to present an ongoing risk management and pharmacovigilance plan, since data from preauthorized clinical studies are usually too limited to identify all potential side effects of the SBP. The safety specification should describe important identified or potential safety issues for the RBP and any that are specific for the SBP.

In summary, the WHO guidelines on evaluating similar biotherapeutic products represent an important step forward in the global harmonization of biosimilar product evaluation and regulation, and provide clear guidance for both regulatory bodies and the pharmaceutical industry.

8.2.2 The European Union

The EU has pioneered the development of a regulatory system for biosimilar products. The EMA began formal consideration of scientific issues presented by biosimilar products at least as early as January 2001 when an *ad hoc* working group discussed the comparability of medicinal products containing biotechnology-derived proteins as active substances (CPMP, 2001). In 2003, the European Commission amended the provisions of the EU's secondary legislation governing requirements for marketing authorization of applications for medicinal products to establish a new category of applications for "similar biological medicinal products" (CD, 2003). In 2005, the EMA issued a general guideline to introduce the concept of similar biological medicinal products, to outline the basic principles to be applied, and to provide applicants with a "user guide" showing where to find relevant scientific information (EMA, 2005). Since then, 14 biosimilar products have been approved by EMA under the pathway. One of the rejected biosimilars is Alpheon (interferon α-2a). It was developed by BioPartners GmbH and designed to become a biosimilar of the reference product Roferon-A for the treatment of adult patients with chronic hepatitis C. The EMA refused the marketing authorization for Alpheon due to the difference identified between Alpheon and the reference product, such as impurities, stability, and side effects.

8.2.2.1 Key Principles and Basic Concept

Unlike the WHO guideline which seems to focus more on recombinant DNA-derived therapeutic proteins, EMA's guidelines clearly indicate that the concept of a "similar biological medicinal product" is applicable to a broad spectrum of products ranging from biotechnology-derived therapeutic proteins to vaccines, blood-derived products, monoclonal antibodies, gene and cell-therapy, and so on. However, comparability exercises to demonstrate similarity are more likely to be applied to highly purified products that can be thoroughly characterized such as biotechnology-derived medicinal products. Considering the amount of data submitted, EMA also requires a full quality dossier, while the comparability exercise at the quality level may allow a reduction of the nonclinical and clinical data requirement compared to a full dossier. In 2011, EMA published a concept paper on the revision of the guideline on similar biological medicinal products (EMA, 2011a). It emphasized another main concept that clinical benefit has already been established by the reference medicinal product and that the aim of a biosimilar development program is to establish similarity to the reference product, not clinical benefit. Besides, a clear definition of "biosimilar," the feasibility of following the generic legal basis for some biological products, and the refinement based on the experience gained are recommended.

8.2.2.2 Reference Biotherapeutic Product

Similar to the regulatory requirement from WHO, EMA requires that the active substance, the pharmaceutical form, strength, and route of administration of the biosimilar should be the same as that of the reference product. The same chosen reference medicinal product should be used throughout the comparability program for quality, safety, and efficacy studies during the development of the biosimilar product. One of the major differences between WHO and EMA in terms of the choice of reference product is that EMA requires the chosen reference medicinal product to be a medicinal product authorized in the EU. Data generated from comparability studies with medicinal products authorized outside the EU may only provide supportive information.

8.2.2.3 Quality

In 2006, the *Guideline on Similar Biological Medicinal Products Containing Biotechnology-Derived Proteins as Active Substance: Quality Issues* was adopted by the Committee for Medicinal Products for Human Use (CHMP) (EMA, 2006b), which addresses the requirements regarding manufacturing processes, the comparability exercises for quality, analytical methods, physicochemical characterization, biological activity, purity, and specifications of the similar biological medicinal products. In 2011, EMA issued a concept paper on the revision of this guideline (EMA, 2011b). This concept paper proposes that the guideline published in 2006 needs refinements, taking into account the evolution of a quality profile during the product life cycle. This is because in the context of a biotherapeutic product claiming or claimed to be similar to another one already marketed, the conclusion of a comparability exercise performed with a reference product at a given time may not hold true from the initial development of the biosimilar, through marketing authorization, until the product's discontinuation.

8.2.2.4 Nonclinical and Clinical Evaluation

The *Guideline on Similar Biological Medicinal Products Containing Biotechnology-Derived Proteins as Active Substance: Non-clinical and Clinical Issues* published in

2006 lays down the nonclinical and clinical requirements for a biological medicinal product claiming to be similar to one already marketed (EMA, 2006e). The nonclinical section of the guideline addresses the pharmacotoxicological assessment, and the clinical section addresses the requirements for PK/PD and efficacy studies. Clinical safety studies as well as the risk management plan, with special emphasis on studying the immunogenicity of the biosimilar products, are also required. In 2011, EMA published a concept paper on the revision of this guideline (EMA, 2011c) that indicates several issues in need of discussion for a potential revision. First, EMA emphasizes the need to follow the 3Rs principles (replacement, reduction, and refinement) with regard to the use of animal experiments. Second, a revised version of the guideline will consider a risk-based approach for the design of an appropriate nonclinical study program. Third, the guideline should be clearer considering the need and acceptance of pharmacodynamics markers, and the measures that should be taken in the event relevant markers are not available. It should be noted, however, that the EMA issued a biosimilarity guideline in October, 2014, making an attempt to address these issues (EMA, 2014).

8.2.2.5 Product Class–Specific Guidelines

The principles of biosimilar drug development discussed earlier in this chapter apply in general to all biological drug products. However, there are no standard datasets that can be applied to the approval of all classes of biosimilars. Each class of biologic varies in its benefit/risk profile, the nature and frequency of adverse events, the breadth of clinical indications, and whether surrogate markers for efficacy are available and validated. Accordingly, the EMA has developed product class–specific guidelines that define the nature of comparative studies. So far, guidance for the development of biosimilar products has been developed for six different product classes: erythropoietins, insulins, growth hormones, alpha interferons, granulocyte-colony stimulating factors, and low-molecular-weight heparins (LMWHs), as well as beta interferons, follicle stimulation hormone, and monoclonal antibodies (EMA, 2006a,b,c,d, 2010a,b, 2011d).

8.2.2.6 European Experiences

As indicated earlier, the EMA has issued scientific guidelines on the quality, nonclinical, and clinical standards for the approval of biosimilars. The EMA has issued product class–specific guidelines (including EPO, G-CSF, insulin, growth hormone, LMW heparin, and interferon-α). According to these product class–specific guidelines, as of December 31, 2010, 14 biosimilar drugs have been approved in Europe. As compared to other regions in the world, Europe holds the highest number of biosimilar approvals. This number is expected to increase as many biologic patents will expire in the near future, which will help increase market size and competition among market participants.

8.2.2.7 Remarks

In summary, the EU has taken a thoughtful and evidence-based approach and has established a well-documented legal and regulatory pathway for the approval of biosimilar products distinct from the generic pathway. In order to grant a biosimilar product, the EMA requires comprehensive and justified comparability studies between the biosimilar and the reference products at the quality, nonclinical, and clinical level, which are explained in detail in the EMA guidelines. The approval pathway of biosimilar products in the EU is based on case-by-case reviews, owing to the complexity and diversity of the biologic products. Therefore, besides the three general guidelines, EMA developed additional product class–specific guidelines on nonclinical and clinical studies. This approval pathway is now held up as one of the gold standards for authorizing biosimilar products.

8.2.3 North America

8.2.3.1 The United States Food and Drug Administration

For the approval of follow-on biologics in the United States, current regulations depend on whether the biological product is approved under the United States Food, Drug, and Cosmetic Act (US FD&C) or it is licensed under the United States Public Health Service Act (US PHS). For those biological drugs marketed under the PHS Act, the BPCI Act passed by the US Congress on March 23, 2010, amends the PHS Act to establish an abbreviated approval pathway for biological products that are highly similar or interchangeable with an FDA-authorized biologic drug, and gives the FDA the authority to approve follow-on biologics under new Section 351(k) of the PHS Act. Some early biological drugs, such as somatropin and insulin, were approved under the FD&C Act. In this case, biosimilar versions can receive approval for new drug applications (NDAs) under Section 505(b) (2) of the FD&C Act.

Following passage of the BPCI Act, in order to obtain input on specific issues and challenges associated with the implementation of the BPCI Act from a broad group of stakeholders, the FDA conducted a two-day public hearing on Approval Pathway for Biosimilar and Interchangeability Biological Product held on November 2–3, 2010, at the FDA office in Silver Spring, Maryland. The scientific issues covered in this public hearing included, but were not limited to, criteria and design for biosimilarity and interchangeability, comparability between manufacturing processes, patient safety and pharmacovigilance, exclusivity, and user fees.

In practice, there is a strong industrial interest and desire for the regulatory agencies to develop review standards and an approval process for biosimilars rather than an *ad hoc* case-by-case review of individual biosimilar applications. As indicated earlier, for this purpose, the FDA has established three committees to ensure consistency in the FDA's regulatory approach of follow-on biologics. The three committees are the Center for Drug Evaluation and Research (CDER)/CBER Biosimilar Implementation Committee (BIC), the CDER Biosimilar Review Committee (BRC), and the CBER Biosimilar Review Committee. The CDER/CBER BRC focus on the cross-center policy issues related to the implementation of the BPCI Act. The CDER BRC and CBER BRC are responsible for considering applicants' requests for advice about proposed development programs for biosimilar products, reviewing biologic license applications (BLAs) that are submitted under Section 351(k) of the PHS Act, and managing related issues. Thus, the review process steps of CDER BRC and CBER BRC include the following: (1) applicant submission of request for advice,

(2) internal review team meeting, (3) internal CDER BRC (CBER BRC) meeting, (4) internal post-BRC meeting, and (5) applicant meeting with CDER (CBER).

Another important issue that arose from the BPCI Act is the interchangeability of biosimilars. Once approved, in many states standard generic drugs can be automatically substituted for the reference product without the intervention of the health care provider. However, the automatic interchangeability cannot be applied to all biosimilars. To meet the higher standard of interchangeability, a sponsor must demonstrate that the biosimilar products can be expected to produce the same clinical result as the reference product in any given patient.

On February 9, 2012, the FDA announced the publication of three draft guidance documents to help industry develop follow-on biologic products, including (1) *Scientific Considerations in Demonstrating Biosimilarity to a Reference Product*, (2) *Quality Considerations in Demonstrating Biosimilarity to a Reference Protein Product*, and (3) *Biosimilars: Questions and Answers Regarding Implementation of the Biologics Price Competition and Innovation Act of 2009*, which were subsequently finalized in 2015 (FDA, 2015a,b,c). Subsequently, the FDA hosted another public hearing on the discussion of these draft guidances at the FDA on May 11, 2012. Similar to the requirements of the WHO and EMA, the FDA considers a number of factors important when assessing applications for biosimilars, including the robustness of the manufacturing process, the demonstrated structural similarity, the extent to which mechanism of action was understood, the existence of valid, mechanistically related pharmacodynamic assays, comparative pharmacokinetics and immunogenicity, and the amount of clinical data and experience available with the original products. The guidances were finalized in early 2015. Even though they do not provide clear standards for assessing biosimilar products, they are the first step toward removing the uncertainties surrounding the biosimilar approval pathway in the United States.

Recently, following the Advisory Committee's recommendation for approval of a proposed biosimilar (by Novartis) to Amgen's Neupogen (filgrastim) on January 7, 2015, the FDA approved the proposed biosimilar on March 26, 2015. Neupogen was originally approved by the FDA in 1991, and its patent expired in December 2013. Neupogen was intended to decrease the incidence of infection, as manifested by febrile neutropenia in patients with nonmyeloid malignancies receiving myelosuppressive anticancer drugs associated with a clinically significant incidence of febrile neutropenia. It should be noted that the approved biosimilar of filgrastim is different from Teva's Neutroval approved by the FDA in August 2012. The approval of Teva's Neutroval was based on a full biologic license application rather than under the FDA's new biosimilar approval pathway which allows Teva to compete directly with Amgen's filgrastim in the United States.

8.2.3.2 Canada (Health Canada)

Health Canada, the federal regulatory authority that evaluates the safety, efficacy, and quality of drugs available in Canada also recognizes that with the expiration of patents for biological drugs, manufacturers may be interested in pursuing subsequent entry versions of these biologic drugs which are called subsequent entry biologics (SEBs) in Canada. In 2010, Health Canada issued *Guidance for Sponsors: Information and Submission Requirements for Subsequent Entry Biologics (SEBs)* whose objective is to provide guidance on how to satisfy the data and regulatory requirements under the Food and Drugs Act and Regulations for the authorization of SEBs in Canada (HC, 2010).

The concept of an SEB applies to all biological drug products, but there are additional criteria to determine whether the product will be eligible to be authorized as SEBs:

1. A suitable reference biological drug exists that was originally authorized based on a complete data package, and has significant safety and efficacy data accumulated.
2. The product can be well characterized by state-of-the-art analytical methods.
3. The SEB can be judged similar to the reference biological drug by meeting an appropriate set of predetermined criteria.

With regard to the similarity of products, Health Canada requires the manufacturer to evaluate the following factors:

1. Relevant physicochemical and biological characterization data.
2. Analysis of relevant samples from the appropriate stages of the manufacturing process.
3. Stability data and impurities data.
4. Data obtained from multiple batches of the SEB and reference to understand the ranges in variability.
5. Nonclinical and clinical data and safety studies.

In addition, Health Canada has stringent postmarketing requirements, including an adverse drug reaction report, periodic safety update reports, suspension or revocation of NOC (notice of compliance). Canada's guidance shares similar concepts and principles, as indicated in the WHO guidelines since the guidance clearly mentions that Health Canada seeks to harmonize as much as possible with other competent regulators and international organizations.

8.2.4 Asian Pacific Region (Japan, Korea, China)

8.2.4.1 Ministry of Health, Labour and Welfare

The Japanese Ministry of Health, Labour and Welfare (MHLW) has also been confronted with the new challenge of regulating biosimilar/follow-on biological products. Based on the similarity concept outlined by the EMA, Japan published a guideline for the quality, safety, and efficacy of biosimilar products in 2009 (MHLW, 2009). The scope of the guideline includes recombinant plasma proteins, recombinant vaccines, PEGylated recombinant proteins, and nonrecombinant proteins that are highly purified and characterized. Unlike the EU, polyglycans such as LMWH have been excluded from the guideline. Another class of product excluded is synthetic peptides, since the desired synthetic peptides can be easily defined by structural analyses and can be defined as generic drugs. Similar to the requirements of the EU, the original biologic should be already approved in Japan. However, there are some differences in the requirements of stability test and toxicology studies for impurities in biosimilars between the EU and Japan. A comparison of the stability of a biosimilar with the reference innovator products as a strategy for development of biosimilars is not always necessary in Japan. In addition, it is not required to evaluate the safety of impurities in the biosimilar product through nonclinical studies without comparison to the original product. According to this guideline, follow-on biologics such as Somatropin and Epoetin alfa BS have been approved in Japan.

8.2.4.2 Korean Food and Drug Administration

In Korea, the Pharmaceutical Affairs Act is the high-level regulation to license all medicines, including biological products. The Korean Food and Drug Administration (KFDA) notifications serve as a lower-level regulation. Biological products and biosimilars are subject to the *Notification of the Regulation on Review and Authorization of Biological Products*. The KFDA actively participates to promote a public dialogue on biosimilar issues. In 2008 and 2009, the KFDA held two public meetings and cosponsored a workshop to gather input on scientific and technical issues. The regulatory framework of biosimilar products in Korea is a three-tiered system: (1) Pharmaceutical Affairs Act; (2) Notification of the Regulation on Review and Authorization of Biological Products; and (3) *Guideline on Evaluation of Biosimilar Products* (KFDA, 2011; Suh and Park, 2011).

As the Korean guideline for biosimilar products was developed along with that of the WHO (WHO, 2009), most of the requirements are similar except for that of the clinical evaluation to demonstrate similarity. The KFDA requires that equivalent rather than noninferior efficacy should be shown in order to open the possibility of extrapolation of efficacy data to other indications of the reference product. Equivalence margins need to be predefined and justified, and should be established within the range that is judged not to be clinically different from reference products in clinical regards.

8.2.4.3 China Food and Drug Administration

The Center for Drug Evaluation (CDE) of the China Food and Drug Administration (CFDA) published a draft guidance for comments on October 29, 2014 (CFDA, 2014). CFDA finalized the draft guidance and published a current trial version on February 28, 2015 (CFDA, 2015). In this version, one significant change is to relax the regulation for the selection of reference product for comparison. In the version for comments, CFDA requires the reference product to be the originator product authorized by CFDA. In the current trial version, CFDA has removed this restriction. In other words, EU-approved and/or US-licensed reference products can be used as the reference products. In addition, for PK studies, the traditional 80%–125% equivalence range is not mandatory. In the current trial version, CFDA indicates that an alternative equivalence range (e.g., scaled average bioequivalence for highly variable drugs) can be used if scientifically justifiable.

8.3 CRITERIA FOR BIOSIMILARITY

As indicated in Chow and Liu (2008), bioequivalence assessment for generic drug products is possible under the Fundamental Bioequivalence Assumption. It states that if two drug products are shown to be bioequivalent in the drug absorption profile (which is measured in terms of the extent and rate of absorption), it is assumed that they will reach the same therapeutic effect or they are therapeutically equivalent. The Fundamental Bioequivalence Assumption assumes that there is an association between pharmacokinetic responses and clinical outcomes. This assumption, however, may not be applicable for the assessment of biosimilarity of biosimilars due to the fundamental differences between generic (small molecular) drug products and similar biological (large molecular) drug products (see Table 8.2).

In what follows, several criteria for the primary assessment of bioequivalence or similarity proposed by the FDA are discussed for their applicability to the assessment of biosimilarity.

8.3.1 Average Bioequivalence (Biosimilarity)

For the assessment of average bioequivalence (ABE) both *in vivo* and *in vitro*, the FDA adopted a one-size-fits-all criterion. That is, for *in vivo* (*in vitro*) bioequivalence assessment, a test drug product is said to be bioequivalent to a reference drug product if the estimated 90% confidence interval for the geometric means ratio (GMR) of the primary pharmacokinetic (PK) parameters (e.g., the area under the blood or plasma concentration-time curve (AUC)) and maximum concentration (C_{max}) is totally within the bioequivalence limits of 80.00%–125.00%. See, for example, Chow and Liu (2008), FDA (2003, 2013), and Chow (2011).

The one-size-fits-all criterion does not take into consideration the therapeutic window and intrasubject variability of a drug which has been identified to have non-negligible impact on the safety and efficacy of generic drug products, as compared to the innovative drug products. In the past several decades, this one-size-fits-all

TABLE 8.2
Fundamental Differences between Generic Drugs and Biosimilars

Generic Drugs	Biological Drugs (Biosimilars)
Made by chemical synthesis	Made by living cells
Defined structure	Heterogeneous structure
	Mixtures of related molecules
Easy to characterize	Difficult to characterize
Relatively stable	Variable
	Sensitive to environmental conditions such as light and temperature
No issue of immunogenicity	Issue of immunogenicity
Usually taken orally	Usually injected
Often prescribed by a general practitioner	Usually prescribed by specialists

criterion has been challenged and criticized by many researchers. It has been suggested that flexible criteria in terms of safety (upper bioequivalence limit) and efficacy (lower bioequivalence limit) should be developed based on the characteristics of the drug, its therapeutic window (TW), and intrasubject variability (ISV) (see Table 8.3).

On the other hand, for orally administered drugs with high within-subject variability and wide therapeutic window (Class D, highly variable drugs; see Table 8.3), the regulatory expectation has become, in some cases, more relaxed. For these drugs, the approach of scaled average bioequivalence has been proposed. This method is, in fact, a special case of the procedure described earlier for individual bioequivalence (BE) when the within-subject variation is high ($\sigma^2_{WR} > \sigma^2_{W0}$). A recent FDA draft guidance makes allowance for this alternative (FDA, 2013).

The assessment of ABE focuses on average bioavailability but ignores the variability associated with the PK responses. Thus, two drug products may fail the evaluation of ABE if the variability associated with the PK responses is large, even though they have identical means. A drug with large variability is considered highly variable. The FDA defines a highly variable drug (HVD) as a drug whose within-subject (or intrasubject) variation is larger than or equal to 30%. This definition based on intrasubject variation, however, is rather arbitrary. One of the problematic aspects of this definition is that the estimated within-subject variability depends on the metrics of pharmacokinetic responses such as AUC and C_{max}. Haidar et al. (2008a) pointed out that HVDs show variable pharmacokinetics as a result of their inherent properties (e.g., distribution, systemic metabolism, and elimination) (see also Davit et al., 2008; Haidar et al., 2008a; Tothfalusi et al., 2009). A drug may have low variability if it is administered intravenously, whereas it can be highly variable after oral administration.

In practice, HVDs often fail to meet current regulatory acceptance criteria for ABE. In the past decade, the topic for evaluation of bioequivalence for HVDs has received much attention. This topic has been discussed several times at regulatory forums and international conferences, but academics and representatives of pharmaceutical industries and regulatory agencies failed to reach a consensus until recently, and an approach of sealed average bioequivalence (SABE) was proposed by Haidar et al. (2008a,b). The SABE approach is briefly described next.

TABLE 8.3
Classification of Drugs

Class	TW	ISV	Example
A	Narrow	High	Cyclosporine
B	Narrow	Low	Theophylline
C	Wide	Low to moderate	Most drugs
D	Wide	High	Chlorpromazine or topical corticosteroids

ISV, intrasubject variability; TW, therapeutic window.

Logarithmic means of the two drug products, denoted by μ_T and μ_R, respectively, are typically compared. The acceptance of bioequivalence is claimed if the difference between the logarithmic means is between prespecified regulatory limits. The limits (θ_A) are generally symmetrical on the logarithmic scale and usually equal $\pm\ln(1.25)$. Thus, the criterion for ABE can be expressed as follows:

$$-\theta_A \le \mu_T - \mu_R < \theta_A$$

In a bioequivalence study, the individual kinetic responses are evaluated from the measured concentrations. The means of the logarithmic responses of the two formulations are calculated. These sample averages estimate the true population means. A variance is also estimated for each kinetic response. It is a measure of the intrasubject variance but not always identical to it. The FDA suggests that the above ABE could be scaled by a standard deviation as follows:

$$-\theta_s \le \frac{(\mu_T - \mu_R)}{\sigma_W} \le \theta_s,$$

where θ_S is the SABE regulatory cutoff. Here the standard deviation (σ_W) is the within-subject standard deviation. In a replicate design, σ_W is generally the within-subject standard deviation of the reference formulation (denoted by σ_{WR}). Thus, the scaling factor of SABE has similar features to the scaling factor of individual bioequivalence (IBE).

8.3.2 Population/Individual Bioequivalence

In the early 1990s, as more generic drug products became available, it was a concern whether the use of generic drug products was safe and whether the approved generic drug products could be used interchangeably. The FDA indicates that an approved generic drug product can be used to substitute the innovative (brand-name) drug product. However, the FDA does not indicate that generic drug products can be used interchangeably. Since generic drug products are approved based on the criterion of the 80/125 rule, there may be a drastic change in blood concentration if one switches from one generic drug to another. For example, if one switches from a drug that was approved on the lower end of the 80/125 rule (say 80%) to another drug that was approved on the higher end of the 80/125 rule (say 120%), then there would be a sudden 50% increase in blood concentration, which may cause a potential safety concern. To address the issue of drug interchangeability in terms of drug prescribability and switchability, between the early 1990s and early 2000s, the FDA suggested using the concepts of population bioequivalence for addressing drug prescribability and individual bioequivalence for addressing drug switchability.

Let y_T be the PK response from the test product and y_R and $y_{R'}$ be two identically distributed PK responses from the reference product. Now consider a measure of the relative difference between the mean squared errors of $y_T - y_R$ and $y_R - y_{R'}$. Thus

$$\theta = \begin{cases} \dfrac{E(y_T - y_R)^2 - E(y_R - y_{R'})^2}{\dfrac{E(y_R - y_{R'})^2}{2}} & \text{if } \dfrac{E(y_R - y_{R'})^2}{2} \geq \sigma_0^2 \\ \dfrac{E(y_T - y_R)^2 - E(y_R - y_{R'})^2}{\sigma_0^2} & \text{if } \dfrac{E(y_R - y_{R'})^2}{2} < \sigma_0^2 \end{cases},$$

where σ_0^2 is a given constant. If y_T, y_R, and $y_{R'}$ are independent observations from different subjects, then the two drug products show population bioequivalence when $\theta < \theta_P$. On the other hand, if y_T, y_R, and $y_{R'}$ are independent observations from the same subject, then the two drug products exhibit individual bioequivalence when $\theta < \theta_I$. Thus, as indicated in Section 8.2.3, for the assessment of individual bioequivalence, the criterion proposed in the FDA guidance (FDA, 2001) can be expressed as

$$\theta_I = \frac{\left(\delta^2 + \sigma_D^2 + \sigma_{WT}^2 - \sigma_{WR}^2\right)}{\max\left\{\sigma_{W0}^2, \sigma_{WR}^2\right\}}, \tag{8.1}$$

where $\delta = \mu_T - \mu_R, \sigma_{WT}^2, \sigma_{WR}^2, \sigma_D^2$ are the true difference in means, intra-subject variabilities of the test product and the reference product, and the variance component due to subject-by-formulation interaction between drug products, respectively. σ_{W0}^2 is the scale parameter specified by the user or regulator. Similarly, the criterion for the assessment of population bioequivalence suggested in the FDA guidance (FDA, 2001) is given by

$$\theta_P = \frac{\left(\delta^2 + \sigma_{TT}^2 - \sigma_{TR}^2\right)}{\max\left\{\sigma_{T0}^2, \sigma_{TR}^2\right\}}, \tag{8.2}$$

where σ_{TT}^2 and σ_{TR}^2 are the total variances for the test product and the reference product, respectively, and σ_{T0}^2 is the scale parameter specified by the user or regulator.

A typical approach is to construct a one-sided 95% confidence interval for $\theta_I(\theta_P)$ for the assessment of individual (population) bioequivalence. If the one-sided 95% upper confidence limit is less than the bioequivalence limit of $\theta_I(\theta_P)$, then we conclude that the test product is bioequivalent to that of the reference product in terms of individual (population) bioequivalence.

8.3.2.1 Masking Effect

The goal for evaluation of bioequivalence is to assess the similarity of the distributions of the PK metrics obtained either from the population or from individuals in the population. Under aggregate criteria, however, different combinations of values for the components of the aggregate criterion can yield the same value. In other words, bioequivalence can be reached by two totally different distributions of PK metrics. This is another artifact of the aggregate criteria. At the 1996 advisory committee meeting, it was reported that a 14% increase in the average (ABE only allows 80%–125%) is offset by a 48% difference in the variability and the test passes IBE but fails ABE. More details regarding individual and population bioequivalence can be found in Chow and Liu (2008) and Chow (1999).

8.3.3 Profile Analysis for *In Vitro* Bioequivalence Testing

As indicated in the FDA draft guidance for *in vitro* bioequivalence testing, profile analysis using a confidence interval approach should be applied to cascade impactor or multistage liquid impinger (MSLI) for particle size distribution. Equivalence may be assessed based on chi-square differences. The idea is to compare the profile difference between test product and reference product samples to the profile variation between reference product samples. More specifically, let y_{ijk} denote the observation from the jth subject's ith stage of the kth treatment. Given a sample (j_0) from test product and two samples (j_0, j_1) from reference products and assuming that there are a total of S stages, the profile distance between test and reference is given by

$$d_{TR} = \sum_{i=1}^{S} \frac{\left(y_{ij_0T} - 0.5\left(y_{ij_1R} + y_{ij_2R}\right)\right)^2}{\left(y_{ij_0T} + 0.5\left(y_{ij_1R} + y_{ij_2R}\right)\right)}.$$

Similarly, the profile variability within reference is defined as

$$d_{RR} = \sum_{i=1}^{S} \frac{\left(y_{ij_1R} - y_{ij_2R}\right)^2}{0.5\left(y_{ij_1R} + y_{ij_2R}\right)}.$$

For a given triplet sample of (Test, Reference 1, Reference 2), the ratio of d_{TR} and d_{RR}, that is,

$$rd = \frac{d_{TR}}{d_{RR}} \tag{8.3}$$

can then be used as a bioequivalence measure for the triplet samples between the two drug products. For a selected sample, the 95% upper confidence bound of $E(rd) = E\left(d_{TR}/d_{RR}\right)$ is then used as a bioequivalence measure for the determination of bioequivalence. In other words, if the 95% upper confidence bound is less than the bioequivalence limit, then we claim that the two products are bioequivalent. The FDA draft guidance recommends a bootstrap procedure to construct the 95% upper bound for $E(rd)$. The procedure is as follows.

Assume that the samples are obtained in a two-stage sampling manner. In other words, for each treatment (test or reference), three lots are randomly sampled. Within each lot, ten samples (e.g., bottles or canisters) are sampled. The following is quoted from the 1999 FDA draft guidance regarding the bootstrap procedure to establish profile bioequivalence. For an experiment consisting of three lots each of test and reference products, and with 10 canisters per lot, the lots can be matched into six different combinations of triplets with two different reference lots in each triplet. The 10 canisters of a test lot can be paired with the 10 canisters of each of the two reference lots in (10 factorial)^2 = (3,628,800)^2 combinations in each of the lot triplets. Hence, a random sample of the N-canister pairing of the six Test–Reference 1–Reference 2 lot triplets is needed. rd is estimated by the sample mean of the rds calculated for the triplets in 10 selected samples of N. Note that the FDA recommends that $N = 500$ be considered.

8.3.4 Similarity Factor for Dissolution Profile Comparison

In vivo bioequivalence studies are surrogate trials for assessing equivalence between test and reference formulations based on the rate and extent of drug absorption in humans to establish similar effectiveness and safety under the fundamental bioequivalence assumption. However, drug absorption depends on the dissolved state of drug product, and dissolution testing provides a rapid *in vitro* assessment of the rate and extent of drug release. Leeson (1995), therefore, suggested that *in vitro* dissolution testing be used as a surrogate for *in vivo* bioequivalence studies to assess equivalence between the test and reference formulations for postapproval changes. For the comparison of dissolution profiles, the FDA guidance suggests considering the assessment of (1) the overall profile similarity and (2) similarity at each sampling time point (FDA, 1997). Since dissolution profiles are curves over time, Chow and Ki (1997) introduced the concepts of local similarity and global similarity. Two dissolution profiles are said to be locally similar at a given time point if their difference or ratio at the given time point is within some equivalence (similarity) limits, denoted by (δ_L, δ_U). Two dissolution profiles are considered globally similar if their differences or ratios are within (δ_L, δ_U) across all time points. Note that global similarity is also known as *uniformly similar*. Chow and Ki (1997) suggested the following similarity limits for comparing dissolution profiles:

$$\delta_L = \frac{Q-\delta}{Q+\delta} \quad \text{and} \quad \delta_U = \frac{Q+\delta}{Q-\delta},$$

where

Q is the desired mean dissolution rate of a drug product as specified in the *United States Pharmacopeia and National Formulary* (USP/NF) individual monograph.

δ is a meaningful difference of scientific importance in mean dissolution profiles of two drug products under consideration.

In practice, δ is usually determined by a pharmaceutical scientist.

In order to achieve these two objectives, based on Moore and Flanner (1996), both the FDA SUPAC (Scale-up and Postapproval Change) guidance (SUPAC-IR, 1995) and guidance on dissolution testing (FDA, 1997) suggest the similarity and difference factor for the assessment of similarity. The similarity factor is then defined as the logarithmic reciprocal square root transformation of 1 plus the mean-squared (the average sum of squares) difference in mean cumulative percentage dissolved between the test and reference formulations over all sampling time points. That is,

$$f_2 = 50 \log\left\{\left[1+\frac{Q}{n}\right]^{-0.5} 100\right\}, \tag{8.4}$$

where

$$Q=\sum_{t=1}^{n}\left(\mu_{R_t}-\mu_{T_t}\right)^2,$$

and log denotes the logarithm based on 10.

On the other hand, the difference factor is the sum of the absolute difference in mean cumulative percentage dissolved between the test and reference formulations divided by the sum of the mean cumulative dissolved of the reference formulation.

$$f_1=\frac{\sum_{t=1}^{n}\left|\mu_{R_t}-\mu_{T_t}\right|}{\sum_{t=1}^{n}\mu_{R_t}} \tag{8.5}$$

It should be noted that the definitions of f_1 and f_2 provided by Moore and Flanner (1996) and in the SUPAC and guidance on dissolution testing are not clear as to whether they are defined based on the population means or the sample averages. However, following the traditional statistical inference with ability for evaluation of error probability, we define both f_1 and f_2 based on the population mean dissolution rates. It follows that f_1 and f_2 are population parameters for the assessment of similarity of dissolution profiles between the test and reference formulations.

The use of the f_2 similarity factor has been discussed and criticized by many researchers (e.g., Liu et al., 1997; Shah et al., 1998; Ma et al., 1999). Chow and Shao (2002) pointed out two main problems in using the f_2 similarity factor for assessing similarity between the dissolution profiles of two drug products. The first problem is its lack of statistical justification. Since f_2 is a statistic and, thus, a random variable, $P(f_2 > 50)$ may be quite large when the two dissolution profiles are not similar. However, $P(f_2 > 50)$ can be very small when the two dissolution profiles are similar. Suppose that the expected value $E(f_2)$ exists and that we can find a 95% lower confidence bound for $E(f_2)$. Then, a reasonable modification to the f_2 similarity factor approach is to replace f_2 with the 95% lower confidence bound for $E(f_2)$. The second problem with using the f_2 similarity factor is that the f_2 similarity factor assesses neither local similarity nor global similarity, owing to the use of the average of the dissolution data.

8.4 STATISTICAL METHODS

For the assessment of biosimilarity of biosimilar products, standard methods for the assessment of bioequivalence for small-molecule drug products are usually considered. These methods include the classical methods such as the confidence interval approach and interval hypotheses testing such as Schuirmann's two one-sided tests procedure, Bayesian methods, and nonparametric methods such as the Wilcoxon–Mann–Whitney two one-sided tests procedure. It should be noted that these methods were derived based on a raw data model under a crossover design, although they can be easily extended to a parallel group design based on log-transformed data. In practice, it is a concern whether these methods are appropriate for the assessment of biosimilarity of biosimilar products due to fundamental differences between small-molecule drug products and large-molecule biological drug products as described in earlier chapters (see also Chow et al., 2011). As a result, the search for biosimilarity criteria, study endpoints (measures of biosimilarity), and statistical methods has become the center of attention for assessment of the biosimilarity of biosimilar products within regulatory agencies.

8.4.1 Interval Hypotheses

In practice, one of the most widely used designs for assessing biosimilarity between biosimilar products and an innovator biological product is probably either a two-sequence, two-period (2 × 2) crossover design or a two-arm parallel group design. Under a valid study design, biosimilarity can then be assessed by means of an equivalence test under the following interval hypotheses

$$H_0: \mu_T - \mu_R \le \theta_L \text{ or } \mu_T - \mu_R \ge \theta_U \text{ vs. } H_a: \theta_L < \mu_T - \mu_R < \theta_U, \tag{8.6}$$

where

(θ_L, θ_U) are prespecified equivalence limits (margins).

μ_T and μ_R are the population means of a biological (test) product and an innovator biological (reference) product, respectively.

That is, biosimilarity is assessed in terms of the *absolute* difference between the two population means. Alternatively, biosimilarity can be assessed in terms of the *relative* difference (i.e., ratio) between the population means. Note that for the assessment of similarity between small-molecule drug products, average bioequivalence, population bioequivalence, and individual bioequivalence are defined in terms of ratios of appropriate parameters under a crossover design. In practice, since many biological products have long half-lives, a crossover design may not be appropriate in these cases for the assessment of biosimilarity. Instead, a parallel group design is considered more appropriate.

This chapter provides a comprehensive summary of standard statistical methods that are commonly used for assessing ABE for small-molecule drug products under a crossover design. In addition, we focus on the discussion of statistical methods proposed by Kang and Chow (2013) under a newly proposed three-arm parallel design for investigation of the biosimilarity of biosimilar products. The statistical analysis methods proposed by Kang and Chow (2013) consider the relative distance based on the absolute mean differences. Under the three-arm design, patients who are randomly assigned to the first group receive a biosimilar (test) product while patients who are randomly assigned to the second and third groups receive the innovator biological (reference) product from different batches. The distance between the test product and the reference product is defined by the absolute mean difference between two products. Similarly, the distance between the reference products from two different batches is defined. The relative distance is the ratio of the two distances whose denominator is the distance between the two reference products from different batches. Under the proposed design, Kang and Chow (2013) claim that the two products are biosimilar if the relative distance is less than a pre-specified margin.

8.4.2 Classic Methods for Assessing Biosimilarity

Under a crossover design, consider the following statistical model for raw data:

$$Y_{ijk} = \mu + S_{ik} + P_j + F_{(j,k)} + C_{(j-1,k)} + e_{ijk}, \tag{8.7}$$

where

Y_{ijk} is the response (e.g., AUC) of the ith subject in the kth sequence at the jth period.
μ is the overall mean.
S_{ik} is the random effect of the ith subject in the kth sequence, where $i = 1, 2, \ldots, g$.
P_j is the fixed effect of the jth period, where $j = 1, \ldots, p$ and $\Sigma_j P_j = 0$.
$F_{(j,k)}$ is the direct fixed effect of the drug product in the kth sequence which is administered at the jth period, and $\Sigma F_{(j,k)} = 0$.
$C_{(j-1,k)}$ is the fixed first-order carryover effect of the drug product in the kth sequence which is administered at the $(j-1)$th period, where $C_{(0,k)} = 0$ and $\Sigma C_{(j-1,k)} = 0$.
e_{ijk} is the (within-subject) random error in observing Y_{ijk}.

It is assumed that $\{S_{ik}\}$ are independently and identically distributed (i.i.d.) with mean 0 and variance σ_s^2, and $\{e_{ijk}\}$ are independently distributed with mean 0 and variances σ_t^2, where $t = 1, 2, \ldots, L$ (the number of formulations to be compared). $\{S_{ik}\}$ and $\{e_{ijk}\}$ are assumed to be mutually independent. The estimate of σ_s^2 is usually used to explain the intersubject variability, and the estimates of σ_t^2 are used to assess the intrasubject variabilities for the tth drug product.

8.4.2.1 Confidence Interval Approach

For bioequivalence assessment of small-molecule drug products, the FDA adopts the 80/125 rule based on log-transformed data. The 80/125 rule states that bioequivalence is concluded if the geometric means ratio (GMR) between the test product and the reference product is within the bioequivalence limits of (80.00%, 125.00%), with a certain statistical assurance. Thus, a typical approach is to consider the method of classic (shortest) confidence interval.

Consider a standard two-sequence, two-period crossover design, under model 8.7, after the log transformation of the data; let $\bar{Y}_T$ and $\bar{Y}_R$ be the respective least squares means for the test and reference formulations which can be obtained from the sequence-by-period means. The classic (or shortest) (1 – 2α) × 100% confidence interval can then be obtained based on the following *t* statistic:

$$T = \frac{(\bar{Y}_T - \bar{Y}_R) - (\mu_T - \mu_R)}{\hat{\sigma}_d \sqrt{\frac{1}{n_1} + \frac{1}{n_2}}}, \tag{8.8}$$

where n_1 and n_2 are the number of subjects in sequences 1 and 2, respectively, and $\hat{\sigma}_d$ is an estimate of the variance of the period differences for each subject within each sequence, which are defined as follows:

$$d_{ik} = \frac{1}{2}(Y_{i2k} - Y_{i1k}), \quad i = 1, 2, \ldots, n_k; \quad k = 1, 2.$$

Thus, $V(d_{ik}) = \sigma_d^2 = \sigma_e^2/2$. Under normality assumptions, T follows a central student t distribution with degrees of freedom $n_1 + n_2 - 2$. Thus, the classic (1 – 2α) × 100% confidence interval for $\mu_T - \mu_R$ can be obtained as follows:

$$L_1 = (\bar{Y}_T - \bar{Y}_R) - t(\alpha, n_1 + n_2 - 2)\hat{\sigma}_d \sqrt{\frac{1}{n_1} + \frac{1}{n_2}},$$

$$U_1 = (\bar{Y}_T - \bar{Y}_R) - t(\alpha, n_1 + n_2 - 2)\hat{\sigma}_d \sqrt{\frac{1}{n_1} + \frac{1}{n_2}}. \tag{8.9}$$

The above (1 – 2α) × 100% confidence interval for $\log(\mu_T) - \log(\mu_R) = \log(\mu_T/\mu_R)$ can be converted into a (1 – 2α) × 100% confidence interval for μ_T/μ_R by taking an antilog transformation.

Note that under a parallel group design, a (1 – 2α) × 100% confidence interval for μ_T/μ_R can be similarly obtained.

8.4.2.2 Schuirmann's Two One-Sided Tests Procedure

The assessment of average bioequivalence is based on the comparison of bioavailability profiles between products. However, in practice, it is recognized that no two drug products will have exactly the same bioavailability profiles. Therefore, if the profiles of the two drug products differ by less than a (clinically) meaningful limit, the profiles of the two drug products may be considered equivalent. Following this concept, Schuirmann (1981) first introduced the use of interval hypotheses 8.6 for assessing average bioequivalence. The concept of interval hypotheses 8.6 shows average bioequivalence by rejecting the null hypothesis of average bio*in*equivalence. In most bioavailability and bioequivalence studies, δ_L and δ_U are often chosen to be $-\theta_L = \theta_U$ = 20% of the reference mean (μ_R). When the natural logarithmic transformation of the data is considered, the hypotheses corresponding to hypotheses 8.6 can be stated as

$$H_0': \frac{\mu_T}{\mu_R} \le \delta_L \text{ or } \frac{\mu_T}{\mu_R} \ge \delta_U \quad \text{vs.} \quad H_a': \delta_L < \frac{\mu_T}{\mu_R} < \delta_U \tag{8.10}$$

where $\delta_L = \exp(\theta_L)$ and $\delta_U = \exp(\theta_U)$. Note that the FDA recommends that $(\delta_L, \delta_U) =$ (80.00%, 125.00%) for assessing average bioequivalence.

Note that the test for hypotheses in Equation 8.10 formulated on the log scale is equivalent to testing for hypotheses 8.6 on the raw scale. The interval hypotheses 8.6 can be decomposed into two sets of one-sided hypotheses

$$H_{01}: \mu_T - \mu_R \le \theta_L \text{ vs. } H_{a1}: \mu_T - \mu_R > \theta_L$$

and

$$H_{02}: \mu_T - \mu_R \ge \theta_U \text{ vs. } H_{a2}: \mu_T - \mu_R < \theta_U. \tag{8.11}$$

The first set of hypotheses is to verify that the average bioavailability of the test formulation is not too low, whereas the second set of hypotheses is to verify that the average bioavailability of the test formulation is not too high. A relatively low (or high) average bioavailability may refer to the concern of efficacy (or safety) of the test formulation. If one concludes that $\theta_L < \mu_T - \mu_R$ (i.e., reject H_{01}) and $\mu_T - \mu_R < \theta_U$ (i.e., reject H_{02}), then it has been concluded that

$$\theta_L < \mu_T - \mu_R < \theta_U.$$

Thus, μ_T and μ_R are equivalent. The rejection of H_{01} and H_{02}, which leads to the conclusion of average bioequivalence, is equivalent to rejecting H_0 in Equation 8.6.

Under hypotheses 8.6, Schuirmann (1987) introduced the two one-sided tests procedure for assessing average bioequivalence between drug products. The proposed two one-sided tests procedure suggests the conclusion of equivalence of μ_T and μ_R at the α level of significance if, and only if, H_{01} and H_{02} in Equation 8.11 are rejected at a predetermined α level of significance. Under the normality assumptions, the two sets of one-sided hypotheses can be tested with ordinary one-sided *t*-tests. We conclude that μ_T and μ_R are average equivalent if

$$T_L = \frac{(\bar{Y}_T - \bar{Y}_R) - \theta_L}{\hat{\sigma}_d \sqrt{\frac{1}{n_1} + \frac{1}{n_2}}} > t(\alpha, n_1 + n_2 - 2)$$

and

$$T_U = \frac{(\bar{Y}_T - \bar{Y}_R) - \theta_U}{\hat{\sigma}_d \sqrt{\frac{1}{n_1} + \frac{1}{n_2}}} < -t(\alpha, n_1 + n_2 - 2). \tag{8.12}$$

The two one-sided t-test procedure is operationally equivalent to the classic (shortest) confidence interval approach; that is, both the classic confidence interval approach and Schuirmann's two one-sided tests procedure will lead to the same conclusion on bioequivalence.

Note that under a parallel group design, Schuirmann's two one-sided tests procedure can be similarly derived with a slight modification from a paired t-test statistic to a two-sample t-test statistic.

8.4.2.3 Bayesian Methods

In previous sections, statistical methods for assessing biosimilarity were derived based on the sampling distribution of the estimate of the parameter of interest, such as the direct drug effect (i.e., $\theta = \mu_T - \mu_R$), which is assumed to be fixed but unknown. Although statistical inference (e.g., confidence interval and interval hypothesis testing) on the unknown direct drug effect can be drawn from the sampling distribution of the estimate, there is little information on the probability of the unknown direct drug effect being within the equivalence limits (θ_L, θ_U). To have a certain assurance on the probability of the direct drug effect being within (θ_L, θ_U), a Bayesian approach (Box and Tiao, 1973), which assumes that the unknown direct drug effect is a random variable and follows a prior distribution, is useful.

In practice, before a biosimilar study is conducted, investigators may have some prior knowledge of the drug product under development. As an example, for a PK biosimilar study, according to past experiments, the investigator may have some information on (1) the intersubject and the intrasubject variabilities and (2) the ranges of AUC or C_{max} for the test and reference products. This information can be used to choose an appropriate prior distribution of the unknown direct drug effect. An appropriate prior distribution can reflect the investigator's belief about the drug products under study. After the study is completed, the observed data can be used to adjust the prior distribution of the direct drug effect, which is called the posterior distribution. Given the posterior distribution, a probability statement on the direct drug effect being within the biosimilarity limits can be made.

A different prior distribution can lead to a different posterior distribution that has an influence on the statistical inference on the direct drug effect. Thus, an important issue in a Bayesian approach is how to choose a prior distribution. Box and Tiao (1973) introduced the use of a locally uniform distribution over a possible range of AUC or C_{max} as a noninformative prior distribution. A noninformative prior distribution assumes that there is an equally likely chance for any two points within the possible range being the true state of the location of the direct drug effect. In this case, the resultant posterior distribution can be used to provide the true state of the location of a direct drug effect. In practice, however, it is also desirable to provide an interval showing a range in which most of the distribution of a direct drug effect will fall. We shall refer to such an interval as a highest posterior density (HPD) interval. The HPD interval is also known as a credible interval (Edwards et al., 1963) and a Bayesian confidence interval (Lindley, 1965). An HPD interval possesses the following properties (Box and Tiao, 1973): (1) the density for every point inside the interval is greater than that for every point outside the interval, and (2) for a given probability distribution, the interval is the shortest. It can be verified that the above two properties imply each other.

In what follows, for illustration purposes, the Bayesian method proposed by Rodda and Davis (1980) is discussed under the following model, which assumes that there are no carryover effects because a washout period of sufficient length can be chosen to completely eliminate the residual effects from one dosing period to the next:

$$Y_{ijk} = \mu + S_{ik} + F_{(j,k)} + P_j + e_{ijk}, \tag{8.13}$$

where Y_{ijk}, μ, S_{ik}, $F_{(j,k)}$, P_j and e_{ijk} were defined in Equation 8.7. Given the results of a biosimilar study, Rodda and Davis (1980) proposed a Bayesian evaluation to estimate the probability of a clinically important difference (i.e., the probability that the true direct drug effect will fall within the bioequivalent limits is estimated). Under the assumption of normality and equal carryover effects, $\bar{d}_{.1}$, $\bar{d}_{.2}$, and $(n_1+n_2-2)\hat{\sigma}_d^2$ are independently distributed as $N(\theta_1, \sigma_d^2/n_1)$, $N(\theta_2, \sigma_d^2/n_2)$, and $\sigma_d^2\chi^2(n_1+n_2-2)$, where

$$\bar{d}_{.k} = \frac{1}{n_k}\sum_{i=1}^{n_k} d_{ik}, \quad k=1, 2.$$

$$\theta_1 = \frac{1}{2}\left[(P_2 - P_1) + (F_T - F_R)\right],$$

$$\theta_2 = \frac{1}{2}\left[(P_2 - P_1) + (F_R - F_T)\right].$$

$$\text{Note that } F = \theta_1 - \theta_2 = (\mu + F_T) - (\mu + F_R)$$

$$= \mu_T - \mu_R$$

Assuming that the noninformative prior distributions for θ_1, θ_2, and log (σ_d) are approximately independent and locally uniformly distributed, then the joint posterior distribution of θ_1, θ_2, and σ_d^2, given data $Y = \{Y_{ijk}, i = 1, 2, \ldots, n_k; j, k = 1, 2\}$, is

$$p(\theta_1, \theta_2, \sigma_d^2 \mid \mathbf{Y}) = p(\theta_1 \mid \sigma_d^2, \bar{d}_{.1})\, p(\theta_2 \mid \sigma_d^2, \bar{d}_{.2})\, p(\sigma_d^2 \mid \hat{\sigma}_d^2), \tag{8.14}$$

$$p(\theta_i \mid \sigma_d^2, \bar{d}_{.i}) = N(\bar{d}_{.i}, \hat{\sigma}_d^2/n_i), \quad i = 1, 2.$$

$$p(\sigma_d^2 \mid \hat{\sigma}_d^2) = (n_1+n_2)\hat{\sigma}_d^2\chi^{-2}(n_1+n_2-2),$$

where χ^{-2} $(n_1 + n_2 - 2)$ is the distribution of the inverse of χ^2 $(n_1 + n_2 - 2)$. Therefore, the joint distribution of $\mu_T - \mu_R$ (=F) and σ_d^2 is given by

$$p(\mu_T - \mu_R, \sigma_d^2 \mid \mathbf{Y}) = p(\mu_T - \mu_R \mid \sigma_d^2, \bar{d}_{.1} - \bar{d}_{.2})\, p(\sigma_d^2 \mid \hat{\sigma}_d^2), \tag{8.15}$$

where

$$p\left(\mu_T-\mu_R \mid \sigma_d^2, \bar{d}_{.1}-\bar{d}_{.2}\right)=N\left[\bar{d}_{.1}-\bar{d}_{.2},\ \hat{\sigma}_d^2\left(\frac{1}{n_1}+\frac{1}{n_2}\right)\right]$$

$$=N\left[\bar{Y}_T-\bar{Y}_R,\ \hat{\sigma}_d^2\left(\frac{1}{n_1}+\frac{1}{n_2}\right)\right].$$

The marginal posterior distribution of F, given data $\mathbf{Y}$, is

$$p\left(\mu_T-\mu_R \mid \mathbf{Y}\right)=\frac{\left(\hat{\sigma}_d^2 m\right)^{-1/2}}{B(1/2, \nu/2)\sqrt{n}}\left\{1+\frac{\left[\left(\mu_T-\mu_R\right)-\left(\bar{Y}_T-\bar{Y}_R\right)\right]^2}{\nu\hat{\sigma}_d^2 m}\right\}^{-(\nu+1)/2}, \tag{8.16}$$

where $m=1/n_1+1/n_2, \nu=n_1+n_2-2$ and $-\infty<\mu_T-\mu_R<\infty$.

Thus, we have

$$T_{RD}=\frac{\left(\mu_T-\mu_R\right)-\left(\bar{Y}_T-\bar{Y}_R\right)}{\hat{\sigma}_d\sqrt{\frac{1}{n_1}+\frac{1}{n_2}}}. \tag{8.17}$$

which has a central student t-distribution with n_1+n_2-2 degrees of freedom. From Equation 8.17, the probability of F being within the biosimilarity limits of (θ_L, θ_U) can be estimated by

$$P_{RD}=P\left\{\theta_L<\mu_T-\mu_R<\theta_U\right\}$$

$$=F_t\left(t_U\right)-F_t\left(t_L\right), \tag{8.18}$$

where F_t is the cumulative distribution function of a central t variable with n_1+n_2-2 degrees of freedom, and

$$t_U=\frac{\theta_U-(\bar{Y}_T-\bar{Y}_R)}{\hat{\sigma}_d\sqrt{\frac{1}{n_1}+\frac{1}{n_2}}},$$

and

$$t_L=\frac{\theta_L-(\bar{Y}_T-\bar{Y}_R)}{\hat{\sigma}_d\sqrt{\frac{1}{n_1}+\frac{1}{n_2}}}. \tag{8.19}$$

The lower and upper limits of the $(1-2\alpha)\times 100\%$ HPD interval are given by

$$L_{RD} = \left(\bar{Y}_T - \bar{Y}_R\right) - t(\alpha, n_1 + n_2 - 2)\hat{\sigma}_d \sqrt{\frac{1}{n_1} + \frac{1}{n_2}},$$

$$U_{RD} = \left(\bar{Y}_T - \bar{Y}_R\right) + t(\alpha, n_1 + n_2 - 2)\hat{\sigma}_d \sqrt{\frac{1}{n_1} + \frac{1}{n_2}}. \quad (8.20)$$

Hence, it is verified that the $(1 - 2\alpha) \times 100\%$ HPD interval in Equation 8.20 is numerically equivalent to the $(1 - 2\alpha) \times 100\%$ classic confidence interval obtained from the sampling theory. However, the interpretation of these two intervals is totally different. For example, a 90% classic confidence interval for F indicates that, in the long run, if the study is repeatedly carried out for numerous times, 90% of the times the interval will contain the unknown direct drug effect $\mu_T - \mu_R$. On the other hand, based on the posterior distribution of $\mu_T - \mu_R$, the chance of $\mu_T - \mu_R$ being within the lower and upper limits of a 90% HPD interval is 90%.

8.4.2.4 Wilcoxon–Mann–Whitney Two One-Sided Tests Procedure

As described in the previous sections, statistical methods for assessing average biosimilarity between drug products were derived under the assumption that $\{S_{ik}\}$ and $\{e_{ijk}\}$ are mutually independent and normally distributed with mean 0 and variance σ_s^2 and σ_e^2. Under these normality assumptions, confidence intervals and tests for interval hypotheses were obtained based on either a two-sample t statistic or an F statistic. In practice, however, one of the difficulties commonly encountered in comparing drug products is whether the assumption of normality (for raw or untransformed data) or log normality (for log-transformed data) is valid. If the normality (or log normality) is seriously violated, the approach based on a two-sample t statistic or an F statistic is no longer justified. In this situation, a distribution-free (or nonparametric) method is useful. In this section, a nonparametric version of the two one-sided tests procedure for testing interval hypotheses—the Wilcoxon–Mann–Whitney two one-sided tests procedure—is discussed. The Hodges–Lehmann estimator associated with the Wilcoxon rank sum test will be used to construct a $(1 - 2\alpha) \times 100\%$ confidence interval for $\mu_T - \mu_R$, the difference in average biosimilarity.

Under the standard 2×2 crossover design consisting of a pair of dual sequences (i.e., RT and TR), a distribution-free rank sum test can be applied directly to the two one-sided tests procedure (Cornell, 1990; Hauschke et al., 1990). We refer to this approach as the Wilcoxon–Mann–Whitney two one-sided tests procedure. Let $\theta = \mu_T - \mu_R$. The two sets of hypotheses in Equation 8.6 can then be rewritten as

$$H_{01}: \theta_L^* \leq 0 \text{ vs. } H_{a1}: \theta_L^* > 0$$

and

$$H_{02}: \theta_U^* \geq 0 \text{ vs. } H_{a2}: \theta_U^* < 0,$$

where $\theta_L^* = \theta - \theta_L$ and $\theta_U^* = \theta - \theta_U$.

Thus, the estimates of θ_L^* and θ_U^* can be obtained as a linear function of period differences d_{ik}, $i = 1, 2, \ldots, n_k$, $k = 1, 2$. Let

$$b_{hik} = \begin{cases} d_{ik} - \theta_h & h = L, U, \text{ for subjects in sequence 1} \\ d_{ik} & \text{for subjects in sequence 2.} \end{cases} \tag{8.21}$$

When there are no carryover effects, the expected value and variance of b_{hik}, where $h = L, U$, $i = 1, 2, \ldots, n_k$, and $k = 1, 2$, are given by

$$E(b_{hik}) = \begin{cases} \frac{1}{2}\left[(P_2 - P_1) + (\theta - 2\theta_h)\right] & \text{for } k = 1 \\ \frac{1}{2}\left[(P_2 - P_1) - \theta\right] & \text{for } k = 2 \end{cases}, \tag{8.22}$$

and

$$V(b_{hik}) = V(d_{ik}) = \sigma_d^2 = \frac{\sigma_e^2}{2}.$$

It can be seen that $E(b_{hi1}) - E(b_{hi2}) = (\theta - \theta_h) = \theta_h^*$.

Thus, for a fixed h, $\{b_{hi1}\}$ and $\{b_{hi2}\}$ have the same distribution except for the difference $\left(= \theta_h^*\right)$ in location of the true formulation effect. Here, the Wilcoxon–Mann–Whitney rank sum test (Wilcoxon, 1945; Mann and Whitney, 1947) for the unpaired two-sample location problem can be directly applied to test each of the two sets of hypotheses given above. Consider the first set of hypotheses that

$$H_{01}: \theta_L^* \leq 0 \text{ vs. } H_{a1}: \theta_L^* > 0.$$

The Wilcoxon–Mann–Whitney test statistic can be derived based on $\{b_{Li1}\}$, $i = 1, 2, \ldots, n_1$ and $\{b_{Li2}\}$, $i = 1, 2, \ldots, n_2$. Let $R(b_{Lik})$ be the rank of b_{Lik} in the combined sample $\{b_{Lik}\}$, $i = 1, 2, \ldots, n_k$, $k = 1, 2$. Also, let R_L be the sum of the ranks of the responses for subjects in sequence 1; that is

$$R_L = \sum_{i=1}^{n_1} R(b_{Li1}).$$

Thus, the Wilcoxon–Mann–Whitney test statistic for H_{01} is given by

$$W_L = R_L - \frac{n_1(n_1 + 1)}{2}.$$

We then reject H_{01} if

$$W_L > w(1 - \alpha) \tag{8.23}$$

where $w(1-\alpha)$ is the $(1-\alpha)$th quantile of the distribution of W_L. Similarly, for the second set of hypotheses that

$$H_{02}: \theta_U^* \geq 0 \text{ vs. } H_{a2}: \theta_U^* < 0,$$

we reject H_{02} if

$$W_U = R_U - \frac{n_1(n_1+1)}{2} < w(\alpha), \tag{8.24}$$

where R_U is the sum of the ranks of $\{b_{Uik}\}$ for subjects in the first sequence. Hence, average bioequivalence is concluded if both H_{01} and H_{02} are rejected; that is,

$$W_L > w(1-\alpha) \text{ and } W_U < w(\alpha). \tag{8.25}$$

The expected values and variances for W_L and W_U under the null hypotheses H_{01} and H_{02}, when there are no ties, are given by

$$E(W_L) = E(W_U) = \frac{n_1 n_2}{2},$$

$$V(W_L) = V(W_U) = \frac{1}{12} n_1 n_2 (n_1 + n_2 + 1). \tag{8.26}$$

When there are ties among observations, average ranks can be assigned to compute W_L and W_U. In this case, however, the expected values and variances of W_L and W_U become

$$E(W_L) = E(W_U) = \frac{n_1 n_2}{2},$$

$$V(W_L) = V(W_U) = \frac{1}{12} n_1 n_2 (n_1 + n_2 + 1 - Q), \tag{8.27}$$

where

$$Q = \frac{1}{(n_1+n_2)(n_1+n_2-1)} \sum_{v=1}^{q} (r_v^3 - r_v),$$

where

q is the number of tied groups.
r_v is the size of the tied group v.

Note that if there are no tied observations, $q = n_1 + n_2$, $r_v = 1$ for $v = 1, 2, \ldots, n$, and $Q = 0$, then Equation 8.27 reduces to Equation 8.26. Since W_L and W_U are symmetric about their mean $(n_1 n_2)/2$, we have

$$w(1-\alpha) = n_1 n_2 - w(\alpha). \tag{8.28}$$

When $n_1 + n_2$, the total number of subjects, is large (say, $n_1 + n_2 > 40$) and the ratio of n_1 and n_2 is close to 1/2, a large sample approximation using the standard normal distribution can be used to approximate for average biosimilarity testing; that is, we may conclude bioequivalence if

$$Z_L > z(\alpha) \text{ and } Z_U < -z(\alpha),$$

where $z(\alpha)$ is the αth quantile of a standard normal distribution and

$$Z_L = \frac{W_L - E(W_L)}{\sqrt{V(W_L)}} = \frac{R_L - \left[\frac{n_1(n_1+n_2+1)}{2}\right]}{\sqrt{\frac{1}{12} n_1 n_2 (n_1+n_2+1)}},$$

$$Z_U = \frac{W_U - E(W_U)}{\sqrt{V(W_U)}} = \frac{R_U - \left[\frac{n_1(n_1+n_2+1)}{2}\right]}{\sqrt{\frac{1}{12} n_1 n_2 (n_1+n_2+1)}}. \tag{8.29}$$

Note that the variances in Z_L and Z_U should be replaced with the variance given in Equation 8.26 if these are ties.

8.5 POWER CALCULATION FOR SAMPLE SIZE

Since power analysis for sample-size calculation for biosimilar studies is similar to that for bioequivalence studies, in this section, we focus on power analysis for sample-size calculation for bioequivalence studies. The statistical procedures and/or formulas for bioequivalence assessment for generic drug products can be directly applied to biosimilarity assessment for biosimilar products.

For bioequivalence trials, a traditional approach for sample-size determination is to conduct a power analysis based on the 80/20 decision rule. This approach, however, is based on point hypotheses rather than interval hypotheses and therefore may not be statistically valid. Phillips (1990) provided a table of sample sizes based on power calculations of Schuirmann's two one-sided tests procedure using the bivariate noncentral t-distribution. However, no formulas are provided. An approximate formula for sample-size calculations was provided in Liu and Chow (1992). Under a standard 2 × 2 crossover design, assuming that $n_1 = n_2 = n$ and $\theta = \mu_T - \mu_R = 0$, the sample size required for achieving $1 - \beta$ power at the α level of significance under interval hypotheses testing (or Schuirmann's two one-sided tests procedure) can be obtained by solving the following equation:

$$n \geq 2\left[t(\alpha, 2n-2) + t\left(\frac{\beta}{2}, 2n-2\right)\right]^2 \left(\frac{\hat{\sigma}_d}{\Delta}\right)^2,$$

where $\Delta = 0.2\mu_R$. If the ±20 rule is used, then the above becomes

$$n \geq \left[t(\alpha, 2n-2) + t\left(\frac{\beta}{2}, 2n-2\right) \right]^2 \left(\frac{CV}{20} \right)^2$$

where $CV = 100 \times \sqrt{2\hat{\sigma}_d^2} / \mu_R$.

In the case where $\theta = \theta_0 \neq 0$, the sample size required for achieving $1 - \beta$ power at the α level of significance under Schuirmann's two one-sided tests procedure can be obtained by solving the following equation:

$$n(\theta_0) \geq 2\left[t(\alpha, 2n-2) + t\left(\frac{\beta}{2}, 2n-2\right) \right]^2 \left(\frac{\hat{\sigma}_d}{\Delta - \theta_0} \right)^2,$$

where $\Delta = 0.2\mu_R$. If the ±20 rule is used, then the above becomes

$$n(\theta_0) \geq \left[t(\alpha, 2n-2) + t\left(\frac{\beta}{2}, 2n-2\right) \right]^2 \left(\frac{CV}{20 - \theta_0'} \right)^2$$

where $\theta_0' = 100 \times \theta_0 / \mu_R$.

In practice, biosimilar studies often utilize higher-order crossover designs such as Balaam's design, the two-sequence dual design, and four-period crossover designs with either two sequences or four sequences. A higher-order crossover design is defined as a crossover design whose number of sequences or whose number of periods is larger than the number of treatments to be compared. Sample-size requirements for a higher-order crossover design can be similarly obtained. Consider the following commonly used high-order crossover designs:

Design 1: Balaam's design—i.e., (TT, RR, RT, TR),
Design 2: Two-sequence, three-period dual design—i.e., (TRR, RTT),
Design 3: Four-period design with two sequences—i.e., (TRRT, RTTR), and
Design 4: Four-period design with four sequences—i.e., (TTRR, RRTT, TRTR, RTTR).

Let n_i, the number of subjects in each sequence i, have the same value n, and F_v denote the cumulative distribution function of the t-distribution with v degrees of freedom. Then the power function, $P_k(\theta)$, of Schuirmann's two one-sided tests at the α nominal level for design k is given as follows (see also Chen et al., 1997):

$$P_k(\theta) = F_{v_k}\left(\left[\frac{\Delta - \theta}{CV\sqrt{\frac{b_i}{n}}} \right] - t(\alpha, v_k) \right) - F_{v_k}\left(t(\alpha, v_k) - \left[\frac{\Delta + \theta}{CV\sqrt{\frac{b_i}{n}}} \right] \right), \quad \text{for } k = 1, 2, 3, 4,$$

where $v_1 = 4n - 3$, $v_2 = 4n - 4$, $v_3 = 6n - 5$, $v_4 = 12n - 5$, $b_1 = 2$, $b_2 = 3/4$, $b_3 = 11/20$, $b_4 = 1/4$.

Similarly, the sample size n required to achieve a $1-\beta$ power at the α nominal level for each corresponding design k after the logarithmic transformation is determined by the following equations:

$$n \geq b_k \left[t(\alpha, v_k) + t\left(\frac{\beta}{2}, v_k\right) \right]^2 \left[\frac{\text{CV}}{\ln(1.25)} \right]^2 \quad \text{if } \delta = 1,$$

$$n \geq b_k \left[t(\alpha, v_k) + t(\beta, v_k) \right]^2 \left[\frac{\text{CV}}{(\ln(1.25) - \ln\delta)} \right]^2 \quad \text{if } 1 < \delta < 1.25,$$

and

$$n \geq b_k \left[t(\alpha, v_k) + t(\beta, v_k) \right]^2 \left[\frac{\text{CV}}{(\Delta - \theta)} \right]^2 \quad \text{if } 0.8 < \delta < 1$$

where ln denotes the natural logarithm and $\text{CV} = \sqrt{\exp(\sigma^2) - 1}$, the coefficient of variation in the multiplicative model, and σ^2 is the residual (within-subject) variance on the log scale. However, because the degrees of freedom are usually unknown, an easy way to find the sample size is to enumerate n.

8.6 UNIFIED APPROACH FOR ASSESSING BIOSIMILARITY

Chow (2009, 2010) proposed the development of a composite index for assessing the biosimilarity of follow-on biologics based on the facts that (1) the concept of biosimilarity for biological products (made of living cells) is very different from that of bioequivalence for drug products and (2) biological products are very sensitive to small changes in the variation during the manufacturing process (i.e., it might have a drastic change in clinical outcome). Although some research on the comparison of moment-based criteria and probability-based criteria for the assessment of (1) average biosimilarity and (2) variability of biosimilarity for some given study endpoints by applying the criteria for bioequivalence is available in the literature (see, e.g., Chow et al., 2010; Hsieh et al., 2010), universally acceptable criteria for biosimilarity are not available in the regulatory guidelines/guidances. Thus, Chow (2009) and Chow et al. (2011) proposed a biosimilarity index based on the concept of the probability of reproducibility as follows:

Step 1. Assess the average biosimilarity between the test product and the reference product based on a given biosimilarity criterion. For illustration purposes, consider the bioequivalence criterion as a biosimilarity criterion. That is, biosimilarity is claimed if the 90% confidence interval of the ratio of means of a given study endpoint falls within the biosimilarity limits of

(80.00%, 125.00%) or (−0.2231, 0.2231) based on raw (original) data or based on log-transformed data.

Step 2. Once the product passes the test for biosimilarity in Step 1, calculate the reproducibility probability based on the observed ratio (or observed mean difference) and variability (Shao and Chow, 2002). Thus, the calculated reproducibility probability will take the variability and the sensitivity of heterogeneity in variances into consideration for the assessment of biosimilarity.

Step 3. We then claim biosimilarity if the calculated 95% confidence lower bound of the reproducibility probability is larger than a prespecified number p_0, which can be obtained based on an estimated reproducibility probability for a study comparing a reference product to itself (the "reference product"). We will refer to such a study as an R–R study. We can then claim (local) biosimilarity if the 95% confidence lower bound of the biosimilarity index is larger than p_0.

In an R–R study, define

$$P_{TR} = P\begin{pmatrix}\text{Concluding average biosimilarity between the test and the} \\ \text{reference products in a future trial given that the average} \\ \text{biosimilarity based on ABE criterion has been established} \\ \text{in first trial}\end{pmatrix} \tag{8.30}$$

A reproducibility probability for evaluating the biosimilarity of the same two reference products, based on the ABE criterion, is defined as:

$$P_{RR} = P\begin{pmatrix}\text{Concluding average biosimilarity of the two same reference} \\ \text{products in a future trial given that the average biosimilarity} \\ \text{based on ABE criterion has been established in first trial}\end{pmatrix} \tag{8.31}$$

Since the idea of the biosimilarity index is to show that the reproducibility probability for comparing "a reference product" with "the reference product" is higher in a study than the comparison of a follow-on biologic with the innovative (reference) product, the criterion of an acceptable reproducibility probability (i.e., p_0) for the assessment of biosimilarity can be obtained based on the R–R study. For example, if the R–R study suggests the reproducibility probability of 90%, that is, $P_{RR} = 90\%$, the criterion of the reproducibility probability for bioequivalence study could be chosen as 80% of the 90% which is $p_0 = 80\% \times P_{RR} = 72\%$.

The biosimilarity index, described above, has the advantages that (1) it is robust with respect to the selected study endpoint, biosimilarity criteria, and study design and (2) the probability of reproducibility will reflect the sensitivity of heterogeneity in variance.

Note that the proposed biosimilarity index can be applied to different functional areas (domains) of biological products such as pharmacokinetics (PK), biological activities, biomarkers (e.g., pharmacodynamics), immunogenicity, manufacturing process, and efficacy. An overall biosimilarity index or totality biosimilarity index across domains can be similarly obtained as follows:

Step 1. Obtain $\hat{p}_i$, the probability of reproducibility for the ith domain, $i = 1, \ldots, K$.

Step 2. Define the biosimilarity index $\hat{p} = \sum_{i=1}^{K} w_i \hat{p}_i$, where w_i is the weight for the ith domain.

Step 3. Claim global biosimilarity if we reject the null hypothesis that $p \leq p_0$, where p_0 is a prespecified acceptable reproducibility probability. Alternatively, we can claim (global) biosimilarity if the 95% confidence lower bound of p is larger than p_0.

Let T and R be the parameters of interest (e.g., pharmacokinetic response) with means of μ_T and μ_R, for a test product and a reference product, respectively. Thus, the interval hypotheses for testing the ABE of two products can be expressed as:

$$H_0: \theta'_L \geq \frac{\mu'_T}{\mu'_R} \text{ or } \theta'_U \leq \frac{\mu'_T}{\mu'_R} \text{ vs. } H_a: \theta'_L < \frac{\mu'_T}{\mu'_R} < \theta'_U$$

where (θ'_L, θ'_U) is the ABE limit. For *in vivo* bioequivalence testing, (θ'_L, θ'_U) is chosen to be (80.00%, 125.00%). The above hypotheses can be reexpressed as:

$$H_0: \theta_L \geq \mu_T - \mu_R \text{ or } \theta_U \leq \mu_T - \mu_R \text{ vs. } H_a: \theta_L < \mu_T - \mu_R < \theta_U$$

where μ_T and μ_R are the means of log-transformed data that are equal to the log-transformed values of μ'_T and μ'_R. (θ_L, θ_U) is (−0.2231, 0.2231), which are equal to the log-transformed values of (80.00%, 125.00%). To calculate the reproducibility probability under the above interval hypotheses, the probability of P_{TR} can be expressed as the following when considering a parallel design, since it is a common design for biological products:

$$P(\delta_L, \delta_U) = P\left(T_L\left(\bar{Y}_T, \bar{Y}_R, s_T, s_R\right) > t_{\alpha,dfp} \text{ and } T_U\left(\bar{Y}_T, \bar{Y}_R, s_T, s_R\right) < -t_{\alpha,dfp} \mid \delta_L, \delta_U\right) \quad (8.32)$$

where

s_T, s_R, n_T, and n_R are the sample standard deviations and sample sizes for the test and reference formulations, respectively.

$T_L(\cdot)$ and $T_U(\cdot)$ are test statistics evaluated at θ_L and θ_U, respectively.

The value of dfp can be calculated by:

$$dfp = \frac{\left(\frac{s_T^2}{n_1} + \frac{s_R^2}{n_2}\right)^2}{\frac{\left(\frac{s_T^2}{n_T}\right)^2}{n_T - 1} + \frac{\left(\frac{s_R^2}{n_T}\right)^2}{n_R - 1}},$$

$$T_L\left(\bar{Y}_T,\bar{Y}_R,s_T,s_R\right)=\frac{\left(\bar{Y}_T-\bar{Y}_R\right)-\theta_L}{\sqrt{\frac{s_T^2}{n_T}+\frac{s_R^2}{n_R}}},$$

$$T_U\left(\bar{Y}_T,\bar{Y}_R,s_T,s_R\right)=\frac{\left(\bar{Y}_T-\bar{Y}_R\right)-\theta_U}{\sqrt{\frac{s_T^2}{n_T}+\frac{s_R^2}{n_R}}},$$

$$\delta_L=\frac{\mu_T-\mu_R-\theta_L}{\sqrt{\frac{\sigma_T^2}{n_T}+\frac{\sigma_R^2}{n_R}}} \text{ and } \delta_U=\frac{\mu_T-\mu_R-\theta_U}{\sigma_d\sqrt{\frac{\sigma_T^2}{n_T}+\frac{\sigma_R^2}{n_R}}} \tag{8.33}$$

σ_T^2 and σ_R^2 are the variances for test and reference formulations, respectively.

The vectors (T_L, T_U) can be shown to follow a bivariate noncentral t distribution with $n_1 + n_2 - 2$ and dfp degrees of freedom, correlation of 1, and noncentrality parameters of δ_L and δ_U (Owen, 1965; Phillips, 1990). Owen (1965) showed that the integral of the above bivariate noncentral t-distribution can be expressed as the difference of the integral between two univariate noncentral t-distributions. Therefore, the power function in Equation 8.33 can be obtained by:

$$P(\delta_L,\delta_U)=Q_f(t_U,\delta_U;0,R)-Q_f(t_L,\delta_L;0,R) \tag{8.34}$$

where

$$Q_f(t,\delta;0,R)=\frac{\sqrt{2\pi}}{\Gamma(f/2)2^{(f-2)/2}}\int_0^R G\left(\frac{tx}{\sqrt{f}-\delta}\right)x^{f-1}G'(x)\,dx$$

$$R=(\delta_L-\delta_U)\frac{\sqrt{f}}{(t_L-t_U)},\ G'(x)=\frac{1}{\sqrt{2\pi}}e^{-x^2/2},\ G(x)=\int_{-\infty}^{x}G'(t)\,dt$$

and

$$t_L=t_{\alpha,dfp},\ t_U=-t_{\alpha,dfp} \text{ and } f=dfp \text{ for parallel design.}$$

Note that when $0<\theta_U=-\theta_L$, $P(\delta_L,\delta_U)=P(-\delta_U,-\delta_L)$.

The reproducibility probabilities increase when sample size increases and the ratio of means is close to 1, while it decreases when variability increases for the same setting of the sample size and means ratio. This shows the impact of variability on reproducibility probabilities. Since true values of δ_L and δ_U are unknown, using the idea of replacing δ_L and δ_U in Equation 8.34 by their estimates based on the sample from the first study, the estimated reproducibility probability can be obtained as:

$$\hat{P}\left(\hat{\delta}_L,\hat{\delta}_U\right)=Q_f\left(t_L,\hat{\delta}_U;0,\hat{R}\right)-Q_f\left(t_U,\hat{\delta}_L;0,\hat{R}\right) \tag{8.35}$$

where

$$\hat{\delta}_L = \frac{\bar{Y}_T - \bar{Y}_L - \theta'_L}{\sqrt{\frac{s_T^2}{n_T} + \frac{s_R^2}{n_R}}},\ \hat{\delta}_U = \frac{\bar{Y}_T - \bar{Y}_L - \theta'_U}{\sqrt{\frac{s_T^2}{n_T} + \frac{s_R^2}{n_R}}},\ \hat{R} = \frac{\left(\hat{\delta}_L - \hat{\delta}_U\right)\sqrt{f}}{\left(t_L - t_U\right)}$$

8.7 CONCLUDING REMARKS

Biological products or medicines are therapeutic agents made of a living system or organisms. As a number of biological products are due to expire in the next few years, the potential opportunity to develop the follow-on products of these originator products may result in the reduction of these products and provide more choices to medical doctors and patients for getting similar treatment care with lower cost. However, the price reductions versus the originator biological products remains to be determined, as the advantage of a slightly cheaper price may be outweighed by the hypothetical increased risk of side-effects from biosimilar molecules that are not exact copies of their originators. Unlike traditional small-molecule drug products, the characteristics and development of biological products are more complicated and sensitive to many factors. Any small change in the manufacturing process may result in a change of therapeutic effect of the biologic products. The traditional bioequivalence criterion for average bioequivalence of small-molecule drug products may not be suitable for evaluation of the biosimilarity of biological products. Therefore, in this chapter, we evaluate the biosimilarity index proposed by Chow et al. (2011) for the assessment of the (average) biosimilarity between the innovator and reference products. Results based on both the estimation and Bayesian approaches demonstrate that the proposed method based on the biosimilarity index can reflect the characteristics and impact of variability on the therapeutic effect of biological products. However, the estimated reproducibility probability based on the Bayesian approach depends on the choice of the prior distributions. If a different prior such as an informative prior is used, a sensitivity analysis may be performed to evaluate the effects of different prior distributions.

The other advantage is that the proposed method can be applied to different functional areas (domains) of biological products such as pharmacokinetics, biological activities, biomarkers (e.g., pharmacodynamics), immunogenicity, the manufacturing process, efficacy, and so on, since it is developed based on the probability of reproducibility. Further research will be employed for development of the statistical testing approach for evaluating biosimilarity across domains.

Current methods for assessing bioequivalence for drug products with identical active ingredients are not applicable to follow-on biologics due to fundamental differences. The assessment of biosimilarity between follow-on biologics and innovator in terms of surrogate endpoints (e.g., pharmacokinetic parameters and/or pharmacodynamic responses) or biomarkers (e.g., genomic markers) requires the establishment of the Fundamental Biosimilarity Assumption in order to bridge the surrogate endpoints and/or biomarker data to clinical safety and efficacy.

Unlike conventional drug products, follow-on biologics are very sensitive to small changes in variation during the manufacturing process, which have been shown to have an impact on the clinical outcome. Thus, it is a concern whether current criteria and regulatory requirements for the assessment of bioequivalence for drugs with small molecules can also be applied to the assessment of biosimilarity of follow-on biologics. It is suggested that current, existing criteria for the evaluation of bioequivalence, similarity, and biosimilarity be scientifically/statistically evaluated in order to choose the most appropriate approach for assessing biosimilarity of follow-on biologics. It is recommended that the selected biosimilarity criteria should be able to address (1) sensitivity due to small variations in both location (bias) and scale (variability) parameters and (2) the degree of similarity which can reflect the assurance for drug interchangeability.

Under the established Fundamental Biosimilarity Assumption and the selected biosimilarity criteria, it is also recommended that appropriate statistical methods (e.g., comparing distributions and the development of biosimilarity index) be developed under valid study designs (e.g., Design 1 and Design 2 described earlier) for achieving the study objectives (e.g., the establishment of biosimilarity at specific domains or drug interchangeability) with a desired statistical inference (e.g., power or confidence interval). To ensure the success of studies conducted for the assessment of biosimilarity of follow-on biologics, regulatory guidelines/guidances need to be developed. Product-specific guidelines/guidances published by the EMA have been criticized for not having standards. Although product-specific guidelines/guidances do not help to establish standards for the assessment of biosimilarity of follow-on biologics, they do provide the opportunity for accumulating valuable experience/information for establishing standards in the future. Thus, numerical studies are recommended, including simulations, meta-analysis, and/or sensitivity analysis, in order to (1) provide a better understanding of these product-specific guidelines/guidances, and (2) check the validity of the established Fundamental Biosimilarity Assumption, which is the legal basis for assessing the biosimilarity of follow-on biologics.

REFERENCES

Box GEP, Tiao GC. (1973) *Bayesian Inference in Statistical Analysis.* Addison-Wesley, Reading, MA.

CD. (2003) Commission Directive 2003/63/EC of 25 June 2003 amending Directive 2001/83/EC of the European Parliament and of the Council on the Community code Relating to medicinal products for human use. *Official Journal of the European Union L* **159**, 46.

CFDA. (2014) Draft guideline on the development and evaluation of biosimilars (Chinese version for public comments). Center for Drug Evaluation, China Food and Drug Administration, October 29, 2014, Beijing, China.

CFDA. (2015) Draft guideline on the development and evaluation of biosimilars (Chinese version for trial purpose). Center for Drug Evaluation, China Food and Drug Administration, February 28, 2015, Beijing, China.

Chen KW, Chow SC, Li G. (1997) A note on sample size determination for bioequivalence studies with higher-order crossover designs. *Journal of Pharmacokinetics and Biopharmaceutics* **25**, 753–765.

Chow SC. (1999) Individual bioequivalence—a review of FDA draft guidance. *Drug Information Journal* **33**, 435–444.
Chow SC. (2009) Criteria for assessing biosimilarity for follow-on biologics. Presented at Current Advanced Statistical Issues in Clinical Trials—Biosimilars. Taipei, Taiwan, October 2, 2009.
Chow SC. (2010) On scientific factors of biosimilarity and interchangeability. Presented at the FDA Public Hearing on Approval Pathway for Biosimilar and Interchangeable Biological Products, Silver Spring, MD, November 2–3, 2010.
Chow SC. (2011) Quantitative evaluation of bioequivalence/biosimilarity. *Journal of Bioequivalence and Bioavailability* **Suppl.1–002**, 1–8, doi:10.4172/jbb.S1-002.
Chow SC. (2013a) *Biosimilars: Design and Analysis of Follow-on Biologics*. CRC Press, Taylor & Francis Group, New York.
Chow SC. (2013b) Assessing biosimilarity and drug interchangeability of biosimilar products. *Statistics in Medicine* **32**, 361–363.
Chow SC, Endrenyi L, Lachenbruch PA, et al. (2011) Scientific factors for assessing biosimilarity and drug interchangeability of follow-on biologics. *Biosimilars* **1**, 13–26.
Chow SC, Hsieh TC, Chi E, Yang J. (2010) A comparison of moment-based and probability-based criteria for assessment of follow-on biologics. *Journal of Biopharmaceutical Statistics* **20**, 31–45.
Chow SC, Ki F. (1997) Statistical comparison between dissolution profiles of drug products. *Journal of Biopharmaceutical Statistics* **7**, 241–258.
Chow SC, Liu JP. (2008) *Design and Analysis of Bioavailability and Bioequivalence Studies*. 3rd edition. Chapman Hall/CRC Press, Taylor & Francis, New York.
Chow SC, Shao J. (2002) A note on statistical methods for assessing therapeutic equivalence. *Controlled Clinical Trials* **23**, 515–520.
Chow SC, Song FY, Endrenyi L. (2015) A note on the Chinese draft guidance on biosimilar products. *Chinese Journal of Pharmaceutical Analysis* **5**, 4–9.
Chow SC, Wang J, Endrenyi L, Lachenbruch PA. (2013) Scientific considerations for assessing biosimilar products. *Statistics in Medicine* **32**, 370–381.
Chow SC, Yang LY, Starr A, Chiu ST. (2013) Statistical methods for assessing interchangeability of biosimilars. *Statistics in Medicine* **32**, 442–448.
Cornell RG. (1990) The evaluation of bioequivalence using nonparametric procedures. *Communications in Statistics—Theory and Methods* **19**, 4153–4169.
CPMP. (2001) Guideline on comparability of medicinal products containing biotechnology-derived proteins as active substance: non-clinical and clinical issues. Committee for Proprietary Medicinal Products. EMEA/CPMP/3097/02/Final8.
Davit BM, Conner DP, Fabian-Fritsch B, et al. (2008) Highly variable drugs: observations from bioequivalence data submitted to the FDA for new generic drug applications. *AAPS Journal* **10**, 148–156.
Edwards W, Lindman H, Savage LJ. (1963) Bayesian statistical inference for psychological research. *Psychology Reviews* **70**, 193.
EMA. (2005) Guideline on similar biological medicinal products. The European Medicines Agency Evaluation of Medicines for Human Use. EMEA/CHMP/437/04, London, UK.
EMA. (2006a) Guideline on similar biological medicinal products. The European Medicines Agency Evaluation of Medicines for Human Use. EMEA/CHMP/437/04, London, UK.
EMA. (2006b) Draft guideline on similar biological medicinal products containing biotechnology-derived proteins as drug substance: quality issues. The European Medicines Agency Evaluation of Medicines for Human Use. EMEA/CHMP/49348/05, London, UK.

EMA. (2006c) Draft annex guideline on similar biological medicinal products containing biotechnology-derived proteins as drug substance—non clinical and clinical issues—guidance on biosimilar medicinal products containing recombinant erythropoietins. The European Medicines Agency Evaluation of Medicines for Human Use. EMEA/CHMP/94526/05, London, UK.

EMA. (2006d) Draft annex guideline on similar biological medicinal products containing biotechnology-derived proteins as drug substance—non clinical and clinical issues—guidance on biosimilar medicinal products containing recombinant granulocyte-colony stimulating factor. The European Medicines Agency Evaluation of Medicines for Human Use. EMEA/CHMP/31329/05, London, UK.

EMA. (2006e) Guideline on similar biological medicinal products containing biotechnology-derived proteins as active substance—non-clinical and clinical issues. EMEA/CHMP/BMWP/42832, London, UK.

EMA. (2010a) Draft guideline on similar biological medicinal products containing monoclonal antibodies. EMA/CHMP/BMWP/403543/2010, London, UK.

EMA. (2010b) Concept paper on similar biological medicinal products containing recombinant follicle stimulation hormone. EMA/CHMP/BMWP/94899/2010, London, UK.

EMA. (2011a) Concept paper on the revision of the guideline on similar biological medicinal products. EMA/CHMP/BMWP/572643, London, UK.

EMA. (2011b) Concept paper on: revision of the guideline on similar biological medicinal products containing biotechnology-derived proteins as active substance: quality issues. EMEA/CHMP/BWP/617111, London, UK.

EMA. (2011c) Concept paper on: revision of the guideline on similar biological medicinal products containing biotechnology-derived proteins as active substance: non-clinical and clinical issues. EMEA/CHMP/BMWP/572828, London, UK.

EMA. (2011d) Guideline on: similar biological medicinal products containing interferon beta. EMA/CHMP/BMWP/652000/2010, London, UK.

EMA. (2014) Guideline on similar biological medicinal products. CHMP/437/04 Rev 1, Committee for Medicinal Products for Human Use (CHMP), London, UK, October 23, 2014.

FDA. (1997) Guidance for industry: dissolution testing of immediate release solid oral dosage forms. Food and Drug Administration, Rockville, MD.

FDA. (2001) Guidance on: statistical approaches to establishing bioequivalence. Center for Drug Evaluation and Research, Food and Drug Administration, Rockville, MD.

FDA. (2003) Guidance on: bioavailability and bioequivalence studies for orally administered drug products—general considerations. Center for Drug Evaluation and Research, Food and Drug Administration, Rockville, MD.

FDA. (2013) Bioequivalence studies with pharmacokinetic endpoints for drugs submitted under an ANDA (draft guidance). United States Food and Drug Administration, Silver Spring, MD.

FDA. (2015a) Scientific considerations in demonstrating biosimilarity to a reference product. United States Food and Drug Administration, Silver Spring, MD.

FDA. (2015b) Quality considerations in demonstrating biosimilarity to a reference protein product. Food and Drug Administration, Silver Spring, MD.

FDA. (2015c) Biosimilars: questions and answers regarding implementation of the Biologics Price Competition and Innovation Act of 2009. Food and Drug Administration, Silver Spring, MD.

Haidar SH, Davit BM, Chen ML, et al. (2008a) Bioequivalence approaches for highly variable drugs and drug products. *Pharmaceutical Research* **25**, 237–241.

Haidar SH, Makhlouf F, Schuirmann DJ, et al. (2008b) Evaluation of a scaling approach for the bioequivalence of highly variable drugs. *The AAPS Journal* **10**, 450–454.

Hauschke D, Steinijans VW, Diletti E. (1990) A distribution-free procedure for the statistical analyses of bioequivalence studies. *International Journal of Clinical Pharmacology, Therapy and Toxicology* **28**, 72–78.

HC. (2010) Guidance for sponsors: information and submission requirements for subsequent entry biologics (SEBs). Health Canada, Ottawa, ON.

Hsieh TC, Chow SC, Liu JP, et al. (2010) Statistical test for evaluation of biosimilarity of follow-on biologics. *Journal of Biopharmaceutical Statistics* **20**, 75–89.

IMS. (2010) Multinational integrated data analysis system (MIDAS). IMS Institute for Healthcare and Informatics, Danbury, CT, December, 2010.

Kang SH, Chow SC. (2013) Statistical assessment of biosimilarity based on relative distance between follow-on biologics. *Statistics in Medicine* **32**, 382–392.

KFDA. (2011) Korean guidelines on the evaluation of similar biotherapeutic products (SBPs). Korean Food and Drug Administration, South Korea.

Leeson LJ. (1995) *In Vitro/in vivo* correlation. *Drug Information Journal* **29**, 903–915.

Lindley DV. (1965) *Introduction to Probability and Statistics for a Bayesian Viewpoint, Part II, Inference*. Cambridge University Press, Cambridge, UK.

Liu JP, Chow SC. (1992) Sample size determination for the two one-sided tests procedure in bioequivalence. *Journal of Pharmacokinetics and Biopharmaceutics* **20**, 101–104.

Liu JP, Ma MC, Chow SC. (1997) Statistical evaluation of similarity factor f_2 as a criterion for assessment of similarity between dissolution profiles. *Drug Information Journal* **31**, 1255–1271.

Ma M, Lin R, Liu JP. (1999) Statistical evaluations of dissolution similarity. *Statistica Sinica* **9**, 1011–1027.

Mann HB, Whitney DR. (1947) On a test of whether one or two random variables is stochastically larger than the other. *Annals of Mathematical Statistics* **18**, 50–60.

MHLW. (2009) Guidelines for the quality, safety and efficacy assurance of follow-on biologics. Yakushoku shinsahatu 0304007, Tokyo, Japan.

Moore JW, Flanner HH. (1996) Mathematical comparison of curves with an emphasis on dissolution profiles. *Pharmaceutical Technology* **20**, 64–74.

Owen DB. (1965) A special case of a noncentral *t* distribution. *Biometrika* **52**, 437–446.

Phillips KF. (1990) Power of the two one-sided tests procedure in bioequivalence. *Journal of Pharmacokinetics and Biopharmaceutics* **18**, 137–144.

Rodda BE, Davis RL. (1980) Determining the probability of an important difference in bioavailability. *Clinical Pharmacology and Therapeutics* **28**, 247–252.

Schuirmann DJ. (1981) On hypothesis testing to determine if the mean of a normal distribution is continued in a known interval. *Biometrics* **37**, 617.

Schuirmann DJ. (1987) A comparison of the two one-sided tests procedure and the power approach for assessing the equivalence of average bioavailability. *Journal of Pharmacokinetics and Biopharmaceutics* **15**, 657–680.

Shah VP, Tsong Y, Sathe P, Liu JP. (1998) In vitro dissolution profile comparison—statistics and analysis of the similarity factor, f2. *Journal of Pharmacology Research* **15,** 889–896.

Shao J, Chow SC. (2002) Reproducibility probability in clinical trials. *Statistics in Medicine* **21**, 1727–1742.

Suh SK, Park Y. (2011) Regulatory guideline for biosimilar products in Korea. *Biologicals* **39**, 336–338.

SUPAC-IR. (1995) Food and Drug Administration guideline on: immediate release solid oral dosage forms: scale-up and post approval changes: chemistry, manufacturing, and controls, in vitro dissolution testing, and in vivo bioequivalence documentation. FDA, Rockville, MD.

Tothfalusi L, Endrenyi L, Arieta AG. (2009) Evaluation of bioequivalence for highly variable drugs with scaled average bioequivalence. *Clinical Pharmacokinetics* **48**, 725–743.

Wang J, Chow SC. (2012) On regulatory approval pathway of biosimilar products. *Pharmaceuticals* **5**, 353–368; doi:10.3390/ph5040353.

WHO. (2009) Guidelines on: evaluation of similar biotherapeutic products (SBPs). Geneva, Switzerland.

Wilcoxon F. (1945) Individual comparisons by ranking methods. *Biometrics* **1**, 80–83.

Woodcock J, Griffin J, Behrman R, et al. (2007) The FDA's assessment of follow-on protein products: a historical perspective. *Nature Reviews Drug Discovery* **6**, 437–442.

9 Extrapolation of Indications for Biosimilars

Opportunity for Developers and Challenges for Regulators

Jian Wang, Wallace Lauzon, Catherine Njue, and Agnes V. Klein
Health Canada

CONTENTS

9.1 INTRODUCTION

Since the approval of the first biosimilar, Omnitrope, by the European Medical Agency in 2006 (EMA, 2006), biosimilars have been authorized in many jurisdictions, including Australia, Canada, Japan, Korea, the US, and many other countries. Subsequent-entry biologic (SEB) is the Canadian term used for a biosimilar that enters the market subsequent to a Canadian biological drug (the reference product) previously authorized in Canada. For the purpose of marketing authorization, a biosimilar relies in part on prior information in the public domain regarding safety and efficacy deemed relevant due to the demonstration of similarity to the reference biologic drug (RBD).

Similarity to the RBD in terms of quality characteristics, biological activity, toxicity, and clinical safety and efficacy can be demonstrated through comprehensive comparability exercises between the biosimilar and the RBD. The side-by-side structural and functional characterization of the biosimilar and the RBD to demonstrate similarity is the foundation of the biosimilar development program. Extrapolation of indications is an established scientific and regulatory process adopted by many regulatory agencies based on clinical data generated in one or two indications (Health Canada, 2010; EMA, 2015; FDA, 2015). Before extrapolation can be considered, biosimilar sponsors must present convincing and compelling similarity data to regulatory agencies. The totality of evidence collected from all comparative studies ultimately determines the marketing authorization of a biosimilar and its authorized indications.

The first biosimilar authorized in Canada was Sandoz's biosimilar growth hormone, Omnitrope, in 2009 (Klein, 2011). To date, Health Canada has issued Notices of Compliance (NOC) for three biosimilars within the product classes of human growth hormone and tumor necrosis factor (TNF)-inhibitor, for use in Canada (see Table 9.1).

TABLE 9.1
Health Canada Authorized Subsequent Entry Biologics/Biosimilars

Product Name	Active Substance	Therapeutic Area	Authorization Date
Omnitrope	Somatropin	Growth hormone deficiency in adults and children	April 20, 2009
Remsima	Infliximab	Ankylosing spondylitis Psoriatic arthritis Psoriasis Rheumatoid arthritis	January 15, 2014
Inflectra	Infliximab	Ankylosing spondylitis Psoriatic arthritis Psoriasis Rheumatoid arthritis	January 15, 2014
Omnitrope	Somatropin	Small for gestational age Small for gestational age Turner syndrome	May 8, 2015

SEBs in Canada are regulated under the same regimen as any new drug. Once market authorization has been granted to a biosimilar, the product becomes a self-standing, independent product, with one caveat: a biosimilar may not be used as an RBD for another biosimilar. The reasons are the following:

1. Canadian regulations are product-specific.
2. Division 8 of the Canadian Food and Drug Regulations indicates that the package submitted in support of a new drug (new drug product) has to satisfy the Minister in terms of safety, efficacy, and quality. This allows for a fair amount of flexibility. At minimum, it allows for the regulator to use science to tailor the information to the product.
3. Generic drug regulations and intellectual property considerations, in the form of the patent linkage regulations, also have a bearing on how biosimilars are regulated, together with data protection provisions.

This chapter deals extensively with the underlying science that was the basis of the Canadian approach to regulating biosimilars. In particular, the present chapter discusses quality, clinical and statistical aspects of biosimilar development, and their importance for the consideration of extrapolation of indications.

9.2 QUALITY CONSIDERATIONS

9.2.1 Complexity of Biologics

For biologics, changes to the starting materials, manufacturing process, equipment, or facility can result in significant unexpected alterations in the intermediate stage product, drug substance, and/or final product. Herein lies the challenge of demonstrating the similarity of a biosimilar to its RBD. This challenge is compounded by the limitations of the analytical methods available to assess the similarity.

The purpose of the analytical and biological similarity exercises is to establish an evidence-based link between the RBD and the biosimilar in order to *leverage the existing body of knowledge available in the public domain regarding the efficacy and safety of the product.* The strength of this link, or perhaps more appropriately, the *residual uncertainty* in the similarity resulting from a thorough evaluation of the results of the analytical and biological similarity exercises drive the breadth and scope of the clinical and nonclinical similarity exercises. It is essential that the analytical and biological similarity exercises demonstrate that the biosimilar and the RBD are highly similar within the limitations of the current state-of-the-art analytical methods for a candidate molecule to be considered biosimilar.

9.2.2 Demonstrating Quality Similarity

The analytical and biological similarity, termed the *quality similarity* for the purposes of this chapter, is not a foreign concept to the regulation of biologics. ICH Q5E Comparability of Biotechnology/Biological Products Subject of Changes in their Manufacturing Process (ICH Q5E, 2004) provides a basis for evaluating the impact of manufacturing changes on a product. While the comparison of two independent products is clearly outside the scope of this guidance, the aim is similar and many of the principles and approaches are applicable. The Canadian SEB guidance (Health Canada, 2010) provides details regarding expectations for the similarity exercises and the applicability of ICH Q5E.

The similarity demonstrated through the quality similarity exercise between the proposed biosimilar and the RBD should be the primary basis for the decision of whether the biosimilar pathway is appropriate for the biosimilar product. The studies that make up the quality similarity exercise should be comprehensive and appropriately justified. The studies should be performed using reliable analytical methods that have been optimized to detect potentially meaningful differences. These studies should include multiple orthogonal methods to characterize the appropriate physical, chemical, and biological attributes. This should include functional assays to evaluate the range of biological activities.

The Canadian biosimilar guidance states that the goal of the determination of similarity between the two products being compared is that they be *highly similar rather than identical* and that any differences observed in quality attributes should have *no adverse impact on safety and efficacy*. This recognizes that the nature of biological drugs precludes the determination of two products being identical and provides the criteria for the determination of highly similar. This definition of highly similar appropriately links the output of the quality similarity exercises to potential clinical outcomes, thus providing a benchmark to filter the noise (differences due to method variability and minor molecular variation) from the signal (clinically meaningful differences).

In practice, the identification of differences that have "no adverse impact on safety and efficacy" is fraught with uncertainty in that the potential impact of differences in the quality similarity of the products is often unknown and there is often considerable uncertainty regarding the mechanisms of action of the molecule. It is incumbent upon the sponsor of the biosimilar product to provide the demonstration that the differences have no adverse impact. Great care should be taken in the design of the quality similarity studies to ensure that all available information regarding the *critical quality attributes (CQAs) and the quality determinants of the safety and efficacy* of the product has been considered. In addition, the results of the quality similarity studies should be discussed fully within the framework of the potential clinical impact, including *justification of assumptions and qualification of the uncertainty.*

The relationship between the observed differences in the quality similarity exercise and the potential clinical outcomes links the determination of similarity to the mechanisms of action of the molecule in the clinic which becomes the basis of the contribution of the quality similarity exercise to the determination of the extrapolation of indications. Considering the extrapolation of indications in this way has two important consequences. First, there is the potential for the determination that a biosimilar meets the definition of highly similar to the RBD in *some, but not all, indications*, if there is a difference in the mechanism of action across indications. Thus, biosimilarity is contextually based rather than an inherent quality of the molecule. Second, the choice of supporting clinical and nonclinical studies needs to consider the gaps in the quality similarity exercise with respect to potential mechanisms of action in order to support the extrapolation of indications. The choice of clinical and nonclinical studies is discussed in detail in the appropriate sections. Regardless of the consequences, the input of the quality similarity assessment to the determination of the extrapolation of indications is the same: *Is there sufficient evidence in the quality similarity package to adequately support the consideration of data extrapolation*? Operationally, this translates into a determination of whether there are specific signals in the data package to indicate a potential safety or efficacy concern with respect to a particular mechanism of action or, more likely, an evaluation of the residual uncertainty regarding the similarity with respect to a particular mechanism of action.

There is a risk in the extrapolation of indications for which there are no primary clinical data. This risk can be mitigated in part by the quality similarity exercise. This mitigation strategy raises high expectations for the quality of the data and the degree of similarity to support biosimilarity. In situations where there is a robust data package that encompasses all of the potential mechanisms of action and all of the quality attributes of the biosimilar product fall within the observed ranges of the reference biological product, the results of the quality similarity exercises clearly support the consideration of the extrapolation of data to other indications. However, the variability of biological products, their labile nature, and the impact of raw materials and the manufacturing process on the CQAs mean that differences are often observed in the quality similarity studies. A number of properties of any biologic may vary without impacting the clinical performance or the safety of the product. Many of these properties are specific to a product class or a specific molecule. In many instances, the relationship between the specific physical chemical characteristics of a molecule and the impact of those characteristics on the clinical efficacy and safety profile are unknown. The relationship between structure and function is explored experimentally both directly and indirectly during the *clinical development of a product*. In the case of innovator products, this clinical experience is often sufficiently robust to predict the impact of certain changes in the characteristics of the product in spite of manufacturing changes.

9.2.3 Dealing with Quality Uncertainty

For biosimilars, which are supported by a comparatively limited clinical program, this product-specific information may not be available. Nonetheless, it is often possible to infer the potential impact of differences in observed characteristics from published product class data to support conclusions regarding the limited potential of the impact of certain differences in some quality characteristics. For instance, there is extensive experience with fully humanized monoclonal antibodies to recognize that a degree of heterogeneity in the C-terminal lysine of human monoclonal antibodies is tolerated without appreciable changes in the efficacy or safety of the molecule. However, in many instances, there may be no reliable information regarding the impact of differences in product-related species. In these cases, the certainty of the quality similarity determination may be impacted, with consequences to potential extrapolation of data. In general, the introduction of a *new molecular species of unknown impact will introduce a greater degree of uncertainty than a variation in the quantity* of a product-related species between the biosimilar and the RBD.

A number of characteristics of biologics are known to have an impact on the function of the molecule and its safety profile. Some of these properties, such as the formation of aggregates in proteins, are generally recognized as being associated with immunogenicity across product types; others, such as the impact of specific glycoforms of monoclonal antibodies on Fc-related functions, are product class specific. Those characteristics that impact the safety profile of the biosimilar in a negative way are much more likely to *impact the appropriateness of the biosimilar pathway.* It is theoretically possible that some characteristics of a molecule impact the safety profile only under certain conditions, such as subcutaneous versus intravenous administration. However, it is difficult to imagine a situation where extrapolation of data would be possible without direct demonstration of the safety profile to mitigate the risk to patient safety.

More interesting with respect to the determination of extrapolation of data to different indications are the differences that impact certain functions of the molecule but not others. For instance, increased *N*-linked fucosylation of the F_C region of human IgG1 is associated with reduced FcγRIIIa binding and lower antibody dependent cellular cytotoxity (ADCC) activity without appreciable differences in antigen-binding qualities (reviewed in Arnold et al., 2007). The key to demonstrating similarity is that the observed differences in the quality similarity package have no adverse impact on the safety or efficacy of the biosimilar. Thus, a difference between the biosimilar and the RBD limited to a significant difference with respect to afucosylation, FcγRIIIa binding, and ADCC would meet that standard for indications which had been demonstrated to rely on mechanisms of action other than ADCC activity but would fail to meet that standard for indications in which ADCC potentially played a role. The biosimilarity would thus be stratified with respect to the indications, and extrapolation *would not be supported by the quality similarity exercise for indications where FcγRIIIa binding and ADCC played a potential role.*

Setting the bar for quality similarity based on the potential impact on clinical efficacy and safety is a logical approach. However, the application of this approach presumes that the level of understanding of the mechanism of action is sufficiently robust to distinguish the contribution of various mechanisms of action to the clinical efficacy. *In instances where this is not the case, it is the responsibility of the sponsor to support the proposed data extrapolation with sufficient experimental evidence and a detailed scientific rationale.* This evidence and the accompanying rationale must be judged as sufficient to address the residual uncertainty identified during the evaluation of the quality similarity package. Depending on the depth of understanding available in the public domain and the experimental evidence available to the sponsor, it may not be possible to address this residual uncertainty without performing additional clinical studies.

9.3 CLINICAL CONSIDERATIONS

9.3.1 Key Components of the Clinical Studies

The extrapolation of indications requires an evaluation of the totality of the evidence. The quality similarity exercise is an important component for identifying potential gaps in the similarity that may impact the clinical efficacy and safety profile of the biosimilar. After side-by-side characterizations of structure and function, and comparative nonclinical studies are completed, a tailored clinical program is planned. The comparative clinical program usually includes studies to investigate clinical pharmacokinetics/pharmacodynamics (PK/PD), clinical efficacy/safety, and immunogenicity between the biosimilar and the RBD. The purpose of the clinical program is not to reestablish the efficacy and safety profile of the biosimilar, since that has been fully established by the reference product and has been assessed at the time for the marketing application of the RBD. Rather, the demonstration of similarity at the analytical and biological level allows a regulatory link to be created that forms the bridge between the innovator's clinical data and the biosimilar product. Thus, the goal of the clinical program is to rule out clinically meaningful differences by resolving the residual uncertainties that remain due to small differences observed during analytical or biological testing or due to the technological limitations of the analytical and biological testing methodologies. If clinically meaningful differences are not observed in well-designed clinical studies, a final determination of similarity can be made and the extrapolation of indications can be considered.

9.3.2 Primary Clinical PK/PD Studies

The first clinical studies that should be conducted are the comparative human PK studies and, when feasible, comparative human PD studies. Comparative PK/PD studies can be used to confirm comparability that has been established through the structural and functional studies, to justify the reduction of subsequent clinical studies (e.g., insulin), and to establish evidence for extrapolation of indications (e.g., cancer vs. RA; adult vs. pediatric). It is not necessary to study the biosimilar in every indication that has been authorized. However, separate comparative PK studies may be required to bridge multiple distinct conditions. The number of studies required depends both on the degree of similarity between the biosimilar and the RBD [ascertained from the chemistry, manufacturing, and control (CMC) data] and on the indications for which the biosimilar is proposed. The design of the comparative PK studies depends on various factors, including clinical context, safety profiles, and PK characteristics of the RBD (target-mediated disposition, linear or nonlinear PK, time dependency, half-life, etc.).

The comparative PK studies should be conducted in a setting that is reflective of the clinical situation and is sensitive enough to detect potential differences if they exist. The single-dose crossover design (short half-life) is generally regarded as the most sensitive assessment of comparative bioavailability and may be conducted in healthy volunteers. However, healthy volunteers may not always adequately reflect the PK parameters in the patient population, since host factors such as receptor expression, receptor internalization rate, and patient status can affect the disposition and clearance of biologics (e.g. target-mediated disposition for mAbs). In addition, route of administration is an important factor to consider in the design and conduct of comparative PK studies. It is highly recommended that the PK of the biosimilar be compared to the RBD using a route that requires an absorption step if such a route is used by the RBD. For biologics, the subcutaneous (sc) route of administration is more discriminative than the intravenous

(iv) route to detect potential PK differences. It is also expected that the sc route is more immunogenic than the iv route (FDA, 2014a), which makes the sc route more appropriate for investigating immunogenicity differences between the biosimilar and the RBD. Immunogenicity is a major issue that triggers concern and debate for developing not only biosimilars, but all biologics. For example, the EMA has authorized at least two biosimilar epoetins for the treatment of renal anemia and chemotherapy-induced anemia. However, one of these epoetins was only studied using the iv route of administration. Thus, it was not authorized by EMA for subcutaneous use since the subcutaneous route of administration had previously been associated with increased immunogenicity leading to the development of antibodies that cross-react with endogenous protein and, subsequently, pure red cell aplasia (PRCA) (EMA, 2008).

When seeking authorization of a biosimilar in Canada, sponsors should consider the principles of study design, statistical methods, and criteria of acceptance as outlined in Health Canada's Guideline *Comparative Bioavailability Standards: Formulations Used for Systemic Effects* as a general guidance for conducting comparative PK studies (Health Canada, 2012). It should be noted that comparative bioavailability criteria differ between the various regulatory agencies, for example, 90% CI of C_{max} is not applicable; AUCt is used; and potency correction may be required for all key parameters by Health Canada (Pen et al., 2015). According to Health Canada's guidance document for PK studies, the 90% CI of AUCt, as well as of the relative mean C_{max} of the biosimilar to the reference product, should be within 80%–125%. At the same time, the FDA recommends that sponsors provide the geometric means, arithmetic means, geometric mean ratios, and 90% CI for AUCt, AUCi, and C_{max} (FDA, 2014b).

It is important to recognize that, unlike generics, biosimilars are not considered to be identical to their respective RBDs. A comparable PK profile between the biosimilar and the RBD does not guarantee that a biosimilar has effects on the body that are comparable to those of the RBD. Therefore, comparative human PD data are desirable and can help to reduce residual uncertainty and support extrapolation of indications. A comparative human PD study can be combined with a PK study to characterize PK/PD relationships. The PD parameters used in comparative studies should be clinically validated and considered as surrogate markers of clinical outcomes, for example, absolute neutrophil count for a biosimilar G-CSF. The PD surrogate should be sensitive enough to detect potential changes induced by the biosimilar and the RBD. Importantly, a therapeutic dose for patients may induce a ceiling PD response in healthy volunteers, thus masking potential differences (dosing sensitivity) between products. In such cases, a lower dose in the steep part of the dose-response curve should be considered (assay sensitivity), and a separate human PD study may be needed. A patient PD study demonstrating similar effects on a clinically relevant PD surrogate could strengthen the claim of biosimilarity. Relevance of the PD surrogate to the mechanism(s) of action of the product is essential in supporting the extrapolation of indications. If pivotal evidence for extrapolation is based on PD parameters and, for the claimed indications, different mechanisms of action are pertinent, then sponsors should provide additional data, either PD or clinical data to cover all claimed clinical indications.

Comparative human PD studies are unlikely to substitute all clinical studies. When PD data from healthy volunteers are used as primary evidence, biosimilar sponsors also need to provide sufficient reassurance of clinical safety, including immunogenicity, in a separate repeat-dose clinical trial. Furthermore, for many biologics, especially for mAbs, there are no relevant PD surrogates available and therefore clinical trials are required.

9.3.3 Primary Clinical Studies

If dose comparative and sensitive PD studies cannot be performed to convincingly demonstrate comparability in a clinically relevant manner, adequately powered, preferably double-blind and randomized clinical trial(s) between the biosimilar and the RBD should be conducted. The main purpose of comparative clinical study/ies is to demonstrate that there are "no clinically meaningful differences" between the products.

The type and number of comparative clinical trials required for biosimilars could be affected by many factors: the nature and complexity of the RBD (such as anticancer mAbs); the limitations of studies comparing structural and functional characteristics; the findings of nonclinical testing; the extent to which differences in structure, function, and nonclinical data can predict clinical outcomes; the degree of understanding of mechanism(s) of action of the RBD and disease pathology; the extent to which human PK/PD can predict clinical outcomes; and the extent of clinical experience with the RBD, including the knowledge base with respect to safety, efficacy, and immunogenicity.

Equivalence trials are generally preferred to noninferiority trials because the suggested superiority in a noninferiority setting would raise questions about the comparability of the two products. This is usually demonstrated by showing that the true treatment difference is likely to lie between prespecified lower and upper equivalence margins that are considered clinically acceptable (ICH E9, 1998). This equivalence margin is the largest difference that can be judged as clinically acceptable for the biosimilar and should be smaller than the effect sizes observed in superiority trials conducted for the RBD. In order to detect differences between the biosimilar and the RBD, equivalence trials should be conducted in at least one sensitive population that is representative of authorized therapeutic indications. In general, a homogeneous population of patients would provide a better chance to detect differences between the biosimilar and the RBD. Ideally, the observed clinical effects should be caused by the direct action of the biosimilar or the RBD without interference of other medications, as concomitant medications may affect or mask differences in PK/PD, efficacy, safety, and/or immunogenicity of the tested products. To validate the effect of the RBD and the sensitivity of the study in the chosen study population, a large body of historical data should be available to justify the selection of the study population and equivalence margin. This is generally done through meta-analysis or systematic review. The mechanism of action in the chosen study population should be well-understood and be representative of the mechanism of action of the RBD in other populations in order to support and justify the extrapolation of indications. In some jurisdictions, clinical studies in an unauthorized indication (e.g., line of therapy, combined therapy, disease severity) may be acceptable to demonstrate "no clinically meaningful differences." Biosimilar sponsors should consult each individual regulatory agency prior to conducting such trials.

The most sensitive study endpoint within a sensitive population should be considered in order to improve the detection of potential differences between products. Potentially, a study could use a clinically relevant and sensitive study endpoint that is different from the innovator's original primary endpoint(s). For instance, the choice of ACR20 versus DAS28 has been debated for the rheumatoid arthritis (RA) population (Dougados et al., 2009), while using ORR or progression-free survival (PFS) as the primary endpoint, instead of overall survival (OS), in oncology trials could be considered for biosimilars (Ahn et al., 2011). Finally, a new surrogate or a more sensitive clinical endpoint or different time points of analysis for traditional study endpoints could all be considered acceptable depending on the proposal (EMA, 2015).

Regardless of the chosen endpoint, clinical studies should demonstrate comparable safety and efficacy profiles. Any differences detected between the efficacy of the biosimilar and the RBD should always be discussed as to whether they are clinically relevant. Generally, the aim of clinical trials is to address minor differences observed during the comparability exercises and to demonstrate that the clinical performances of the biosimilar and the RBD are comparable. However, *it is important to note that clinical data cannot be used to justify substantial differences in quality attributes.*

Extrapolation of efficacy is acceptable when a biosimilar has demonstrated high comparability to the RBD in physicochemical and functional characteristics; in nonclinical studies obtained from relevant species; and in clinical PK/PD and clinical efficacy studies in population sensitive enough to show differences, and if all concerns with respect to the principle of indication extrapolation have been addressed.

9.3.4 Safety and Immunogenicity

For a particular product, safety profiles may differ among the various indications due to a number of factors. For example, therapeutic dose, concurrent conditions, and/or concomitant medicines may have an impact. Although there are concerns with the practice of extrapolating safety profiles among indications, safety comparability demonstrated in an indication with high sensitivity to detect differences may support extrapolation to other indications.

Most biologics induce some level of antidrug antibodies (ADAs), and these ADAs may have an undesirable clinical effect on pharmacokinetics, efficacy, and/or safety. Particularly, a change in the production process of a biologic might influence the immunogenic potential of that biologic (Shankar et al., 2006). Unwanted immunogenicity is currently difficult to predict from the analytical and nonclinical data in terms of incidence, characteristics, clinical consequences, and significance. Immunogenicity evaluation should be part of clinical efficacy and safety studies for biosimilars. To support extrapolation of indications, clinical studies should be carefully planned, and experimental data should be systematically collected from a sufficiently large number of patients and sufficient study duration in at least one clinical study to characterize the variability in immunogenicity response. The biosimilar has to demonstrate that immunogenicity is not increased and that the type of immunogenicity is not changed in a sensitive population.

Dose selection is also important. For instance, some mAbs are thought to inhibit antibody formation when administered at high doses; therefore, studies have to be conducted with low dose (Brinks, 2013). Immunogenicity testing of the biosimilar and the RBD should be conducted within the biosimilar comparability exercise using the same assay format and sampling schedule. The assay used to detect antibodies is an important consideration during the clinical development of a biosimilar and should meet all current standards. Comparison of data obtained for the biosimilar with historical data obtained for the RBD is generally not considered appropriate (Giezen and Schneider, 2014). This is because RBDs may have historical immunogenicity data based on out-of-date assays, which, by today's standards, would be considered to have inadequate sensitivity. Thus, any comparison of immunogenicity needs a side-by-side test of the biosimilar and its RBD to ensure valid comparison. Without a side-by-side comparison, sensitive immunogenicity assays used currently may show higher antibody positive results with the biosimilar.

9.3.5 Extrapolation of Indications

Extrapolation of data is already an established scientific and regulatory principle for biosimilars, which has been exercised, by regulatory agencies, since the first biosimilar approvals. Extrapolation guidance exists in the biosimilar guidelines prepared by the Canada, EMA, US (Health Canada, 2010; EMA, 2015; FDA, 2015), and many other countries. In Canada, biosimilar sponsors are permitted to apply for one or more clinical indications granted to the Canadian RBD. The biosimilar sponsor is not required to submit complete clinical data for every requested indication as long as clinical data and/or scientific rationales are provided that can address the principles that Health Canada uses to determine whether extrapolation is appropriate. Based on the totality of evidence obtained from detailed and comprehensive comparative structural and functional characterizations, nonclinical studies, human PK/PD studies, and pivotal clinical trials, extrapolation of indications could be justified based on: mechanism(s) of action that play(s) a role in each of the indicated conditions for which a sponsor applies; pathophysiological mechanism(s) of the indicated diseases, which is an important determinant in the extrapolation assessment as some mAbs hold indications for the treatment of diseases that bear little resemblance to each other; a safety profile in the respective conditions and/or populations; and clinical experience with the reference biologic drug. In each case, a detailed scientific rationale that appropriately addresses the benefits and risks of such a proposal should be provided to adequately support the data extrapolation (Health Canada, 2010).

It is noted that if the mechanism of action of the drug substance and the target receptor(s) involved in the tested and in the extrapolated indication(s) are the same, extrapolation of indications is usually not problematic. However, when the mode of action is complex and involves multiple receptors or binding sites, the contribution may differ between indications or may not be well known (Weise et al., 2014). Thus, additional data are necessary to provide further reassurance that the biosimilar and the RBD will behave alike, and also in the extrapolated indications. The regulatory flexibility of current regulations in Canada, particularly those for new drug products such as biosimilars and expressed as "the information to be provided on the quality, safety and efficacy of a drug product as acceptable to the Minister (of Health)" allowed Health Canada to define the following principles:

1. Similarity must be demonstrated by comprehensive comparative characterization of the two products involved: the RBD and the biosimilar.
2. Even when two products are generally comparable, and there are minor, usually considered unimportant, differences based on side-by-side comparability between two active ingredients, it is entirely plausible that these differences might have an impact on the mechanism/s of action of a product. In that event, extrapolation is precluded.
3. Differences in mechanisms of action for each condition or use preclude extrapolation, as had been clearly demonstrated between the original infliximab and the biosimilars (Scott et al., 2014).
4. Differences in pathophysiological mechanism/s of the disease/s (indications and uses) and differences or similarities between them preclude extrapolation.
5. Differences in clinical experience and the lack of information in the public domain compared to the reference drug chosen preclude extrapolation and require full clinical trials to support those indications. The drug is considered as an entirely new product for that indication or use.
6. The type and design of trials using sensitive populations and endpoints must be capable of detecting changes in the endpoints chosen.
7. Other considerations: for example: including route of administration; posology; and PK/PD profiles in each indication or route of administration.

The primary element deciding whether or not extrapolation can be considered is the analytical characterization of the biosimilar. Variations to the production process, even for well-characterized products, might alter their efficacy and safety properties. For example, a biosimilar infliximab was filed to Health Canada as a biosimilar to the Canadian authorized infliximab. The Canadian regulatory decision was based on a critical assessment of the CMC, nonclinical, and clinical data of the biosimilar infliximab in the Canadian data package. After a careful assessment, Health Canada recommended extrapolation from rheumatoid arthritis (RA) and ankylosing spondylitis (AS) (diseases in which the biosimilar was studied) to psoriatic arthritis (PsA) and plaque psoriasis (Ps); however, extrapolations to adult and pediatric Crohn's disease (CD) and ulcerative colitis (UC) were not granted, because the comparability between the RBD and the biosimilar infliximab indicated insufficient similarity between the two products with respect to afucosylation, FcγRIIIa receptor binding, and *in vitro* ADCC (Health Canada, 2014). Thus, the biosimilar infliximab was deemed to be comparable to the reference in situations where the mechanism of action depends exclusively on its binding to TNFα.

In indications for which the mechanism of action is not clearly defined or where there may be a role for ADCC in the mechanism of action, it could not be conclusively determined that the molecules were comparable. Furthermore, from a clinical perspective, it was considered that monoclonal antibodies can function through multiple mechanisms of action and that the mechanisms involved in one disease may differ from those involved in another. ADCC may be an active mechanism of action for infliximab in the setting of CD and UC, but not in the setting of the studied RA and AS. Therefore, extrapolation from RA and AS to inflammatory bowel diseases (IBDs) cannot be recommended. For the same biosimilar product, EMA granted all indications. EMA believes that a lower amount of afucosylation species, which translates in a lower binding affinity to FcγRIIIa receptors, is not considered clinically relevant because it does not affect the activities of biosimilar infliximab in the experimental models that are most relevant to the pathophysiological conditions. Furthermore, the contribution of ADCC to the mode of action of infliximab, or any TNF antagonist, has not been established in patients (EMA, 2013).

In order to support extrapolation of indications, additional factors that may influence the decision to allow extrapolation have to be considered (Scott et al., 2014). The studied population (age, sex, and ethnic origin, etc.) should be similar to and representative for those being extrapolated. The primary clinical trial's duration of treatment, route of administration, and dosage range should be similar: extrapolation to a different route (IV to SC) or from a low dosage to a high dosage would unlikely be permissible if there is no PK/PD bridging study. Similarly, extrapolation between two indications with very different immunogenic profiles would not be likely. Monotherapy is recommended to support extrapolation of indications, if a well-established monotherapy has been used by the originator's product in a population that is considered to be sensitive to detect differences and is representative of the target populations intended for clinical practice. Since monotherapy would provide better comparative efficacy, safety, and immunogenicity profiles without the interference of concomitant medications, biosimilar sponsors should make an effort to select such a patient population that allows for comparison and extrapolation.

In general, extrapolation is based on the totality of evidence that could be scientifically verifiable and rationalized; uncertainty that may cause efficacy and safety concerns needs to be addressed prior to marketing authorization of biosimilars; and if similarity cannot be sufficiently demonstrated, sponsors should pursue a standalone authorization pathway.

In order to extrapolate the data generated with a biosimilar in one indication to other indications held by the RBD, it will be critical to design and conduct a high-quality equivalence trial with carefully selected equivalence margins. The trial should be shown to have assay sensitivity, and there should be some confidence that the constancy assumption is valid in order to facilitate in the proper interpretation of the trial results.

9.4 STATISTICAL CONSIDERATIONS

9.4.1 Designing a Meaningful Trial to Support Additional Indications

Although equivalence or noninferiority studies may be acceptable for the comparative clinical studies of the biosimilar to the RBD, equivalence trials are generally preferred (Health Canada, 2010; WHO, 2010; EMA, 2015; FDA, 2015). A demonstration of equivalence, as opposed to noninferiority, is especially important when extrapolation to other indications is one of the goals of the development program for the biosimilar. Noninferiority trials are one sided and hence do not exclude the possibility that the investigational product could be found to be superior to the comparator agent. In the biosimilar setting, *post-hoc* justification that a finding of superiority is not clinically relevant can be difficult. For example, one concern would be whether a finding of superiority might be associated with an increased rate of adverse events if the biosimilar is used at the same dose as the RBD (WHO, 2010). If such an association is found, then the new product would not be considered similar to the RBD and would need to be developed as a stand-alone product. Such a finding is clearly undesirable at the completion of the comparative clinical study that marks the end of the comparability exercise for the biosimilar (Njue, 2011).

On the other hand, a finding of superiority is not an issue for equivalence trials as they are two sided and are designed to provide statistical evidence that the biosimilar is neither inferior to the RBD by more than a specified margin nor superior to the RBD by more than a specified margin (FDA, 2015). As such, equivalence is demonstrated only when the confidence interval for the estimated treatment effect falls entirely within the lower and upper clinically acceptable equivalence limits, and superiority of the biosimilar to the RBD would lead to a failed equivalence trial.

An important consideration during the planning and design of the equivalence trial for the biosimilar is that it should be conducted with a sensitive and well-established clinical model regarding both the study population and the study endpoints in order to ensure assay sensitivity, which is important for all equivalence trials. In the biosimilar context, it would provide some confidence that the trial, as planned and designed, will have the ability to detect differences between the biosimilar and the RBD if such differences exist (ICH E10, 2000). A trial that lacks sensitivity could lead to the erroneous conclusion of equivalence of the biosimilar to the RBD if the biosimilar is inferior (Chow and Liu, 2004). In the absence of a direct comparison to a placebo arm in the equivalence trial which compares the biosimilar to the RBD, assay sensitivity has to be determined indirectly. Hence, historical evidence should be provided which shows that appropriately designed and conducted trials with the RBD against placebo have reliably demonstrated the superiority of the RBD over placebo (ICH E10, 2000). Measures should also be put in place to minimize any factors that would reduce the sensitivity of the study, including missing data, patient withdrawals, and protocol deviations.

The constancy assumption, which assumes that the historical effect size of the RBD against placebo is preserved in the current trial, is another important consideration for equivalence trials. The statistical methods (discussed later) used for the analysis of equivalence trials are based on the constancy assumption, and the methods may not be valid if this assumption is violated (Chow and Liu, 2004). In order to ensure the validity of this assumption, the biosimilar equivalence trial should be designed and conducted in a similar manner to the trials that established the efficacy of the RBD with regards to study factors such as patient population, endpoints, concomitant therapies, trial duration, and the methods and timing for the assessment of study endpoints (ICH E10, 2000). Relevant scientific advances should also be taken into account in the design of the biosimilar equivalence study and may be used to justify differences in key study factors between the current trial and the historical trials with the RBD. For example, if a different primary endpoint from the one used in the trials that established the efficacy of the RBD is used, then a justification for this endpoint should be provided. The justification provided should include historical evidence to support the sensitivity of the selected endpoint as well as the determination of the equivalence margins.

9.4.2 Determining the Equivalence Margins

An equivalence trial is designed with the primary objective of showing that the response to two or more treatments differs by an amount that is clinically unimportant (ICH E9, 1998). The clinically unimportant difference is referred to as the equivalence margin and has to be determined during the planning and design phase of the study. The appropriate determination of the margins is critical as both the determination of the required sample size and the interpretation of study results depend on it. Historical evidence of the effect size of the RBD compared to placebo for the selected endpoint and population should be provided to support the proposed margins and should take into account the magnitude and variability of the effect size. The final choice of the margins is based on a combination of statistical evidence from historical trials and clinical judgment. Typically, symmetric margins to rule out inferiority and superiority are used, and the use of asymmetric margins with a larger margin to rule out superiority compared with the margin to rule out inferiority can be justified in certain cases (FDA, 2015). To date, the biosimilars approved by Health Canada have utilized symmetric margins.

Several equivalence margins can be used to test the hypothesis of equivalence of the biosimilar to the RBD (Chow and Liu, 2004). The first margin is simply the smallest effect size that the reference can be expected to have relative to a placebo control, preferably based on a meta-analysis of historical trials. The second margin is a fraction of the first and is selected because it is considered clinically important to ensure that the biosimilar retains a substantial fraction of the reference. A third limit, which transforms the equivalence hypothesis into a superiority hypothesis, is only used in cases when the active control has not consistently been shown to be superior to placebo and is not considered relevant for biosimilars. In the biosimilar setting, the efficacy of the RBD compared to placebo has already been demonstrated, and the second margin that preserves a proportion of the effect size of the reference is the most relevant and commonly used margin.

The fraction of the effect size of the RBD that must be retained by the biosimilar should be clearly justified in each case and should take into account the smallest clinically important difference in a given setting. A commonly used value is the 50% rule in which 50% of the effect size of the RBD is preserved, and its wide use appears to be historical. For example, the equivalence trial for the biosimilar infliximab that was filed to Health Canada as a biosimilar to the RBD used the 50% rule to determine the equivalence margin based on historical data with the RBD. However, the 50% rule is not expected to be justifiable in all cases, as it could result in a margin that is larger than what is considered the smallest clinically important difference. Once the margin has been selected, the determination of the required sample size should be based on methods specifically designed for equivalence trials (Hwang, 2005).

9.4.3 Data Analysis and Interpretation of Trial Results

Two methods are typically used for the statistical analysis of data from equivalence trials, and both methods can be applied to biosimilars. The most commonly used method is based on the indirect confidence interval comparison that requires specification of the equivalence limits (Chow and Liu, 2004; FDA, 2010). Equivalence is demonstrated when the confidence interval for the selected metric of the treatment effect falls entirely within the lower and upper equivalence limits. If a p-value approach is used, then the p-values should be computed based on the two-one sided t-tests (TOST) procedure testing simultaneously the null hypotheses of inferiority and superiority (Schuirmann, 1987). In using the TOST procedure, equivalence is demonstrated when the p-values obtained are less than the significance level used. An error that is commonly made is concluding equivalence based on a statistically nonsignificant p-value from testing the null hypothesis of no difference. In general, concluding equivalence based on observing a statistically nonsignificant p-value from testing the null hypothesis of no difference between the biosimilar and the RBD is incorrect (ICH E9, 1998). This observation is especially applicable to secondary endpoints for which comparability margins are not typically predefined. In such cases, it is erroneous to conclude similarity on the secondary endpoints based on statistically nonsignificant p-values from the test of a difference between the biosimilar and the RBD.

The second method that is sometimes used for analyzing equivalence trial data is the virtual comparison method that synthesizes the estimate of the treatment effect relative to the control in the current equivalence trial with the estimate of the treatment effect of the control relative to placebo from historical trial (Chow and Liu, 2004; FDA, 2010). The resulting statistics are treated as if they were derived from the same trial. This method is unique as it does not require the prespecification of equivalence limits. The method can also be used to provide an estimate of the fraction of the effect size of the active control that is retained by the test drug product, and this can be useful in the overall interpretation of the study results. To date, the biosimilars that have been approved by Health Canada used the indirect confidence interval comparison method because there appears to be more familiarity with this approach, in general, relative to the synthesis method.

With regard to analysis sets, analysis based on the intention to treat (ITT) approach in an equivalence trial might bias the results toward equivalence (ICH E9, 1998). The per-protocol (PP) approach, which excludes subjects who either do not meet study eligibility criteria or have major protocol violations, can also lead to biased results. Bias is especially a concern when a large proportion of subjects are excluded from analysis. As a result, equivalence trials should be analyzed using ITT and per-protocol approaches, and both approaches should support equivalence (Snapinn, 2000). In using each approach, a careful assessment of the potential for bias should be performed. Health Canada's approach to this issue has been to assess the results from analyses based on both approaches in order to ensure that both sets of results support the conclusion of equivalence of the biosimilar to the RBD.

Finally, proper interpretation of the results from an equivalence trial is critical. Overall, the lack of a placebo concurrent control creates challenges in the interpretation of the results of any equivalence trial, and this is no exception for biosimilar equivalence trials (Chow and Liu, 2004). To facilitate the proper interpretation of the study results, it is important to show that the trial has assay sensitivity. A trial of poor quality is likely to lack assay sensitivity and may lead to an erroneous conclusion of equivalence of the biosimilar to the RBD (Snapinn, 2000). Another important consideration in the interpretation of the study results is the constancy assumption, as previously discussed (Chow and Liu, 2004). Verification of the assumption can be difficult and typically requires the external validation of the results of the current trial relative to historical trials with the RBD.

9.5 CONCLUSIONS

The extrapolation of indications is the leveraging of safety and efficacy data from clinical studies in one indication to support the authorization of other indications in which the biosimilar has not been studied but for which the reference product is authorized. After review of all quality data, nonclinical data, and PK/PD study/ies and clinical trial(s), a biosimilar product may or may not be authorized for all routes of administration, doses, and indications for which the reference product is authorized. The biosimilar guidance documents from regulatory authorities for biosimilars indicate that it may be possible to extrapolate clinical data to other indications (in which the biosimilar has not been studied) where rationales are sufficiently persuasive. The decision to extrapolate should be based primarily on the demonstration of similarity through extensive comparability studies that compare the physicochemical attributes and the biological activity between the biosimilar and reference product.

The clinical development for biosimilars consists of complex processes associated with regulatory and scientific issues. A biosimilar has to demonstrate that its structure and function(s) are (highly) similar to the reference product prior to conducting clinical trials in sensitive populations. Extrapolation of indications is possible when the appropriate data and rationales are provided. However, extrapolation would be unlikely if there is evidence that an observed difference may impact a mechanism that could be important in the treatment of an indication to which extrapolation is required. Based on the same dataset, regulatory agencies may render different regulatory decisions; this is the same as for any other therapeutic product. Further dialogue and harmonization among regulatory agencies on their decision making should be considered as experience with the scientific and regulatory issues related to biosimilars increases.

Recently, a new amendment to the Food and Drugs Act, Bill C-17, also called Vanessa's law, has provided the regulator with added authorities such that it is now possible to take a lifecycle approach to the regulation of any drug product in Canada, including biosimilars.

REFERENCES

Ahn C, Lee SC, Yongu UT. (2011) Statistical considerations in the design of biosimilar cancer clinical trials. *Korean Journal of Applied Statistics* **24(3)**, 495–503.

Arnold JN, Wormald MR, Sim RB, et al. (2007) The impact of glycosylation on the biological function and structure of human immunoglobulins. *Annual Review of Immunology* **25**, 21–50.

Brinks V. (2013) Immunogenicity of biosimilar monoclonal antibodies. *Generics and Biosimilars Initiative Journal* **2(4)**, 188–193.

Chow SC, Liu JP. (2004) *Design and Analysis of Clinical Trials: Concepts and Methodologies*. John Wiley & Sons, Hoboken, NJ.

Dougados M, Schmidely N, Le Bars M, et al. (2009) Evaluation of different methods used to assess disease activity in rheumatoid arthritis: analyses of abatacept clinical trial data. *Annals of Rheumatic Diseases* **68(4)**, 484–489.

EMA. (2006) Assessment report for omnitrope. Available from: http://www.ema.europa.eu/docs/en_GB/document_library/EPAR_-_Assessment_Report_-_Variation/human/000607/WC500137237.pdf.

EMA. (2008) Assessment report for epoetin alfa hexal, Procedure no. EMEA/H/C/725/II/0006. Available from: http://www.ema.europa.eu/docs/en_GB/document_library/EPAR_-_Assessment_Report_-_Variation/human/000726/WC500028283.pdf.

EMA. (2013) Assessment report for remsima. Available from: http://www.ema.europa.eu/docs/en_GB/document_library/EPAR_-_Public_assessment_ report/human/002576/WC500151486.pdf.

EMA. (2013) European public assessment report for inflectra (infliximab) EMA/402688/2013 EMEA/H/C/ 002778. Available from: http://www.ema.europa.eu/ema/index.jsp?curl¼pages/ medicines/human/medicines/002778/human_med_001677.jsp& S130.

EMA. (2015) Guideline on similar biological medicinal products containing biotechnology-derived proteins as active substance: non-clinical and clinical issues. Available from: http://www.ema.europa.eu/docs/en_GB/document_library/Scientific_guideline/2015/01/WC500180219. pdf.

FDA. (2010) Draft guidance for industry: non-inferiority clinical trials. Available from: http://www.fda.gov/downloads/drugs/guidancecomplianceregulatoryinformation/guidances/ucm070951.pdf.

FDA. (2014a) Guidance for industry: immunogenicity assessment for therapeutic protein products. Available from: http://www.fda.gov/downloads/drugs/guidancecompliance regulatoryinformation/guidances/ucm338856. pdf.

FDA. (2014b) Draft guidance for industry: clinical pharmacology data to support a demonstration of biosimilarity to a reference product. Available from: http://www.fda.gov/downloads/Drugs/GuidanceComplianceRegulatoryInformation/Guidandes/UCM397017.pdf.

FDA. (2015) Guidance for industry: scientific considerations in demonstrating biosimilarity to a reference product. Available from: http://www.fda.gov/downloads/Drugs/GuidanceComplianceRegulatoryInformation/Guidances/UCM291128.pdf.

Giezen TJ, Schneider CK. (2014) Safety assessment of biosimilars in Europe: a regulatory perspective. *Generics and Biosimilars Initiative Journal* **3(4)**, 180–183.

Health Canada. (2010) Guidance for sponsors: information and submission requirements for subsequent entry biologics (SEBs). Available from: http://www.hc-sc.gc.ca/dhp-mps/brgtherap/applic-demande/guides/seb-pbu/seb-pbu_2010-eng.php.

Health Canada. (2012) Guidance document: comparative bioavailability standards: formulations used for systemic effects. Available from: http://www.hc-sc.gc.ca/dhp-mps/alt_formats/pdf/prodpharma/applic-demande/guide-ld/bio/gd_standards_ld_normes-eng.pdf.

Health Canada. (2014) Summary basis of decision (SBD) for inflectra. Available from: http://www.hc-sc.gc.ca/dhp-mps/prodpharma/sbd-smd/drug-med/sbd_smd_2014_inflectra_159493-eng.php.

Hwang IK. (2005) Active-controlled noninferiority/equivalence trials: methods and practice. In: *Statistics in the Pharmaceutical Industry*, Ralph BC, Tsay JY, eds., 193–230. Chapman and Hall, New York.

ICH E9. (1998) Statistical principles for clinical trials. Available from: http://www.ich.org/fileadmin/Public_Web_Site/ICH_Products/Guidelines/Efficacy/E9/Step4/E9_Guideline.pdfp4/Q5E_Guideline.pdf.

ICH E10. (2000) Choice of control group and related issues in clinical trials. Available from: http://www.ich.org/fileadmin/Public_Web_Site/ICH_Products/Guidelines/Efficacy/E10/Step4/E10_Guideline.pdf.

ICH Q5E. (2004) Comparability of biotechnological/biological products subject to changes in their manufacturing process. Available from: http://www.ich.org/fileadmin/Public_Web_Site/ICH_Products/Guidelines/Quality/Q5E/Step4/Q5E_Guideline.pdf.

Klein AV. (2011) The first subsequent entry biologic authorized for market in Canada: the story of Omnitrope, a recombinant human growth hormone. *Biologicals* **39(5)**, 278–281.

Njue C. (2011) Statistical considerations for confirmatory clinical trials for similar biotherapeutic products. *Biologicals* **39**, 266–269.

Pen A, Klein AV, Wang J. (2015) Health Canada's perspective on the clinical development of biosimilars and related scientific and regulatory challenges. *Generics and Biosimilars Initiative Journal* **4(1)**, 36–41.

Schuirmann DJ. (1987) A comparison of the two one-sided tests procedure and the power approach for assessing the equivalence of average bioavailability. *Journal of Pharmacokinetics and Biopharmaceutics* **15(6)**, 657–680.

Scott BJ, Klein AV, Wang J. (2014) Biosimilar monoclonal antibodies: a Canadian regulatory perspective on the assessment of clinically relevant differences and indication extrapolation. *Journal of Clinical Pharmacology* **55(S3)**, S123–132.

Shankar G, Shores E, Wagner C, et al. (2006) Scientific and regulatory considerations on the immunogenicity of biologics. *Trends in Biotechnology* **24(6)**, 272–280.

Snapinn SM. (2000) Noninferiority trials. *Current Control Trials in Cardiovascular Medicine* **1(1)**, 19–21.

Weise M, Kurki P, Wolff-Holz E, et al. (2014) Biosimilars: the science of extrapolation. *Blood* **124(22)**, 3191–3196.

WHO. (2010) Guidelines on evaluation of similar biotherapeutic products (SBPs). Available from: http://www.who.int/biologicals/areas/biological_therapeutics/BIOTHERAPEUTICS_FOR_WEB_22APRIL2010.pdf.

10 Interchangeability, Switchability, and Substitution of Biosimilar Products

Paul Declerck
University of Leuven

Laszlo Endrenyi
University of Toronto

Shein-Chung Chow
Duke University School of Medicine

CONTENTS

10.1 INTRODUCTION

The US Food and Drug Administration (FDA) was authorized in 1984 to approve generic drug products under the Drug Price Competition and Patent Term Restoration Act, which is known as the Hatch–Waxman Act. For the approval of small-molecule generic drug products, the FDA requires that evidence on the *average* of bioavailability in terms of the rate and extent of drug absorption be provided. The assessment of bioequivalence as a surrogate endpoint for quantitative evaluation of drug safety and efficacy is based on the Fundamental Bioequivalence Assumption. This assumption states that if two drug products are shown to be bioequivalent in average bioavailability, it is assumed that they will reach the same therapeutic effect, or they are therapeutically equivalent and hence can be used interchangeably. The assumption is expressed in the preface of the "Orange Book" of the FDA (FDA, 2016) as well as in a guideline of the European Medicines Agency (EMA) (EMA, 2010). Under the Fundamental Bioequivalence Assumption, regulatory requirements, study design, criteria, and statistical methods for assessment of bioequivalence have been well established (see, e.g., Chow and Liu, 2008; EMA, 2010; FDA, 2001, 2013; Schuirmann, 1987; WHO, 2005).

For large-molecule biosimilar products, on the other hand, in 2009 the US Congress passed the Biologics Price Competition and Innovation (BPCI) Act, which authorized the FDA to approve follow-on biologics (biosimilars) (Federal Register, 2009). Formal guidelines were finalized in 2015, the same year that the first biosimilar was approved in the United States. Europe was the first region to establish guidelines for biosimilar development (EMA, 2006). Since then, these guidelines have been regularly adapted as the field is evolving rapidly. Other guidelines have been developed by the World Health Organization, Health Canada, Australia, South Korea, and Japan. Overall, these different guidelines are very similar and are based on a stepwise approach, of which the extensive comparative evaluation of the biochemical, physicochemical, and functional properties of the biosimilar with those of the reference product forms a solid foundation (see Chapters 14 and 15 in this book). Subsequently, clinical safety and efficacy of the proposed biosimilar products as compared to the licensed biological product follows an abbreviated pathway, which is not designed to prove clinical benefit per se but merely to confirm clinical similarity. Whereas in Europe no firm statement is made on interchangeability of biosimilars, the FDA has made it clear that granting biosimilarity does not imply interchangeability. However, it is not clear whether the Fundamental Bioequivalence Assumption for generic approval is applicable for the approval of biosimilar products due to fundamental differences between small-molecule generic drug products and large-molecule biosimilars. As a result, study design, criteria, and statistical methods for assessment of interchangeability have not yet been established (see, e.g., Chow, 2013; FDA, 2015a,b,c; see also Chapter 11 in this book).

When various products of a given drug are on the market, one of the important questions is if they can be substituted with each other. The answer to the question is different for small-molecule drugs and for large-molecule biologics. Since a number of biological products' patents have expired and many more are due to expire in the next few years, the subsequent production of biosimilar products has aroused considerable interest within the pharmaceutical/biotechnology industry as biosimilar manufacturers strive to obtain part of an already large and rapidly growing market. The potential opportunity for price reductions versus the originator biological products remains to be determined, as the advantage of a slightly cheaper price may be outweighed by the uncertainty regarding the safety of repeated switching between reference and biosimilar products and between biosimilars. Thus, it is a concern whether approved biosimilar products can be used interchangeably. It should also be realized that the speed of uptake of biosimilars may depend on whether or not they can be used interchangeably with the reference product. Therefore, apparent scientific discussions on this aspect are often tarnished by economic concerns.

On the other hand, one must realize that granting the status of "interchangeability" between two drugs will automatically lead to subsequent practices regarding switching and substitution. Before substitution of small molecules can be considered, regulators must declare that sufficient similarity of the safety and efficacy of the drug products has been demonstrated. With small-molecule, chemical drugs, *sufficient similarity* is usually based on a clearly defined, rigorous statement of the *bioequivalence* between the pharmacokinetic profiles of the contrasted drug products. The comparison typically involves the generic and the reference formulation. Thus, clear, simple regulatory requirements and criteria have been declared for establishing the bioequivalence between a small-molecule, generic formulation and a reference product as set out in regulatory guidances (EMA, 2010; FDA, 2001, 2013).

Within the context of biological drugs, this issue is much more complicated. For instance, biological drugs have much higher molecular weights with convoluted structures, they are produced biotechnologically by host cell lines, and they are less stable and have complex physicochemical properties. Their properties are susceptible to small differences in the environment (e.g., temperature, light, oxygen) and the manufacturing process. During their production processes, there is a possibility of contaminants, which can be difficult to detect and, at times, impossible to remove. Even though techniques for the characterization of biologicals have been very much advanced and lead to analytical fingerprinting of the molecule (see Chapter 2 in this book), differences between the biosimilar and reference product do exist and are either detectable or may remain elusive as exemplified in the European Public Assessment Reports (EPAR, EMA website) of approved biosimilars. The most difficult issue involves learning which differences may be clinically relevant. In addition, biologicals are immunogenic, and this property may be impacted through frequent switching between highly similar products. Indeed, immunological pathways indicate that switching a stable patient to a strongly related molecule may cause a challenge to the immunologically tolerized state versus the first therapeutic. Well-established tolerization mechanisms include clonal deletion, receptor editing, clonal anergy, blockade of memory response, and competitive tolerance (Stewart et al., 1997). Immunological response to biological therapeutics varies between patients. Select patient subsets develop antidrug antibodies (ADAs), with clinical impact ranging from no impact at all to secondary loss of response. The immune response to biologicals is a dynamic process, and, importantly, ADAs may develop and disappear over time. Whether a stable patient subjected to a nonmedical switch between the reference and a biosimilar (or vice versa) can maintain tolerance is not known.

As a consequence of the differing features of small-molecule generics and biosimilars, regulatory considerations and approaches to the interchangeability of two products are very different after their bioequivalence has been declared or when their biosimilarity has been stated.

As noted earlier, a regulatory statement of bioequivalence of a small-molecule generic formulation to a reference product generally implies their interchangeability within patients. In contrast, a regulatory declaration of biosimilarity between a biosimilar and its reference product does *not* imply interchangeability; the two conditions are sharply different. Therefore, clear criteria need to be set to grant the property "interchangeability," and in particular, conditions for its application in a real-life setting are needed to ensure the continuing safe use of biologicals. Thus, in many jurisdictions, including the United States, approved biosimilar preparations may, without further permission, not be switched from and to their reference drug.

There is also a clear separation between *prescribability* and *switchability*. Anderson and Hauck (1990) distinguished between these two terms even for small-molecule generic products. Under the first scenario, a patient had no prior exposure to a drug in any of its forms; that is, he or she has been naïve to the drug. In such situations, any of the approved products are *prescribable*, provided that their safety and efficacy are satisfactory and, therefore, they have been approved by regulators. However, after the patient has started to receive one of the products, it may not be substituted by another formulation. Under the second scenario, the patient is already taking the drug but opts for a cheaper alternative for economic reasons. In this case, the different products must be therapeutically equivalent. Therefore, the products should be *switchable* within individuals. The switch may then occur either from the reference product to the biosimilar or from the biosimilar to the reference product (R $\rightarrow$ B or B $\rightarrow$ R).

In the event that two or more biological drug products can be considered interchangeable, switching between these biologicals may become more general. Directions of switches that need to be considered may not be restricted to R $\rightarrow$ B and B $\rightarrow$ R. The question that arises is whether or not other independently developed biosimilars (e.g., B_2 and B_3) will also be involved in that "loop" allowing for instance $R \rightarrow B_1 \rightarrow B_2 \rightarrow B_1 \rightarrow B_3 \rightarrow R \rightarrow B_2$. It is obvious that this would require clinical studies that demonstrate unambiguously the interchangeability of the various products.

Depending on the condition, two test models were proposed. To describe prescribability quantitatively the so-called population bioequivalence (PBE) model was developed (Schall and Luus, 1993). This required the comparison of the population distributions for the relevant pharmacokinetic parameters. A simple parallel-group design study was considered adequate to establish PBE. In contrast, interchangeability could be demonstrated by satisfying the requirements of a model for individual bioequivalence (IBE) (Schall and Luus, 1993; Sheiner, 1993). The implementation of IBE calls for a more complex study design: IBE requires investigations with a crossover study design in which both drug products are measured (at least) twice within individuals (Endrenyi et al., 2013; FDA, 2001).

Population and individual bioequivalence are discussed in detail in Chapter 11 of this book.

10.2 GENERAL QUESTIONS AND CONCERNS ABOUT THE INTERCHANGEABILITY OF BIOSIMILARS

Klein et al. (2014) noted several questions for the careful consideration of substitution and interchangeability between biological products. They observed that the "complexity and impurity profile of an SEB [subsequent-entry biological = biosimilar] means that automatic interchangeability of SEBs, or even of originator biologicals, has a real potential for different, and sometimes unexpected, clinical consequences."

Another possible concern is that switching between biological products can exacerbate immunogenicity, with potentially negative effects. It has been noted that the immunogenicity of biosimilars "cannot be fully predicted using preclinical/clinical studies as the process is abbreviated and the number of patients/subjects in clinical studies is relatively limited" (Klein et al., 2014). For example, loss of tolerance and loss of efficacy were observed within 1 year when patients, having Crohn's disease and maintained with infliximab, were switched to adalimumab as compared to patients who remained on infliximab therapy (Van Assche et al., 2012).

Professional organizations raised concerns about switching patients from an effective and safe biological drug to a biosimilar. For instance, the British Society for Rheumatology (2015) advised against the summarily switching of such patients. A committee of the European Society for Paediatric Gastroenterology, Hepatology, and Nutrition recommended that children with inflammatory bowel disease not be switched at present to a biosimilar (Ridder et al., 2015).

Another observation is that some quality attributes of biologicals (reference as well as biosimilar) may drift in time (Schiestl et al., 2011). Biosimilarity only needs to be demonstrated at the time of approval, and both reference and biosimilars can make their own independent manufacturing changes. Therefore, it is not unlikely that throughout the life cycle of the two products differences may be introduced that affect their degree of similarity. Consequently, biosimilarity between reference and biosimilar can never be guaranteed over time. Obviously, this also has an impact on the duration of the validity of interchangeability if granted.

10.3 INTERCHANGEABILITY IN THE UNITED STATES

The BPCI Act of 2009 clearly defines biosimilarity and interchangeability separately (Federal Register, 2009). The BPCI Act defines the biosimilar (or biosimilarity) as a biological product that is *highly similar* to the reference product (notwithstanding minor differences in clinically inactive components), and states that there are *no clinically meaningful differences* between the biosimilar product and the reference product in terms of safety, purity, and potency. Important guidances were recently issued on the implementation of the biosimilar pathway (FDA, 2015a,b,c). The BPCI Act also provides explicit definitions and conditions for interchangeability. "The terms *interchangeable* and *interchangeability* mean that: 1) the biological product is *biosimilar* to the reference product; 2) the biological product can be *expected* to produce the *same* clinical result as the reference product in *any* given patient; 3) for a product administered more than once, the *risks* of safety and reduced efficacy *of alternating and switching* are not greater than with the use of the reference product without alternating or switching" (emphasis added).

Thus, there is a clear and strong distinction between *biosimilarity* and *interchangeability*. Biosimilarity is a precondition of interchangeability, and the regulatory approval of the biosimilarity of two products does not at all imply their interchangeability. If two biological products have already been approved and are marketed, an additional application must be submitted which would demonstrate that conditions for their interchangeability are also satisfied.

The strong distinction and hierarchy between the biosimilarity and interchangeability of biological products are in sharp contrast to the close parallel between the bioequivalence and interchangeability of small-molecule drug products.

However, while the BPCI Act distinguishes between biosimilarity and interchangeability, such a distinction and criteria for demonstrating interchangeability is still to be made more explicit in the forthcoming FDA guidelines.

10.3.1 Automatic Substitution in the United States

The BPCI Act also decrees the important consequences of the approval of interchangeability between biologic drug products (Endrenyi et al., 2013). The BPCI Act states that the interchangeable "biological product may be substituted for the reference product without the intervention of the health care provider who prescribed the reference product" (Federal Register, 2009).

Consequently, according to the BPCI Act, if a test product is judged to be interchangeable with the reference product, then it may be substituted, even alternated, without a possible intervention, or even notification, of the prescribing physician.

This is a very permissive stipulation and has raised concerns at the level of state legislation. The basis of these concerns is that substitution of a drug product by another could give rise to changes of the safety and efficacy in individuals. It is assumed, however, that when interchangeability has been demonstrated, the risks associated with these changes are minimal.

For instance, when a patient is switched from a reference product to a biosimilar, R → B, then the associated risk should not be higher than when the subject continues to be exposed to the reference product: R → R. Scenarios of switching and alternating have not been stated yet by the FDA. On the other hand, as of January 4, 2016, at least 31 states had considered enabling legislation that establishes state standards for substitution of a biosimilar prescription product to replace the original biological product (Cauchi, 2016). It has been signed into law in 18 states and Puerto Rico.

The language of the legislation can be confusing. The laws generally refer to the substitution of a prescribed biological product by an alternative biological product, as in California. This may mean alternative prescription (i.e., prescribability for bio-naïve patients) but may imply also interchanging (i.e., switchability). It is therefore reassuring that the legislation generally contains various safeguards and conditions (Cauchi, 2016). For example, they often refer to the precondition of the approval of interchangeability by the FDA. Some require that the prescriber and/or the patient be notified. The prescriber's active or passive ("does not object") approval of substitution is generally sought. Some provide liability protections for pharmacists who substitute biosimilars.

However, there is pressure to extend automatic substitution. For example, in 2015 the Generic Pharmaceutical Association stated: "The Generic Pharmaceutical Association (GPhA) and its Biosimilars Council applaud the enactment of legislation in five states to allow automatic substitution for Food and Drug Administration (FDA) approved interchangeable biologic products." (GPhA, 2015). The statement appears to reach somewhat beyond the intent and content of the legislation: interchangeability is not automatic but is subject to conditions and constraints.

The conditions and constraints of the state legislation are more cautious (for instance, when they refer to the approval of substitution by the prescriber) than the federal BPCI Act which, as seen above, calls for substitution by biosimilars without the intervention of the prescriber. The differing views may require legal resolution.

10.3.2 Switching and Alternating

As noted above, the BPCI Act refers to the switching and alternating of drug products. Switching refers to changing between the reference and biosimilar products, R → B and B → R. In an initial phase, upon introduction of an approved biosimilar primarily the former (R → B) will occur. However, subsequent to competitive incentives, over time B → R may also be considered, subsequently, also including alternating between R and B. However, in the long term, multiple biosimilars may be available and result in the alternating of multiple products as described above (e.g., $R \rightarrow B_1 \rightarrow B_2 \rightarrow B_1 \rightarrow B_3 \rightarrow R \rightarrow B_2$). The expectation of controlling the risks of safety and efficacy applies to every switching component in this chain of alternation.

One should raise the question of whether interchangeability (for patients under existing therapy) between different, independently developed biosimilars should even be considered. Indeed, since biosimilarity (a prerequisite for considering interchangeability according to the FDA) has not been demonstrated between two biosimilars, interchangeability for two or more products cannot even be considered unless biosimilarity between both products is shown followed by extensive clinical studies to prove interchangeability. It is obvious that in situations where multiple biosimilars are involved, the chances of impacts of drifts in quality attributes are much more pronounced.

10.4 INTERCHANGEABILITY IN CANADA

In Canada, the term *substitutability* is applied to two products that can be used in lieu of the other during and within the same treatment period, that is, for interchangeability as discussed in this chapter. The general guidance of Health Canada on biosimilars (or subsequent-entry biologics) does not consider explicitly interchangeability (Health Canada, 2010a). Still, the position of Health Canada on the substitution of biological products is clear but rather soft: "Health Canada does not support the automatic substitution of a subsequent-entry biologic for its reference biologic drug. Health Canada therefore recommends that physicians make only well-informed decisions regarding therapeutic interchange" (Health Canada, 2010b).

Health Canada's position does not have legal authority. Funds for the reimbursement of pharmaceutical expenses are dispensed by the provinces. They will be keen to promote introduction of new biosimilars, that is, their prescribability. Furthermore, they will also seek the status of interchangeability for biosimilar products. Thus, in Canada the positions on interchangeability of biologic products and the related conditions will be diverse and controversial.

10.5 INTERCHANGEABILITY IN THE EUROPEAN UNION

The EU was the first region in the world to set up a legal framework and regulatory pathways for biosimilars. In fact, the word "biosimilars" was coined during this legislative process.

The concept of a "similar biological medicinal product" was adopted in EU pharmaceutical legislation in 2004 and came into effect in 2005. The EMA was the first to lay down an abbreviated regulatory pathway for biosimilars. As of April 2016, 23 biosimilars were approved within the product classes of human growth hormone, granulocyte colony-stimulating factor, erythropoietin, follitropin, TNF-inhibitor, and insulin for use in the EU; some products were withdrawn (Declerck et al., 2016). The EMA has issued many guidelines for biosimilars, including several product class–specific guidelines. An updated general guideline on biosimilars was issued in 2014 (EMA, 2014).

Only the EMA can grant marketing authorization for biosimilars in the European Union. More precisely, the European Commission issues decisions concerning the authorization of these medicinal products on the basis of the scientific opinions from the EMA. The resulting marketing authorization is valid in all EU member states.

For granting approval for a biosimilar product, the EMA requires clinical trials, including comparability studies to the originator product, to demonstrate safety and efficacy. Additionally, the prospective market authorization holder also must demonstrate lack or at least comparable immunogenicity in long-term clinical trials. Guidelines have been established to provide further details on specific needs to demonstrate biosimilarity for nine primary product classes.

Interchangeability issues are not discussed by the EMA but are a matter for the national competent authorities (Weise et al., 2012). Long-term clinical investigations and systematic monitoring of the efficacy and tolerability of biosimilars in all indications are still needed (Declerck et al., 2010), and even then establishing the interchangeability of biosimilar and reference products will be difficult. One should also take into account that some biosimilars may not carry the same indications as those for which the reference drug is approved. Furthermore, individual regions do exercise their own discretion in regard to the approval of biosimilars for all the indications assigned to the originator drug. For example, in January 2014, Health Canada approved two brands of the monoclonal antibody infliximab as "subsequent-entry biologics" for some (but not all) approved indications of the existing brand (Declerck et al., 2015; Scott et al., 2015). For a recently approved biosimilar of etanercept, the use in children is excluded because low-dose formulations are not available (EMA, 2015). It is beyond any doubt that such situations complicate the real-life application of any "interchangeability" consideration.

The EMA has made it clear that a biosimilar is not the same as a generic drug and has handed the interchangeability issue over to individual countries. As an EMA document states (EMA, 2009): "The EMA evaluates biosimilar medicines for authorization purposes. The Agency's evaluations do not include recommendations on whether a biosimilar should be used interchangeably with its reference medicine. For questions related to switching from one biological medicine to another, patients should speak to their doctor or pharmacist."

In the absence of central European directives, the 28 member states follow diverse directions. In several countries, either legislation has been passed or regulations prohibit automatic substitution by pharmacists (Thimmaraju et al., 2015). Given that most currently approved biosimilars are applicable only in hospital settings, the impact of such a strict prohibition is moderate. At the other extreme, some countries (e.g., Poland and Bulgaria) have no relevant law or guidelines and permit automatic substitution. Substitution is conditional in several other countries (Thimmaraju et al., 2015) and is often restricted to treatment-naïve patients.

Equally important is the reimbursement policy (Declerck and Simoens, 2012), where there are many national variations. One of the options is that new patients need to be treated with a biosimilar product if it is available; otherwise the treatment is not reimbursed. But patients already treated with a brand-name biological have the right to continue receiving the same biological brand. Consequently, the prescribability, but not the interchangeability, of biosimilar products is recommended.

Even though it has been reported that there has been some experience in Europe with switching (Ebbers et al., 2012), it needs to be stressed that adequate data are scarce. The studies reported so far suffer from a number of important drawbacks. First, all studies referred to in Ebbers et al. (2012) deal with epoetins, G-CSFs, or human growth hormones. These are relatively small biologicals characterized as having low immunogenicity. Second, most of these studies have included a small number of patients or volunteers, had a short duration, and investigated a switch in only one direction. Therefore, these data do not reflect potential clinical practice. Third, some of these studies did not study switches related to biosimilars but rather switches between different originator molecules. Fourth, the majority of these studies were designed to evaluate comparable efficacy, and thus most if not all were strongly underpowered to detect changes in safety profiles upon switching. In addition, any postapproval data analysis is also not valid for drawing firm conclusions on the safety of switching since such databases (1) are not set up to detect adverse events specifically related to switches at an individual level and (2) typically suffer from underreporting.

Currently, the Norwegian Medicines Agency is encouraging a substitution culture and has initiated a clinical study to investigate the safety and efficacy of switching between IFX (Remicade) and the biosimilar Remsima in patients with rheumatoid arthritis, spondyloarthritis, psoriatic arthritis, ulcerative colitis, Crohn's disease, and chronic plaque psoriasis (NOR-SWITCH study, 2015). The primary endpoint of the NOR-SWITCH study is the occurrence of disease worsening in the indications being studied. However these endpoints cannot actually be analyzed together, and the sample size in each disease is underpowered to show a difference if it exists (Declerck et al., 2015).

In France, a new law (but yet to be implemented by decree) states that pharmacists will be legally permitted to substitute a biosimilar for the prescribed reference biological medicine as long as the prescribing physician has not marked the prescription as "nonsubstitutable" (Allen & Overy, 2014). Importantly, substitution will be allowed only for treatment-naïve patients. It could therefore be noted that the law actually considers the "replacement" of the reference drug by a biosimilar product (i.e., their parallel prescribability) and not their substitution (i.e., within-subject switching) as understood by the European Consensus Group. If the pharmacist decides to substitute/replace a biosimilar for the prescribed biological, the brand name of the dispensed product should be written on the prescription and the prescribing physician should be informed in order to maintain an accurate medical record.

10.6 CONCLUDING REMARKS

Regulatory approval of both small-molecule generics and biological biosimilars means directly an agreement only that they are prescribable—that is, that the drug products may be provided to naïve patients who have not taken the drug until this point in any form.

However, considerations are very different for the two kinds of drug products. For small-molecule generics, it is generally assumed that regulatory approval of bioequivalence with a reference drug indicates (or, better, implies) that the two drug products are therapeutically equivalent. Consequently, their free interchangeability is generally also assumed. This also enables ready substitution of the various products in pharmacies. Since small-molecule drugs are generally safe and effective, these assumptions and practices usually work. However, as outlined in the present contribution, there are important exceptions.

The situation is entirely different in considerations of interchangeability of biological products. First, the terminology has become regrettably confusing. In the United States, the BPCI Act introduced the terms of switching and alternating without providing clear interpretation. Switching presumably refers to changing between the reference and biosimilar products, R → B and B → R, primarily the former. Alternating could involve multisource products, thus also a change of B → B_2. Interchangeability would then cover both switching and alternating. All these changes are conducted within subjects. It is important to emphasize this because the terms are sometimes used incorrectly in the literature and regulations of the states, when they imply changes between subjects—that is, when they refer to the condition of prescribability. The situation is different, but not less confusing, in the European Union. The intent of the proposed terms of switching, interchanging, and substitution may be that they be considered as changes within patients (Ebbers and Chamberlain, 2014). However, they are often used and implemented for changes between subjects, that is, again in the sense of prescribability. Box 10.1 presents a summary of definitions, based on various proposed characterizations in the literature and practical applications in the field.

BOX 10.1

Interchangeability: Is a property of two drugs that refers to the fact that changing one drug for the other is expected to achieve the same clinical effect in a given clinical setting and in any patient. Interchangeability implies that alternating (i.e., back-and-forth switching) between the two drugs does not affect either the safety or the efficacy in any patient.

Switching: The medical practice, by the treating physician, of exchanging one drug for another drug with the same therapeutic intent *in patients who are undergoing treatment.* Switching may also imply alternating (i.e., repeated switching) between various drugs (including reference and different biosimilars). Safe switching requires that the drugs have been formally demonstrated to be fully interchangeable.

Substitution: The practice of dispensing a drug different from the drug that was prescribed without consulting the prescriber. The substituted drug should be equivalent to the originally prescribed drug. A distinction needs to be made between treatment-naïve patients (biosimilarity required; prescribability required) and patients already under treatment (interchangeability required).

Making such distinctions between the postapproval use of biosimilars may be needed to describe the different possibilities. But reloading the meaning of these common English words, which in some countries have been used as synonyms, could lead to confusion. The word "switching" appears to be a particularly bad choice of wording because native English speakers seem to associate switching with the inadvertent exchange of medicines. Moreover, confusion could arise, and has arisen, because the terms could be interpreted for the conditions of both prescribability (between subjects) and switchability (within subjects) (Tothfalusi et al., 2014).

The EMA has issued product-related guidelines for biosimilars, as well as a general guideline in 2014, and has approved a large number of biosimilars. In some cases, these were multisource products, that is, independently developed biosimilars referring to the same reference drug. However, EMA left the decision on interchanging to member states. As a consequence, the regulations have spanned a wide range from those enabling automatic substitution to laws forbidding substitution. This is a very regrettable situation. It is in part due to the lack of solid evidence regarding safety in case of accepting interchangeability as a general property of biosimilars. Indeed, even if biosimilar data were to be available for a particular biosimilar, it should be realized that these data should not be generalized to other biosimilars. It is appreciated that the FDA has followed a more cautious approach. Relying on the BPCI Act, it has issued general guidelines on various aspects of the development and regulation of biosimilars. The consideration of interchangeability requires additional demonstration; the conditions for these have not yet been clarified. The FDA approved one product (Zarxio) until March 2016, for which the condition of interchangeability was not requested and not granted.

It is almost impossible for a sponsor to claim or demonstrate interchangeability according to the definition given in the BPCI Act. Alternatively, it is possible that a meta-analysis that combines all data given in the submissions be conducted (by the regulatory agency) to evaluate the relative risks of switching and alternating of drug interchangeability. In practice, meta-analyses can be conducted for safety monitoring of approved biosimilar products. This could be important especially when there are a number of biosimilar products in the marketplace.

REFERENCES

Allen & Overy. (2014) Biosimilar substitution in France: no way back? *The Lawyer*, January 10, 2014. Available from: http://www.thelawyer.com/briefings/biosimilar-substitution-in-france-no-way-back/3014733.article.

Anderson S, Hauck WW. (1990) Consideration of individual bioequivalence. *Journal of Pharmacokinetics and Biopharmaceutics* **18(3)**, 259–273.

British Society for Rheumatology. (2015) Position statement on biosimilar medicines. Available from: http://www.rheumatology.org.uk/about_bsr/press_releases/bsr_supports_the_use_of_biolsimilars_but_recommends_measures_to_monitor_safety.aspx.

Cauchi R. (2016) State laws and legislation related to biological medications and substitutions of biosimilars. National Conference of State Legislatures (NCSL), Washington, DC. Available from: http://www.ncsl.org/research/health/state-laws-and-legislation-related-to-biologic-medications-and-substitution-of-biosimilars.aspx#Mandatory.

Chow SC. (2013) *Design and Analysis of Biosimilar Studies.* Chapman and Hall/CRC Press, Taylor & Francis, New York.

Chow SC, Liu JP. (2008) *Design and Analysis of Bioavailability and Bioequivalence Studies*, 3rd edition. Chapman and Hall/CRC Press, Taylor & Francis, New York.

Declerck PJ, Darendeliler F, Góth M, et al. (2010) Biosimilars: controversies as illustrated by rhGH. *Current Medical Research and Opinion* **26**, 1219–1229.

Declerck P, Farouk-Rezk M, Rudd PM. (2016) Biosimilarity versus manufacturing change: two distinct concepts. *Pharmaceutical Research* **33**, 261–268.

Declerck P, Mellstedt H, Danese S. (2015) Biosimilars: terms of use. *Current Medical Research and Opinion* **31**, 2325–2330.

Declerck PJ, Simoens S. (2012) European perspective on the market accessibility of biosimilars. *Biosimilars* **2**, 33–40.

Ebbers HC, Chamberlain P. (2014) Interchangeability. An insurmountable fifth hurdle? *Generics and Biosimilars Initiative Journal* **3(2)**, 88–93.

Ebbers HC, Muenzberg M, Schellekens H. (2012) The safety of switching between therapeutic proteins. *Expert Opinion in Biology and Therapy* **12(11)**, 1473–1485.

EMA. (2006) Guideline on similar biological medicinal products. The European Medicines Agency, evaluation of medicines for human use. EMEA/CHMP/437/04, London, UK. Available from: http://www.ema.europa.eu/docs/en_GB/document_library/Scientific_guideline/2009/09/WC500003517.pdf.

EMA. (2009) Questions and answers on biosimilar medicines (similar biological medicinal products), London, UK. Available from: http://www.ema.europa.eu/docs/en_GB/document_library/Medicine_QA/2009/12/WC500020062.pdf.

EMA. (2010) Guideline on the investigation of bioequivalence. European Medicines Agency, London, UK. January 20, 2010. Available from: http://www.ema.europa.eu/docs/en_GB/document_library/Scientific_guideline/2010/01/WC500070039.pdf.

EMA. (2014) Guideline on similar biological medicinal products. London, UK, October 23, 2014. Available from: http://www.ema.europa.eu/docs/en_GB/document_library/Scientific_guideline/2014/10/WC500176768.pdf.

EMA. (2015) EPAR summary for the public. Available from: http://www.ema.europa.eu/docs/en_GB/document_library/EPAR_-_Summary_for_the_public/human/004007/WC500200381.pdf.

Endrenyi L, Chang C, Chow SC, Tothfalusi L. (2013) On the interchangeability of biologic drug products. *Statistics in Medicine* **32(3)**, 434–441.

FDA. (2001) Guidance for industry: statistical approaches to establishing bioequivalence. U.S. Food and Drug Administration, Center for Drug Evaluation and Research, Rockville, MD. Available from: http://www.fda.gov/downloads/Drugs/GuidanceComplianceRegulatoryInformation/Guidances/ucm070244.pdf.

FDA. (2013) Draft guidance for industry: bioequivalence studies with pharmacokinetic endpoints for drugs submitted under an ANDA. U.S. Food and Drug Administration, Center for Drug Evaluation and Research, Silver Spring, MD. Available from: http://www.fda.gov/downloads/Drugs/GuidanceComplianceRegulatoryInformation/Guidances/UCM377465.pdf.

FDA. (2015a) Scientific considerations in demonstrating biosimilarity to a reference product. United States Food and Drug Administration, Silver Spring, MD. Available from: http://www.fda.gov/downloads/Drugs/GuidanceComplianceRegulatoryInformation/Guidances/UCM291128.pdf.

FDA. (2015b) Quality considerations in demonstrating biosimilarity to a reference protein product. United States Food and Drug Administration, Silver Spring, MD. Available from: http://www.fda.gov/downloads/Drugs/GuidanceComplianceRegulatoryInformation/Guidances/UCM291134.pdf.

FDA. (2015c) Biosimilars: questions and answers regarding implementation of the Biologics Price Competition and Innovation Act of 2009. United States Food and Drug Administration, Silver Spring, MD. Available from: http://www.fda.gov/downloads/Drugs/GuidanceComplianceRegulatoryInformation/Guidances/UCM444661.pdf.

FDA. (2016) Approved drug products with therapeutic equivalence evaluations (Orange Book). Available from: http://www.fda.gov/Drugs/DevelopmentApprovalProcess/ucm079068.htm.

Federal Register. (2009) *Biologics Price Competition and Innovation Act.* §7002(a)(2). Available from: http://www.fda.gov/downloads/Drugs/GuidanceComplianceRegulatoryInformation/ucm216146.htm.

GPhA. (2015) GPhA and biosimilars council praise new state laws to allow automatic substitution for interchangeable biologics. Press release of Generic Pharmaceutical Association, Washington, DC, May 13, 2015. Available from: http://www.gphaonline.org/gpha-media/press/gpha-and-biosimilars-council-praise-new-state-laws-to-allow-automatic-substitution-for-interchangeable-biologics.

Health Canada. (2010a) Guidance for sponsors: information and submission requirements for subsequent entry biologics (SEBs), Ottawa, ON. Available from: http://www.hc-sc.gc.ca/dhp-mps/alt_formats/pdf/brgtherap/applic-demande/guides/seb-pbu/seb-pbu-2010-eng.pdf.

Health Canada. (2010b) Questions and answers to accompany the final guidance for sponsors: information and submission requirements for subsequent entry biologics (SEBs). Question 15. Ottawa, ON, 2010. Available from: http://www.hc-sc.gc.ca/dhp-mps/brgtherap/applic-demande/guides/seb-pbu/notice-avis_seb-pbu_2010-eng.php.

Klein AV, Wang J, Bedford P. (2014) Subsequent entry biologics (biosimilars) in Canada: approaches to interchangeability and the extrapolation of indications and uses. *Generics and Biosimilars Initiative Journal* **3(3)**, 150–154.

NOR-SWITCH study. (2015) Clinical trials database, last updated November 25, 2015. Available from: http://clinicaltrials.gov/ct2/show/record/NCT02148640.

Ridder L, Waterman M, Turner D, et al. (2015) Use of biosimilars in paediatric inflammatory bowel disease: a position statement *Journal of Pediatric Gastroenterology and Nutrition* **61(4)**, 503–508.

Schall R, Luus H. (1993) On population and individual bioequivalence. *Statistics in Medicine* **12**, 1109–1124.

Schiestl M, Stangler T, Torella C, et al. (2011) Acceptable changes in quality attributes of glycosylated biopharmaceuticals. *Nature Biotechnology* **29**, 310–312.

Schuirmann DJ. (1987) A comparison of the two one-sided tests procedure and the power approach for assessing the equivalence of average bioavailability. *Journal of Pharmacokinetics and Biopharmaceutics* **15**, 657–680.

Scott BJ, Klein AV, Wang J. (2015) Biosimilar monoclonal antibodies: a Canadian regulatory perspective on the assessment of clinically relevant differences and indication extrapolation. *Journal of Clinical Pharmacology* **55(Suppl. 3)**, S123–S132.

Sheiner LB. (1993) Bioequivalence revisited. *Statistics in Medicine* **30**, 1777–1788.

Stewart JJ, Agosto H, Litwin S, et al. (1997) A solution to the rheumatoid factor paradox: pathologic rheumatoid factors can be tolerized by competition with natural rheumatic factors. *The Journal of Immunology* **159**, 1728–1738.

Thimmaraju PK, Rakshambikai R, Farista R, Jurulu K. (2015) Legislations on biosimilar interchangeability in the US and EU: developments far from visibility. *GaBI Online* posted: June 1, 2015. Available from: http://www.gabionline.net/Sponsored-Articles/Legislations-on-biosimilar-interchangeability-in-the-US-and-EU-developments-far-from-visibility.

Tothfalusi L, Endrenyi L, Chow SC. (2014) Statistical and regulatory considerations in assessments of interchangeability of biological drug products. *European Journal of Health Economics* **15 (Suppl. 1)**, S5–S11.

Van Assche G, Vermeire S, Ballet V, et al. (2012) Switch to adalimumab in patients with Crohn's disease controlled by maintenance infliximab: prospective randomised SWITCH trial. *Gut* **61(2)**, 229–234.

Weise M, Bielsky MC, De Smet K, et al. (2012) Biosimilars: what clinicians should know. *Blood* **120**, 5111–5117.

WHO. (2005) Draft revision on multisource (generic) pharmaceutical products: guidelines on registration requirements to establish interchangeability. World Health Organization, Geneva, Switzerland.

11 Design and Analysis of Studies for Assessing Interchangeability

Shein-Chung Chow
Duke University School of Medicine

Laszlo Endrenyi
University of Toronto

CONTENTS

11.1 INTRODUCTION

When the patent of a brand-name drug expires, other drug products usually seek market authorization. Chemically based drugs having small molecules, the alternatives are generic products; with biologically based drugs having large molecules, the alternative products are biosimilars (or termed subsequent-entry biologics in Canada). Regulatory authorities compare the relevant properties of the generic or biosimilar product with those of the original, brand-name drug. As a result of the comparison, bioequivalence of the generic formulation or biosimilarity of the biosimilar product with the brand-name drug could be declared, and market authorization could be granted (Chow, 2011, 2013).

After an alternative drug product (either generic or biosimilar) has received market authorization, the question that arises is if it is interchangeable with the original brand-name drug. In the case of generic formulations, a statement of bioequivalence usually (but not always) implies therapeutic equivalence and therefore interchangeability. With biosimilar products the issue is much more complicated. It is discussed in another chapter of this book.

The present chapter considers the design and analysis of studies for the assessment of interchangeability of biological products. In order to provide a background, Section 11.2 discusses criteria for drug interchangeability in terms of population and individual bioequivalence for generic drug products. It also presents a newly proposed scaled criterion for drug interchangeability that could be applied to both generics and biosimilars. Section 11.3 summarizes various study designs for assessing the risk of switching, alternating, and both switching and alternating. A unified approach for assessing drug interchangeability of biosimilar products is given in Section 11.4. Section 11.5 provides concluding remarks.

11.2 CRITERIA FOR INTERCHANGEABILITY

11.2.1 Criteria for Population and Individual Similarity

A generic drug can generally be used to substitute for the brand-name drug if it has been shown to be bioequivalent to that drug. The FDA does not indicate that two generic copies of the same brand-name drug can be used interchangeably, even though they are bioequivalent to the same brand-name drug. In practice, bioequivalence between generic copies of a brand-name drug is not required. However, as more generic drug products become available, it is a concern whether the approved generic drug products have the same quality and therapeutic effect as the brand-name drug product and whether they can be used safely and interchangeably. The concept of drug interchangeability for small-molecule drug products includes drug prescribability and drug switchability. To evaluate whether the generic drug products can be used safely and interchangeably, the FDA suggests that population bioequivalence and individual bioequivalence be assessed for addressing the drug prescribability and drug switchability of approved generic drug products, respectively (FDA, 2001, 2003a,b).

11.2.1.1 Population Bioequivalence

Drug prescribability is referred to as the physician's choice for prescribing an appropriate drug for his or her patients between the brand-name drug and its generic copies. To address drug prescribability, the FDA recommends that population bioequivalence (PBE) be assessed. In addition to average of bioavailability, PBE focuses on variability of bioavailability. The 2001 FDA guidance recommends that the following criterion be used for assessing PBE (FDA, 2001, 2003b):

$$\theta_P = \frac{\left(\delta^2 + \sigma_{TT}^2 - \sigma_{TR}^2\right)}{\max\left\{\sigma_{T0}^2, \sigma_{TR}^2\right\}}, \tag{11.1}$$

where

$\sigma_{TT}^2, \sigma_{TR}^2$ are the total variances for the test product and the reference product, respectively.

σ_{T0}^2 is the scale parameter specified by the regulatory agency or the sponsor.

PBE can be claimed if the one-sided 95% upper confidence bound for θ_p is less than a prespecified bioequivalence limit. In view of the above PBE criterion, PBE can be claimed if the null hypothesis in

$$H_0: \lambda \geq 0 \text{ vs. } H_a: \lambda < 0 \tag{11.2a}$$

is rejected at the 5% level of significance and the observed geometric means ratio (GMR) is within the limits of 80% and 125%, where

$$\lambda = \delta^2 + \sigma_{TT}^2 - \sigma_{TR}^2 - \theta_{\text{PBE}} \max\left(\sigma_{TR}^2, \sigma_0^2\right) \tag{11.2b}$$

and θ_{PBE} is a constant specified in the 2001 FDA draft guidance. Under a 2 × 2 crossover design, the one-sided 95% upper confidence bound for θ_P can be obtained under the following model:

$$y_{ijk} = \mu + F_l + P_j + Q_k + S_{ikl} + \epsilon_{ijk} \tag{11.3}$$

where μ is the overall mean, P_j is the fixed effect of the jth period, Q_k is the fixed effect of the kth sequence, F_l is the fixed effect of the lth drug product, S_{ikl} is the random effect of the ith subject in the kth sequence under the lth drug product, and ϵ_{ijk}'s are independent random errors distributed as $N\left(0, \sigma_{Wl}^2\right)$. It is assumed that S_{ikl}'s and ϵ_{ijk}'s are mutually independent. It can be verified that (S_{ikT}, S_{ikR}), $i = 1, 2, \ldots, n_k$; $k = 1, 2$ are independent and identically distributed bivariate normal random vectors with mean 0 and an unknown covariance matrix

$$\begin{pmatrix} \sigma_{BT}^2 & \rho\sigma_{BT}\sigma_{BR} \\ \rho\sigma_{BT}\sigma_{BR} & \sigma_{BR}^2 \end{pmatrix},$$

where σ_{Bl}^2 denotes the between-subject variability for the lth drug product, where $l = T$ or $l = R$. Thus, we have

$$\sigma_{TT}^2 = \sigma_{BT}^2 + \sigma_{WT}^2 \quad \text{and} \quad \sigma_{TR}^2 = \sigma_{BR}^2 + \sigma_{WR}^2.$$

Under model 11.3, unbiased estimators for δ, σ_{TT}^2, and σ_{TR}^2 can be obtained as follows:

$$\hat{\delta} = \frac{\bar{y}_{11} - \bar{y}_{12} - \bar{y}_{21} + \bar{y}_{22}}{2} \sim N\left(\delta, \frac{\sigma_{1,1}^2}{4}\left(\frac{1}{n_1} + \frac{1}{n_2}\right)\right).$$

Here $\bar{y}_{jk}$ is the sample mean of the observations in the kth sequence at the jth period, and $\sigma_{1,1}^2$ is $\sigma_{a,b}^2 = \sigma_D^2 + a\sigma_{WT}^2 + b\sigma_{WR}^2$ with $a = 1$ and $b = 1$. Commonly considered unbiased estimators for σ_{TT}^2 and σ_{TR}^2 are given by

$$\hat{\sigma}_{TT}^2 = \frac{1}{n_1 + n_2 - 2}\left[\sum_{i=1}^{n_1} (y_{i11} - \bar{y}_{11})^2 + \sum_{i=1}^{n_2} (y_{i22} - \bar{y}_{22})^2\right]$$

$$\sim \frac{\sigma_{TT}^2 \lambda_{n_1+n_2-2}^2}{n_1 + n_2 - 2}$$

and

$$\hat{\sigma}_{TR}^2 = \frac{1}{n_1 + n_2 - 2}\left[\sum_{i=1}^{n_1} (y_{i21} - \bar{y}_{21})^2 + \sum_{i=1}^{n_2} (y_{i12} - \bar{y}_{12})^2\right]$$

$$\sim \frac{\sigma_{TR}^2 \lambda_{n_1+n_2-2}^2}{n_1 + n_2 - 2}.$$

By Chow et al. (2002), the following approximate 95% upper confidence bound for λ when $\sigma_{TR}^2 \geq \sigma_0^2$ can be obtained:

$$\hat{\lambda}_U = \hat{\delta}^2 + \hat{\sigma}_{TT}^2 - (1 + \theta_{\text{PBE}})\hat{\sigma}_{TR}^2 + t_{0.05, n_1+n_2-2}\sqrt{V}, \tag{11.4}$$

where V is an estimated variance of $\hat{\delta}^2 + \hat{\sigma}_{TT}^2 - (1 + \theta_{\text{PBE}})\hat{\sigma}_{TR}^2$ of the form

$$V = \left(2\hat{\delta}, 1, -(1 + \theta_{\text{PBE}})\right) C \left(2\hat{\delta}, 1, -(1 + \theta_{\text{PBE}})\right)'$$

and C is an estimated variance-covariance matrix of $\left(\hat{\delta}, \hat{\sigma}_{TT}^2, \hat{\sigma}_{TR}^2\right)$. Since $\hat{\delta}$ and $\left(\hat{\sigma}_{TT}^2, \hat{\sigma}_{TR}^2\right)$ are independent, C is given by

$$C = \begin{pmatrix} \frac{\sigma_{1,1}^2}{4}\left(\frac{1}{n_1} + \frac{1}{n_2}\right) & (0,0) \\ (0,0)' & \frac{(n_1 - 1)C_1}{(n_1 + n_2 - 2)^2} + \frac{(n_2 - 1)C_2}{(n_1 + n_2 - 2)^2} \end{pmatrix},$$

where

C_1 is the sample covariance matrix of $\left(\left(y_{i11}-\bar{y}_{11}\right)^2,\left(y_{i21}-\bar{y}_{21}\right)^2\right)$, $i=1,2,\ldots,n_1$,

C_2 is the sample covariance matrix of $\left(\left(y_{i22}-\bar{y}_{22}\right)^2,\left(y_{i12}-\bar{y}_{12}\right)^2\right)$, $i=1,2,\ldots,n_2$.

On the other hand, when $\sigma_{TR}^2<\sigma_0^2$, the upper confidence bound for λ should be modified as follows:

$$\hat{\lambda}_U=\hat{\delta}^2+\hat{\sigma}_{TT}^2-(1+\theta_{\text{PBE}})\hat{\sigma}_0^2+t_{0.05,n_1+n_2-2}\sqrt{V_0}, \tag{11.4a}$$

where $V_0=(2\hat{\delta},\ 1,\ -1)C(2\hat{\delta},\ 1,-1)'$.

11.2.1.2 Individual Bioequivalence

Drug switchability is defined as the switch from one drug (e.g., a brand-name drug) to another (e.g., a generic copy) within the same patient whose concentration of the drug has been titrated to a steady, efficacious, and safe level. To address drug switchability, the FDA suggests that individual bioequivalence (IBE) be assessed under replicated crossover designs such as a replicated 2 × 2 crossover design, that is, (TRTR, RTRT), or a 2 × 3 two-sequence dual design, such as (TRT, RTR). In addition to average of bioavailability, IBE focuses on the within-subject variability of bioavailability and variability due to subject-by-drug interaction. The 2001 FDA draft guidance recommends that the following criterion be used for assessing IBE (see also, Patnaik et al., 1997; Chow, 1999):

$$\theta_I=\frac{\left(\delta^2+\sigma_D^2+\sigma_{WT}^2-\sigma_{WR}^2\right)}{\max\left\{\sigma_{W0}^2,\sigma_{WR}^2\right\}},$$

where $\delta=\mu_T-\mu_R, \sigma_{WT}^2, \sigma_{WR}^2, \sigma_D^2$ are the true difference between means, intrasubject variabilities of the test product and the reference product, and the variance component due to subject-by-formulation interaction between drug products, respectively. σ_{W0}^2 is the scale parameter specified by the regulatory agency or the sponsor (FDA, 2001). In view of the above IBE criterion, IBE can be claimed if the null hypothesis in

$$H_0:\gamma\geq 0 \text{ vs. } H_a:\gamma<0 \tag{11.5a}$$

is rejected at the 5% level of significance and the observed GMR is within the limits of 80% and 125%, where

$$\gamma=\delta^2+\sigma_D^2+\sigma_{WT}^2-\sigma_{WR}^2-\theta_{\text{IBE}}\max\left(\sigma_{WR}^2,\sigma_{W0}^2\right) \tag{11.5b}$$

and θ_{IBE} is a constant specified in the 2001 FDA draft guidance.

For the assessment of IBE, the FDA recommends that a replicated 2 × 2 crossover design, that is, (TRTR, RTRT) or (RTRT, TRTR) be used. Under the 2 × 2 replicated crossover design, the one-sided 95% upper confidence bound for θ_I can be obtained under the following statistical model:

$$y_{ijk}=\mu+F_l+W_{ljk}+S_{ikl}+\epsilon_{ijk} \tag{11.6}$$

where μ is the overall mean, F_l is the fixed effect of the lth drug product, W_{ljk}'s are fixed period, sequence, and interaction effects, S_{ikl} is the random effect of the ith subject in the kth sequence under the lth drug product, and ϵ_{ijk}'s are independent random errors distributed as $N\left(0,\sigma_{Wl}^2\right)$. It is assumed that S_{ikl}'s and ϵ_{ijk}'s are mutually independent. Under model 11.6, σ_D^2 is given by

$$\sigma_D^2 = \sigma_{BT}^2 + \sigma_{BR}^2 - 2\rho\sigma_{BT}\sigma_{BR},$$

which is the variance of $S_{ikT} - S_{ikR}$. Note that σ_D^2 is usually referred to as the variance component due to the subject-by-drug interaction. It can be verified that when $\sigma_{WR}^2 \geq \sigma_{W0}^2$, the linearized criterion γ can be decomposed as follows:

$$\gamma = \delta^2 + \sigma_{0.5,0.5}^2 + 0.5\sigma_{WT}^2 - (1.5 + \theta_{\text{IBE}})\sigma_{WR}^2. \tag{11.7}$$

Now, under model 11.6, for subject i in sequence k, let x_{ilk} and z_{ilk} be the average and the difference, respectively, of the observations from drug product l and let $\bar{x}_{lk}$ and $\bar{z}_{lk}$ be, respectively, the sample mean based on x_{ilk}'s and z_{ilk}'s. Thus, under model 11.6, unbiased estimators for δ, $\sigma_{0.5,0.5}^2$, and σ_{WR}^2 can be obtained as follows:

$$\hat{\delta} = \frac{\bar{x}_{T1} - \bar{x}_{R1} + \bar{x}_{T2} - \bar{x}_{R2}}{2} \sim N\left(\delta, \frac{\sigma_{0.5,0.5}^2}{4}\left(\frac{1}{n_1} + \frac{1}{n_2}\right)\right),$$

$$\hat{\sigma}_{0.5,0.5}^2 = \frac{(n_1 - 1)s_{d1}^2 + (n_2 - 1)s_{d2}^2}{n_1 + n_2 - 2} \sim \frac{\sigma_{0.5,0.5}^2 \lambda_{n_1+n_2-2}^2}{n_1 + n_2 - 2},$$

where s_{dk}^2 is the sample variance based on $x_{iTk} - x_{iRk}$, $i = 1, 2, \ldots, n_k$; an unbiased estimator of σ_{WT}^2 is given by

$$\hat{\sigma}_{WT}^2 = \frac{(n_1 - 1)s_{T1}^2 + (n_2 - 1)s_{T2}^2}{n_1 + n_2 - 2} \sim \frac{\sigma_{WT}^2 \lambda_{n_1+n_2-2}^2}{n_1 + n_2 - 2},$$

where s_{Tk}^2 is the sample variance based on z_{iTk}, $i = 1, 2, \ldots, n_k$; an unbiased estimator of σ_{WR}^2 is given by

$$\hat{\sigma}_{WR}^2 = \frac{(n_1 - 1)s_{R_1}^2 + (n_2 - 1)s_{R_2}^2}{n_1 + n_2 - 2} \sim \frac{\sigma_{WR}^2 \lambda_{n_1+n_2-2}^2}{n_1 + n_2 - 2},$$

where s_{Rk}^2 is the sample variance based on z_{iRk}, $i = 1, 2, \ldots, n_k$. Furthermore, since $\hat{\delta}$, $\hat{\sigma}_{0.5,0.5}^2$, $\hat{\sigma}_{WT}^2$, and $\hat{\sigma}_{WR}^2$ are independent, when $\sigma_{WR}^2 \geq \sigma_{W0}^2$ an approximate 95% confidence upper bound for γ can be obtained as follows:

$$\hat{\gamma}_U = \hat{\delta}^2 + \hat{\sigma}_{0.5,0.5}^2 + 0.5\hat{\sigma}_{WT}^2 - (1.5 + \theta_{IBE})\hat{\sigma}_{WR}^2 + \sqrt{U}, \tag{11.8}$$

where U is the sum of the following four quantities:

$$\left[\left(\left|\hat{\delta}\right| + t_{0.05,n_1+n_2-2}\frac{\hat{\sigma}_{0.5,0.5}}{2}\sqrt{\frac{1}{n_1}+\frac{1}{n_2}}\right)^2 - \hat{\delta}^2\right]^2,$$

$$\hat{\sigma}^4_{0.5,0.5}\left(\frac{n_1+n_2-2}{\lambda^2_{0.05,n_1+n_2-2}}-1\right)^2,$$

$$0.5^2\hat{\sigma}^4_{WT}\left(\frac{n_1+n_2-2}{\lambda^2_{0.05,n_1+n_2-2}}-1\right)^2,$$

and

$$(1.5+\theta_{\text{IBE}})^2\hat{\sigma}^4_{WR}\left(\frac{n_1+n_2-2}{\lambda^2_{0.05,n_1+n_2-2}}-1\right)^2.$$

When $\sigma^2_{WR} < \sigma^2_{W0}$, an approximate 95% confidence upper bound for γ is given by

$$\hat{\gamma}_U = \hat{\delta}^2 + \hat{\sigma}^2_{0.5,0.5} + 0.5\hat{\sigma}^2_{WT} - 1.5\hat{\sigma}^2_{WR} - \theta_{\text{IPE}}\sigma^2_{W0} + \sqrt{U_0}, \tag{11.8a}$$

where U_0 is a sum like U except that the four quantities should be replaced by

$$1.5^2\hat{\sigma}^4_{WR}\left(\frac{n_1+n_2-2}{\lambda^2_{0.05,n_1+n_2-2}}-1\right)^2.$$

11.2.1.3 Remarks

Both criteria for PBE and IBE are aggregated, moment-based criteria that involve several variance components, including intersubject and intrasubject variabilities. Since the criteria are nonlinear functions of the direct drug effect, intersubject and intrasubject variabilities for the test product and the reference product, and the variability due to subject-by-drug interaction (for IBE criterion), a typical approach is to linearize the criteria and then apply the method of modified large sample (MLS) or extended MLS for obtaining an approximate 95% upper confidence bound of the linearized criteria (see, e.g., Hyslop et al., 2000; Lee et al., 2004). The key is to decompose the linearized criteria into several components and obtain independent and unbiased estimators of these components for obtaining a valid approximate upper confidence bound.

Alternatively, one may consider the method of generalized pivotal quantity (GPQ) to assess PBE and/or IBE (see, e.g., Chiu et al., 2010, 2013). The idea of GPQ is briefly described below. Suppose that Y is a random variable whose distribution depends on a vector of unknown parameters $\zeta = (\theta, \eta)$, where θ is a parameter of interest and η is a vector of nuisance parameters. Let Y be a random sample from Y and y be the observed value of Y. Furthermore, let $R = R(Y; y, \zeta)$

be a function of Y, y, and ζ. The random quantity R is referred to as a GPQ, which satisfies the following two conditions:

1. The distribution of R does not depend on any unknown parameters;
2. The observed value of R, say $r = R(y; y, \zeta)$ is free of the vector of nuisance parameters η.

In other words, r is only a function of (y, θ). Thus, a $(1-\alpha) \times 100\%$ generalized upper confidence limit for θ is given by $R_{1-\alpha}$, where $R_{1-\alpha}$ is the $100(1-\alpha)$th percentile of the distribution of R. The percentile can be analytically estimated using a Monte Carlo algorithm.

11.2.2 Scaled Criteria for Drug Interchangeability

While the criterion for the determination of average biosimilarity is based on the BE requirement of (80.00%, 125.00%) using log-transformed data, the criterion for the assessment of interchangeability is not clear. As indicated by Chen et al. (2000), variability due to subject-by-drug formulation interaction will have an impact on drug interchangeability. Let σ_D^2 be the variance component due to the subject-by-formulation interaction. Chow et al. (2015) proposed a scaled criterion for interchangeability (SCDI) which adjusts for both intrasubject variability and the variability due to subject-by-formulation variability. The current BE criterion adjusted for intrasubject variability leads to the so-called scaled average bioequivalence (SABE) criterion, which is considered suitable for highly variable drug products (see, e.g., Tothfalusi and Endrenyi, 2003; Benet, 2006; Davit et al., 2008; Haidar et al., 2008; Endrenyi and Tothfalusi, 2009; Tothfalusi et al., 2009; Davit and Conner, 2010; Tothfalusi and Endrenyi, 2011; Davit et al., 2012). In addition to adjusting intrasubject variability, following the idea of individual bioequivalence, Chow et al. (2015) proposed further adjusting the BE criterion with respect to the variability due to subject-by-formulation variability in order to have a more accurate and reliable assessment of interchangeability. As indicated in the 2001 FDA guidance on bioequivalence, the criterion for assessing individual bioequivalence (IBE) is given by:

$$\theta = \frac{\left(\mu_T - \mu_R\right)^2 + \sigma_D^2 + \left(\sigma_{WR}^2 - \sigma_{WT}^2\right)}{\max\left(\sigma_0^2, \sigma_{WR}^2\right)}, \tag{11.9}$$

where σ_{WR}^2 and σ_{WT}^2 are intrasubject variances for the reference product and the test product, respectively, and σ_0^2 is a regulatory constant.

Chow et al.'s (2015) proposed criterion is based on the first two components of the criterion for individual bioequivalence, which consists of (1) the criterion for average biosimilarity adjusted for intrasubject variability of the reference product (i.e., SABE), and (2) a correction for the variability due to subject-by-product variability (i.e., σ_D^2). The proposed criterion for assessing interchangeability (i.e., switching and alternating) is briefly derived in the following:

Step 1: Unscaled ABE criterion

Let BEL be the BE limit which generally equals 1.25. Thus, biosimilarity requires that

$$\frac{1}{\text{BEL}} \le \text{GMR} \le \text{BEL} \tag{11.10}$$

This implies

$$-\log(\text{BEL}) \le \log(\text{GMR}) \le \log(\text{BEL}), \tag{11.10a}$$

or

$$-\log(\text{BEL}) \le \mu_T - \mu_R \le \log(\text{BEL}), \tag{11.10b}$$

where μ_T and μ_R are logarithmic means.

Step 2: Scaled ABE (SABE) criterion

The difference between logarithmic means is adjusted for the intrasubject variability as follows:

$$-\log(\text{BELS}) \le \frac{\mu_T - \mu_R}{\sigma_W} \le \log(\text{BELS}), \tag{11.11a}$$

or

$$-\log(\text{BELS})\sigma_W \le \mu_T - \mu_R \le \log(\text{BELS})\sigma_W, \tag{11.11b}$$

where σ_W^2 is a within-subject variation and BELS is the BE limit for SABE. In practice, σ_{WR}^2, the within-subject variation of the reference product is often considered.

Step 3: Proposed scaled criterion for SCDI

Considering the first two components of the individual bioequivalence criterion, we have the following relationship:

$$\frac{(\mu_T - \mu_R)^2 + \sigma_D^2}{\sigma_W^2} = \frac{2\delta\sigma_D + (\delta - \sigma_D)^2}{\sigma_W^2},$$

where $\delta = \mu_T - \mu_R$. When δ and σ_D are close, we observe that

$$\frac{\delta^2 + \sigma_D^2}{\sigma_W^2} \approx \frac{2\delta\sigma_D}{\sigma_W^2}.$$

The assumption is reasonable when both δ and σ_D are small.
Thus, the proposed SCDI is:

$$-\log(\text{BELS}) \leq \left(\frac{\mu_T - \mu_R}{\sigma_W}\right)\left(\frac{2\sigma_D}{\sigma_W}\right) \leq \log(\text{BELS}). \tag{11.12a}$$

Now, let $f = \sigma_W/(2\sigma_D)$, a correction factor for drug interchangeability. Then, the proposed SCDI criterion is given by:

$$-\log(\text{BELS}) f\sigma_W \leq \mu_T - \mu_R \leq \log(\text{BELS}) f\sigma_W. \tag{11.12b}$$

Note that the statistical properties and finite sample performance need further research.

Following the concept of the criterion for individual bioequivalence and the idea of SABE, the proposed SCDI criterion is developed in order to adjust the usual one-size-fits-all approach for both intrasubject variability of the reference product and variability due to subject-by-product interaction. As compared to SABE, SCDI may result in wider or narrower limits depending on the correction factor f, which is a measure of the relative magnitude between σ_{WR} and σ_D.

The proposed SCDI criterion for drug interchangeability depends on the selection of regulatory constants for σ_{WR} and σ_D. In practice, the observed variabilities may deviate far from the regulatory constants. Thus, it is suggested that the following hypotheses be tested before the use of the SCDI criterion:

$$H_{01}: \sigma_{WR} > \sigma_{W0} \text{ vs. } H_{a1}: \sigma_{WR} \leq \sigma_{W0}, \tag{11.13a}$$

and

$$H_{02}: \sigma_D > \sigma_{D0} \text{ vs. } H_{a2}: \sigma_D \leq \sigma_{D0} \tag{11.13b}$$

If we fail to reject the null hypotheses H_{01} or H_{02}, then we will stick with the suggested individual regulatory constants; otherwise estimates of σ_{WR} and/or σ_D should be used in the SCDI criterion.

Lower and upper limits for SCDI and SABE at various values of σ_{WR} and σ_D are provided in Table 11.1.

It should be noted, however, that statistical properties and/or the finite sample performance of SCDI with estimates of σ_{WR} and/or σ_D are not well established. Further research is needed.

11.3 STUDY DESIGNS

For the assessment of bioequivalence for chemical drug products, a standard two-sequence, two-period (2 × 2) crossover design is often considered, except for drug products with relatively long half-lives. Since most biosimilar products have relatively long half-lives, it is suggested that a parallel group design should be considered. However, the parallel group design does not provide independent estimates of

TABLE 11.1
Lower and Upper Limits for SABE and SCDI Given σ_D and σ_{WR}

				SABE		SCDI	
σ_D	σ_{WR}	σ_D/σ_{WR}	f	Lower	Upper	Lower	Upper
0.15	0.15	1	0.5	−1.488	1.488	−2.975	2.975
	0.2	0.75	0.667	−1.116	1.116	−1.674	1.674
	0.25	0.6	0.833	−0.893	0.893	−1.071	1.071
	0.3	0.5	1	−0.744	0.744	*−0.744*	*0.744*
	0.4	0.375	1.333	−0.558	0.558	−0.418	0.418
0.2	0.15	1.333	0.375			−3.967	3.967
	0.2	1	0.5			−2.231	2.231
	0.25	0.8	0.625	Same as above		−1.428	1.428
	0.3	0.667	0.75			−0.992	0.992
	0.4	0.5	1			*−0.558*	*0.558*
0.25	0.15	1.667	0.3			−4.959	4.959
	0.2	1.25	0.4			−2.789	2.789
	0.25	1	0.5	Same as above		−1.785	1.785
	0.3	0.833	0.6			−1.24	1.24
	0.4	0.625	0.8			−0.697	0.697
0.3	0.15	2	0.25			−5.95	5.95
	0.2	1.5	0.333			−3.347	3.347
	0.25	1.2	0.417	Same as above		−2.142	2.142
	0.3	1	0.5			−1.488	1.488
	0.4	0.75	0.667			−0.837	0.837
0.4	0.15	2.667	0.187			−7.934	7.934
	0.2	2	0.25			−4.463	4.463
	0.25	1.6	0.312	Same as above		−2.856	2.856
	0.3	1.333	0.375			−1.983	1.983
	0.4	1	0.5			−1.116	1.116

Note: Highlighted rows have accounted for regulation constants ($\sigma_{D0} = 0.15$; $\sigma_{W0} = 0.2, 0.25$); italicized SCDI limits are identical to the SABE limits at the given σ_D and σ_W.

variance components such as inter- and intrasubject variabilities and the variability due to subject-by-product interaction. Thus, it is a major challenge to assess biosimilarity (especially to assess drug interchangeability) under parallel group designs since each subject will receive the same product only once.

As indicated in the BPCI Act, for a biological product that is administered more than once to an individual, the risk in terms of safety or diminished efficacy of alternating or switching between use of the biological product and the reference product should not be greater than the risk of using the reference product without such alternation or switch. Thus, for assessing drug interchangeability, an appropriate study design should be chosen in order to address (1) the risk in terms of safety or diminished efficacy of alternating or switching between use of the biological product and the reference product, (2) the risk of using the reference product without such alternation or switch, and (3) the relative risk between switching/alternating and without switching/alternating. In this section, several useful designs for addressing switching and alternation of biosimilar products are discussed.

11.3.1 Designs for Switching

Consider the broader sense of switchability. In this case, the concept of switching includes (1) switch from "R to T," (2) switch from "T to R," (3) switch from "T to T," and (4) switch from "R to R." Thus, in order to assess the interchangeability of switching, a valid study design should be able to assess biosimilarity between "R and T," "T and R," "T and T," and "R and R" based on some biosimilarity criteria. For this purpose, the following study designs are useful.

11.3.1.1 Balaam's Design

Balaam's design is a 4 × 2 crossover design, denoted by (TT, RR, TR, RT). Under a 4 × 2 Balaam's design, qualified subjects will be randomly assigned to receive one of the four sequences of treatments: TT, RR, TR, and RT. For example, subjects in the sequence 1 of TT, will receive the test (biosimilar) product first and then crossover to receive the reference (innovative biological) product after a sufficient length of washout time. In practice, a Balaam design is considered the combination of a parallel design (the first two sequences) and a crossover design (sequences #3 and #4). The purpose of part of the parallel design is to obtain independent estimates of intrasubject variabilities for the test product and the reference product. In the interest of assigning more subjects to the crossover phase, an unequal treatment assignment is usually employed. For example, we may consider a 1:2 allocation to the parallel phase and the crossover phase. In this case, for a sample size of $N = 24$, 8 subjects will be assigned to the parallel phase and 16 subjects will be assigned to the crossover phase. As a result, 4 subjects will be assigned to both sequences #1 and #2, while 8 subjects will be assigned to sequences #3 and #4, assuming that there is a 1:1 ratio treatment allocation within each phase.

As can be seen from Figure 11.1, the first sequence provides not only an independent estimate of the intrasubject variability of the test product, but also the assessment for "switch from T to T," while the second sequence provides an independent estimate of the intrasubject variability of the reference product and compares the difference between "R and R." The other two sequences assess the similarity for the "switch from T to R" and the "switch from R to T," respectively. Under the 4 × 2 Balaam design, the following comparisons are usually assessed:

1. Comparisons by sequence;
2. Comparisons by period;
3. T versus R based on sequences #3 and #4—this is equivalent to a typical 2 × 2 crossover design;
4. T versus R given T based on sequences #1 and #3;
5. R versus T given R based on sequences #2 and #4;
6. The comparison between (1) and (3) for the assessment of treatment-by-period interaction.

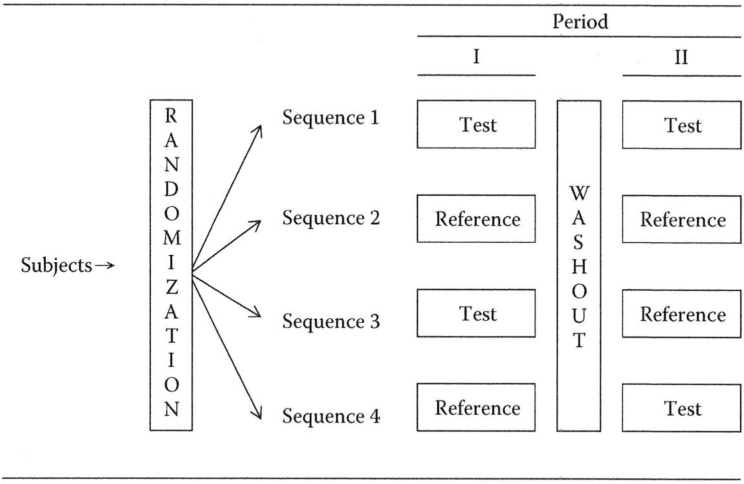

FIGURE 11.1 Balaam's design. *Note:* Balaam's design is a 4 × 2 crossover design.

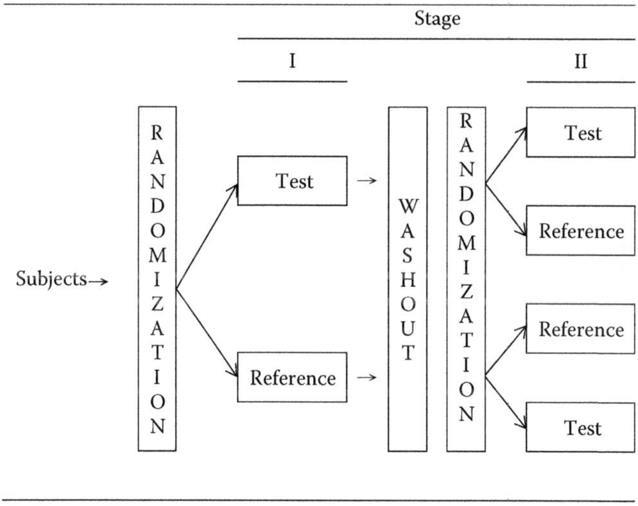

FIGURE 11.2 Two-stage design. *Note:* Stage 2 is nested within Stage 1.

Interpretations of the above comparisons are different. More information regarding statistical methods for data analysis of the Balaam design can be found in Chow and Liu (2008).

11.3.1.2 Two-Stage Design

Alternatively, a two-stage crossover design described in Figure 11.2 may be useful for addressing the interchangeability of switching. Under the two-stage design, qualified subjects are randomly assigned to receive either the test product or the reference product at the first stage. At the second stage, after a sufficient length of washout, subjects are randomly assigned to receive either the test product or the reference product with either equal or unequal ratio of treatment allocation. At the end of the study, the two-stage design will lead to four sequences of treatments, that is, TT, TR, RT, and RR similar to those in Balaam's design.

Note that the above-mentioned two-stage design that consists of a parallel phase (stage 1) and a crossover phase (stage 2) is similar to a placebo-challenging design proposed by Chow et al. (2000). As a result, statistical methods proposed by Chow et al. (2000) are useful for a valid analysis of data collected from a two-stage design described above. Similarly, under the two-stage design, the above comparisons (1)–(6) can also be made based on the methods proposed by Chow et al. (2000).

11.3.2 Designs for Alternating

For addressing the concept of alternating, an appropriate study design should allow the assessment of differences between "R to T" and "T to R" for alternating of "R to T to R" to determine whether the drug effect has returned to the baseline after the second switch.

For this purpose, the following study designs are useful.

11.3.2.1 Two-Sequence Dual Design

Two-sequence dual design is a 2 × 3 higher-order crossover design consisting of two dual sequences, namely TRT and RTR. Under the two-sequence dual design, qualified subjects will be randomly assigned to receive either the sequence of TRT or the sequence of RTR. Of course, there is a sufficient length of washout between dosing periods. Under the two-sequence dual design, we will be able to evaluate the relative risk of alternating between use of the biological product and the reference product and the risk of using the reference product without such alternating (Figure 11.3).

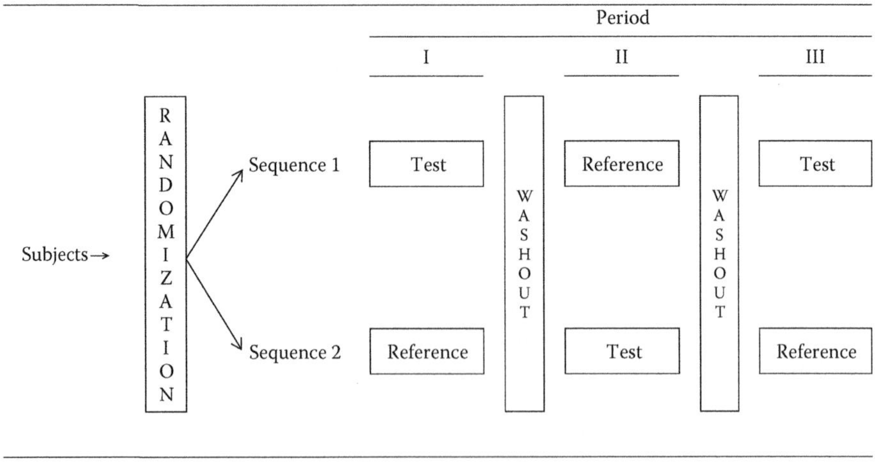

FIGURE 11.3 Two-sequence dual design. *Note:* Two-sequence dual design is a 2 × 3 crossover design.

Note that expected values of the sequence-by-period means, analysis of variance table, and statistical methods (e.g., the assessment of average biosimilarity, inference on carryover effect, and the assessment of intrasubject variabilities) for the analysis of data collected from a two-sequence dual design are given in Chow and Liu (2008). In case there are missing data (i.e., incomplete data), statistical methods proposed by Chow and Shao (2002) are useful.

11.3.2.2 Williams Design

For a broader sense of alternation involving more than two biologics, for example, two biosimilars T_1 and T_2 and one innovative product R, there are six possible sequences: (R T_2 T_1), (T_1 R T_2), (T_2 T_1 R), (T_1 T_2 R), (T_2 R T_1), and (R T_1 T_2). In this case, a 6 × 3 Williams design for comparing three products is useful (see also Chow and Liu, 2008). A Williams design is a variance-balanced design that consists of six sequences and three periods. Under the 6 × 3 Williams design, qualified subjects are randomly assigned to receive one of the six sequences. Within each sequence, a sufficient length of wash is applied between dosing periods (see also Figure 11.4).

Detailed information regarding (1) construction of a Williams design, (2) analysis of variance table, and (3) statistical methods for analysis of data collected from a 6 × 3 Williams design adjusted for carryover effects, in the absence of unequal carryover effects, and adjusted for drug effect can be found in Chow and Liu (2008).

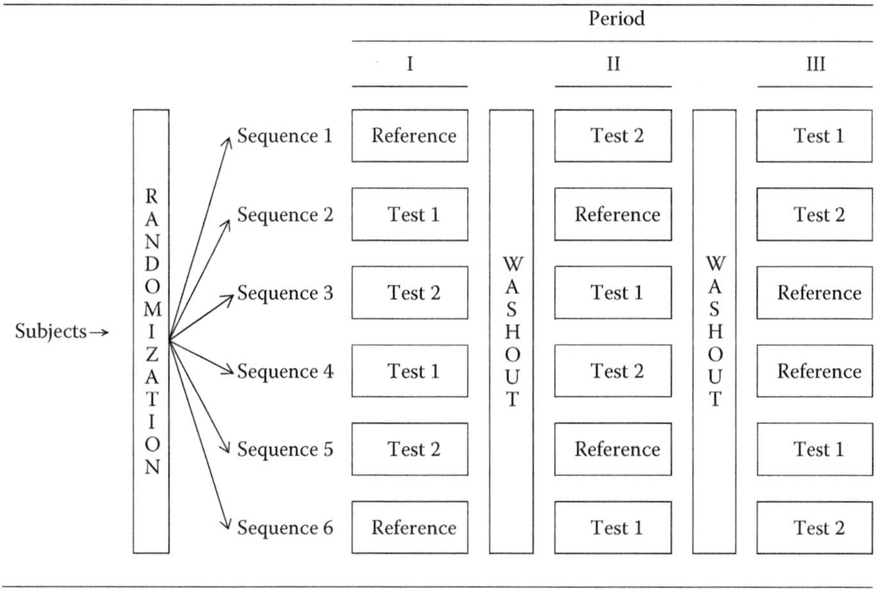

FIGURE 11.4 Williams design. *Note:* Test 1 and Test 2 are two different biosimilar products, and Reference is the reference (innovative) product.

11.3.3 Designs for Switching and Alternating

In the previous two subsections, useful study designs for addressing switching and alternating of drug interchangeability are discussed, respectively. In practice, however, it is of interest to have a study design that can address both switching and alternating. In this case, an intuitive study design is to combine a switching design with an alternating design. Along this line, in this section, several useful designs for addressing the switching and alternating of drug interchangeability are introduced.

11.3.3.1 Modified Balaam's Design

As indicated earlier, Balaam's design is useful for addressing switching, while a two-sequence dual design is appropriate for addressing alternating. In the interest of addressing both switching and alternating in a single trial, we may combine the two study designs as follows: (TT, RR, TRT, RTR), which consists of a parallel design (the first two sequences) and a two-sequence dual design (the last two sequences). We will refer to this design as a modified Balaam design, which is illustrated in Figure 11.5.

The figure illustrates that data collected from the first two dosing periods (which are identical to Balaam's design) can be used to address switching, while data collected from sequences #3 and #4 can be used to assess the relative risks of alternating.

11.3.3.2 Complete Design

The modified Balaam's design is not a balanced design in terms of the number of dosing periods. In the interest of balance in dosing periods, it is suggested that the modified Balaam's design be further modified as (TTT, RRR, TRT, RTR). We will

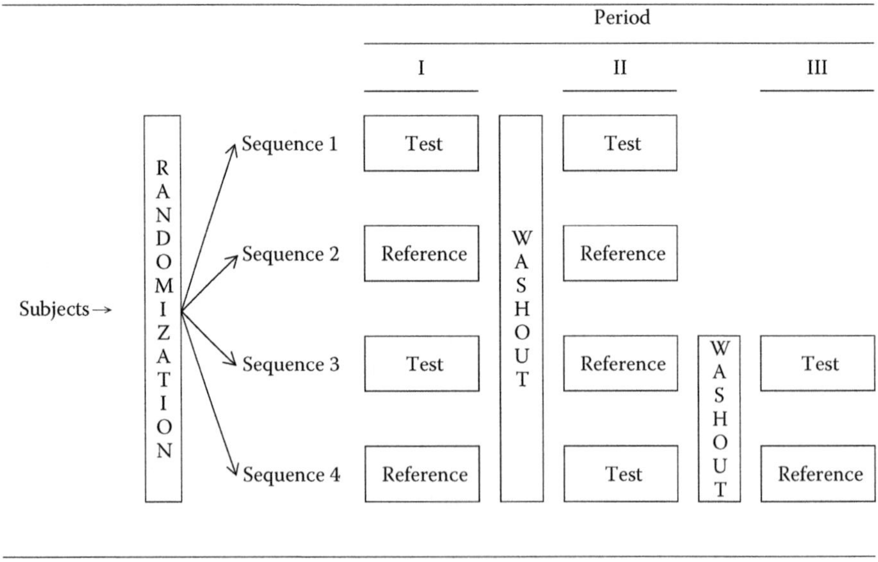

FIGURE 11.5 Modified Balaam's design. *Note:* Sequences #3 and #4 constitute a two-sequence dual design.

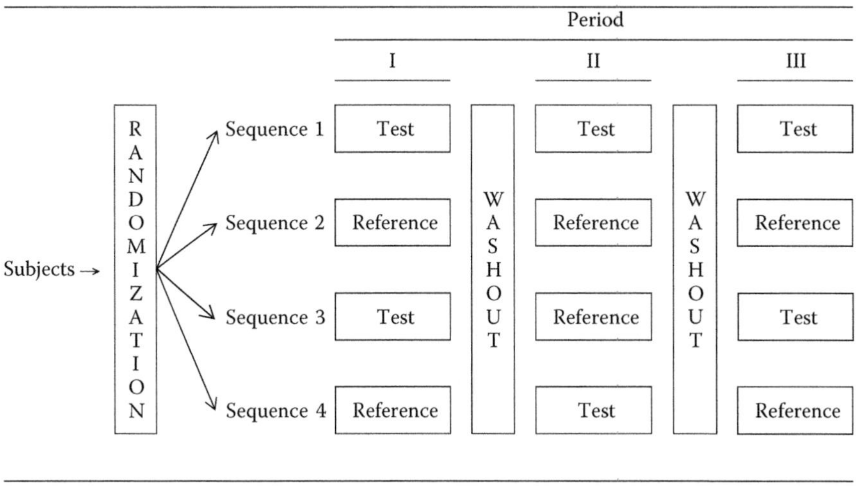

FIGURE 11.6 Complete design. *Note:* "Complete" in the sense that the treatments at the second dosing period one are repeated for sequences #1 and #2.

refer to this design as a complete design. The difference between the complete design and the modified Balaam's design is that the treatments are repeated at the third dosing period for sequences #1 and #2. Data collected from sequence #1 will provide a more accurate and reliable assessment of intrasubject variability, while data collected from sequence #2 is useful in establishing baseline for the reference product (Figure 11.6).

Note that statistical methods for analysis of data collected from the complete design are similar to those under the modified Balaam's design.

11.3.3.3 Alternative Design

For the assessment of individual bioequivalence under a replicated design, Chow et al. (2002) indicated that the optimal design among 2 × 3 crossover designs is a so-called extra-reference design, which is given by (TRR, RTR). Thus, an alternative design is to combine a parallel design (TTT, RRR) and a 2 × 3 extra-reference design for addressing both switching and alternating. The resultant study design is then given by (TTT, RRR, RTR, TRR) (Figure 11.7).

11.3.3.4 Adaptive Designs

In recent years, the use of adaptive design methods in clinical research has become very popular due to its flexibility and efficiency for identifying any (or optimal) clinical benefits of the test treatment under investigation (see, e.g., Chow and Chang, 2011). Similar ideas can be applied to assessment of the biosimilarity and interchangeability of biosimilar products. For example, a two-stage adaptive design that combines two independent studies into a single trial may be useful. Some adaptations (modifications or changes) can be implemented at the end of the first stage after the review of accumulated data collected from the first stage. More information regarding various adaptive trial designs can be found in Chow and Chang (2011).

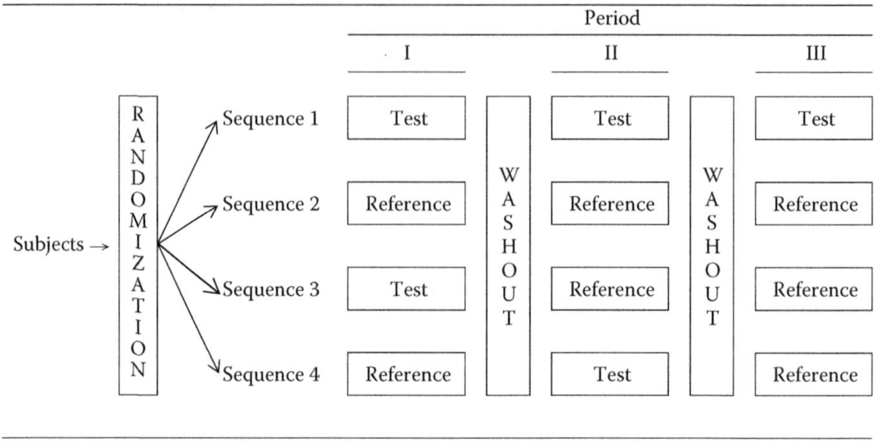

FIGURE 11.7 Alternative design. *Note:* Sequences #3 and #4 comprise an extra-reference design.

11.4 UNIFIED APPROACH FOR ASSESSING INTERCHANGEABILITY

In practice, switching and alternating can only be assessed after the biosimilar products under study have been shown to be highly similar to the innovative biological drug product. Based on a similar idea for development of the biosimilarity index (Shao and Chow, 2002; Chow et al., 2011), a general approach for development of switching index and/or alternating index for addressing switching and/or alternating can be obtained.

11.4.1 Totality Biosimilarity Index

For a given criterion for biosimilarity and under a valid study design, the biosimilarity index for a given functional area or domain can be obtained by the following steps (see also Chow et al., 2013):

Step 1: Assess average biosimilarity based on a given criterion, for example, (80%, 125%) based on log-transformed data;
Step 2: Calculate the local biosimilarity index (i.e., reproducibility) based on the observed ratio and variability;
Step 3: Claim local biosimilarity if the 95% confidence lower bound of p is larger than p_0, a prespecified number, where p_0 can be obtained based on an estimate of reproducibility probability for a study comparing a reference product to itself (the reference product), that is, an R–R study.

A totality biosimilarity index can be derived across all functional areas, or domains can be obtained by the following steps:

Step 1: Obtain $\hat{p}_i$, the biosimilarity index for the ith domain;

Step 2: Define the totality biosimilarity index as $\hat{p}_T = \sum_{i=1}^{K} w_i \hat{p}_i$, where w_i is the weight for the ith domain, where i = 1,2,..., K (number of domains or functional areas);

Step 3: Claim biosimilarity if the 95% confidence lower bound of p_T is larger than a prespecified value p_{T0}, which can be determined based on an estimate of the totality biosimilarity index for studies comparing a reference product to itself (the reference product).

The totality biosimilarity index described above has the advantages that (1) it is robust with respect to the selected study endpoint, biosimilarity criteria, and study design; (2) it takes variability into consideration (one of the major criticisms in the assessment of average bioequivalence); (3) it allows the definition and assessment of the degree of similarity (in other words, it provides a partial answer to the question "how similar is considered similar?"); and (4) the use of the biosimilarity index or totality biosimilarity index will reflect the sensitivity of heterogeneity in variance.

11.4.2 Switching Index (SI)

A similar idea can be applied to develop a switching index under an appropriate study design such as the 4 × 2 Balaam's crossover design described earlier. Thus, biosimilarity for "R to T," "T to R," "T to T," and "R to R" needs to be assessed for addressing the issue of switching.

Define $\hat{p}_{Ti}$ the totality biosimilarity index for the ith switch, where i = 1 (switch from R to R), 2 (switch from T to T), 3 (switch from R to T), and 4 (switch from T to R). As a result, the switching index (SI) can be obtained as follows:

Step 1: Obtain $\hat{p}_{Ti}$, i = 1,..., 4;

Step 2: Define the switching index as $\text{SI} = \min_i \{\hat{p}_{Ti}\}$, i = 1,..., 4; which is the largest order of the biosimilarity indices;

Step 3: Claim switchability if the 95% confidence lower bound of $\min_i\{p_{Ti}\}$, i = 1,..., 4, is larger than a pre-specified value p_{S0}.

Let $P_{T_1}, P_{T_2}, \ldots, P_{T_4}$ be a random sample from a continuous distribution with a probability density function $f(p)$ and cumulative distribution function $F(p)$, and let $P_{T(1)}, P_{T(2)}, \ldots, P_{T(4)}$ be the order statistics obtained from the above sample. Thus, the probability density function of the defined switching index $\text{SI} = P_{T(4)}$ is given by

$$f_{\text{SI}}(p) = \frac{4!}{3!}(F(p))^3 (1 - F(p))^0 f(p) = 4F(p)^3 f(p)$$

and the expected value and variance of SI can be given by

$$\mu_{\text{SI}} = E(\text{SI}) = 4\int p(F(p))^3 f(p)dp,$$

and

$$\mathrm{Var(SI)} = E\left(\mathrm{SI}^2\right) - (\mu_{\mathrm{SI}})^2,$$

where $E\left(\mathrm{SI}^2\right) = \int p^2 (F(p))^3 f(p)\, dp$ denoted as the second raw moment of SI.

With a given distribution function $F(p)$, the expected value and the variance of order statistics could be derived (see David and Nagaraja, 2003). However, the population distribution may be unknown or difficult to determine. Several results of nonparametric bounds for the moments of order statistics have been provided. David (1981) summarized the distribution-free bounds on the expected values of the order statistics when the observations $p_{T_1}, p_{T_2}, \ldots, p_{T_4}$ are i.i.d. from a population with expectation μ and variance σ^2. The earliest result provided in Gumbel (1954) and Hartley and David (1954) concerns the minimum,

$$\mu_{\mathrm{SI}} \le \mu + \sigma(n-1)(2n-1)^{-1/2} = \mu + 1.1339\sigma, \tag{11.14}$$

where the sample size is $n = 4$.

However, the observations may be dependent and/or are from a different distribution. It could be obtained in Arnold and Groeneveld (1979) that the bound in Equation 11.14 becomes

$$\mu_{\mathrm{SI}} \le \mu + \sigma(n-1)^{1/2} = \mu + 1.73205\sigma \tag{11.15}$$

when independence cannot be assumed. On the other hand, for the variance of order statistics, the upper bounds for the variance of order statistics derived by Papadatos (1995) can be used. That is,

$$\mathrm{Var(SI)} < n\sigma^2 = 4\sigma^2. \tag{11.16}$$

In order to obtain the estimates of expectation and variance of SI, the sample mean and sample variance of the observations $p_{T_1}, p_{T_2}, \ldots, p_{T_4}$ could be used to replace μ and σ^2, respectively, in the bound 11.14 or 11.15 of μ_{SI} and the bound 11.16 of Var(SI).

As a result, the 95% confidence low bound of SI can be obtained. We then claim switching if the 95% confidence low bound for SI is larger than p_{S_0}.

11.4.3 Alternating Index (AI)

A similar idea can be applied to develop an alternating index under an appropriate study design. Under the modified Balaam's crossover design of (TT, RR, TRT, RTR), biosimilarity for "R to T to R" and "T to R to T" needs to be assessed for the evaluation of alternating. For example, the assessment of differences between "R to T" and "T to R" for alternating of "R to T to R" needs to be evaluated in order to determine whether the drug effect has returned to the baseline after the second switch.

Define p_{Ti} as the totality biosimilarity index for the ith switch, where $i = 1$ (switch from R to R), 2 (switch from T to T), 3 (switch from R to T), and 4 (switch from T to R). As a result, the alternating index (AI) can be obtained as follows:

Step 1: Obtain $\hat{p}_{Ti}, i = 1,\ldots,4$;
Step 2: Define the range of these indices, $\text{AI} = \max_i\{\hat{p}_{Ti}\} - \min_i\{\hat{p}_{Ti}\},\ i = 1,\ldots,4$, as the alternating index;
Step 3: Claim alternation if the 95% confidence lower bound of $\max_i\{p_{Ti}\} - \min_i\{p_{Ti}\},\ i = 1,\ldots,4$, is larger than a prespecified value p_{A0}.

The estimates of expected values and variance of AI could be similarly obtained following the process of the confidence lower bound of SI. Suppose that $P_{T_1}, P_{T_2},\ldots,P_{T_4}$ is a random sample from a continuous distribution with probability density function $f(p)$ and cumulative distribution function $F(p)$, and $P_{T(1)}, P_{T(2)},\ldots,P_{T(4)}$ are the order statistics obtained from the above sample. Thus, the joint density function of $\max_i\{\hat{p}_{Ti}\}$ and $\min_i\{\hat{p}_{Ti}\}$, denoted by $f_{(1,4)}\left(p_{T(1)}, p_{T(4)}\right)$, is given by

$$f_{(1,4)}(p_{T(1)}, p_{T(4)}) = \frac{4!}{2!} f(p_{T(1)}) f(p_{T(1)}) \left[F(p_{T(1)}) - F(p_{T(1)})\right]^2$$

$$= 12 f(p_{T(1)}) f(p_{T(1)}) \left[F(p_{T(1)}) - F(p_{T(1)})\right]^2$$

and the expected value and the variance of AI can be given by

$$\mu_{AI} = E(\text{AI}) = 12 \iint (p_{T(1)} - p_{T(1)}) f(p_{T(1)}) f(p_{T(1)}) \left[F(p_{T(1)}) - F(p_{T(1)})\right]^2 dp_{T(1)}\, dp_{T(4)},$$

and

$$\text{Var(AI)} = E(\text{AI}^2) - (\mu_{AI})^2,$$

where

$$E(\text{SI}^2) = 12 \iint (p_{T(1)} - p_{T(1)})^2 f(p_{T(1)}) f(p_{T(1)}) \left[F(p_{T(1)}) - F(p_{T(1)})\right]^2 dp_{T(1)}\, dp_{T(4)}$$

is denoted as the second moment of the sample range.

Based on the results for dependent distribution-free samples in Arnold and Groeneveld (1979), the bound on the expected value for the range of order statistics is given by

$$\mu_{AI} = E\,\underset{i}{\boldsymbol{max}}\{p_{Ti}\} - E\,\underset{i}{\boldsymbol{min}}\{p_{Ti}\}$$

$$\leq \sigma \left\{\frac{n(n - k_2 + 1 + k_1)}{((n - k_2 + 1)k_1)}\right\}^{1/2} = 2.82\sigma,$$

where $k_1 = 1$ and $k_2 = 4$ denote the orders of the difference. On the other hand, the upper bound for the variance of AI could be obtained by

$$\begin{aligned}\text{Var(AI)} &= \text{Var}\left(P_{T(1)}\right) + \text{Var}\left(P_{T(4)}\right) + 2\text{Cov}\left(P_{T(1)}, P_{T(4)}\right) \\ &< \text{Var}\left(P_{T(1)}\right) + \text{Var}\left(P_{T(4)}\right) + 2\text{Var}\left(P_{T(1)}\right)^{1/2} \text{Var}\left(P_{T(4)}\right)^{1/2} \\ &< n\sigma^2 + n\sigma^2 + 2n\sigma^2 = 16\sigma^2,\end{aligned}$$

which is derived based on the Cauchy-Schwarz inequality (Casella and Berger, 2002) and the result for the upper bound of the variance of order statistics (Papadatos, 1997). We could estimate μ and σ^2 by the sample mean and sample variance in order to construct the confidence lower bound for AI. Thus, we then claim switching if the 95% confidence lower bound for AI is larger than p_{A_0}. Therefore, we may claim interchangeability if both switching and alternating are concluded.

11.4.4 Remarks

The above biosimilarity index (totality biosimilarity index) for the assessment of biosimilarity and the switching index and/or alternating index for the assessment of interchangeability are developed based on the reproducibility probability. Hence, they are probability-based indices. In practice, we may consider moment-based indices for the assessment of biosimilarity and interchangeability. For example, we may consider

$$\hat{z}_d = \frac{\hat{\mu}_T - \hat{\mu}_R}{\hat{\sigma}_d},$$

a standardized score for measuring the distance between the test (T) and reference (R) products. In this case, the biosimilarity index can be defined as $\text{BI} = \hat{z}_d$ or $\text{BI} = \Phi(\hat{z}_d)$.

11.5 CONCLUDING REMARKS

With small-molecule drug products, bioequivalence generally reflects therapeutic equivalence. Drug prescribability, switching, and alternating are generally considered reasonable. With biological products, however, variations are often higher (other than pharmacokinetic factors may be sensitive to small changes in conditions). Thus, often only parallel-group design rather than crossover kinetic studies can be performed. It should be noted that with follow-on biologics, biosimilarity does *not* reflect therapeutic comparability. Therefore, switching and alternating should be pursued with extreme caution.

The concept of drug interchangeability in terms of prescribability and switchability for small-molecule drug products is similar but different from that for large-molecule biological products as defined in the BPCI Act. Thus, the usual methods for addressing drug interchangeability through the assessment of population/

individual bioequivalence cannot be directly applied for the assessment of drug interchangeability for biosimilar products. For biosimilar products, the assessment of drug interchangeability in terms of the concepts of switching and alternating involves (1) the assessment of biosimilarity and (2) the evaluation of the relative risk of switching and alternating. It should be noted that there is a clear distinction between the assessment of biosimilarity and the evaluation of drug interchangeability. In other words, the demonstration of biosimilarity does not imply drug interchangeability.

Based on the totality biosimilarity index for the assessment of biosimilarity, the switching index and alternating index for addressing drug interchangeability of biosimilar products can be obtained under an appropriate switching design and alternating design, respectively. The proposed switching and alternating indices have the advantages that (1) they can be applied regardless of the criteria for biosimilarity and study design used; (2) the assessment is made based on relative difference with the reference product [i.e., the relative difference between (T vs. R) and (R vs. R)]; (3) it can address the commonly asked question that "how similar is considered highly similar?" "the degree of similarity," and "interchangeability in terms of switching and alternating" in terms of the degree of *reproducibility*; and most importantly (4) the proposed method is in compliance with current regulatory thinking (i.e., totality of the evidence, relative risk of switching and alternating for interchangeability). It should be noted, however, that the proposed totality biosimilarity index and/or switching and alternating indices depend on the selection of weights for each domain or functional area for achieving the totality of the evidence for assessment of biosimilarity and/or interchangeability. The performances of the proposed totality biosimilarity index, switching index, and alternating index are currently being studied via clinical trial simulations by Zhang et al. (2013).

In practice, it is a concern regarding how many subjects are needed for providing totality of the evidence across different functional areas such as PK/PD, clinical efficacy/safety, manufacturing process, and addressing relative risks of switching and alternating for interchangeability. Based on the proposed totality biosimilarity index (for the assessment of biosimilarity) and switching and alternating indices (for the assessment of interchangeability), the sample size required for achieving certain statistical inference (assurance) can be obtained following the procedure described in Chow et al. (2008).

As indicated earlier, the broader sense of the concepts of switching and alternating could involve a number of biosimilar products, for example, T_i, $i = 1, \ldots, K$, which have been shown to be biosimilar to the same innovative (reference) drug product. Under the broader sense of interchangeability, it is almost impossible for a sponsor to claim or demonstrate *interchangeability* according to the definition as given in the BPCI Act. Alternatively, it is suggested that a meta-analysis that combines all of the data given in the submissions be conducted (by the regulatory agency) to evaluate the relative risks of the switching and alternating of drug interchangeability. In practice, meta-analysis can be conducted for safety monitoring of approved biosimilar products. This is extremely important, especially when there are a number of biosimilar products in the marketplace.

REFERENCES

Arnold B, Groeneveld R. (1979) Bound on expectations of linear systematic statistics based on dependent samples. *Annals of Statistics* **7**, 220–223.

Benet L. (2006) Why highly variable drugs are safer. Meeting of FDA Committee for Pharmaceutical Science, October 6, 2006. http://www.fda.gov/ohrms/dockets/ac/06/slides/2006-4241s2_2.htm.

Casella G, Berger R. (2002) *Statistical Inference*. 2nd edition. Duxbury, Pacific Grove, CA.

Chen ML, Patnaik R, Hauck WW, et al. (2000) An individual bioequivalence criterion: regulatory considerations. *Statistics in Medicine* **19**, 2821–2842.

Chiu ST, Chen C, Chow SC, Chi E. (2013) Assessing biosimilarity using the method of generalized pivotal quantities. *Generics and Biosimilars Initiative Journal* **3(2)**, 130–135.

Chiu ST, Tsai PY, Liu JP. (2010) Statistical evaluation of non-profile analyses for the *in vitro* bioequivalence. *Journal of Chemometrics* **24**, 617–625.

Chow SC. (1999) Individual bioequivalence: a review of FDA draft guidance. *Drug Information Journal* **33**, 435–444.

Chow SC. (2011) Quantitative evaluation of bioequivalence/biosimilarity. *Journal of Bioequivalence and Bioavailability* **Suppl. 1–002**, 1–8.

Chow SC. (2013) *Design and Analysis of Biosimilar Studies*. Chapman and Hall/CRC Press, Taylor & Francis, New York.

Chow SC, Chang M. (2011) *Adaptive Design Methods in Clinical Trials*. 2nd edition, Chapman and Hall/CRC Press, Taylor & Francis, New York.

Chow SC, Endrenyi L, Lachenbruch PA, et al. (2011) Scientific factors for assessing biosimilarity and drug interchangeability of follow-on biologics. *Biosimilars* **1**, 13–26.

Chow SC, Liu JP. (2008) *Design and Analysis of Bioavailability and Bioequivalence Studies*. 3rd edition, Chapman and Hall/CRC Press, Taylor & Francis, New York.

Chow SC, Shao J, Ho HC. (2000) Statistical analysis for placebo-challenging design in clinical trials. *Statistics in Medicine* **19**, 1029–1037.

Chow SC, Shao J. (2002) A note on statistical methods for assessing therapeutic equivalence. *Controlled Clinical Trials* **23**, 515–520.

Chow SC, Shao J, Wang H. (2002) Individual bioequivalence testing under 2 × 3 crossover designs. *Statistics in Medicine* **21**, 629–648.

Chow SC, Shao J, Wang H. (2008) *Sample Size Calculation in Clinical Research*. 2nd edition, Taylor & Francis, New York.

Chow SC, Xu H, Endrenyi L, Song FY. (2015) A new scaled criterion for drug interchangeability. *Chinese Journal of Pharmaceutical Analysis* **35(5)**, 844–848.

Chow SC, Yang LY, Starr A, Chiu ST. (2013) Statistical methods for assessing interchangeability of biosimilars. *Statistics in Medicine* **32**, 442–448.

David H. (1981) *Order Statistics*. Wiley, New York.

David H, Nagaraja H. (2003) *Order Statistics*. John Wiley & Sons, New York.

Davit BM, Chen ML, Conner DP, et al. (2012) Implementation of a reference-scaled average bioequivalence approach for highly variable generic drug products by the US Food and Drug administration. *AAPS Journal* **14**, 915–924.

Davit BM, Conner D. (2010) Reference-scaled average bioequivalence approach. In: Kanfer I, Shargel L, eds., *Generic Drug Product Development: International Regulatory Requirements for Bioequivalence*, 271–272. Informa Healthcare, New York.

Davit BM, Conner DP, Fabian-Fritsch B, et al. (2008) Highly variable drugs: observations from bioequivalence data submitted to the FDA for new generic drug applications. *AAPS Journal* **10**, 148–156.

Endrenyi L, Tothfalusi L. (2009) Regulatory conditions for the determination of bioequivalence of highly variable drugs. *Journal of Pharmacy and Pharmaceutical Science* **12**, 138–149.

FDA. (2001) Guidance for industry: statistical approaches to establishing bioequivalence. Center for Drug Evaluation and Research, U.S. Food and Drug Administration, Rockville, MD.

FDA. (2003a) Guidance on bioavailability and bioequivalence studies for orally administered drug products: general considerations. Center for Drug Evaluation and Research, U.S. Food and Drug Administration, Rockville, MD.

FDA. (2003b) Guidance on bioavailability and bioequivalence studies for nasal aerosols and nasal sprays for local action. Center for Drug Evaluation and Research, U.S. Food and Drug Administration, Rockville, MD.

Gumbel EJ. (1954) The maxima of the mean largest value and of the range. *Annals of Mathematical Statistics* **25**, 76–84.

Haidar SH, Davit B, Chen ML, et al. (2008) Bioequivalence approaches for highly variable drugs and drug products. *Pharmaceutical Research* **25**, 237–241.

Hartley HO, David HA. (1954) Universal bounds for mean range and extreme observation. *Annals of Mathematical Statistics* **25**, 85–99.

Hyslop T, Hsuan F, Holder DJ. (2000) A small sample confidence interval approach to assess individual bioequivalence. *Statistics in Medicine* **19**, 2885–2897.

Lee Y, Shao J, Chow SC. (2004) The modified large sample confidence intervals for linear combinations of variance components: extension, theory and application. *Journal of American Statistical Association* **99**, 467–478.

Papadatos N. (1995) Maximum variance of order statistics. *Annals of the Institute of Statistical Mathematics* **47**, 185–193.

Patnaik RN, Lesko LJ, Chen ML, Williams RL. (1997) Individual bioequivalence: new concepts in the statistical assessment of bioequivalence metrics. *Clinical Pharmacokinetics* **33**, 1–6.

Shao J, Chow SC. (2002) Reproducibility probability in clinical trials. *Statistics in Medicine* **21**, 1727–1742.

Tothfalusi L, Endrenyi L. (2003) Limits for the scaled average bioequivalence of highly variable drugs and drug products. *Pharmaceutical Research* **20**, 382–389.

Tothfalusi L, Endrenyi L. (2011) Sample sizes for designing bioequivalence studies for highly variable drugs. *Journal of Pharmacy and Pharmaceutical Science* **15**, 73–84.

Tothfalusi L, Endrenyi L, Garcia Areta A. (2009) Evaluation of bioequivalence for highly-variable drugs with scaled average bioequivalence. *Clinical Pharmacokinetics* **48**, 725–743.

Zhang A, Tzeng JY, Chow SC. (2013) Statistical considerations in biosimilar assessment using biosimilarity index. *Journal of Bioavailability & Bioequivalence* **5**, 209–214.

12 The Role of the Immunogenicity Evaluation for Biosimilars

Paul Chamberlain
NDA Advisory Services Ltd.

CONTENTS

12.1 INTRODUCTION

The role of the immunogenicity evaluation of biosimilar candidates, in direct comparison with the reference product, is to provide assurance of absence of clinically relevant differences in undesirable immune responses to the different versions of a given biological medicinal product.

The evaluation of the potential impact of immunogenicity on the clinical benefit-risk of a biosimilar candidate, relative to its reference product, is based on principles common to other aspects of the biosimilarity exercise, namely:

- Immunogenicity is evaluated in a direct, side-by-side comparison using the most sensitive conditions to detect a difference that could have a clinical impact;
- Results are interpreted in relation to differences detected at the product quality level;
- Findings in one therapeutic setting can be extrapolated to other indications if adequately justified; although caution is applied where there is uncertainty associated with the nature of potential risks for the reference product, and where immunogenicity could have a higher impact in populations not evaluated in preauthorization clinical studies;
- It is not necessary to define benefit-risk per se; rather, the emphasis is on applying the most sensitive test to detect differences;
- The product quality dossier provides the primary evidence of biosimilarity.

As will be explained in this chapter, differences in the immunogenic potential of biosimilar candidates are directly related to measurable product quality attributes: accordingly, risk of higher immunogenicity can be mitigated by effective control at the manufacturing, analytical, formulation and storage levels. The biosimilar candidate could have a lower level of immunogenicity than the reference product if this translated into an improved safety profile, and a significant and clinically relevant increase in efficacy can be excluded.

Ideally, the manufacturing process would be locked prior to performing any comparative clinical studies, and the drug product formulation–primary container combination to be marketed would be used for the clinical immunogenicity evaluation. Analytical data from comparative stability studies (real-time, accelerated degradation, and stress conditions) are important to identify the potential for differences that could appear in the drug product, for example, formation of aggregates and subvisible particles, following quality control testing and batch release.

The most challenging practical aspect for the immunogenicity evaluation is the development and validation of assays to detect antidrug antibodies (ADAs) to the product versions being compared, since there are many potentially confounding factors for data interpretation. While the principles of the bioanalytical strategy for evaluation of ADA formation in the biosimilarity exercise share many common features with those applied to the development of novel biopharmaceutical entities, there is an additional need to demonstrate equivalent specificity and sensitivity to detect ADAs induced by each of the product versions. Then because clinically relevant drug levels can interfere with detection of ADAs, it may be necessary to include a sample pretreatment step to optimize sensitivity to detect potential differences in ADA response across treatment groups.

But how should the comparative clinical study be designed relative to the dynamics of the immune response? What is an acceptable difference? Why is it reasonable to extrapolate findings across different indications, where patients to be treated may be more or less immune responsive? And how reliable are the results from a randomized controlled clinical trial for identifying immunogenicity-related risks that might manifest when the product is used to treat patients in the real world?

The primary aims of this chapter are to share experience gained in the evaluation of comparative immunogenicity of biosimilar and reference products to meet EU and US regulatory standards, to suggest strategies for overcoming some of the challenges, and to reflect on how unresolved issues might be addressed.

12.2 DEFINITION OF IMMUNOGENICITY

The term *immunogenicity*, as used in this chapter, refers to undesirable immune responses associated with the administration of a medicinal product. This may encompass one or more mechanisms, including:

- Binding of the medicinal product to preexisting antibodies that recognize epitopes associated with defined molecular features of the drug product, including primary sequence, posttranslational modifications, or process-derived impurities or excipients;
- Induction of treatment-emergent antibodies reactive with the drug product or with endogenous counterparts;
- Stimulation of innate immune effector cells, for example, via engagement by the drug product of Fc receptors or pattern recognition receptors, with consequent release of pro-inflammatory cytokines and/or upregulation of adaptive immune responses.

Immune responses induced by the medicinal product may have no clinical impact or may be associated with changes to the pharmacokinetic or pharmacodynamic properties of the product and/or induction of treatment-emergent hypersensitivity reactions. In cases where the drug product shares structural and functional properties with endogenous factors, ADA might also have potential to cross-react with, and interfere with, the function of the endogenous factor.

Evaluation of relative immunogenicity of biosimilar candidates should be designed to take into account the pertinent underlying causes and consequences of immune responses induced by the particular product, in a manner that is relevant for the intended clinical use. In practice, measurement of drug-reactive antibodies is used as the primary index of induction of a host immune response in the recipient; it is this index that is most commonly referred to when comparing the relative immunogenicity of biosimilar and reference products. However, it is important to recognize that this is an indirect *in vitro* measure of a complex *in vivo* response that may be more or less biased by the assay conditions used—as well as by a dynamic relationship between immune stimulatory and tolerance mechanisms that depend on patient-related factors and conditions of use (route of administration, dose regimen, concurrent medications, etc.).

For this reason, the relative immunogenicity of biosimilar and reference versions is assessed by correlation of the surrogate bioanalytical indices (ADA incidence and titer) with clinical parameters; the clinical endpoints should reflect the PK, PD, therapeutic efficacy, and safety dimensions in the populations to be treated. The comparison of the bioanalytical signals should also incorporate analysis of the relative magnitude and chronology of the immune response induced by the product versions being compared, requiring sequential sampling to enable correlation of the magnitude of the evolving immune response with relevant clinical parameters.

12.3 EU AND US REGULATORY CONTEXT COMPARED

The main features of EU (EMA, 2012, 2014a,b,c, 2015) and FDA guidelines (FDA, 2014a,b, 2015, 2016) for evaluation of the immunogenicity of biosimilar candidates are compared in Table 12.1.

In practice, sponsors have been able to apply a broadly similar approach in designing the immunogenicity evaluation to meet EU and FDA regulatory expectations for biosimilarity.

The main considerations for the regulator are to ensure that:

- Immunogenicity-related risks have been adequately controlled from the product quality perspective, taking into account the risks identified for the reference products and remaining uncertainties (including longer-term impact on overall benefit-risk);
- Suitably specific and sensitive bioanalytical methods have been applied to monitor ADA formation in a directly comparative (biosimilar vs. reference product) manner;
- Sources of bioanalytical and clinical bias have been minimized to an extent that enables an objective comparison;

TABLE 12.1
EU versus US Regulatory Recommendations for Evaluation of Immunogenicity of Biosimilars

Feature	EU	US

Guidance for immunogenicity evaluation of biosimilars	• Overarching biosimilars guidelines:
	  • General principles (EMA, 2014b)
	  • Quality considerations (EMA, 2014a)
	  • Nonclinical and clinical (EMA, 2014c)
	• Guideline on immunogenicity assessment of biotechnology-derived therapeutic proteins (EMA, 2015)
	• Guideline on immunogenicity assessment of monoclonal antibodies intended for *in vivo* clinical use (EMA, 2012)
		• Scientific considerations in demonstrating biosimilarity to a reference product (FDA, 2015)
		• Draft guidance for industry: clinical pharmacology data to support a demonstration of biosimilarity to a reference product (FDA, 2014a)
		• Draft guidance for industry: assay development for immunogenicity testing of therapeutic proteins (FDA, 2016)
		• Guidance for industry: immunogenicity assessment of therapeutic protein products (FDA, 2014b)

Objective of immunogenicity evaluation in biosimilarity exercise	To identify differences in relative immunogenicity of the biosimilar versus reference product that could affect overall clinical benefit/risk; a lower immunogenicity of the biosimilar product could be acceptable if equivalent efficacy is demonstrated	To exclude a clinically meaningful difference in the immunogenicity of the biosimilar relative to its reference product; differences in immune responses between a proposed product and the reference product in the absence of observed clinical sequelae may be of concern and may warrant further evaluation (e.g., extended period of follow-up evaluation)
Types of clinical studies	Relative immunogenicity should be monitored in all clinical studies in the biosimilar program, including comparative PK/PD and therapeutic studies. In the case that no therapeutic equivalence study is performed, a separate study of comparative immunogenicity may be required prior to authorization	
Clinical measures of immunogenicity	Direct (head-to-head) comparison of incidence, magnitude, and dynamics of antidrug antibody formation (neutralizing and nonneutralizing) correlated with clinically relevant parameters, including PK/PD, efficacy and incidence, severity and timing of immune-mediated adverse events	
Extent of similarity expected	Detected difference in ADA formation does not have a measurable impact on relevant clinical outcomes. (Biosimilar version could be less immunogenic if equivalent efficacy but more favorable safety profile were demonstrated.)	
Extrapolation	Extrapolation of immunogenicity findings across therapeutic indications should be adequately justified; additional clinical evaluation of higher risk therapeutic uses may be required, for example, where the consequences could be severe	Evaluate immunogenicity using a study population and treatment regimen that are adequately sensitive for predicting a difference across the conditions of use; often, this will be the population and regimen for the reference product for which development of immune responses *with adverse outcomes* is most likely to occur

(Continued)

TABLE 12.1 (*Continued*)
EU versus US Regulatory Recommendations for Evaluation of Immunogenicity of Biosimilars

<table>
<tr><th>Feature</th><th>EU</th><th>US</th></tr>
<tr><td>Flexibility in choice of bioanalytical methodology</td><td colspan="2">Sponsors have flexibility to justify choice of technical platform
A single or dual ADA screening/confirmatory assay format can be used for testing clinical samples from subjects treated with either biosimilar or reference products; this may be performed using only the biosimilar as labeled antigen in the single-assay approach; or two separate assays may be configured, one using the biosimilar as labeled antigen and the other using the reference product as labeled antigen. Samples should be tested in an operator-blinded manner. Signal specificity should be confirmed in parallel using both the biosimilar and reference versions as competing antigen in ADA assay
A cell-based assay is preferred for measurement of the neutralizing capacity of ADAs in the case where the product is an agonist and/or binds to a tissue target; a CLBA assay can be used if the product is an antagonist of a soluble target</td></tr>
<tr><td>Role of in vivo nonclinical ADA data</td><td>Not considered to be instructive; collect and store samples in case retrospective testing is requested</td><td>Measurement of antitherapeutic protein antibody responses in animals may provide useful information; differences observed may reflect potential structural or functional differences between the two products not captured by other analytical methods</td></tr>
<tr><td>EU or US reference product to be used for clinical immunogenicity comparison</td><td colspan="2">For the main therapeutic equivalence study, a single reference product (either EU or US source) may be used if comparability demonstrated at the analytical level and formulation/presentation is equivalent
For comparative PK/PD studies, use of both the EU- and US-approved reference products has usually been required; but a single reference product might be allowed if appropriately justified by the sponsor</td></tr>
<tr><td>Duration and extent of ADA monitoring in preauthorization clinical studies</td><td>Depends on perceived risk for product, as well as timing of onset and persistence of ADA response
Six to twelve months for chronic administration products approved to date</td><td>Highly dependent on product; need to agree on study designs with agency
Six-month evaluation for chronic administration products may be sufficient; 12-month posttreatment follow-up of responders to chronic use products</td></tr>
<tr><td>Switching studies</td><td>Not requested</td><td>Rerandomization and transition from reference product to biosimilar has been requested for different biosimilar candidates</td></tr>
<tr><td>Postauthorization risk management</td><td colspan="2">Heightened monitoring for immunogenicity-related risks via different provisions</td></tr>
</table>

ADA, antidrug antibody.

- There is a sufficient weight of evidence to make an informed decision that there is no increased risk of clinically impactful immunogenicity for the biosimilar candidate; or, in the case of lower immunogenicity of the biosimilar compared to the reference product, that the different versions can be expected to have the same efficacy in all licensed indications;
- The biosimilar product formulation–primary container combination that is to be placed on the market has been appropriately tested, that is, including clinical evaluation in direct comparison to the reference product obtained from a suitable geographic source;
- Postmarketing surveillance and, if warranted, postauthorization studies, are performed to monitor risks that were not fully evaluated at the time of marketing authorization;
- Physicians and patients can consider the biosimilar version as a therapeutic alternative to the reference product, based on a robust scientific evaluation by concerned regulatory agencies.

While the different regulatory agencies may apply substantially common approaches, individual agencies might look at the data in different ways and arrive at different conclusions about the benefit versus risk proposition. Regulatory guidance documents are intended to be enabling, rather than restrictive, and allow for considerable flexibility if the sponsor provides adequate justification. Regulatory guidance is not legally binding—but sponsors should seek endorsement, via Scientific Advice procedures, for adopting their preferred approach.

An interactive approach is particularly important for addressing questions such as:

- *Could clinical immunogenicity data from studies performed with reference product sourced from other markets or regions be acceptable as primary evidence, or only useful in a supportive role?*
- *How should alternative presentations be evaluated, for example, prefilled syringe versus auto-injector, for potential differences in immunogenicity-related risks?*

There is no standard response to these questions because the nature of the risk, as well as the level of uncertainty about the influence of interacting variables, can be quite different depending on the product and the population(s) to be treated.

12.4 IDENTIFYING RISKS BASED ON PROFILE OF REFERENCE MEDICINAL PRODUCT

The starting point for the immunogenicity assessment of a biosimilar candidate is the existing knowledge of the risks identified for the reference product. The adverse outcomes of undesirable immunogenicity can vary considerably between different products, as summarized in Table 12.2.

This prior knowledge informs the design of the immunogenicity evaluation by aligning known structure–immunogenicity relationships with the extent of analytical

TABLE 12.2
Identified Immunogenicity-Related Risks for Reference Products

Product	Clinical Impact of Immunogenicity/ADA	Published References
Epoetin alfa	Cross-reactive neutralizing ADAs causing amPRCA	Casadevall et al. (1996)
Cetuximab	Severe allergic reactions in presensitized subjects	Chung et al. (2008)
Infliximab	Immune complex-related hypersensitivity and loss of efficacy	Baert et al. (2003) and Bendtzen et al. (2006)
Adalimumab	Loss of efficacy and increased incidence of injection site reactions	Bartelds et al. (2007) and Murdace et al. (2013)
Rituximab	Loss of efficacy in patients with severe pemphigus and rare cases of hypersensitivity reactions	Schmidt et al. (2009) and Ataca et al. (2015)
Somatropin	Possible reduction in PK/PD/efficacy in very rare cases	Pfizer (2014)
Insulin glargine	Possible reduction in PK/PD/efficacy in very rare cases	Fineberg et al. (2007)
Follitropin-alfa	Negative impact not identified	Loumaye et al. (1998)
Bevacizumab	Negative impact not identified	EPAR for Avastin
Trastuzumab	Negative impact not identified	EPAR for Herceptin
Omalizumab	Negative impact not identified	Somerville et al. (2014)
Filgrastim	Negative impact not identified	Amgen (2015b)
Pegfilgrastim	Negative impact not identified	Amgen (2015a)

ADA, antidrug antibody; amPRCA, antibody-mediated pure red cell aplasia.

characterization, rigor of process and product quality control, and choice of formulation and primary container for the biosimilar candidate.

It should be evident that the clinical impact associated with undesirable immunogenicity of the originator products varies across a wide scale; for 6 out of 13 reference product versions listed in Table 12.2, there was no detectable clinical impact associated with undesirable immunogenicity. In addition, the rate of occurrence of these risks can be extremely low for the most severe risks, for example, antibody-mediated pure red cell aplasia (amPRCA) associated with originator versions of erythroid stimulating agents (Casadevall, 2005). This implies that different levels of risk evaluation and mitigation are likely to be required for different products, according to the balance between severity, probability, and detectability of adverse outcomes.

Risks related to undesirable immunogenicity should also be viewed in relation to the severity of dose-limiting toxicity. For example, rituximab can induce host antidrug antibodies, but these tend to have rather less adverse clinical impact than the release of pro-inflammatory cytokine induced by lysis of the target cells

(Mok, 2014). Moreover, the mechanism of action of rituximab has a self-limiting effect on ADA formation as a consequence of the intended B-lymphocyte-depleting effect. This may explain why ADAs were detected in only a relatively low proportion of rheumatoid arthritis subjects receiving long-term treatment with rituximab, without a clear correlation with ADA-related adverse effects (van Vollenhoven et al., 2010).

12.5 POTENTIAL INCREMENTAL RISKS ASSOCIATED WITH INDEPENDENT MANUFACTURING PROCESS

An essential requirement for biosimilarity is that the expression construct for a proposed product will encode the same primary amino acid sequence as its reference product (EMA, 2014a; FDA, 2015). This means that the primary structure of the major proportion of the expressed biosimilar product should have the same intrinsic immunogenic potential as the reference product in terms of the peptide sequences that can be presented by antigen-presenting cells to T-cell receptors in an MHC (major histocompatibility complex) class II-restricted manner.

Accordingly, by definition, incremental immunogenicity of a biosimilar candidate would then be a consequence of one or more of the following factors:

- Instability of the active substance in the particular drug product formulation–primary container combination that resulted in product-related variants with higher intrinsic potential to be presented by MHC class II to T-cell receptors;
- Conformational differences in the active substance, for example, oligomers, aggregates, or subvisible particles that could alter recognition by B-lymphocytes and/or mass-balance of antigen uptake and processing by antigen-presenting cells;
- Differences in posttranslational modification, particularly the qualitative and quantitative content of nonhuman glycans that could alter recognition by preexisting antibodies in recipients sensitized by prior exposure;
- Altered pattern of positional isomerization of polyethylene glycol (PEG)-conjugated protein, if this resulted in altered antigen uptake, processing, or presentation;
- Different qualitative or quantitative composition of host cell-derived, or other process-derived, impurities that can modify the adaptive immune response to the active substance via stimulation of innate immune effector cells, or that may directly induce hypersensitivity reactions.

Although patient-related variables—genotypic polymorphisms, immune tolerance, immune competence, autoimmune status, and prior sensitization to nonhuman glycans or to PEG—would be expected to impact the extent of the manifestations of intrinsic immunogenic potential of the therapeutic protein, in principle these variables would apply equally to the biosimilar and reference versions if the clinical populations were balanced with respect to these variables.

12.6 CONTROL OF CRITICAL PRODUCT QUALITY ATTRIBUTES FOR IMMUNOGENICITY-RELATED RISKS

Immunogenicity-related risks for biosimilar candidates need to be effectively controlled at the manufacturing and product quality testing levels. With the advance in analytical technologies, the risk of incremental immunogenicity can be evaluated and, to a large extent, avoided by analytical characterization, batch release testing, and stability testing. Clinical evaluation with appropriate bioanalytical testing then provides confirmatory evidence of comparative immunogenicity.

Thus, although the starting materials, cell substrate, manufacturing process, product formulation, and primary container might all be different from those used to produce the reference product, the impact of these variables on identified risk factors for undesirable immunogenicity of the molecule can be detected if an appropriate combination of methods are applied for analytical characterization and stability testing.

The evaluation and mitigation measures for potential immunogenicity-related risk factors associated with the product quality of biosimilar candidates are summarized in Table 12.3.

TABLE 12.3
Product Quality-Related Factors for Differential Immunogenicity of Biosimilars

Risk Factor	Risk Evaluation	Risk Mitigation
Instability of the active substance	Formulation development studies	Adequate justification for any differences in product formulation
	Compatibility of drug product with primary container	Choice of suitable primary container
	Definition of stability-indicating quality attributes	Overlapping analytical profiles for relevant quality attributes of biosimilar versus reference product
	Comparative stability of commercial drug product formulation–primary container combination under forced degradation conditions linked to analytical characterization of relevant quality attributes	No increase in clinically impactful ADA for biosimilar drug product to be marketed
	Extended analytical characterization to include methods sensitive to detect differences in primary and higher-order structure and thermal stability	Clear instructions on handling and storage conditions, allied to adequate supervision of supply chain
	Comparative clinical studies to measure ADA response using commercial drug product formulation–primary container combination	

(Continued)

TABLE 12.3 (*Continued*)
Product Quality-Related Factors for Differential Immunogenicity of Biosimilars

Risk Factor	Risk Evaluation	Risk Mitigation
Conformational differences in the active substance	Peptide mapping to confirm integrity of expected intramolecular disulfide bonds	Comparable higher-order structure
	Extended analytical characterization to include methods sensitive to detect differences in higher-order structure	Overlapping conformational epitope-sensitive, antibody binding profiles for biosimilar versus reference products
	In vitro binding of antibodies reactive with conformational epitopes	No marked increase in tendency for aggregate or subvisible particle formation during storage of the drug product in the formulation–primary container combination to be commercialized
	Orthogonal methods to define profile of aggregates and subvisible particles in submicron to low-micron range: applied to batch release and stability testing	No increase in clinically impactful ADA for biosimilar drug product to be marketed
	Demonstration of antigenic equivalence for binding to the positive control antibody used for validation of the ADA assays	Routine batch release testing of drug product includes methods of suitable sensitivity to detect increased levels of product-related aggregates
	Comparative clinical studies to measure ADA response using commercial drug product formulation–primary container combination	Clear instructions on handling and storage conditions, allied to adequate supervision of supply chain
Differences in posttranslational modification	Understanding of dependence of posttranslational heterogeneity on manufacturing process conditions	Choice of host cell line that does not increase risk of undesirable immunogenicity due to nonhuman posttranslational modification
	Comparative, high-resolution, analysis of posttranslational modification profile, including qualitative and quantitative definition of content of nonhuman glycans (e.g., Neu5Gc, Gal-α-1, 3-Gal, hyper-mannosylation)	Detected levels of posttranslational modification do not increase clinically impactful immunogenicity of biosimilar, as measured in a comparative clinical study versus reference product
	Comparative clinical studies to measure ADA response using commercial drug product formulation–primary container combination	Manufacturing process control to ensure adequate consistency of quality profile
	Cross-reactivity testing of ADA signals detected in pretreatment samples from clinical studies	

(*Continued*)

TABLE 12.3 (*Continued*)
Product Quality-Related Factors for Differential Immunogenicity of Biosimilars

Risk Factor	Risk Evaluation	Risk Mitigation
Altered pattern of positional isomerization of polyethylene glycol-conjugated protein	Comparative peptide mapping to define quantitative profile of positional isomers	Analytical characterization demonstrates highly similar structural profile, including distribution of positional isomers
	Process variables influencing site-specific conjugation	Adequate manufacturing process control to ensure consistency of quality profile
	Comparative clinical studies to measure ADA response using commercial drug product formulation–primary container combination	No increase in clinically impactful ADA to any component of the drug product
	ADA monitoring includes suitable method to detect polyethylene glycol-reactive antibodies in pre- and posttreatment samples	
Differences in qualitative or quantitative composition of host cell-derived, or other process-derived, impurities	Two-dimensional gel electrophoresis (charge + size separation) with silver staining and Western blotting detection to identify host cell-derived protein profile	Analytical assays validated for suitable specificity and sensitivity to detect process-related impurities
	Application of process-specific host cell protein assay; assays for other impurities as relevant	Downstream process design enables effective removal of process-related impurities
	Demonstration of effective clearance of process-related impurities by downstream process	Suitable drug substance/drug product release limits for process-related impurities
	Comparative clinical studies to measure ADA response using commercial drug product formulation–primary container combination	No increase in clinically impactful ADA to any component of the drug product
	Cross-reactivity testing of ADA signals detected in pretreatment samples from clinical studies	

ADA, antidrug antibody; Neu5Gc, *N*-glycolyl neuraminic acid; Gal, galactose.

12.7 IMPLICATIONS OF PRODUCT QUALITY VARIABLES FOR REGULATORY ASSESSMENT

Any detectable differences in variables that represent identified risk factors in Table 12.3 should be justified by demonstrating an absence of a difference in *clinically impactful* immunogenicity of the biosimilar candidate relative to the reference product.

This is performed in a comparative clinical study in which the relationship of the ADA response to relevant clinical parameters is evaluated over a suitable treatment period.

In justified cases, that is, where no impact on clinical properties is detected when tested under suitably sensitive conditions, it could be acceptable for the biosimilar to have higher or lower amounts of particular product-related variants. For example, the relative level of C-terminal clipped-human IgG is not expected to influence the immunogenicity of therapeutic monoclonal antibodies. In general, the low levels of nonhuman glycans, including *N*-glycolyl neuraminic acid (Neu5Gc) added by some mammalian cell substrates, has not been associated with enhanced immunogenicity of therapeutic proteins—even though a relatively high proportion of human subjects possess preexisting antibodies reactive with Neu5Gc (Amon et al., 2014).

In the case of cetuximab, however, a relatively high level of Gal-α-1,3-Gal linked *N*-glycans, as well as of Neu5Gc, is associated with each of the Fab arms of the molecule (Qian et al., 2007). The quantity and spatial disposition of the Gal-α-1,3-Gal linked *N*-glycans in cetuximab have been attributed as the causal factor of severe IgE-mediated hypersensitivity reactions in subjects who have been previously sensitized by environmental or dietary exposure (Chung et al., 2008). As a consequence of the immunogenicity-related risk identified for the reference product, it might be justifiable for a biosimilar cetuximab candidate to be manufactured in a different cell line from that used for the reference product (e.g., CHO instead of Sp2/0), on the basis that equivalent efficacy could be achieved with an improved safety profile (i.e., lower incidence of severe immune-mediated adverse events) and that adequate similarity for other quality parameters was demonstrated.

The recognition that product aggregates were a plausible causal factor for increased incidence of amPRCA in renal anemia patients treated with subcutaneous administration of an originator version of epoetin alfa (Rossert et al., 2004) was instrumental in increasing rigor of analysis and control of levels of oligomers, aggregates, and subvisible particles. There is accumulating evidence (Joubert et al., 2012; Rombach-Riegraf et al., 2014) to support a role for aggregates in increasing the mass balance of antigen uptake and processing by antigen-presenting cells, as well as for directly stimulating B-lymphocytes to bypass B-cell tolerance. It appears that even relatively low levels of subvisible particles are able to provide co-stimulatory signals to enhance antigen-specific T-cell responsiveness, and that such additional stimulation may contribute to enhancement of immunogenicity, depending on the intrinsic immunogenic potential of the molecule (Ahmadi et al., 2015).

Choice of primary container can represent an influential variable because different groups have demonstrated an association between residual tungsten particles in glass prefilled syringes and an increased tendency for protein aggregate formation (Bee et al., 2009; Jiang et al., 2009; Liu et al., 2010). Accordingly, risk mitigation for biosimilar products should include careful selection of the primary container and demonstration of comparative (vs. reference product) stability of the drug product–primary container formulation to be commercialized.

Residual host cell-derived protein levels cannot be directly compared for the biosimilar versus reference product because quantitation depends on the availability of process-specific assay reagents. Nevertheless, the product quality dossier for the biosimilar product will need to demonstrate effective clearance of host cell-derived proteins, and no increase in clinically impactful immunogenicity that could be ascribable to an unacceptably high level of residual host cell-derived protein.

The potential for formulation differences to influence drug product stability in a manner that might not be detected by analytical methods applied for routine drug product batch release testing increases the importance of conducting (as early as possible in the biosimilar product development program) comparative stability studies using forced degradation conditions. In the author's experience, it is possible that, even when the same qualitative and quantitative composition of excipients is used to formulate the biosimilar product compared to the reference product, there may be a measurable difference in pH range—necessitating a slightly different specified pH range for the biosimilar drug product version. Acceptability of such a difference should also be justified by comparative stability studies and exclusion of the potential for increased clinically impactful immunogenicity.

12.8 EXAMPLES OF POTENTIAL PRODUCT QUALITY ISSUES IDENTIFIED IN THE PREAUTHORIZATION PHASE

Some examples of potential immunogenicity-related concerns arising from analytical differences detected in the preauthorization phase for different biosimilar candidates are summarized in Table 12.4.

In all cases, the regulatory process was effective in identifying and mitigating a potential increase in the immunogenicity-related risk. For Binocrit, this involved additional clinical evidence prior to approval for subcutaneous use in renal anemia; the purification process for the approved version of Omnitrop was modified to

TABLE 12.4
Impact of Quality-Related Risk Factors Identified in Preauthorization Phase for Candidate Biosimilars

Detected Quality Difference	Impact
HMW variants associated tungsten residue (Binocrit) Ref: Seidl et al. (2012)	Possible association with induction of neutralizing ADA in two subjects; one confirmed case of amPRCA (CKD SC route only); change to low-tungsten prefilled syringe
E. coli HCP impurity (early version of somatropin) Ref: EPAR for Omnitrop	Treatment-emergent antibodies to HCP and reported enhancement of ADA reactive with somatropin; additional purification step to remove HCP impurity
Higher level of Neu5Gc Ref: EPAR for Ovaleap	No impact of preexisting Neu5Gc-reactive antibodies on PK
Different product-related impurity profile Ref: EPAR for Alpheon	No apparent difference in immunogenicity; not authorized because analytical comparability not adequately demonstrated

HMW, high molecular weight; amPRCA, antibody-mediated pure red cell aplasia; CKD, chronic kidney disease; SC, subcutaneous; HCP, host cell protein; ADA, antidrug antibody; Neu5Gc, *N*-glycolyl neuraminic acid; EPAR, European Public Assessment Report.

remove the host cell-derived protein that co-purified with somatropin; clinical data for Ovaleap showed no impact of preexisting Neu5Gc-reactive antibodies on clinical parameters; and Alpheon was not approved due to uncertainty about biosimilarity, mainly in relation to quality aspects (Chamberlain, 2014).

12.9 REGULATORY REQUIREMENTS FOR ADA DETECTION IN THE BIOSIMILARITY EXERCISE

Regulatory guidance (EMA, 2012, 2015; USP, 2015; FDA, 2016) defining standards for bioanalytical assays to be applied to ADA assays for biotechnology-derived products in general are also applicable to the bioanalysis of nonclinical and clinical samples for the biosimilarity exercise. However, the need to design and validate assays to provide a reliable measure of the *relative* ADA response induced by *independently manufactured* versions of a therapeutic protein does create additional considerations for regulators and sponsors alike.

While sponsors are at liberty to select their preferred assay format, there is a regulatory need to ensure that the choice of assay format does not bias the detection of ADA induced by either of the product versions being compared, and has adequate sensitivity to detect ADA in the presence of clinically relevant drug concentrations.

Both the FDA and the Committee for Human Medicinal Products (CHMP) will accept a single- or a dual-assay approach in which (1) the ADA assay format is configured using only the biosimilar product as the labeled antigen ("single-assay approach"); or (2) separate assays are configured using the biosimilar product as the labeled antigen in one assay and the reference product as the labeled antigen in the second assay ("dual-assay approach"). In both cases, clinical samples should be tested in an operator-blinded manner, and the specificity of any detected positive samples should be confirmed by competitive inhibition of the signal by both the unlabeled biosimilar and unlabeled reference products.

In practice, sponsors have tended to prefer the single-assay option because this minimizes variability associated with using multiple assays and labeled antigens, and avoids redundant testing and the possibility of generating conflicting results for the same sample tested in different assays. Validation of two assays not only doubles the amount of experimental work but creates a second set of assay cut points and drug tolerance limits.

Given the abbreviated nature of the clinical program for biosimilars, sponsors should plan to have completed assay validation prior to commencing clinical studies—even though this is not a strict regulatory requirement. It is wise to seek endorsement via Scientific Advice, prior to initiation of clinical studies, for the proposed bioanalytical strategy, including technical details of assay format, schedule of clinical sampling and data analysis. This is particularly relevant for the choice of assay format to measure the neutralizing capacity of ADAs detected in the screening/confirmatory assays because for products that act as antagonists of soluble targets it is usually acceptable to apply a competitive ligand binding assay (CLBA) format in place of a cell-based bioassay—the latter being required in the case of a product whose primary mechanism of action depends on engagement of a ligand expressed on the cell surface.

Because it is necessary to correlate measures of drug-reactive antibodies with PK parameters, appropriately rigorous validation of the impact of ADA on the accuracy of the assay ("PK assay") used for quantitation of drug concentration should also be considered (e.g., see Sailstad et al., 2014). Although industry-driven "White Papers" can provide a useful source of technical options, it should be remembered that these have no official regulatory jurisdiction, and sponsors have flexibility to apply alternative approaches.

Finally, it is important to emphasize that the absence of standardized positive control antibody reagents precludes objective benchmarking of assay sensitivity: comparison to historical bioanalytical data is rarely possible because different assay conditions are used; and it is not feasible to define clinically impactful ADA levels prior to analyzing clinical samples with product/application-specific assays.

12.10 ADA ASSAY SPECIFICITY

The immunogenicity of therapeutic proteins is measured indirectly, using an *in vitro* assay that reflects binding of a mixed (in terms of binding specificity, amount, and avidity) antibody population in the test article with a fixed amount of antigen (typically labeled). Treatment-emergent immunogenicity is inferred by a difference in the signals for the pre- versus posttreatment samples.

Since versions of the same molecule produced by two independent manufacturing processes are to be compared, the author has found it helpful to include a demonstration of antigenic equivalence (or "bioanalytical similarity") of the respective product versions in the validation of assay specificity. Such demonstration (described below) then provides the rationale for applying a single assay format to measure clinical samples from subjects receiving either the biosimilar candidate or the reference product. This also demonstrates that chemical modification of the labeled antigen(s) used in the assay has not biased the relative binding in the solution phase of a positive control antibody to the different unlabeled versions.

A more important source of bias for detection of ADA induced by therapeutic monoclonal antibodies has been interference by residual circulating drug, which can lead to substantial underestimation of the true incidence of ADA formation. Sample pretreatment involving acid-dissociation, often in combination with affinity capture (Bourdage et al., 2007; Smith et al., 2007), has been widely used to achieve suitable assay sensitivity to detect ADA in the presence of clinically relevant levels of therapeutic monoclonal antibody.

Sponsors should also evaluate and minimize potential interference associated with nonspecific binding factors (e.g., Rheumatoid Factor, antihuman IgG Fc allelic antibodies, human antimouse antibody, as relevant) and target ligand present in samples from the different patient populations to be treated in clinical studies. Ideally, this aspect should be included in the initial assay development experiments to avoid having to redevelop and revalidate the ADA assay at a later stage. Inclusion of an excess of nonspecific human IgG in the assay buffer may be a useful strategy to overcome much of the nonspecific binding that can be observed in plasma or serum obtained from certain patient populations, for example, autoimmune subjects.

12.11 ADA ASSAY SENSITIVITY

Although the FDA has traditionally defined a target sensitivity for ADA assays to be used for clinical sample analysis, it can be quite challenging to translate this target to different products because the reported sensitivity can be heavily biased by the choice of positive control antibody reagent, as well as by many other variables. Therefore, the objective should be to optimize sensitivity to detect differences in levels of clinically impactful antibodies. Recent FDA draft guidance (FDA, 2016) provides cross-references to examples indicating that ADA levels as low as 100 ng/mL may have a clinical impact.

On the other hand, historical data to establish a "clinically impactful" ADA level in terms of a universally applicable, positive control antibody reagent tested under standardized assay conditions with the relevant antigen(s) are rarely available. As a consequence, attempting to define a clinically relevant ADA level in advance of performing the clinical study is unlikely to be feasible for most biosimilar candidates at the present time. For the same reasons, predefinition of acceptance criteria for a difference in the detectable ADA response between the biosimilar and reference products is unlikely to provide a meaningful test of a clinically relevant difference—which is the objective of the comparative immunogenicity assessment for biosimilarity.

Accordingly, sponsors will need to develop and validate assays that have optimal sensitivity to detect one or more positive control antibodies that bind to relevant (for the human population) epitopes on the active substance. Then, the suitability of the reported sensitivity may be assessed by comparing ADA signals obtained in the screening assay with *in vitro* neutralizing capacity, *in vivo* drug levels, pharmacodynamics, efficacy, and safety.

12.12 CHOICE OF POSITIVE CONTROL ANTIBODY

One or more positive control antibody reagents can be used to benchmark assay sensitivity during method development and validation, and then to serve as a quality control reagent for monitoring consistent performance of the assay during clinical sample analysis.

From the EU regulatory perspective, it is not necessary to generate in-house positive control antibody reagents using either/both the biosimilar and reference products as immunogens; a commercially sourced-reagent could be perfectly suitable. Some FDA reviewers have requested that in-house positive control antibody reagents be generated (in animals) against both the biosimilar and the reference products, and that these be used to demonstrate bioanalytical similarity of the product versions in the ADA assay. However, since the reagents will be generated in nonhuman species, differences are likely to be of equivocal relevance for the human immune response. Moreover, they will almost certainly differ in terms of amount and avidity of antibody due to interanimal variability in the immune response—rather than reflecting real structural differences between the product versions. Therefore, in the author's experience, it is preferable to use a single positive control reagent and to demonstrate that this reacts in an equivalent manner with the biosimilar and reference products (as discussed below).

For human therapeutic monoclonal antibody products, fractionation of a polyclonal antiserum raised in animals may be necessary to remove antispecies reactive antibodies that are not representative of the immune response of humans to the same product. Alternatively, animals could be immunized with the $F(ab)_2$ domain of the therapeutic monoclonal antibody, or an anti-idiotypic monoclonal antibody reagent could be used for products in which the human immune response is known to be directed primarily to the complementarity-determining regions (CDRs) (e.g., adalimumab, van Schouwenburg et al., 2013; or infliximab, van Schie et al., 2015). In the case of fusion proteins or conjugates, assay validation should include positive control antibodies reactive with the different moieties of the molecule.

12.12.1 Validation of Antigenic Equivalence ("Bioanalytical Similarity")

Although the positive control antibody reagent can only represent a surrogate index of the ADA response to be measured in the assay, the author has found it useful to demonstrate that the positive control antibody reacts in an equivalent manner with the biosimilar and reference products. Antigenic equivalence can be demonstrated by using a competitive binding format of the screening assay in which a wide (e.g., 100-fold) concentration-range of unlabeled versions of the biosimilar and reference products are added as competing antigens to inhibit the signal for binding of the positive control antibody to the labeled biosimilar antigen in the assay. Visual overlap of the inhibition curves (or, if considered necessary, a statistical test of equivalence) may then be used as evidence that there are no detectable differences—at least when using a surrogate antibody—in the antigenicity of the biosimilar and reference products. In the case that the curves do not overlap, or fail a statistical test of equivalence, biosimilarity might be questioned.

12.13 CONFIRMING SPECIFICITY OF POSITIVE SIGNALS

Since clinical samples are required to be tested in an operator-blinded manner, preferably in a single assay format that uses the biosimilar product as the antigen, it becomes relevant to confirm specificity of screened positive samples by competitive inhibition with unlabeled biosimilar and reference product tested in parallel. In any case, the screening assay cut point is set to yield a 5% false positive rate, so the confirmatory assay step is essential to correctly distinguish true from false positives. The confirmatory assay cut point should be established in a statistically rigorous manner to enable reliable classification of sample results as positive or negative. Different regulatory agencies may recommend different levels of stringency according to the risks associated with the product, such that sponsors will need to justify the approach actually applied.

Typically, a high molar excess of competing antigen is used in the confirmatory step of ADA assays, potentially undermining the value of the test to discriminate differential reactivity of the ADA with the biosimilar and reference antigens. An alternative approach using two concentrations of competing antigen—one representing submaximal inhibition of the positive control signal in the assay and the other representing a just maximal inhibitory level—might provide a more instructive comparison that human ADA responses to either product were reacting in a comparable manner to the respective antigens. Although this alternative approach does not represent a regulatory expectation, it could provide supporting evidence that the measurement of antigenicity, used to infer immunogenicity, is not biased by the choice of antigen for the ADA assay; and that the individual human polyclonal immune responses to the treatment appear to react in solution phase in an equivalent manner with the respective product versions.

For some biosimilar products (e.g., insulin glargine), it is relevant to test the cross-reactivity of confirmed positive signals with related products (other insulins to which pretreated subjects may have been exposed to) and the endogenous molecule (native human insulin). Indeed, this was part of the bioanalytical evidence presented to support the EU approval of Abasaglar (EPAR for Abasaglar).

12.14 ADA TESTING STRATEGY

The multitier testing scheme that is recommended for bioanalysis of any biological medicinal product is equally applicable to the development of biosimilars. This multitier assay strategy is illustrated in Figure 12.1.

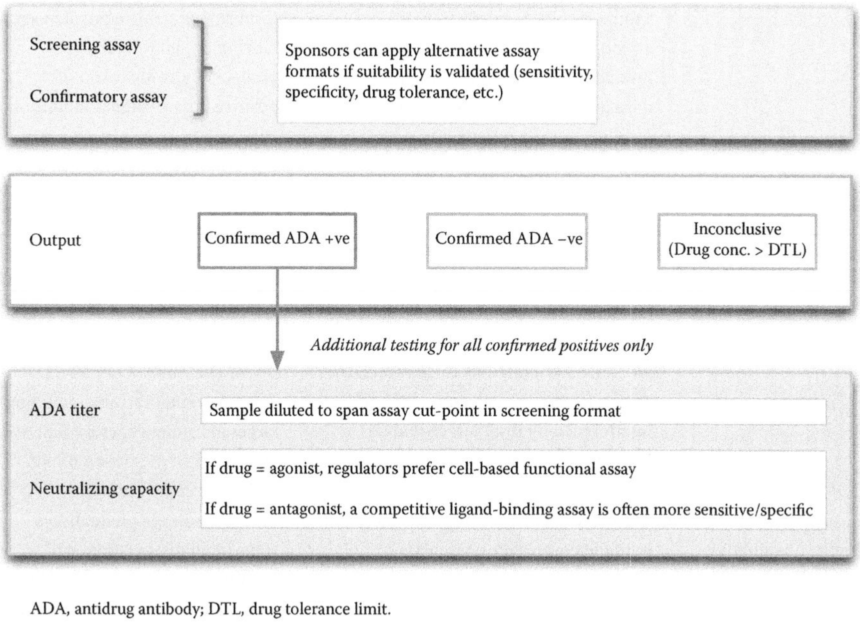

FIGURE 12.1 Multitier test scheme for detection of ADA. All clinical samples are tested in the screening assay to detect the presence of ADA; samples with signals above the cut-point are then tested to confirm reactivity with solution-phase competing antigen; only the confirmed positive samples are further tested to estimate ADA titer (reciprocal of minimum dilution yielding a signal above the assay cut-point) and capacity to neutralize a relevant biological function of the drug. In the case that the drug level in the sample exceeds the validated drug tolerance limit, samples are classified as "inconclusive" for ADA.

12.15 ADA ASSAY VALIDATION

Particularly relevant features of the method validation exercise for ADA assays to be applied to the biosimilarity exercise, based on the author's experience, are summarized in Table 12.5. These would be addressed in addition to other parameters that are relevant for the validation of suitability of ADA assays for biotechnology-derived products in general. Ultimately, there are no legally binding requirements

TABLE 12.5
Priorities for Validation of ADA Assay to Be Used in Biosimilarity Exercise

Assay Characteristic	Objective	Strategy
Specificity (qualitative nature of antigen binding)	• Demonstrate specificity of the assay to detect antibodies reactive with all potentially immunogenic components (including product-related variants and process-derived impurities) of the drug product	• Confirm binding of positive control antibodies and lack of binding of negative control matrix • Test competitive inhibition of binding with all potentially relevant competing antigens
	• Minimize bias for detection of antibodies reactive to the particular product version administered to individual subjects, be it the biosimilar candidate or the reference product	• Confirm antigenic equivalence of biosimilar and reference products by comparison of positive control signal inhibition curves • Confirm lack of impact of antigen labeling on reactivity with positive controls
Matrix interference	• Generate reliable comparative (quasi-quantitative) profiles of incidence, magnitude, and neutralizing capacity of drug reactive antibodies, for all clinical populations to be evaluated, that is, taking into account potential interference by drug target-related antigen, cross-reactive, and nonspecific binding factors	• Evaluate detection of low-quality control level spiked into individual samples representing all clinical populations to be tested • Investigate impact of patient/disease-related factors, including target interference, cross-reactive antibodies in pre-dose sample, and nonspecific binding
Sensitivity (quasi-quantitative level of detectability) in the presence of clinically relevant drug levels	• Minimize impact of drug tolerance and immune complexes on quasi-quantitative estimation of ADA incidence and titer	• Evaluate sample pretreatment steps (acid dissociation and partitioning) to ensure that drug tolerance limit exceeds expected drug level at time of sample collection

Note: Other characteristics, including precision, dilutional linearity, robustness, and sample stability should also be validated.

ADA, antidrug antibody.

for the bioanalytical features of the comparative immunogenicity evaluation—and individual sponsors have the opportunity to justify different approaches on the basis of scientific soundness.

12.16 ROLE OF NONCLINICAL STUDIES

The value of nonclinical *in vivo* studies for evaluation of the relative immunogenicity of related versions of any given therapeutic protein depends on the nature of the molecule and the extent of structural and functional homology to endogenous counterparts. EU and FDA guidelines (FDA, 2014b, 2015; EMA, 2015) acknowledge a potential role for nonclinical studies as part of the evaluation of the immunogenicity of biotechnology-derived products in general because such studies may have relevance for hazard identification (i.e., qualitative risks) for human subjects. For example, comparative immunogenicity studies in wild-type animals have yielded data to indicate whether related versions of a given molecule could have a markedly different quantitative profile in terms of extent of ADA formation, in a manner that can be correlated with defined differences (e.g., levels of subvisible particles) in product quality (Jeandidier et al., 2002; Fradkin et al., 2009).

However, in the case of the biosimilarity exercise involving product versions that have demonstrated highly similar structural properties, ADA data from comparative nonclinical toxicology studies was not judged to be instructive for the approval of the initial wave of biosimilars approved in the EU, even in cases where some differences were evident (Chamberlain, 2014).

For example, while filgrastim does not appear to induce ADAs in humans—due to effective immune tolerance to the exogenous protein—filgrastim is immunogenic in rats (EPAR for Zarzio) because of interspecies sequence differences in the G-CSF molecule. Epoetin alfa also induces ADAs in nonhuman species (EPAR for Binocrit and EPAR for Silapo) for the same reason. For therapeutic monoclonal antibody products, a major part of the ADA response in animals will be directed against the human IgG Fc region (van Meer et al., 2013). Therefore, differences in the immune responsiveness of animals to relevant (for humans) structural features of biosimilar and reference versions of the same molecule may be difficult to detect or interpret in the context of a stronger immune response against xenogeneic structural motifs. In addition, relatively small experimental group sizes increase the influence of interanimal variability on the assessment of relative immunogenicity.

12.17 CLINICAL IMMUNOGENICITY EVALUATION

Although immunogenicity is caused by factors that can be detected and controlled at the manufacturing and product quality testing levels, there remains uncertainty about how extrinsic (manufacturing, formulation, storage conditions, dose regimen, and patient heterogeneity) and intrinsic (B- and T-cell epitopes) factors may interact in the biological context to influence the extent of the immune response in individual subjects.

For this reason, a directly comparative, clinical immunogenicity evaluation to detect differences in the incidence and severity of immune responses induced by the biosimilar candidate and the reference product is normally expected (FDA, 2015).

EU regulatory guidance indicates that in certain cases, it may be possible to justify omission of directly comparative evaluation of clinical immunogenicity (EMA, 2014c); this might be the case for less complex recombinant proteins (e.g., teriparatide) that can be adequately characterized by analytical methods and have not been associated with clinically impactful immunogenicity.

If a sponsor is seeking to extrapolate immunogenicity findings for one condition of use to other conditions/therapeutic indications, the sponsor should consider using a study population and treatment regimen that are sensitive enough to predict a difference in the *adverse effects* associated with immune responses to the proposed product and the reference product across the different conditions of use. Usually, this will be the population and regimen for the reference product for which development of immune responses with adverse outcomes is most likely to occur (FDA, 2015). In the case of insulin, treatment-naïve subjects are likely to provide a more sensitive test of relative immunogenicity (and its clinical impact) than previously treated patients, explaining why both populations were included in the preapproval studies for Abasaglar (EPAR for Abasaglar). Thus, a different study population could be used for comparing immunogenicity from that used to demonstrate comparable efficacy.

For chronically administered agents, EU guidance recommends monitoring antibody formation during a comparative treatment period of 6–12 months, depending on the product. FDA guidance recommends that the extent of monitoring be based on the identified risks for the particular product. In the case of chronically administered agents, however, the posttreatment follow-up monitoring period for ADA positive subjects should be one year, unless a shorter duration can be justified.

In the EU, a biosimilar candidate could have lower immunogenicity than the reference product, provided that this did not result in a significant and clinically relevant increase in efficacy. FDA guidance (FDA, 2015) cautions that differences in immune responses between a proposed product and the reference product in the absence of observed clinical sequelae may be of concern and may warrant further evaluation (e.g., extended period of follow-up evaluation).

As emphasized already, interpretation of bioanalytical results for ADA formation should always be correlated with clinical measures because (1) bioanalytical assays are more or less confounded; and (2) there is no clearly established relationship between ADA response dynamics and clinical impact (due to a lack of bioanalytical assay standardization and objective assay controls) for any product.

12.18 WHAT TO MEASURE

The clinical immunogenicity evaluation should seek to measure the following comparative indices of the humoral response to the biosimilar and reference products:

- Confirmed ADA positive versus ADA negative incidence in pre- versus posttreatment samples, distinguishing between transient versus persistent antibodies if relevant;
- ADA titer or other index of magnitude (%B/T for insulin, which represents proportion of total antigen binding in radioimmunoassay, a measure of anti-drug antibody formation);

- Neutralizing capacity of confirmed ADA;
- Relative time-course of ADA (neutralizing and nonneutralizing) detection;
- Cross-reactive capacity of ADA (e.g., insulin glargine vs. endogenous insulin);
- ADA reactive with process-related impurities (e.g., host cell-derived proteins, if relevant).

Although monitoring of cell-mediated immune responses has not been a feature of the immunogenicity evaluation for the biosimilar products approved to date, there might be value in comparing the *in vitro* responses of innate and adaptive immune effector cells (e.g., to exclude a potential influence of differences in drug product formulation on the stability of a therapeutic protein). However, this is not a current regulatory requirement.

12.19 PRESENTATION OF DATA

Questions from regulatory agencies about the suitability of ADA assay methodology have been very common for the biosimilar applications reviewed to date. EU regulatory guidance has been updated (EMA, 2015) to encourage sponsors to submit an integrated summary of immunogenicity as part of the application for marketing authorization. The author's format for a biosimilar candidate includes the following subheadings:

1. Identified risks and uncertainty for reference product
2. Control of product quality-related risks
3. Suitability of bioanalytical methodology
4. Results of comparative clinical evaluation
5. Conclusions about relative immunogenicity
6. Recommendations for ongoing risk management.

Under subheading (3), the rationale for selecting particular methods and assay controls can be presented in relation to a critical discussion of the validated method performance characteristics. The author has found it helpful to include this integrated summary within CTD Section 5.3.5.3, since this provides more flexibility to include more tabular and graphical data than is possible in Module 2.7.2.4.

12.20 EXTENT OF CLINICAL EVALUATION

Data on ADA incidence and titer from both comparative PK (Phase 1) and therapeutic equivalence (Phase 3) studies have contributed to the assessment of immunogenicity for most of the biosimilar products approved in the EU (Chamberlain, 2014). Exceptions were filgrastim, for which therapeutic equivalence was established in comparative PK/PD studies, supported by a safety study (EPAR for Zarzio), and biosimilar insulin glargine (Abasaglar), where the EPAR reports immunogenicity data from the two, 52-week duration, Phase 3 studies, but not from the Phase 1 PK/PD studies.

Monitoring of ADA during Phase 1 and Phase 3 studies has enabled assessment of potential differences in the incidence and magnitude of ADA formation on sensitive PK parameters and on (less sensitive) clinical efficacy and safety endpoints. While correlation with PD endpoints has been important for cytokine products, often there are rather limited PD correlates for therapeutic monoclonal antibodies. An additional opportunity of 6- and 12-month duration therapeutic equivalence studies is the comparative evaluation of the dynamics of antibody formation in direct relation to clinical response and to steady-state drug concentration.

FDA guidance recommends that clinical evidence of comparative immunogenicity should be generated in addition to comparative PK/PD, most likely in a separate clinical study. Thus, the current practice is to include immunogenicity-related endpoints in separate studies that evaluate: (1) comparative PK (and PD if suitable markers are available); *and* (2) therapeutic equivalence during repeated administration for 6–12 months. The FDA has, in addition, required evaluation of incremental ADA formation following transition of subjects from the reference medicinal product to the biosimilar, for example, during an open-label extension period of the therapeutic equivalence study.

12.21 COMPARATIVE PK STUDIES

Comparative, parallel-group, PK studies in healthy volunteers and patients, powered to demonstrate bioequivalence, have yielded valuable information on relative immunogenicity for different therapeutic monoclonal antibody products, including adalimumab (Kaur et al., 2014), trastuzumab (Yin et al., 2014), and rituximab (Florez et al., 2014).

In the case of adalimumab, a single subcutaneous administration to healthy volunteers induced detectable ADA formation in 50%–70% of subjects. This enabled a correlation of the magnitude of the ADA response with primary and secondary PK parameters—which, arguably, represent the most sensitive clinical measures of an impact of ADA formation for adalimumab. Thus, for products for which there is an identified risk of clinical impactful immunogenicity associated with altered PK, correlation of ADA formation with PK parameters in a comparative PK study could enable assessment of the relative clinically impactful immunogenicity of biosimilar candidate with the reference product.

Neither trastuzumab nor rituximab was found to induce a detectable ADA response in these comparative PK studies; and bioequivalence was established from the highly similar PK parameters between the biosimilar and reference versions.

The primary statistical analysis for the comparative PK study via subcutaneous administration should include C_{max}, and AUC0-infinity as primary endpoints; $T_{1/2}$ should also be included, as well as C_{trough} for a repeated administration dose schedule. A secondary descriptive analysis would then compare relevant PK parameters for the ADA positive versus ADA negative subjects, and upper- versus lower quartile ADA titer subpopulations, in each treatment group; supportive graphical displays (e.g., scatterplots) could be included to illustrate the distribution of individual values in the respective treatment groups/subpopulations.

In the context of a global development program, it has been necessary for sponsors to include the reference product from different regional sources (EU and US), as well as testing the biosimilar product formulation–container combination that is intended for commercialization.

12.22 THERAPEUTIC EQUIVALENCE STUDIES

The primary evidence for the relative immunogenicity of the EU-approved biosimilar products has been obtained from therapeutic equivalence studies, rather than from the comparative PK studies discussed in the preceding subsection (Chamberlain, 2014). Current EU regulatory guidance (EMA, 2014c) recommends that therapeutic equivalence studies be performed for products for which there are no surrogate markers for efficacy. The study population should be sensitive for detecting the clinical impact of potential differences between the biosimilar and reference products, as well as representing an approved therapeutic indication of the reference product.

As stated above, FDA guidance indicates that a separate clinical study from the comparative PK/PD (Phase 1) study may be required to evaluate the immunogenicity of the biosimilar candidate in direct comparison to the reference product. Nevertheless, the ADA response should be measured in all clinical studies.

The protocol for therapeutic equivalence studies should specify a descriptive data analysis of ADA formation; this could include a description of the incidence and median titer of ADAs, the proportion of subjects with neutralizing antibodies (ideally, subclassified by low/medium/high titer), and a correlation of ADA positive versus ADA negative status with drug trough concentration, efficacy, and frequency of adverse events. To check for systematic bias, the proportion of false positives (positive in screening assay but negative in confirmatory assay) in each treatment group should be compared. If the residual drug concentration exceeds the validated drug tolerance limit in some subjects, ADA results should also be evaluated separately for the subpopulations with drug concentration below or above the drug tolerance level.

Sample time-points should be selected to indicate the dynamics of antibody formation in relation to treatment outcome, taking into account the need to minimize the potential for interference in the ADA assay by residual circulating drug. The ADA sampling schedule used for the comparative study of Remsima versus Remicade in ankylosing spondylitis is summarized in Figure 12.2. A similar ADA sample schedule was used in the therapeutic equivalence study, CT-P13 1.3, performed in rheumatoid arthritis patients (EPAR for Remsima).

This sampling schedule enabled a demonstration of a highly similar profile of neutralizing ADA incidence and magnitude across the two treatment groups in both clinical studies (EPAR for Remsima), as illustrated in Figure 12.3a and b.

A biosimilar version of etanercept, SB4 (Benepali), was reported to have a lower incidence of detectable ADA than for the reference product (Enbrel) during the initial 24-week treatment period of a Phase 3 comparative efficacy study (Emery et al., 2015). CHMP questioned whether insufficient drug tolerance of the ADA assay, allied to higher drug levels in some subjects, may have contributed to a higher number of false negatives in the biosimilar treatment arm at the 8-week time-point (EPAR for Benepali). Nevertheless, because no difference in incidence of ADA signals detected after the

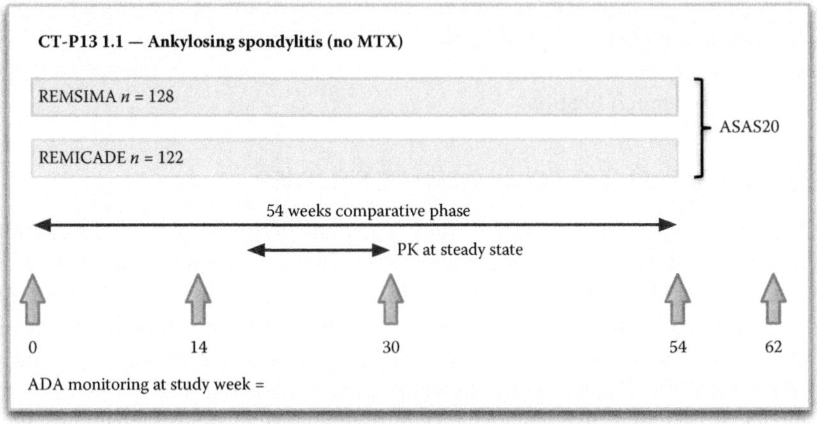

MTX, methotrexate; ASAS20, assessments in ankylosing spondylitis, 20% improvement; ADA, antidrug antibody; PK, pharmacokinetics

FIGURE 12.2 ADA sampling times in Remsima study CT-P13 1.1. Study CT-P13 1.1 was a randomized, double-blind, parallel group study performed in ankylosing spondylitis patients who were not receiving concomitant methotrexate. Patients received multiple iv infusions of 5 mg/kg of Remsima or Remicade during a 54-week period. The primary objective was to demonstrate comparable PK at steady state (between Weeks 22 and 30); samples for ADA testing were collected at the indicated time-points.

8-week time-point and equivalent therapeutic efficacy to Enbrel was demonstrated, the Agency judged that the transient difference did not compromise biosimilarity.

12.23 CHOICE OF SENSITIVE STUDY POPULATION IN RELATION TO EXTRAPOLATION

The FDA recommends that if a sponsor is seeking to extrapolate immunogenicity findings for one condition of use to other conditions of use, the sponsor should consider using a study population and treatment regimen that are adequately sensitive for predicting a difference in immune responses between the proposed product and the reference product across the different conditions of use (FDA, 2015). The FDA guidance goes on to explain that, usually, this will be the population and regimen for the reference product for which development of immune responses with *adverse outcomes* is most likely to occur.

Extrapolation of immunogenicity data across therapeutic indications has been permitted for most biosimilar products approved to date in the EU, namely, somatropin, filgrastim, follitropin-alfa, insulin glargine, infliximab, and etanercept. The single exception to date was for a different route of administration for epoetin alfa, based on prior evidence of an incremental risk of immunogenicity for the subcutaneous route for treatment of renal anemia patients with the reference product (Casadevall, 2005). Additional comparative clinical efficacy, safety, and immunogenicity data were considered necessary to support approval of biosimilar versions for

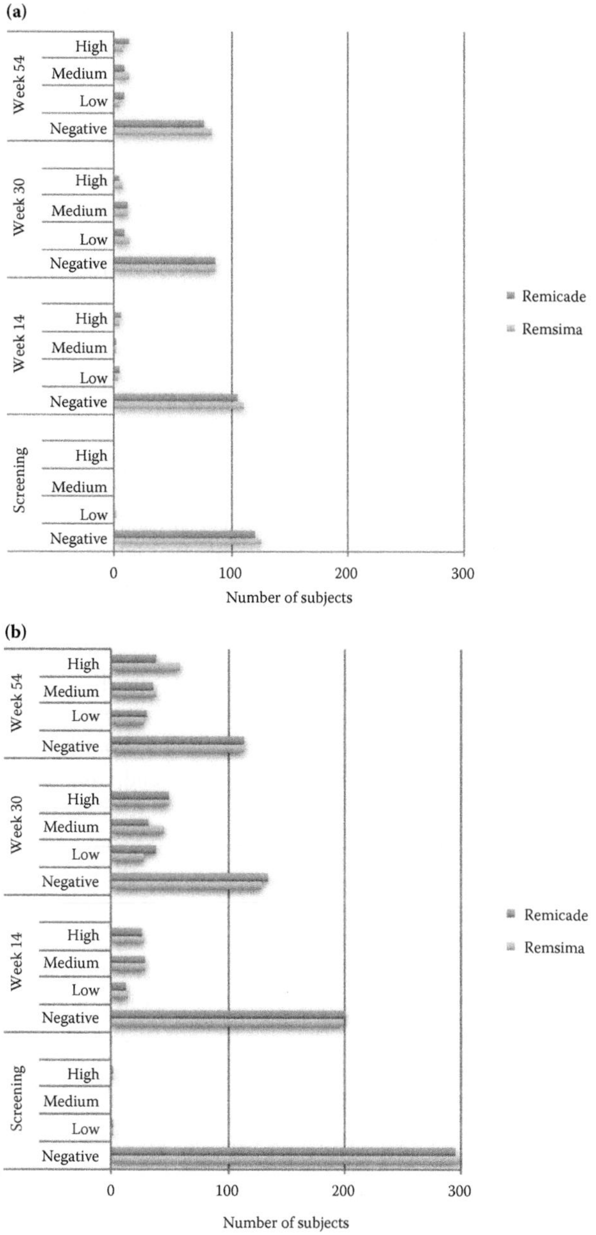

FIGURE 12.3 **(See color insert.)** ADA response; number of subjects (x-axis) vs. treatment time/titer category from screening visit to Week 54 of treatment period (y-axis) (based on EPAR for Remsima). (a) Study CT-P13 1.1 (ankylosing spondylitis). (b) Study CT-P13 3.1 (rheumatoid arthritis). Y-axis shows distribution of neutralizing antibody titer (negative/low/medium/high) at each sample time-point in Studies CT-P13 1.1 (ankylosing spondylitis) and CT-P13 3.1 (rheumatoid arthritis); bars on x-axis represent number of subjects treated with Remicade (light gray) or Remsima (dark gray).

subcutaneous administration of epoetin alfa in renal anemia, and a separate noncomparative safety study was performed using subcutaneous administration in oncology patients (Chamberlain, 2014).

Regulatory agencies may be more reluctant to extrapolate conclusions about relative immunogenicity from the oncology to autoimmune disease setting for rituximab, although long-term safety studies have not revealed a clear negative impact associated with ADA formation during chronic administration of rituximab to treat rheumatoid arthritis (van Vollenhoven et al., 2010).

Since healthy volunteers will be fully immune competent, measurement of AUC0-infinity following a single administration of adalimumab to healthy volunteers could provide the most sensitive endpoint for detecting a difference in clinically impactful ADA formation.

Concomitant use of immunosuppressive medication is not the only consideration for assessing the suitability of the clinical population(s) to be evaluated: in the case of infliximab (EPAR for Remsima), the ankylosing spondylitis population (no concomitant MTX) had a lower detected incidence of ADA compared to the rheumatoid arthritis population (treated with concomitant MTX), possibly due to higher residual drug interference in the ADA assay associated with the higher dose level used in the ankylosing spondylitis indication. Thus, detectability of the ADA response as well as the relationship of the ADA response to adverse outcomes will need to be considered in choosing the most suitable conditions under which to evaluate immunogenicity.

Chronic administration therapeutic efficacy studies may have relatively low discriminatory power for detecting a potential influence of lower/higher relative immunogenicity on efficacy if the drug is dosed at a supramaximal level in terms of the dose-efficacy response curve, and/or the clinical efficacy endpoints are associated with relatively high intersubject variability in pharmacological responsiveness. In addition, therapeutic equivalence studies are not powered to enable a statistically rigorous comparison of safety signals—which may be more accurately estimated in the longer-term postauthorization setting (e.g., via interventional observational cohort studies).

In summary, the experience gained to date for the EU-approved biosimilar products has strongly substantiated the practice of extrapolating immunogenicity findings across all therapeutic indications authorized for the reference product; or, in the case of epoetin alfa, of justifying the requirement for additional evidence prior to authorization for use in higher risk settings.

12.24 SWITCHING

An elective decision by the supervising physician to switch patients between medications that can be expected to achieve the same therapeutic effect is part of routine medical practice. An extensive review of the experience gained from switching patients between different versions of the same biological product, or even between different biological products licensed for the same therapeutic indication (Ebbers et al., 2012), did not reveal evidence for an increase in immunogenicity-related risks that were associated with switching of medications. In the EU, switching experience has been accumulated as a result of local tendering processes leading to a decision to procure a biosimilar version of somatropin, filgrastim, epoetin alfa, insulin glargine, or infliximab, thereby causing a nonelective (either for the physician or patient) switch. Such switches have not yet identified an increased immunogenicity-related risk for the EU-approved biosimilar products, and there was not a single signal in the EudraVigilance serious adverse event database as of May 2015 (FIMEA, 2015).

Switching studies are not part of the current EU regulatory requirements for approval of biosimilar products, as there is an implicit understanding that a candidate that meets the rigorous approval standards for biosimilarity *can* be regarded as a therapeutic alternative to the reference product (Weise et al., 2012). The regulator's decision to approve the biosimilar version serves as an adequate basis to allow physicians to make the choice on whether to switch a patient from a reference to a biosimilar version, or vice versa. This consideration applies to the efficacy, safety, and immunogenicity dimensions of the overall clinical benefit-risk assessment. To date, this position has been publicly affirmed by the authorities of four EU member states, namely, Finland, Germany, Ireland, and The Netherlands (Ekman, 2016).

Biosimilar development programs for some chronic administration products have included controlled transition between treatments, with the intent of reinforcing prescriber confidence in the biosimilar version as well as to satisfy scientific advice received from the FDA. However, the discriminatory sensitivity of such study designs is rather questionable, not least because of the potentially confounding influences associated with the persistence of therapeutic effects and immune responses to the previous treatment (Ebbers and Chamberlain, 2014). This severely limits the feasibility of designing clinical studies to demonstrate interchangeability as required by US legal statute, if measures of clinically relevant immunogenicity of the different product versions were to be included as part of the evidence. Without very long (6–18 months, depending on the product) drug washout periods—which may be considered to be unethical—the clinically relevant parameters would be confounded by effects (pharmacodynamics and immune response) associated with earlier treatment. Low feasibility could explain why there are so few examples of well-controlled switching studies for different versions of therapeutic proteins.

One soundly controlled study that evaluated switching between two *distinct* human Factor VIII products (with different primary amino acid sequences, i.e., not biosimilar) showed that switching had no impact—even in the case of intentionally modified versions of a molecule that has identified immunogenicity-related risks (Hay et al., 2015).

The directly comparative switching experience gained for infliximab, in which subjects treated for 52 weeks with one version of infliximab were either switched to a different version or maintained on the same version (Park et al., 2013; Yoo et al., 2013) for a further 12 months, revealed a remarkably similar level of ADA incidence in rheumatoid arthritis patients, and only a marginally higher numerical difference in ankylosing patients (Figure 12.4).

The comparative design of the NOR-SWITCH study (NCT02148640) incorporates measures of relative anti-infliximab antibody formation in some 500 patients stabilized on the reference product, who are then randomized 1-to-1 for continuing treatment on either Remicade ($n = 250$) or Remsima ($n = 250$) for a 52-week period. Results from this study indicate a comparable incidence of treatment-emergent ADA for patients switched to Remsima compared to those remaining on Remicade (Jørgensen et al., 2016).

Rheumatoid arthritis PLANETRA extension	Time-point	Maintained on CT-P13 ($n = 159$)	Switched from Remicade to CT-P13 in extension phase ($n = 143$)
Yoo DH et al. *Arthritis Rheum.* 2013		% ADA positive	
	Week 54	49.1	49.3
	Week 78	50.4	49.6
	Week 102	46.4	49.6

Ankylosing spondylitis PLANETAS extension	Time-point	Maintained on CT-P13 ($n = 90$)	Switched from Remicade to CT-P13 in extension phase ($n = 84$)
Park W et al. *Arthritis Rheum.* 2013		% ADA positive	
	Week 54	22.2	26.2
	Week 78	24.4	31.3
	Week 102	25.0	30.7

ADA, antidrug antibody

FIGURE 12.4 Impact of switching from Remicade to Remsima (CT-P13). ADA positive frequency of patients completing the initial 54-week double-blind treatment period with Remsima or Remicade in Studies, CT-P13 1.1 (ankylosing spondylitis) or CT-P13 3.1 (rheumatoid arthritis), followed by 48-week Open-Label extension periods of continuing treatment: in the PLANETRA extension study, 158 patients were maintained on Remsima, while 144 patients were switched from Remicade to Remsima; in the PLANETAS extension study, 88 patients were maintained on Remsima, while 86 patients were switched from Remicade to Remsima.

Ultimately, it is not possible definitively to prove the negative, that is, that there is zero incremental immunogenicity-related risk associated with switching patients between different biological medicinal products, be they originator or biosimilar products. Preauthorization switching studies might not provide suitably diverse populations for identifying an incremental risk for "real-world" use. Moreover, it is difficult to identify parameters for objective comparison of therapeutic outcome that would not be confounded (in terms of both treatment effect and immune response) by earlier treatment. Arguably, this might place more emphasis on the role of *post*-authorization monitoring of immunogenicity-related risks, as discussed in the next subsection.

The interchangeability, switching, and substitution of biosimilars are discussed further in Chapter 10 of this book.

12.25 POSTAUTHORIZATION EVALUATION

Accumulation of additional information on longer-term safety, including immunogenicity, is a standard element of the EU Risk Management Plan to be applied for postauthorization monitoring of biosimilars.

EU approval of a biosimilar product is conditional on a heightened level of post-authorization risk management provisions: most notably, warnings and precautions in prescribing information and product labeling, focused pharmacovigilance, and, in some cases, commitments to perform postapproval interventional clinical studies and/or establish patient registries to monitor particular safety concerns.

While postauthorization studies are unlikely to provide the same quality of data as randomized controlled preauthorization studies, they do have potential to provide evidence of sustained benefit-risk under real-world conditions, that is, where patients to be treated are not preselected to meet stringent inclusion and exclusion criteria. In the case of immunogenicity, observational cohort studies have contributed highly instructive information on the relationship between sustained efficacy, steady-state drug concentration, and ADA formation for adalimumab (Bartelds et al., 2011).

Thus, a prospectively designed, noncomparative, observational cohort study could provide postauthorization evidence of reduced drug concentration and loss of efficacy causally related to formation of ADAs—which may then be compared with historical information for the reference product. This would require a "low-intervention" type study, in the sense that periodic blood samples would be collected for measurement of drug concentration and ADA, and patients would also be monitored for sustainability of efficacy and for adverse events of particular relevance, for example, injection-site and hypersensitivity reactions.

Information on the relative incidence of serious adverse events associated with treatment by different versions of the same molecule, that is, the reference product and different approved biosimilars, could be accumulated from patient registries. While patient registries can be useful to monitor the incidence of targeted serious adverse events and, in some cases, sustained benefit (via "drug survival" rates), immunogenicity is rarely measured. Thus, the prospectively designed, post-authorization, observational cohort study would seem to offer the best opportunity for monitoring the impact of the long-term immunogenicity of approved biosimilar products—albeit in a noncontrolled (noncomparative) sense.

All approved products are subject to periodic reevaluation of benefit-risk by regulatory agencies. This would include a review of spontaneous adverse event reports and published literature to identify immune-mediated adverse events.

One potential gap is incomplete transparency of data accumulated during the postauthorization phase for different biosimilar products, which could be addressed by periodic updating of the detailed information provided in the EPARs issued at the time of the initial approval, to include outcomes of provisions in the Risk Management Plan. In addition, lack of harmonization across EU member states for regulatory requirements for conduct of postauthorization, observational cohort studies involving relatively low interventional impact could discourage such studies. The implementation of a "low-intervention" designation for prospectively designed observational studies of approved products, involving periodic blood sampling to measure steady-state drug concentration relative to ADA formation, and monitoring of sustained efficacy and targeted adverse events, may create a possibility for such studies to be approved only by the concerned Ethics Committees—without a need for national regulatory agency approval.

12.26 MANAGING POTENTIAL RISKS ASSOCIATED WITH POSTAUTHORIZATION MANUFACTURING CHANGES

In both the EU and US regulatory systems, postapproval manufacturing changes for biosimilar products are evaluated according to the ICH Q5E guideline (ICH, 2005). The prechange and postchange drug substance and drug product are directly compared by extended analytical characterization and stability testing, and, in rare cases, nonclinical or clinical data might be requested to address known risk factors that cannot be fully evaluated by analytical testing alone. Currently, there is no requirement to repeat the demonstration of biosimilarity in direct comparison to the reference product (although the FDA expectations for interchangeability designation are yet to be defined).

If the particular product was known to induce a clinically impactful ADA response and if risk of incremental immunogenicity associated with the particular change(s) could not be excluded by analytical testing alone, a comparative (pre- vs. postchange drug product) clinical study might be performed to evaluate relative immunogenicity. On the other hand, the utility of such a study would depend on the balance between severity of the negative outcome and probability of detection of a clinically relevant difference.

In the case of epoetin alfa, a change in the formulation–primary container combination for the originator product was associated with an increase in risk of amPRCA in chronic renal anemia subjects treated by subcutaneous administration. Causality was most likely due to a confluence of factors (Bennett et al., 2004), including suboptimal storage of the product in the outpatient setting, which may have resulted in elevated levels of product-related aggregates. However, the incidence of treatment-induced amPRCA, even when the risk is heightened, may be too low to detect within the scale of randomized controlled clinical studies that are feasible in the postmarketing setting. Thus, the risk might be more effectively controlled at the product quality level, that is, by rigorous selection of the primary container to exclude identified risk factors (i.e., residual tungsten particles) allied to extensive orthogonal analytical characterization of aggregate formation under worst-case handling conditions (Seidl et al., 2012).

For originator products, it is quite uncommon for clinical data to be required to support the comparability demonstration for postapproval manufacturing changes. The regulatory practice reflects the ability to exclude an incremental immunogenicity-related risk for most products via analytical characterization and comparative stability studies that demonstrate adequate control of critical quality attributes. Since the risk factors for incremental immunogenicity are the same for authorized biosimilars and originator products, it is reasonable to apply the same regulatory standard to assessing postauthorization changes.

12.27 UNRESOLVED ISSUES/FUTURE PERSPECTIVES

Although initially considered as a major uncertainty for approval of biosimilars, immunogenicity-related risks have not yet proven to be a critical issue in the postauthorization setting for candidates meeting rigorous product quality standards.

Nevertheless, longer-term experience is required to understand whether this level of assurance is translatable to all products in all indications.

As more experience with approved biosimilar products is accumulated, it is conceivable that the value of the comparative immunogenicity evaluation during chronic (>3 months) administration may be increasingly questioned for detection of differences between product versions that have been shown to meet EU and US standards of similarity at the analytical level. Although this might presently be perceived as a controversial notion, a comparative PK study supported by postauthorization monitoring could provide an adequately sensitive approach for detecting real differences in immunogenicity between highly similar versions of the same molecule.

Postauthorization monitoring of immunogenicity-related risks for biotherapeutic products in general could be supplemented by more effective use of prospectively designed observational cohort studies that include “low-intervention” measures, such as periodic blood sampling for measurement of drug concentration and ADA, which are not typically part of patient registries. Availability of standardized control reagents for ADA assays would facilitate comparison across studies performed with different product versions (van Schouwenburg et al., 2016), thereby potentially obviating a need for directly comparative postauthorization studies.

Maintenance of biosimilarity following independent changes to the manufacturing process, formulation, or primary container of the respective versions of the product depends on the application of the well-established regulatory standards for establishing the comparability of pre- versus postchange product (ICH Q5E). In most cases, comparability is demonstrated without clinical studies. This raises potential concern about a risk of failure to detect changes in immunogenicity that could impact overall clinical benefit versus risk in authorized therapeutic indications. Arguably, clinical immunogenicity evaluation might be required if there were uncertainty about the impact of proposed manufacturing or formulation changes for products with an identified severe clinical outcome; but application of a more rigorous standard for biosimilar versions compared to originator products does not seem justified by the safety experience accumulated since the approval of the first biosimilars in Europe more than 10 years ago.

12.28 SUMMARY

Assessment of the impact of undesirable immunogenicity on overall clinical benefit and risk is an essential part of the regulatory authorization process for biosimilars, as it is for biotechnology-derived products in general. Uncertainty about the way identified risk factors can interact to manifest as adverse outcomes has, in practice, been managed effectively to date (up until June 2016) by a rigorous standard of control of the manufacturing process and by extensive comparative analytical characterization of the biosimilar and reference products. Risk mitigation has included comparative clinical evaluation in the preauthorization phase, using optimized bioanalytical methods to analyze humoral responses that can be correlated with relevant clinical endpoints of PK, efficacy, and safety. Heightened monitoring of immunogenicity-related risks is a standard part of the postapproval risk management plan. In the real-world setting, neither the regulatory approach of extrapolation of immunogenicity findings across different therapeutic indications, nor the clinical practice of switching patients from treatment with the reference product to treatment with a biosimilar version, has yet revealed an incremental immunogenicity-related risk for any of the biosimilar products approved in the EU since 2006.

ACKNOWLEDGMENTS

I would like to express my gratitude for the very helpful peer-review comments provided by Dr. Frits Lekkerkerker and Dr. Markku Toivonen (both of NDA Advisory Board), as well as the constructive advice and diligent editorial work of Professor Paul Declerck (University of Leuven).

REFERENCES

Ahmadi M, Bryson CJ, Cloake EA, et al. (2015) Small amounts of sub-visible aggregates enhance the immunogenic potential of monoclonal antibody therapeutics. *Pharmaceutical Research* **32**, 1383–1394.

Amgen. (2015a) Neulasta® prescribing information, September 2015. Available from: http://pi.amgen.com/united_states/neulasta/neulasta_pi_hcp_english.pdf.

Amgen. (2015b) Neupogen® prescribing information, September 2015. Available from: http://pi.amgen.com/united_states/neupogen/neupogen_pi_hcp_english.pdf.

Amon R, Reuven EM, Ben-Arye SL, Padler-Karavani V. (2014) Glycans in immune recognition and response. *Carbohydrate Research* **389**, 115–122.

Ataca P, Atilla E, Kendir R, et al. (2015) Successful desensitization of a patient with rituximab hypersensitivity. *Case Reports in Immunology* **2015**, Article ID 524507, 4 pages.

Baert F, Noman M, Vermeire S, et al. (2003) Influence of immunogenicity on the long-term efficacy of infliximab in Crohn's disease. *New England Journal of Medicine* **348(7)**, 601–608.

Bartelds GM, Krieckaert CL, Nurmohamed MT, et al. (2011) Development of antidrug antibodies against adalimumab and association with disease activity and treatment failure during long-term follow-up. *Journal of the American Medical Association* **305(14)**, 1460–1468.

Bartelds GM, Wijbrandts CA, Nurmohamed MT, et al. (2007) Clinical response to adalimumab: relationship to anti-adalimumab concentrations in rheumatoid arthritis. *Annals of the Rheumatic Diseases* **66**, 921–926.

Bee JS, Nelson SA, Freund E, et al. (2009) Precipitation of a monoclonal antibody by soluble tungsten. *Journal of Pharmaceutical Sciences* **98**, 3290–3301.

Bendtzen K, Geborek P, Svenson M, et al. (2006) Individualized monitoring of drug bioavailability and immunogenicity in rheumatoid arthritis patients treated with the tumor necrosis factor α inhibitor infliximab. *Arthritis and Rheumatism* **54(12)**, 3782–3789.

Bennett CL, Luminari S, Nissenson AR, et al. (2004) Pure red-cell aplasia and epoetin therapy. *New England Journal of Medicine* **351**, 1403–1408.

Bourdage JS, Cook CA, Farrington DL, et al. (2007) Affinity capture elution (ACE) assay. *Journal of Immunological Methods* **327**, 10–17.

Casadevall N. (2005) What is antibody-mediated pure red cell aplasia (PRCA)? *Nephrology, Dialysis, Transplantation* **20(Suppl. 4)**, iv3–iv8.

Casadevall N, Dupuy E, Molho-Sabatier P, et al. (1996) Brief report: autoantibodies against erythropoietin in a patient with pure red-cell aplasia. *New England Journal of Medicine* **334(10)**, 630–633.

Chamberlain PD. (2014) Multidisciplinary approach to evaluating immunogenicity of biosimilars: lessons learnt and open questions based on 10 years' experience of the European Union regulatory pathway. *Biosimilars* **4**, 23–43.

Chung CH, Mirakhur B, Chan E, et al. (2008) Cetuximab-induced anaphylaxis and IgE specific for galactose-α-1,3-galactose. *New England Journal of Medicine* **358(11)**, 1109–1117.

Ebbers HC, Chamberlain P. (2014) Interchangeability: an insurmountable 5th hurdle? *Generics and Biosimilars Initiative Journal* **3(2)**, 88–93. Available at: http://www.gabi-journal.net.

Ebbers HC, Muenzberg M, Schellekens H. (2012) The safety of switching between therapeutic proteins. *Expert Opinion on Biological Therapy* **12(11)**, 1473–1485.

Ekman N. (2016) Biosimilars from the perspective of an EU regulator. Presentation to the European Association of Hospital Pharmacists Congress, March 17, 2016. Available from: http://www.medicinesforeurope.com/wp-content/uploads/2016/03/20160317-EAHP-satellite-on-Biosimilar-medicine-Medicines-for-Europe-Ekman.pdf.

EMA. (2012) Guideline on immunogenicity assessment of monoclonal antibodies intended for in vivo use, EMA/CHMP/BMWP/14327/2006, May 2012.

EMA. (2014a) Guideline on similar biological medicinal products containing biotechnology-derived proteins as active substance: quality issues, EMA/CHMP/BWP/247713/2012, Rev. 1, May 2014.

EMA. (2014b) Guideline on similar biological medicinal products, EMA/CHMP/437/04, Rev. 1, October 2014.

EMA. (2014c) Guideline on similar biological medicinal products containing biotechnology-derived proteins as active substance: non-clinical and clinical issues, EMA/CHMP/BMWP/42832/2005. Rev. 1, December 2014.

EMA. (2015) Guideline on immunogenicity assessment of biotechnology-derived therapeutic proteins, EMA/CHMP/BMWP/14327/2006, Rev. 1, September 2015.

Emery P, Vencovsky J, Sylwestrzak A, et al. (2015) A phase III randomised, double-blind, parallel-group study comparing SB4 with etanercept reference product in patients with active rheumatoid arthritis despite methotrexate therapy. *Annals of the Rheumatic Diseases*, doi:10.1136/annrheumdis-2015-207588.

European Public Assessment Reports (EPAR) for different products (Abasaglar®, Alpheon®, Avastin®, Benepali®, Herceptin®, Nivestim®, Omnitrop®, Ovaleap®, Remsima®, Silapo®, & Zarzio®). Available from: http://www.ema.europa.eu.

FDA. (2014a) Draft guidance for industry: Clinical pharmacology data to support a demonstration of biosimilarity to a reference product. Food and Drug Administration, CDER/CBER, May 2014.

FDA. (2014b) Guidance for industry: immunogenicity assessment of therapeutic protein products. Food and Drug Administration, CDER/CBER, August 2014.

FDA. (2015) Scientific considerations in demonstrating biosimilarity to a reference product. Food and Drug Administration, CDER/CBER, April 2015.

FDA. (2016) Draft guidance for industry: assay development for immunogenicity testing of therapeutic proteins, Rev. 1. Food and Drug Administration, CDER/CBER, April 2016.

FIMEA. (2015) Position statement, May 22, 2015. Finnish Medicines Agency. Available at: http://www.fimea.fi/whats_new/1/0/are_biosimilars_interchangeable?

Fineberg SE, Kawabata TT, Finco-Kent D, et al. (2007) Immunological responses to exogenous insulin. *Endocrine Reviews* **28(6)**, 625–652.

Florez A, et al. (2014) Clinical pharmacokinetic (PK) and safety (immunogenicity) of rituximab biosimilar RTXM83 in combination with chemotherapy CHOP in patients with diffuse large B-cell lymphoma (DLBCL). *Blood* **124**, 5472.

Fradkin AH, Carpenter JF, Randolph TW. (2009) Immunogenicity of aggregates of recombinant growth hormone in mouse models. *Journal of Pharmaceutical Sciences* **98(9)**, 3247–3264.

Hay CR, Palmer BP, Chalmers EA, et al. (2015) The incidence of factor VIII inhibitors in severe haemophilia A following a major switch from full-length to B-domain-deleted Factor VIII: a prospective cohort comparison. *Haemophilia* **21(2)**, 219–226.

ICH Q5E. (2005) Note for guidance on biotechnological/biological products subject to changes in their manufacturing process. Available at http://www.ema.europa.eu.

Jeandidier N, Boullu S, Busch-Brafin M-S, et al. (2002) Comparison of antigenicity of Hoechst 21PH Insulin using either implantable intraperitoneal pump or subcutaneous external pump infusion in type 1 diabetic patients. *Diabetes Care* **25(1)**, 84–88.

Jiang Y, Nashed-Samuel Y, Li C, et al. (2009) Tungsten-induced protein aggregation: solution behavior. *Journal of Pharmaceutical Sciences* **98**, 4695–4710.

Jørgensen K, et al. (2016) Biosimilar infliximab (ct-p13) is not inferior to originator infliximab: results from the 52-week randomized NOR-SWITCH trial. United European Gastroenterology Week (UEGW), LB15.

Joubert MK, Hokom M, Eakin C, et al. (2012) Highly aggregated antibody therapeutics can enhance the *in vitro* innate and late-stage T-cell immune responses. *Journal of Biological Chemistry* **287(30)**, 25266–25279.

Kaur P, et al. (2014) FRI0264 a randomized, single-blind, single-dose, three-arm, parallel group study in healthy subjects to demonstrate pharmacokinetic equivalence of ABP 501 and adalimumab: results of comparison with adalimumab (EU). *Annals of Rheumatic Disease* **73(Suppl. 2)**, 479.

Liu W, Swift R, Torraca G, et al. (2010) Root cause analysis of tungsten-induced protein aggregation in pre-filled syringes. *PDA Journal of Pharmaceutical Science and Technology* **64**, 11–19.

Loumaye E, Dreano M, Galazka A, et al. (1998) Recombinant follicle stimulating hormone: development of the first biotechnology product for the treatment of fertility. *Human Reproduction Update* **4(6)**, 862–881.

Mok CC. (2014) Rituximab for the treatment of rheumatoid arthritis: an update. *Drug Design, Development and Therapy* **8**, 87–100.

Murdace G, Spanò F, Puppo F. (2013) Selective TNF-α inhibitor-induced injection site reactions. *Expert Opinion on Drug Safety* **12(2)**, 187–193.

NOR-SWITCH study, NCT02148640. Available from: https://clinicaltrials.gov/ct2/show/NCT02148640.

Park W, Miranda P, Brzosko M, et al. (2013) Efficacy and safety of CT-P13 (infliximab biosimilar) over two years in patients with ankylosing spondylitis: comparison between continuing with CT-P13 and switching from infliximab to CT-P13. *Arthritis and Rheumatism* **65(12)**, 3326.

Pfizer. (2014) Genotropin® prescribing information, September 2014. Available from: http://labeling.pfizer.com/showlabeling.aspx?id=577.

Qian J, Liu T, Yang L, et al. (2007) Structural characterization of N-linked oligosaccharides on monoclonal antibody cetuximab by the combination of orthogonal matrix-assisted laser desorption/ionization hybrid quadrupole-quadrupole time-of-flight tandem mass spectrometry and sequential enzymatic digestion. *Analytical Biochemistry* **363**, 8–18.

Rombach-Riegraf V, Karle AC, Wolf B, et al. (2014) Aggregation of human recombinant monoclonal antibodies influences the capacity of dendritic cells to stimulate adaptive T-cell responses *in vitro*. *PLoS One* **9**, e86322.

Rossert J, Casadevall N, Eckardt KU. (2004) Anti-erythropoietin antibodies and pure red cell aplasia. *Journal of American Society of Nephrology* **15**, 398–406.

Sailstad JM, Amaravadi L, Clements-Egan A, et al. (2014) A white paper: consensus and recommendations of a global harmonization team on assessing the impact of immunogenicity on pharmacokinetic measurements. *AAPS Journal* **16(3)**, 488–498.

Schmidt E, Hennig K, Mengede C, et al. (2009) Immunogenicity of rituximab in patients with severe pemphigus. *Clinical Immunology* **132(3)**, 334–341.

Seidl A, Hainzl O, Richter M, et al. (2012) Tungsten-induced denaturation and aggregation of epoetin alfa during primary packaging as a cause of immunogenicity. *Pharmaceutical. Research* **29(6)**, 1454–1467.

Smith HW, Butterfield A, Sun D. (2007) Solid-phase extraction with acid dissociation (SPEAD) sample treatment prior to ELISA. *Regulatory Toxicology and Pharmacology* **49**, 230–237.

Somerville L, Bardelas J, Viegas A, et al. (2014) Immunogenicity and safety of omalizumab in pre-filled syringes in patients with allergic (IgE-mediated) asthma. *Current Medical Research Opinion* **30(1)**, 59–66.

USP. (2015) Immunogenicity assays—design and validation of immunoassays to detect anti-drug antibodies, Chapter 1106.

van Meer PJK, Kooijman M, Brinks V, et al. (2013) Immunogenicity of mAbs in non-human primates during non-clinical safety assessment. *mAbs* **5(5)**, 810–816.

van Schie KA, Hart MH, de Groot ER, et al. (2015) The antibody response against human and chimeric anti-TNF therapeutic antibodies primarily targets the TNF binding region. *Annals of the Rheumatic Diseases* **74(1)**, 311–314.

van Schouwenburg PA, Kruithof S, Wolbink G, et al. (2016) Using monoclonal antibodies as an international standard for the measurement of anti-adalimumab antibodies. *Journal of Pharmaceutical and Biomedical Analysis* **120**, 198–201.

van Schouwenburg PA, van de Stadt LA, de Jong RN, et al. (2013) Adalimumab elicits a restricted anti-idiotypic antibody response in autoimmune patients resulting in functional neutralisation. *Annals of the Rheumatic Diseases* **72(1)**, 104–109.

van Vollenhoven RF, Emery P, Bingham CO 3rd, et al. (2010) Long-term safety of patients receiving rituximab in rheumatoid arthritis trials. *Journal of Rheumatology* **37(3)**, 1–10.

Weise M, Bielsky M-C, De Smet K, et al. (2012) Biosimilars: what clinicians should know. *Blood* **120(26)**, 5111–5117.

Yin D, Barker KB, Li R, et al. (2014) A randomized phase 1 pharmacokinetic trial comparing the potential biosimilar PF-05280014 with trastuzumab in healthy volunteers (REFLECTIONS B327-01). *British Journal of Clinical Pharmacology* **78(6)**, 1281–1290.

Yoo DH, Prodanovic N, Jaworski J, et al. (2013) Efficacy and safety of CT-P13 (infliximab biosimilar) over two years in patients with rheumatoid arthritis: Comparison between continued CT-P13 and switching from infliximab to CT-P13. *Arthritis and Rheumatism* **65(12)**, 3319.

13 Pharmacovigilance of Biosimilars

Shehla Hashim, Souleh Semalulu,
Felix Omara, and Duc Vu
Health Canada

CONTENTS

13.1 INTRODUCTION

Medicinal products are initially approved based largely on the safety and efficacy data from randomized clinical trials. Such trials are often conducted in selected target populations, where the sample sizes are small and the treatment duration is short; therefore, rare events and adverse events with long lag periods are not fully characterized, and safety in the wider general population is unknown. Thus, there are uncertainties with regards to the entire safety profile of the drug at the time of authorization. Once authorized, the safety of drugs is dependent on several factors such as how they are used in the "real world" under the authorized indications, the nature of the drug, characteristics of individual patients, and the quality of the product. This necessitates systematic and continuous monitoring of drug safety throughout its life cycle, in order to detect any new adverse events and to minimize both known and previously unknown safety concerns that may arise in the postmarket setting in order to maintain a positive benefit–risk profile of the drug.

13.1.1 ROLE OF PHARMACOVIGILANCE IN DRUG SAFETY

Pharmacovigilance as a discipline originated following the medical disaster with thalidomide in 1961. Thalidomide was used by pregnant women, and it resulted in congenital abnormalities in the infants born to these mothers. Following these tragedies, regulations in general and national pharmacovigilance centers were set up to follow up drugs once they were authorized and marketed.

The World Health Organization (WHO) defines pharmacovigilance as "the science and activities relating to the detection, assessment, understanding and prevention of adverse effects or any other drug-related problem. The goals of pharmacovigilance are to identify and evaluate safety signals and to mitigate the risks of each product, in a timely manner" (WHO, 2002).

A WHO article summarizes the specific aims of pharmacovigilance as being to "improve patient care and safety in relation to the use of medicines and all medical and paramedical interventions, improve public health and safety in relation to the use of medicines, contribute to the assessment of benefit, harm, effectiveness and risk of medicines, encouraging their safe, rational and more effective (including cost effective) use, and to promote understanding, education and clinical training in pharmacovigilance and its effective communication to the public" (WHO, 2004).

Pharmacovigilance can be passive, such as the routine monitoring of spontaneous reports, or it can be active, which requires putting in place appropriate risk management activities for specific risks associated with use of the product. Most of the benefit–risk information obtained in the postmarket setting comes from spontaneous reports of adverse events, which may be provided by patients, caregivers, health care professionals, and market authorization holders. Some of the databases that collect reports of adverse events include the EudraVigilance Database of the European Medicines Agency, the Uppsala Monitoring Center of the WHO, the FDA Adverse Event Reporting System (FDA AERS), and the Canada Vigilance database. Identification of signal(s) from these reports involves a process of hypothesis generation, assessment of the available information for causality, and derivation of conclusions based on the scientific evidence. However, there are certain drawbacks with spontaneous adverse event reports such as the quality of reporting, details of the information provided, and patient-related confounding factors that make adequate assessment of causal association of an adverse event to the drug difficult, or sometimes impossible.

Underreporting is a major limitation of spontaneous adverse event reporting systems, as many adverse events may go undetected, or there may not be enough evidence to link the drug to the adverse event of concern. New data-mining techniques and the analytical methods that are now available to the pharmacovigilance specialist have greatly enhanced the way pharmacovigilance is done (Meyboom et al., 1997; ICH, 2004; Poluzzi et al., 2012). In recent years, there has been a shift to more proactive pharmacovigilance. This has led to the introduction of risk management plans as tools for planning pharmacovigilance and risk minimization activities. Risk management plans are used to address some of the uncertainties, especially those related to the potential risks that exist at the time a drug is authorized for market. Some of the activities that may be requested as part of a pharmacovigilance plan are postauthorization studies and/or registries, in order to better characterize the benefit–risk profile of the drug throughout its life cycle.

13.1.2 Biopharmaceuticals and Biosimilars

The increase in the development of biological products or biopharmaceuticals presents unique opportunities for monitoring and pharmacovigilance. Biopharmaceuticals are proteins that are generally used to treat a variety of severe and life-threatening diseases. They are fundamentally different from the usual generic products, owing to the size and complexity of the active agents and the manufacturing process. In some jurisdictions, some of these products are authorized for rare diseases as orphan drugs; therefore, information on their safety profile at the time of authorization may sometimes be very limited due to small clinical trial sample size. Although biological products are considered targeted therapeutic drugs, they are associated with several adverse events. The most common safety concerns with biological products include the potential risk of infectious disease transmission, lot-to-lot variability, issues around immunogenicity, hypersensitivity reactions, alterations of immune function; which may lead to increased risks of infection, autoimmunity and/or cancer development, long biological half-lives leading to prolonged pharmacodynamic effects even after cessation of treatment, and unknown short- and long-term risks.

The most frequently associated risks with biopharmaceuticals are attributable to their immune-modulatory actions. A number of these responses are difficult to identify in preclinical studies since such reactions tend to be species-specific and cannot always be extrapolated to humans. An example of such a reaction was seen with anti-CD28 antibody TGN1412 in healthy volunteers. All six volunteers who received TGN1412 had a systemic inflammatory response, characterized by a rapid induction of pro-inflammatory cytokines, accompanied by headache, myalgia, nausea, diarrhea, erythema, vasodilatation, and hypotension (Suntharalingam et al., 2006; Attarwala, 2010). These reactions were, however, not recorded in the preclinical studies of TGN1412. Due to the complexity of biotherapeutics and the difficulties in predicting the adverse reactions that may be seen when these products are introduced to the market, their postmarket surveillance is crucial. Postmarket surveillance for biological products poses many challenges. With the introduction of biosimilar products, which are biologicals that have demonstrated biosimilarity to the reference product with respect to structure, function, animal toxicity, human pharmacokinetics and pharmacodynamics, clinical immunogenicity, and clinical safety and effectiveness, the pharmacovigilance of these products has become even more challenging and complex (Zuniga and Calvo, 2010; Weise et al., 2012; Casadevall et al., 2013).

Biosimilars [also called subsequent-entry biologics (SEBs) or follow-on biologics] are biotherapeutics that have demonstrated similarity, but that are not identical to the approved reference biologics or biotechnology derived products. They are not considered identical because of differences in the inherent complexity of the protein molecule itself, the complex, multistep manufacturing processes that utilize live cells, and the difficulties in accurate characterization of the reference biological product. In addition, cell lines used in the manufacture of a given biotherapeutic constitute proprietary information about the reference biological product, which is not available to the manufacturer of the biosimilar product. It is recognized that even slight changes in the protein molecule may have an unexpected clinical impact and can alter the benefit–risk profile of the drug. Changes in the manufacturing process may also result in alterations to the product that may or may not be clinically significant. Changes, such as differences in the primary amino acid sequence, may lead to significant heterogeneity. Likewise, modifications to amino acids, sugar moieties (due to glycosylation), or other side chains can affect the potency or degradation of the biotherapeutic, the antigenic sites, and the solubility of the protein. Generation of new antigenic epitopes in the protein may increase the antigenicity of the product. Also, changes in the tertiary folding of the protein may result in protein aggregation, oxidation, and deamidation, leading to an increased immunogenic response to the drug. Such changes may be introduced by several factors, including the manufacturing process and environmental conditions such as exposure to light, changes in temperature, moisture content, shear, or the material used for packaging. In addition, impurities introduced in the final product, due either to the manufacturing process or to the product itself, may increase the probability of a severe immune reaction to a protein product (De Groot and Scott, 2007; Zuniga and Calvo, 2010; Weise et al., 2012; Socinski et al., 2015).

13.2 SCIENTIFIC GUIDELINES FOR THE REGULATION OF BIOSIMILARS

Generic versions of chemically synthesized medicines follow well-established scientific standards that have been put in place by regulatory authorities worldwide. Authorization of a generic medicine is based on proving its chemical identity and its bioequivalence to the reference product. If the test product and reference product are comparable in dosage forms, contain identical amounts of the identical medicinal ingredient(s), and display similar pharmacokinetic profiles, they are said to be bioequivalent. Generic medicines are usually authorized on the basis of abbreviated applications. Nonclinical and clinical studies that are required for reference drug products are usually not required for generic products. However, regulators are aware that the prototype bioequivalence guidance that has worked well for generic products is not applicable to complex, biologically derived drugs. Therefore, specific guidelines have been developed by various international regulatory authorities to deal with the challenges related to the authorization and pharmacovigilance of biosimilars.

In principle, the guidance document on technical requirements for registration of pharmaceuticals for human use developed by the International Council of Harmonization (ICH) is applicable for biosimilar products as well. The WHO released guidelines on the evaluation of similar biotherapeutic products (SBPs) in 2009. These guidelines are intended for regulatory agencies to provide guidance for the evaluation, authorization, and regulation of biosimilars (WHO, 2009).

The European Union (EU) was the first to develop such a guidance document as an annex to Directive 2001/83/EC. It details a new process for the authorization of biological medicines. In this guidance, the agency added the requirement for safety and efficacy data, in addition to the studies to demonstrate the similarity between reference and biosimilar product. In the EU, the general principles for regulation of biosimilars are given in the overarching guidance document. In addition, the EU has implemented specific guidance documents to address the unique safety issues that may be associated with a particular product class. These include the guidelines for granulocyte-colony stimulating factor, somatropins, recombinant interferon alpha, low-molecular-weight heparins, recombinant erythropoietin, monoclonal antibodies, follicle stimulating hormones, interferon beta, and recombinant human soluble insulin (EMA, 2006a,b,c, 2009a,b, 2010, 2012, 2013a,b, 2015).

In Canada, the guidance for biosimilars or SEBs was developed around the same time as the WHO document and was implemented in 2010 to provide manufacturers and stakeholders with information regarding the quality, safety, and efficacy requirements for approval of biosimilar products (Health Canada, 2010).

The US Food and Drug Administration (US FDA) allows for abbreviated licensing of biosimilars under the Biologics Price Competition and Innovation Act (BPCIA) of 2009. The licensure pathway permits a biosimilar biological product to be licensed under Section 351(k) of the Public Health Service Act (PHS Act). As per the guidance document, the biosimilar products should be shown to be similar to or interchangeable with a previously licensed reference product. "A biological product is considered similar to the reference product if it is highly similar to the reference product, and if there are no clinically meaningful differences between the products in terms of safety and efficacy" (US FDA, 2015a).

13.3 CHALLENGES IN PHARMACOVIGILANCE OF BIOSIMILARS

13.3.1 Uncertainties with Safety Profile of the Product at Authorization

At the time of authorization of a biosimilar drug product, a reduced nonclinical and clinical data package is acceptable if similarity between the reference product and the biosimilar product is clearly demonstrated. Information on issues such as immunogenicity, which may lead to hypersensitivity, to infusion related reactions, and the like, have to be characterized for biosimilar products before approval.

A slight change in the manufacturing process can alter the safety profile of the protein product, resulting in the formation of antidrug antibodies (ADAs) that could alter the overall benefit–risk profile of the product. Therefore, safety issues, such as immunogenicity, are a major concern for biotherapeutic products (Chirino et al., 2004; Schellekens, 2002; Sharma, 2007). For example, such changes leading to unexpected adverse events were seen with the epoetin product EPREX and, more recently, with interferon beta, REBIF. EPREX (epoetin α) is a synthetic erythropoietin used to treat anemia in patients with chronic kidney failure. EPREX was first authorized in Europe in 1988. In 1998, there was a marked increase in cases of pure red cell aplasia (PRCA) reported with epoetin (EPO) alpha products, with the majority being in chronic kidney disease patients taking subcutaneous EPREX. PRCA is a rare disorder that manifests as sudden severe onset of anemia characterized by complete absence of red cell precursors in the bone marrow and a reticulocyte count below 10×10^9/L. Serum analyses of patients with PRCA showed the presence of

EPO antibodies. Analysis of case series showed that the spike in cases of PRCA was temporally related to the replacement of human serum albumin by polysorbate 80 as stabilizer. It was suggested that polysorbate 80 could have increased the immunogenicity of EPREX by either forming micelles with epoetin or reacting with uncoated rubber stoppers to form leachates. The manufacturer also noted that the use of polysorbate 80 made the formulation less stable and therefore more vulnerable to degradation, which may have also resulted in increased immunogenicity (Boven et al., 2005; McKoy et al., 2008).

More recently, an increase in reports of thrombotic microangiopathy was observed with use of REBIF, an interferon beta product, in patients with multiple sclerosis (Hunt, 2014). Four cases of serious, some fatal thrombotic microangiopathy (TMA) were reported in patients in south Scotland. Six additional cases of TMA were reported to regulatory authorities in the United Kingdom (UK). There were very few cases reported globally for the initial 9 years from the launch of the product. It was noted that the increase in cases of TMA was seen in only those countries that used the same serum-free formulation as that being used in the UK formulation of REBIF (Giovannoni et al., 2009; Hunt, 2014). In this study, a strong correlation was seen to one manufacturing source and one formulation, thereby emphasizing that change in formulations can alter the safety profile of a product in such a way that may have clinically significant effects.

In order to address the limitations of the clinical trials in terms of the safety profile of biotherapeutics, including biosimilar drugs, regulators have made it mandatory to submit a risk management plan that has to be agreed upon by the regulatory authorities and by the manufacturer at the time of authorization of the drug. The risk management plan includes safety specifications, and based on these specifications, a detailed pharmacovigilance plan and risk minimization plan are developed by the marketing authorization holder (MAH). Based on the risk profile of the drug, the pharmacovigilance plan may include routine activities, which involve the collection, analysis, and reporting of spontaneous adverse event reports, or certain additional safety studies may be requested postauthorization to address efficacy or specific safety concerns. A risk minimization plan includes steps to be taken by the MAH to provide all relevant information regarding benefits and risks associated with the product to the stakeholders. In Canada, this information is disseminated as part of the product label in the Canadian product monographs and also as patient leaflets for every authorized drug. In cases where additional risk minimization activities are deemed necessary to mitigate specific risks, the MAH may be required to provide educational material and alert cards. Adverse drug reactions have to be reported as required by the Food and Drug Regulations in the same way as for any other medicinal product. Once the product is authorized, the MAH is required to provide periodic safety update reports (PSURs) as per ICH E2E guidelines in order to follow up on the already known and/or new or rare adverse events and to assess the overall benefit risk profile of the drug.

In the EU, specific guidance documents for biosimilars have been developed, based on the complexity of the molecule, to address the safety issues that are unique to that product class and to focus on the requirement for testing and monitoring of the product (EMA, 2006b,c, 2009a,b, 2010, 2012, 2013a,b, 2015). Any specific safety monitoring imposed on the reference product is required for biosimilar products as well. For example, all biosimilars for epoetin are required to monitor the incidence of PRCA. In the case of monoclonal antibodies, especially those authorized for cancer indication, the EU guidance states that additional long-term immunogenicity and safety data are required postauthorization. Health Canada recommends similar monitoring for specific safety concerns with biosimilar products.

In the EU, the MAH may also be required to conduct postauthorization safety studies (PASSs) in order to identify and characterize a safety hazard, confirm the safety profile of the medicines, and to assess the effectiveness of the risk management measures. Drug utilization studies (DUSs) are also used by regulatory authorities to identify the prescribing trends in clinical practice. In Canada, with the introduction of Bill C17, also known as Vanessa's Law, Health Canada now has the regulatory authority to order additional tests and studies from sponsors. Bill C17 is crucial in giving the authority to enhance the safety oversight of drugs throughout their life cycle (Health Canada, 2014).

13.3.2 Issues with Interchangeability/Substitutability/Switching

Another challenge in the pharmacovigilance of SEBs is interchangeability. As per the consensus document sponsored by the EU, the terms *switching*, *interchanging*, and *substitution* have distinct meanings. Switching refers to the treating physician's decision to exchange one medicine for another without the consent of the patient. Interchangeability is changing the reference product for the biosimilar by a health care professional, with the expectation of the same clinical outcome without the knowledge or consent of the patient. Substitution is the practice of dispensing one medicine instead of another that is equivalent, without the knowledge of the physician or the patient (European Commission, 2013). The initial clinical trials conducted to obtain authorization for a product are usually not powered to assess differences in adverse drug reactions between biosimilar product and reference product. As already described above, the complexity of the manufacturing process and changes in the formulations can lead to differences within batches of the same product. Not only are there differences between the reference product and the biosimilar product, but different lots of the biosimilar products may also vary. Therefore, tracking of each lot is important to ensure the traceability of the product, especially in cases where a particular batch may be associated with the adverse reaction (European Commission, 2013; Ebbers and Chamberlain, 2014).

There is a theoretical possibility that interchanging an innovator product with a biosimilar product may increase the immunogenicity to the drug product, which could potentially alter the efficacy and safety of both products. If the drugs are switched or substituted automatically, it is difficult to ascertain whether the adverse drug reactions are related to the innovator product or to the biosimilar product. Changes in formulation postauthorization can lead to substantially different product characteristics than the one approved. The studies required at the postapproval stage are not designed to be comparative to the reference product and therefore can create challenges for postmarket surveillance (Ebbers and Chamberlain, 2014). In Canada, the guidance document states that SEBs are not "generic biologics" and "many characteristics associated with the authorization process and marketed use for generic drugs do not apply." Furthermore, it states that "authorization of biosimilar is not a declaration of pharmaceutical or therapeutic equivalence to the reference biologic drug." Therefore, automatic substitution and interchangeability is not advised (Health Canada, 2010). However, in Canada interchangeability and substitutability are both under provincial jurisdiction (Klein et al., 2014).

In the EU, it is acknowledged that in clinical practice at the national level, "switching" and "interchanging" of medicines might occur. Interchangeability is not directly addressed in the EMA guidance, but the guidance does mention that biosimilars are not generic products and, therefore, the decision to treat a patient with innovator product or a biosimilar product should be made following the opinion of a qualified health care professional.

According to the FDA guidance, substitution is determined at the state level in accordance with the state pharmacy laws. The product is considered interchangeable with its reference product if the MAH can demonstrate biosimilarity of the reference product and if the biosimilar product produces the same clinical result as the reference product in patients. For products that have multiple uses, the MAH is required to demonstrate that the risk in terms of safety or diminished efficacy of alternating between the use of the biosimilar product and the use of the reference product is not greater than the risk of using the reference product without such alternating or switching. The US FDA has recently approved Zarxio, a biosimilar product to Neupogen and emphasizes that it is not an interchangeable product since the company did not request approval of Zarxio as an interchangeable product. A biological product that is approved as an interchangeable product in the US may be substituted for the reference product without the intervention of the health care provider (US FDA, 2015a).

Health care professionals play a crucial role in how medicines are prescribed. In order to deal with the issues around substitutability, educational programs may be needed. As more information on biosimilar safety is gathered over the years, the issue of substitution between biosimilar product and reference product will need to be revisited.

Interchangeability, switchability, and substitution of biosimilars are discussed in Chapter 10 of this book.

13.3.3 Challenges with Naming and Tracking of Biosimilars

The introduction of biosimilars also poses a challenge for naming and tracking adverse events associated with the biosimilar and innovator product. The international nonproprietary (INN) system of naming was established by the World Health Organization in 1953 to promote "clear identification, safe prescription and dispensing of medicines to patients, and for communication and exchange of information among health professionals and scientists worldwide." All drugs are given an INN name according to a standard procedure that provides information regarding the active ingredient in the drug (WHO, 1997). The manufacturers are allowed to choose their own unique brand names. Biosimilars have unique brand names but share nonproprietary names with the respective innovator products. For example, Remicade, which is the reference product for Inflectra as well as Remsima, share the same international nonproprietary name infliximab. At present, INN names distinguish between glycosylated and nonglycosylated biosimilar products by adding a Greek letter suffix for glycosylated products as with epoetin (epoetin α, epoetin β, epoetin θ, and epoetin ζ) (US FDA, 2015b). The INN system has been very effective in the pharmacovigilance of drugs but poses a challenge in product identification and tracking of biosimilars. To prevent inadvertent substitution of products sharing the same INN name and to accurately track the biological products once they are marketed, the naming of biosimilar products and how they are prescribed becomes crucial. With several biosimilar products for a same reference product, it becomes essential that these are clearly identified by their brand names in order to clearly distinguish them from the innovator as well as other biosimilars. An adverse event associated or reported with the biosimilar product identified only by nonproprietary name may be incorrectly attributed to the reference product or to another product that is a biosimilar.

The INN for biosimilars is a topic of much ongoing discussion. The WHO is working to develop a harmonized set of product identifiers that can be used to trace the products more effectively. Health Canada will follow WHO's guidance on the INN naming of the product once it is released.

So far, various regulatory jurisdictions have some guidance in place to address the issues pertaining to naming and tracking the adverse drug reactions associated with biosimilar products. In Canada, the guidance document mentions that biosimilar products have to be prescribed by their brand names which are unique, in order to trace the adverse event to the correct product. In the EU, it is recommended that the product be prescribed by its brand name and batch number. In the US, the FDA is recommending a routine use of designated suffixes to identify biological products in order to improve pharmacovigilance and to differentiate among biological products that have not been determined to be interchangeable (US FDA, 2015a). In a notice issued in August 2015, the FDA proposed that for biological products, a nonproprietary name that includes a unique suffix composed of four lowercase letters will be designated. This naming convention is applicable to both previously licensed and newly licensed biological products. The FDA has implemented this rule for some of the filgrastim products, epoetin alfa, and infliximab. Based on this new naming convention the official names and proper names of these products would be filgrastim-bflm, filgrastim-jcwp, filgrastim-vkzt, pegfilgrastim-ljfd, epoetin alfa-cgkn, and infliximab-hjmt (US FDA, 2015b).

The WHO guidelines mention that, as for all biotherapeutics, an adequate system is necessary to ensure specific identification of the subsequent biologic products (i.e., traceability). It further states that the "national regulatory authority (NRA) shall provide a legal framework for proper pharmacovigilance surveillance and ensure the ability to identify any biotherapeutics marketed in their territory which is the subject of adverse reaction reports." This implies that an adverse reaction report for any biotherapeutic should include, in addition to the INN, other important indicators such as proprietary (brand) name, manufacturer's name, lot number, and country of origin (WHO, 2009).

Vermeer et al. published a cross-sectional study using the FDA's FAERS and EudraVigilance and found that, in the spontaneous reports available in these databases, batch numbers are not recorded for 76% and 79% of reports, respectively. Furthermore for 96.2% of the biopharmaceuticals for which a biosimilar was available in the EU, the product name could be identified. It has been found that the traceability of biopharmaceuticals, especially for those that have a biosimilar product, was much higher than that for generic products due to higher reporting of batch numbers (Vermeer et al., 2013).

How the biosimilar products are prescribed (i.e., by brand name or by INN name) and whether the lot number is recorded on the prescription will be critical in monitoring the adverse events associated with these products. Educational material for the health care providers informing them of the need to include brand names and lot numbers in the prescription will go a long way to address this issue and to maintain the standard of patient safety.

13.4 CONCLUSIONS

Postmarket surveillance of biotherapeutics is essential to monitor the safety and effectiveness of drugs under real-world conditions. Introduction of biosimilars has presented a whole new set of challenges for pharmacovigilance. Biosimilars are biological products that are similar to previously approved respective reference biotherapeutics. Biosimilars are authorized based on a reduced nonclinical and clinical data package compared to those of respective reference biologic products; therefore, their benefit–risk profiles may not be fully characterized. It is known that even minor changes in the manufacturing process of biologicals can significantly affect their benefit–risk profiles, and, of greatest concern for biosimilars, is the risk of immunogenicity. Authorized biosimilars are regulated as new drugs and have the same pharmacovigilance requirements as their respective reference products. A risk management plan is required at the time of marketing authorization application. Depending on the known safety profile of the reference product and the data provided for the biosimilar product, the MAH may be required to include additional pharmacovigilance activities such as certain postmarket safety studies, including registries. Once authorized, the manufacturer of the biosimilar product is required to submit PSURs on a regular basis. Since the safety profile of the reference product is usually well established, it is expected that the safety information included in the RMP for the biosimilar would mirror that of the reference product, in addition to the safety data generated for the biosimilar during clinical trials, to reflect a more comprehensive safety profile of the biosimilar product.

Due to the naming convention, difficulties may arise if the products are prescribed by the INN names and where the lot numbers of the products are not adequately captured. Guidelines to address this issue and the issues pertaining to naming biosimilars are still under consideration by the WHO. It is very important that biosimilar products are clearly identified in order to specifically distinguish them from the respective reference products and other biosimilar products in the class, so that the adverse reactions associated with each product can be clearly identified or traced.

Lastly, a proactive risk management approach to establish the benefit–risk profile of biosimilar products will be the cornerstone of pharmacovigilance. As new information is gained, new innovative safety monitoring tools will have to be developed to address the challenges brought about by the unique characteristics of the different biosimilars, especially issues of immunogenicity, interchangeability, and substitutability. Meanwhile, rigorous safety monitoring for biosimilar products is required in the postmarket setting to ensure their ongoing safety and efficacy.

REFERENCES

Attarwala H. (2010) TGN1412: from discovery to disaster. *Journal of Young Pharmacists* **2(3)**, 332–336.

Boven K, Stryker S, Knight J, et al. (2005) The increased incidence of pure red cell aplasia with EPREX formulation in uncoated rubber stopper syringes. *Kidney International* **67(6)**, 2346–2353.

Casadevall N, Edwards IR, Felix T, et al. (2013) Pharmacovigilance and biosimilars: considerations, needs and challenges. *Expert Opinion on Biological Therapy* **13(7)**, 1039–1047.

Chirino AJ, Ary, ML, Marshall SA. (2004) Minimizing the immunogenicity of protein therapeutics. *Drug Discovery Today* **9(2)**, 82–90.

De Groot AS, Scott DW. (2007) Immunogenicity of protein therapeutics. *Trends in Immunology* **28(11)**, 482–490.

Ebbers HC, Chamberlain P. (2014) Interchangeability: an insurmountable fifth hurdle? *Generics and Biosimilars Initiative Journal* **3(2)**, 88–93.

EMA. (2006a) Guidelines for similar biological products containing biotechnology-derived proteins as active substance: non-clinical and clinical issues. June 1, 2006. Available from: http://www.ema.europa.eu/docs/en_GB/document_library/Scientific_guideline/2009/09/WC500003920.pdf.

EMA. (2006b) Annex to guidelines for similar biological medicinal products containing biotechnology-derived proteins as active substance: non-clinical and clinical issues: guidelines on similar medicinal products containing recombinant granulocyte-colony stimulating factor. June 1, 2006. Available from: http://www.ema.europa.eu/docs/en_GB/document_library/Scientific_guideline/2009/09/WC500003955.pdf.

EMA. (2006c) Annex to guidelines for similar biological medicinal products containing biotechnology-derived proteins as active substance: non-clinical and clinical issues: guidelines on similar medicinal products containing somatropin. June 1, 2006. Available from: http://www.ema.europa.eu/docs/en_GB/document_library/Scientific_guideline/2009/09/WC500003956.pdf.

EMA. (2009a) Guideline on similar biological medicinal products containing interferon alpha. April, 2009. Available from: http://www.ema.europa.eu/docs/en_GB/document_library/Scientific_guideline/2009/09/WC500003930.pdf.

EMA. (2009b) Guideline on non-clinical and clinical development of similar biological medicinal products containing low-molecular-weight-heparins. October 2009. Available from: http://www.ema.europa.eu/docs/en_GB/document_library/Scientific_guideline/2009/09/WC500003927.pdf.

EMA. (2010) Guideline on non-clinical and clinical development of similar biological medicinal products containing recombinant erythropoietins. October 1, 2010. Available from: http://www.ema.europa.eu/docs/en_GB/document_library/Scientific_guideline/2010/04/WC500089474.pdf.

EMA. (2012) Guideline on similar biological medicinal products containing monoclonal antibodies—non-clinical and clinical issues. December 1, 2012. Available from: http://www.ema.europa.eu/docs/en_GB/document_library/Scientific_guideline/2012/06/WC500128686.pdf.

EMA. (2013a) Guideline on non-clinical and clinical development of similar biological medicinal products containing recombinant human follicle stimulating hormone (r-hFSH). September 1, 2013. Available from: http://www.ema.europa.eu/docs/en_GB/document_library/Scientific_guideline/2013/03/WC500139624.pdf.

EMA. (2013b) Guideline on similar biological medicinal products containing interferon beta. September 1, 2013. Available from: http://www.ema.europa.eu/docs/en_GB/document_library/Scientific_guideline/2013/03/WC500139622.pdf.

EMA. (2015) Guideline on non-clinical and clinical development of similar biological medicinal products containing recombinant human insulin and insulin analogues. September 1, 2015. Available from: http://www.ema.europa.eu/docs/en_GB/document_library/Scientific_guideline/2015/03/WC500184161.pdf.

European Commission. (2013) What you need to know about biosimilar medicinal products: a consensus information paper. Available from: http://ec.europa.eu/DocsRoom/documents/8242/attachments/1/translations/en/renditions/pdf.

Giovannoni G, Barbarash O, Semanaz Casser F, et al. (2009) Safety and immunogenicity of a new formulation of interferon β-1a (REBIF new formulation) in a phase IIIb study in patients with relapsing multiple sclerosis: 96 week results. *Multiple Sclerosis* **15(2)**, 219–228.

Health Canada. (2010) Guidance for sponsors information and submission requirements for subsequent entry biologics. March 5, 2010. Available from: http://www.hc-sc.gc.ca/dhp-mps/alt_formats/pdf/brgtherap/applic-demande/guides/seb-pbu/seb-pbu-2010-eng.pdf.

Health Canada. (2014) Protecting Canadians from Unsafe Drugs Act (Vanessa's Law). November 6, 2014. Available from: http://www.hc-sc.gc.ca/dhp-mps/legislation/unsafedrugs-droguesdangereuses-eng.php.

Hunt D. (2014) Thrombotic microangiopathy associated with interferon beta. *New England Journal of Medicine* **370**, 1270–1271.

ICH E2E. (2004) Guideline on pharmacovigilance planning. November 18, 2004. Available from: http://www.ich.org/fileadmin/Public_Web_Site/ICH_Products/Guidelines/Efficacy/E2E/Step4/E2E_Guideline.pdf.

Klein AV, Wang J, Bedford P. (2014) Subsequent entry biologics (biosimilars) in Canada: approaches to interchangeability and the extrapolation of indications and uses. *Generics and Biosimilars Initiative Journal* **3(3)**, 150–154.

McKoy JM, Stonecash RE, Cournoyer D, et al. (2008) Epoetin associated pure red cell aplasia: past, present and future considerations. *Transfusion* **48(8)**, 1754–1762.

Meyboom RH, Egberts AC, Edwards IR, et al. (1997) Principles of signal detection in pharmacovigilance. *Drug Safety* **16(6)**, 355–365.

Poluzzi E, Rasch, E, Piccinni C, De Ponti F. (2012) Data mining techniques in pharmacovigilance: analysis of the publicly accessible FDA Adverse Event Reporting System (AERS). In: *Data Mining Applications in Engineering and Medicine*, Karahoca A, Ed., Available from: http://www.intechopen.com/books/data-mining-applications-in-engineering-and-medicine/data-mining-techniques-in-pharmacovigilance-analysis-of-the-publicly-accessible-fda-adverse-event-re.

Schellekens H. (2002) Bioequivalence and the immunogenicity of biopharmaceuticals. *Nature Reviews Drug Discovery* **1(6)**, 457–462.

Sharma B. (2007) Immunogenicity of therapeutic proteins. Part 3: impact of manufacturing changes *Biotechnology Advances* **25(3)**, 325–331.

Socinski MA, Curigliano G, Jacobs I, et al. (2015) Clinical considerations for the development of biosimilars in oncology. *Monoclonal Antibodies* **7(2)**, 286–293.

Suntharalingam G, Perry MR, Ward S, et al. (2006) Cytokine storm in a phase 1 trial of the anti-CD28 monoclonal antibody TGN1412. *New England Journal of Medicine* **355**, 1018–1028.

US FDA. (2015a) Scientific considerations in demonstrating biosimilarity to a reference product guidance for industry. April 2015. Available from: http://www.fda.gov/downloads/Drugs/GuidanceComplianceRegulatoryInformation/Guidances/UCM291128.pdf.

US FDA. (2015b) Nonproprietary naming of biologic products guidance for industry. August 2015. Available from: http://www.fda.gov/downloads/Drugs/GuidanceCompliance RegulatoryInformation/Guidances/UCM459987.pdf.

Vermeer NS, Straus SM, Mantel-Teeuwisse AK, et al. (2013) Traceability of biopharmaceuticals in spontaneous reporting system: a cross sectional study in FDA Adverse Event Reporting System (FAERS) and EudraVigilance databases. *Drug Safety* **36(8)**, 617–625.

Weise M, Bielsky M-C, Smet K, et al. (2012) Biosimilars: what clinicians should know. *Blood* **120(26)**, 5111–5117.

WHO. (1997) Guidelines on use of international nonproprietary names (INNs) for pharmaceutical substances. Available from: http://apps.who.int/iris/bitstream/10665/63779/1/who_pharm_s_nom_1570.pdf.

WHO. (2002) The importance of pharmacovigilance: safety monitoring of medicinal products. Available from: http://apps.who.int/medicinedocs/pdf/s4893e/s4893e.pdf.

WHO. (2004) Pharmacovigilance: ensuring the safe use of medicines. WHO policy perspectives on medicines. October 2004. Available from: http://apps.who.int/medicinedocs/pdf/s6164e/s6164e.pdf.

WHO. (2009) Guidelines on evaluation of similar biotherapeutic products (SBPs) expert committee on biological standardization. October 2009. Available from: http://www.who.int/biologicals/areas/biological_therapeutics/biotherapeutics_for_web_22april2010.pdf.

Zuniga L, Calvo B. (2010) Biosimilars: pharmacovigilance and risk management. *Pharmacoepidemiology and Drug Safety* **19(7)**, 661–669.

14 Patent Exclusivities Affecting Biosimilars in the United States, Canada, and Europe

Noel Courage
Bereskin & Parr LLP

Lynn C. Tyler
Barnes & Thornburg LLP

CONTENTS

14.1 INTRODUCTION

A biosimilar drug can reference a pioneering drug's clinical and safety package to reduce the burden for regulatory approval. Meeting the health and safety requirements for approval is only part of the concern for biosimilar manufacturers. They must also navigate exclusivities in relation to the pioneering drug. These exclusivities may be either the patent protection or a regulatory exclusivity on the pioneering drug. In some countries, resolution of patent issues can be linked to regulatory approval, as a precondition to the government providing marketing authorization. This chapter reviews the patent exclusivities affecting the biosimilar approval process. Now that the regulatory pathways are becoming more clearly laid out, patent owners with biological drugs on the market are attempting to improve their patent barriers, while biosimilar manufacturers are trying to find freedom-to-operate with respect to patented drugs. As an example, Celltrion and Hospira had to navigate freedom-to-operate against Johnson & Johnson's patents in order to get a biosimilar infliximab antibody approved by regulators in the US (2016), Canada (2014), and the European Union (EU, 2013; Loftus, 2016).

A brief overview of the regulatory process in Canada, the US, and Europe will be provided, followed by an analysis of the pertinent patent issues affecting biosimilars. It will be seen that patent owners and biosimilar manufacturers have to approach patent issues on a region-by-region basis because each jurisdiction differs in the way that patents may be linked, if at all, to the biosimilar regulatory approval process.

14.2 PATENT OVERVIEW

A brief overview of patents and patent enforcement is provided below.

14.2.1 Common Characteristics of a Patent

Patents provide the exclusive right to exploit an invention. The precise nature of the rights varies from country to country, but they generally provide the exclusive right to make, construct, use, offer to sell, and sell the invention. For a biological drug, examples of patentable inventions can include the compound per se, pharmaceutical formulations, processes of manufacture, and medical uses. A patent has a maximum term of 20 years from filing date (see, e.g., Canadian Patent Act; U.S. Code: Title 35). The patent term may be extended under local law in some regions. The US has patent term extension available for certain regulatory delays by the Food and Drug Administration in reviewing and approving the drug. The US also has patent term extension for certain delays by the US Patent and Trademark Office (USPTO). Europe has patent term extension available for certain regulatory delays, called a Supplementary Protection Certificate (SPC). Canada does not have patent term extension, but will implement a maximum 2-year extension for the future following the Canada–EU free trade agreement reached in 2013.

In order to receive the exclusive right, the patent applicant has to file a patent application that contains a description sufficient to show others skilled in that area how to make and use the invention. Most countries examine patent applications as a precondition to granting a patent to ensure the invention meets the required criteria for patentability. These criteria include novelty and inventiveness, at a minimum. The patent application must also claim patentable subject matter and have proper support and description of the invention. The specific terminology and standards vary between regions. If a patent applicant is able to successfully negotiate patent claim scope with the patent office, then a patent can be granted. There is no guarantee of obtaining an issued patent.

There is a risk of "submarine patents" in the US and Canada, more so in the US. A small number of older patent filings are confidential until issuance, so their progress cannot be monitored. Under the old patent term rules, these patents have their term calculated from their patent grant date, not their filing date. Canada changed its old rules for those patent applications filed on or after October 1, 1989. The US changed its corresponding rules on patent term in June 1996. Therefore, submarine patents are a concern at present but will be less of a problem as time goes on and fewer patents remain pending under the old rules. In 2011, Hoffmann-La Roche and Amgen obtained new US patents covering the drug Enbrel (etanercept), which was a product believed to be nearing patent expiry (Graeser, 2012). At a minimum, the issuance of US patents extending the protection for Enbrel to 2028 surprised Sandoz which, according to the Complaint it filed, had taken "significant steps in preparation for commercializing [a biosimilar] etanercept in reliance on the expectation it would be able to begin marketing immediately upon FDA approval and without the cloud of potential infringement claims" (*Sandoz vs. Amgen*, 2013).

Therefore, some biosimilar developers of older drugs face a risk that after investing many millions of dollars on the development of a biosimilar, a new patent will issue in the US or Canada and preclude commercialization of the biosimilar for many more years. As recently as January 2014, there were still 450 confidential patent applications at the USPTO that had been filed before June 8, 1995, and thus will receive 17 years of patent term if issued as patents (Crouch, 2014). Twenty patents were issued on such applications in 2013, and five of those were in the biotech space (US Patents 8,357,513, 8,399,250, 8,507,196, 8,557,768, and 8,603,777, 2013).

14.2.2 Patent Opposition/Postgrant Review

Prior art that could affect patentability can typically be filed at patent offices in Canada, the US, and Europe, while a patent application is pending. The patent office examiner will consider the relevance of this information during the examination process. The rules vary between countries on what can be filed, and when.

Certain patent offices allow an approved patent application to be opposed prior to grant, or shortly after grant. There are limited time windows for opposition to be filed. Most countries have some sort of postgrant patent review process, such as reexamination, where, at a minimum, new prior art and substantial new issues of patentability may be raised.

14.2.3 Patent Litigation

After patent grant, the infringer may be sued in court for patent infringement and held liable for damages caused by the infringement. An infringer can also be liable to pay reasonable compensation for its activities for the period between publication and before patent grant. (For an example of such a law, see the Canadian case, *Baker Petrolite Corp. vs. Canwell Enviro-Industries Ltd.*, 2002.) In some countries, such as Canada and the UK, the patent owner can elect between damages and an accounting of profits (i.e., disgorgement of infringer's profits). The accounting of profits is an unusual remedy in that the patent owner can take the infringer's profits arising from the infringement, regardless of whether the patent owner suffered any damages. Punitive damages may be requested, but the likelihood of such an award varies between jurisdictions. The US courts can award punitive damages up to three times the assessed amount of damages for willful infringement (35 U.S. Code § 284—Damages). Punitive damages are infrequently awarded in Canada and are relatively insubstantial compared to the US (*Bell Helicopter Textron Canada Limitée vs. Eurocopter*, 2013).

An injunction and delivery up to the patent owner of the infringing drug may also be ordered after a trial. An injunction may also be requested before trial but is unlikely to be granted in the US, Europe, or Canada in pharmaceutical patent cases. Injunctions are infrequent because pharmaceutical cases are complex, and the damage caused by infringement is usually compensable by damages so unlikely to cause irreparable harm to the patent owner.

The biosimilar company may challenge the validity of a patent, for example, that the patent claims lack novelty, inventiveness, or support. A patent is presumed valid during litigation, and the onus is on the challenger to invalidate a patent. Each country has different laws and standards on patent validity issues. At least a couple of years (often much longer) elapse before there is a trial decision on infringement and validity.

14.2.4 Linkage of Patents to Biosimilar Approval Process

Biosimilar approval by a regulatory agency can be linked to patent issues in some jurisdictions. "Linkage" rules require biosimilar drug companies to establish freedom-to-operate (clearance) with respect to certain patents as a *precondition* to market authorization. A patent owner with a product on the market can sue to try to stop the biosimilar manufacturer from getting its marketing authorization.

This is in sharp contrast to conventional patent enforcement litigation, which typically begins only *after* marketing authorization is granted and a competitor drug is launched on the market. The focus of conventional patent litigation is not blocking the government from issuing the marketing approval, but suing for patent infringement and requesting remedies such as an injunction and damages.

14.2.5 Settlement

Companies may settle patent issues between themselves, without litigation. For example, the biosimilar developer may enter an agreement with the reference product sponsor in which the reference product sponsor agrees not to challenge the entry of the biosimilar into the market. The parties are free to make a settlement agreement, subject to compliance with antitrust laws (*FTC vs. Actavis, Inc.*, 2013).

14.3 CANADIAN BIOSIMILAR APPROVAL PROCESS

Biosimilars are regulated in Canada under Health Canada's preexisting powers granted by the Food and Drugs Act and Regulations (Food and Drugs Act, 1985; Food and Drug Regulations, 2016). Biosimilars are reviewed and approved via the same pathway that has been used for many years for assessing new small-molecule pharmaceutical drugs. This pathway involves a new drug submission (NDS) showing the drug's safety, effectiveness, and quality. Health Canada approves drugs for marketing by issuing a marketing authorization, called a Notice of Compliance (NOC). The NOC is issued only after a drug manufacturer's NDS is approved. Biosimilars do *not* proceed via an abbreviated new drug submission, which is the pathway used for conventional generic pharmaceutical drugs. Biosimilars are not bioequivalent to the reference product and have other characteristics considered different enough that Health Canada will usually not approve them via an abbreviated new drug submission (ANDS). The biosimilar manufacturer may nonetheless benefit from a reduced regulatory burden by relying on the previously approved drug's clinical and safety data.

A high-level overview of the regulatory approval requirements for biosimilar drugs is provided in the Health Canada guidance document for sponsors: *Information and Submission Requirements for Subsequent Entry Biologics (SEBs)* (Health Canada, 2010). Health Canada also recommends that biosimilar manufacturers review Europe's extensive published guidance since it is often consistent with the Canadian approach. To date, at least a somatropin, filgrastim, and monoclonal antibody (infliximab) have been approved in Canada. Omnitrope (somatropin; human growth hormone) is the earliest example of a biosimilar that has been approved in Canada (Health Canada, 2009) by relying on clinical data for Pfizer's innovator somatropin product.

14.3.1 Linkage of Canadian Patents to Biosimilar Approval Process

Biosimilar approval can be linked to patent issues in Canada. There is a specialized litigation option available to protect certain patented biosimilar drugs on the market. The process may block a second-entry biological drug that is referencing a patent owner's clinical trial data. This unique Canadian system is informally called the NOC Regulations [*Patented Medicines (Notice of Compliance) Regulations*], and it is well established from its use in relation to conventional pharmaceutical drugs.

14.3.2 The Health Canada Patent Register

In connection with the NOC Regulations, Health Canada maintains its own Patent Register, independent from the Canadian Intellectual Property Office, relating to medicines and their use. However, Health Canada does not have any involvement in examining or issuing patents—it keeps a list of relevant already-issued patents. The Patent Register is somewhat analogous to the US "Orange Book" for conventional pharmaceutical litigation.

The sole purpose of the Register is to prevent patent infringement in certain circumstances. Only a company that applies for, or already has, an NOC for an approved, patented drug may be eligible to take advantage of this specialized process. While only certain types of patents qualify, Health Canada clearly stated that biologic patents are eligible for the Register (Health Canada, 2015). To be eligible for listing, a patent must claim either an approved:

- Medicinal ingredient;
- Formulation that contains the medicinal ingredient;
- Dosage form; or
- Use of the medicinal ingredient.

The Patent Register also has strict, nonextendable time limits. A patent can be listed at one of two times:

- A patent that is issued at the time of filing an NDS (or supplementary NDS) must be listed at the time of filing the submission; or
- A patent application that has a filing date before the filing of the NDS (or supplementary NDS), must be listed within 30 days after the issuance of the patent.

Health Canada vets all requests to list patents on the Register. If a patent is refused listing, Health Canada gives the requestor the opportunity to make written arguments in favor of listing.

14.3.3 Patent Register Procedure

In order to receive marketing authorization, the biosimilar company must address freedom-to-operate with respect to patents on the Health Canada Patent Register. If there are no relevant patents on the Patent Register, then Health Canada will not have to hold up drug approval pending resolution of patent issues. The Patent Register provides significant benefits to a patent owner by keeping a competitor from entering the market. In contrast, prior to the creation of the Patent Register, the patent owner often had to use conventional patent litigation to chase a competitor for patent infringement *after* the medicine was already on the market. This created inefficiencies since interlocutory injunctions are difficult to obtain and it takes a long time to bring a patent infringement case to a trial. The brand-name company would lose significant market share in the meantime.

The preemptive NOC Proceeding is a faster and cheaper way to keep a competitor off the market than a patent infringement trial. The filing of the NOC Proceeding starts an automatic 24-month stay of NOC issuance to the biosimilar company (it is, in effect, like an injunction). The NOC may only be issued when the patent expires or the biosimilar company wins the NOC Proceeding. In contrast, as mentioned above, it is typically very difficult to obtain an interlocutory injunction against a competitor in Canada in a patent infringement lawsuit.

A disadvantage of filing the NOC Proceeding is in the event that the generic or biosimilar company wins, the patent owner may be liable to the competitor for costs and damages for delaying biosimilar drug entry into the market. The NOC Regulations have been extensively litigated in the context of conventional, small-molecule pharmaceuticals. The NOC Regulations will become an important patent protection tool as biosimilar development increases.

An early biosimilar NOC Proceeding involved Teva's biosimilar version of filgrastim (*Amgen Canada Inc. et al. vs. Teva Pharmaceutical Industries Ltd. et al.*, 2012). Amgen had the first approved filgrastim product in Canada, under the brand name Neupogen, whose patent expires on July 31, 2024. Amgen started an NOC Proceeding in an attempt to block Teva's biosimilar filgrastim. In defense, Teva had alleged that claims of the Amgen patent in issue were either invalid, infringed, or not relevant. The case was settled in August 2013. The terms of the Canadian settlement are unknown since neither company appears to have issued a press release on its terms. In particular, it was not publicly stated whether Teva agreed to keep its filgrastim product off the market for a period of time. In the US patent litigation on the filgrastim product (which is completely independent from the Canadian litigation), Teva had earlier admitted in a settlement that certain Amgen US patent claims were infringed, valid, and enforceable, and a court injunction was issued that would keep Teva off the market until December 2013. Teva did not have its Canadian NOC as of April 1, 2016. Apotex won its NOC Proceeding with Amgen and has its NOC for filgrastim and launched its product (*Amgen Canada Inc. vs. Apotex Inc.*, 2015). Amgen Canada Inc. and Samsung Bioepis Co., Ltd. are currently in an NOC Proceeding regarding Samsung's enteracept, which involves infringement and validity issues (*Amgen Canada Inc. et al. vs. Samsung Bioepis Co., Ltd. and the Minister of Health*, 2015).

14.3.4 Interaction of NOC Regulations with Conventional Patent Infringement Lawsuit

The NOC Proceeding is not a patent infringement or validity action; rather, it is a summary application proceeding, the outcome of which is a decision on whether or not the biosimilar manufacturer may get its NOC and enter the marketplace prior to patent expiry. Patent issues are taken into account in the NOC Proceeding, but there is no final determination of infringement or validity. The Canadian government intends to amend laws to turn NOC Proceedings into full actions.

A conventional patent infringement lawsuit may be brought by a patent owner, irrespective of whether it has engaged in an NOC Proceeding. The patent owner can commence the patent infringement lawsuit in Federal Court, even if it loses in an NOC Proceeding. Likewise, the biosimilar company may lose in the NOC Proceeding but later establish at a trial that the patent is invalid or not infringed. Canadian patent infringement litigation is more extensive and permits discovery and trial testimony.

14.3.5 Risk Mitigation: Postgrant Patent Office Challenges in Canada

The grounds to oppose a patent application or patent are quite limited in Canada. Printed prior art publications may be filed any time during the pendency of an application, along with a statement of pertinency (Patent Act, s. 34). Another basis of challenge is reexamination (Patent Act, s. 48.1 to 48.5). Either the patent owner or a third party may request reexamination of a granted patent at the Canadian Patent Office. The outcome of reexamination could be maintenance of the same claim scope, narrowed patent claim scope, or patent refusal. The patentee has the option to submit amended claims, if necessary, to try to salvage its patent. The person requesting reexamination must provide printed prior art and show its pertinency and the existence of a substantial new question of patentability. Reexamination is not commonly used to try to revoke a patent because the challenger is not allowed to participate beyond its first submission of comments on relevance. The patent owner may then engage in multiple rounds of argument and claim amendments with the Patent Office. So reexamination is typically risky for the challenger unless the prior art is very close and likely to defeat novelty.

14.4 THE US BIOSIMILAR APPROVAL PROCESS

The US federal government has created a specially designed, abbreviated biosimilar approval pathway. The Biologics Price Competition and Innovation Act (BPCIA) was part of the Patient Protection and Affordable Care Act (aka "ObamaCare") (H.R. 3590, 111th Congress, 2010) and created the regulatory process by which competitive versions of a previously licensed biological product could be brought to market (see § 7002). Prior to the BPCIA system, certain biosimilar manufacturers could attempt to obtain biosimilar approvals through Section 505(b)(2) of the Federal Food, Drug, and Cosmetic Act or a standard new biologic license application (BLA) (Public Health Service Act—Regulation of Biological Products, Section 351). These avenues are still open, but the BPCIA is expected to become the primary process for biosimilar approval.

As amended, Section 351(i) of the Public Health Service (PHS) Act defines *biosimilarity* to mean "that the biological product is highly similar to the reference product notwithstanding minor differences in clinically inactive components" and that "there are no clinically meaningful differences between the biological product and the reference product in terms of the safety, purity, and potency of the product." It was not until March 2015 that the first biosimilar, Sandoz's filgrastim (Zarxio), was approved in the US by the BPCIA pathway. The product did not launch, however, until September 2015 as a result of litigation with Amgen, discussed below. The FDA assigned the product a temporary generic name "filgrastim-sndz" and still developing its policies on biosimilar naming. It recently issued a draft guidance on the naming of biosimilars (US FDA, 2015a) that proposes to use an arbitrary, four-letter suffix on the names of biosimilars to distinguish them from their reference product counterparts.

Although more properly characterized as the absence of a reward or incentive rather than the presence of a risk, it is nonetheless worth noting that under the BPCIA the first biosimilar to market does not receive any period of marketing exclusivity, unlike a generic drug. Rather, the BPCIA gives a 1-year exclusivity only to a follow-on biologic that qualifies as interchangeable with the reference product (42 U.S. Code § 262—Regulation of biological products).

The BPCIA laid out the biosimilar pathway in general terms. The FDA then issued guidance documents elaborating on the pathway in April 2015 (US FDA, 2015b,c,d). The FDA published scientific guidance, quality guidance, and a list of questions and answers. Much remains to be learned about the regulatory pathway. Some recent court decisions, discussed below, are starting to fill in gaps as to how the regulatory regime will interact with patent issues.

14.4.1 Linkage of US Patents to Biosimilar Approval Process

The BPCIA includes several provisions directed to resolving patent disputes through either settlement or litigation. The BPCIA has some conceptual similarity to the Drug Price Competition and Patent Term Restoration Act of 1984 (the Hatch–Waxman Act), which provides a path for market entry to generic small-molecule pharmaceuticals. Patent issues were not previously linked to the biologic regulatory approval system, so this is a significant change of the new law. There are also very distinct differences in the patent litigation pathways under the BPCIA and the Hatch–Waxman Act. For example, there is no Orange Book for listing patents under the BPCIA. Instead, there is a process for a complex, private exchange of information between competitors. The patent owner may be informed of at least a portion of the biosimilar company's confidential manufacturing process, and the biosimilar manufacturer is informed of the innovator's patents. This process also produces patent litigation options, as discussed in more detail below.

The BPCIA litigation provisions contain several "patent" ambiguities in key areas, particularly whether various lists of patents to be litigated are exclusive. Courts will have to resolve these issues.

Some review of the litigation provisions is necessary to understand the more important areas of contention. The Act prescribes a rather elaborate interaction, informally referred to by some in the industry as a "dance," in which reference product sponsors and biosimilar applicants (referred to as "subsection (k) applicants" based on the subsection of the statute that creates biosimilar applications) may engage before commencing any patent infringement litigation (42 U.S.C. § 262(l), 2013). The dance can begin when the applicant submits to the FDA an application for approval of a biosimilar drug. The statute states that "[w]hen a subsection (k) applicant submits an application" to the FDA, the applicant will give a copy of the application to one in-house lawyer for the reference product sponsor and to outside counsel for the sponsor, subject to certain confidentiality restrictions. Later, the statute states that the copy of the application "shall" be provided to the sponsor "[n]ot later than 20 days after the Secretary [through the FDA] notifies the subsection (k) applicant that the application has been accepted for review." In addition, at that point the applicant must also provide "such other information that describes the process or processes used to manufacture the biological product that is the subject of the application."

Although the language appears to require disclosure of the application, a Court recently decided that it was optional. In *Amgen Inc. vs. Sandoz Inc.* (2016), Sandoz offered to provide portions of its abbreviated biologics license application (aBLA) for Zarxio to the reference product manufacturer, Amgen. In reply, Amgen demanded the entire application. Amgen and Sandoz did not agree on the extent of disclosure of the aBLA required by the BPCIA. The Court decided that disclosure of the aBLA was not mandatory. The word "shall" in paragraph (l)(2)(A) was not read in isolation, and when read in the context of the statute, it did not mean "must." The Court stated that it was not mandatory because the statute also contained an option for the reference product sponsor to take to penalize the biosimilar applicant for noncompliance with the statute (42 U.S. Code § 262—Regulation of biological products and 35 U.S. Code § 271—Infringement of patent). Therefore, even though the BPCIA prescribes an elaborate information exchange and patent review process, the first biosimilar was approved under the BPCIA without going through that full process.

The FDA will also not get involved between the parties to require the biosimilar company to disclose its drug application and manufacturing process. This was decided when Amgen filed a Citizen's Petition requesting that the FDA require Momenta to disclose its application to Amgen before the FDA begins its review process. In a decision on March 25, 2015, the FDA denied the Citizen's Petition. The FDA's position was that the BPCIA does not contemplate FDA involvement in monitoring or enforcing the exchange of information between the reference product and biosimilar manufacturers.

Although Sandoz and Amgen bypassed the BPCIA patent litigation process, there is a case pending, *Amgen Inc. vs. Apotex Inc.* (2015), in which the BPCIA patent information exchange step has been completed by mutual agreement. A brief overview of the patent information exchange will be provided before discussion of the case. Within 60 days after the receipt of the biosimilar applicant's application, the sponsor (patent owner) must provide the applicant with a list of patents that the sponsor believes "could reasonably be asserted" and identify any that are available for license. Sixty days after receiving the sponsor's list of patents, the applicant must provide the sponsor with (1) its own list of patents that it believes could be asserted, and either (2) a detailed statement, on a claim by claim basis, of the factual and legal basis why each patent on the sponsor's and applicant's (if any) list(s) is invalid, unenforceable, or would not be infringed, or (3) a statement that the applicant does not intend to market the product before the patent expires. The applicant must also provide a response to the sponsor's indication of patents that are available for license. The final step in this phase is that, within 60 days of receiving the applicant's detailed statement, the sponsor must provide its own detailed statement, again on a claim-by-claim basis, of the factual and legal basis why each patent will be infringed and a response to the applicant's statement on validity and enforceability.

If the parties cannot agree on patents to be litigated, the statute goes on to prescribe another set of steps in the prelitigation dance. The first of these steps is that the applicant notifies the sponsor of the number of patents the applicant will include on a list of patents to be litigated. Five days later, the parties simultaneously exchange lists of patents that each believes "should be the subject of an action for patent infringement." The number of patents on the sponsor's list cannot exceed the number on the applicant's list, unless the applicant's list does not include any patents, in which case the sponsor can list one patent.

In the *Amgen vs. Apotex* case, Apotex submitted a BLA to the FDA for a biosimilar version of Amgen's Neulasta (pegfilgrastim). Beginning in December 2014, the parties exchanged information and statements as required by the BPCIA, and subsequently two US patents were designated for inclusion in the patent action. Amgen had standing to bring the patent infringement action because it was an act of potential infringement by Sandoz to submit a BLA to the FDA seeking approval of its pegfilgrastim when Amgen, the reference product owner, had patents listed with the FDA under the BPCIA.

Here are the steps in the patent dance as they occurred between Amgen and Apotex, as stated in the Amgen Complaint:

> "...46. On December 16, 2014, Amgen received a letter from in-house counsel for Apotex Inc., notifying Amgen that the Apotex BLA had been accepted for review by FDA and that Apotex intended to provide Amgen the Apotex BLA pursuant to 42 U.S.C. § 262(*l*)(2).
>
> 47. Subsequently, Amgen received a copy of the Apotex BLA under the confidentiality provisions set forth in 42 U.S.C. § 262(*l*)(1).
>
> 48. Pursuant to 42 U.S.C. § 262(*l*)(3)(A), on February 27, 2015, Amgen provided Apotex a list of patents for which it believed a claim of patent infringement could reasonably be asserted against the Apotex Pegfilgrastim Product ("Amgen's (*l*)(3)(A) list"). Amgen's (*l*)(3)(A) list included the Patents in Suit.
>
> 49. On April 17, 2015, Apotex provided Amgen with its statements designated as being in accordance with 42 U.S.C. § 262(*l*)(3)(B).
>
> 50. On June 16, 2015, Amgen provided Apotex with a detailed statement, pursuant to 42 U.S.C. § 262(*l*)(3)(C).
>
> 51. Between June 22, 2015 and July 7, 2015, Amgen and Apotex engaged in good faith negotiations, pursuant to 42 U.S.C. § 262(*l*)(4). On July 7, 2015, Amgen and Apotex agreed that the Patents in Suit should be the subject of any patent infringement action brought pursuant to 42 U.S.C. § 262(*l*)(6)(A)..."

Subsequent to the Complaint, Amgen filed its Answer with Counterclaims in October 2015. Although Apotex agreed with Amgen on the patents to be included in BPCIA litigation, Apotex denied infringing and asserted that the patents were invalid. Apotex also asserted that Amgen was violating the Sherman Act (US competition law) by launching a suit for one of its method of manufacture patents. Apotex alleged that it provided Amgen with clear and irrefutable proof of noninfringement during the information exchange. Apotex believes that competition law applies if Amgen included this patent for the purpose of significantly delaying resolution of patent litigation and Apotex's entry into the market. Apotex also raised a number of issues about BPCIA procedures. It requested a declaration that a notice of commercial marketing to Amgen (42 U.S. Code § 262—Regulation of biological products) is not mandatory, so that noncompliance cannot be basis for an injunction. Apotex did acknowledge that other penalties potentially apply under 42 U.S.C. § 262(1)(9)(B). Apotex lost on the notice issue in an initial court decision and appealed.

The Court of Appeals affirmed the injunction, however, stating that the 180-day notice of commercial marketing is mandatory and cannot be given until the FDA has licensed the biosimilar. According to the Court, the 180-day notice provision was intended to give the trial court and parties time to complete a preliminary injunction hearing in an orderly manner. The Court rejected Apotex's argument that making the 180-day notice requirement mandatory extended the statutory exclusivity for the reference product by 6 months. The Court reasoned that, as time passes, reference products will be newer and biosimilar applicants will be able to file their applications long before the 12-year statutory exclusivity expires. As a result, the applicant will receive its license and can give its notice well within that period. Finally, the Court rejected Apotex's argument that the sole remedy for failure to give the 180-day notice is for the reference product sponsor to file a declaratory judgment action. The statute does not contain any language making the declaratory judgment action an exclusive remedy, and the Court found the heavy burden to infer exclusivity was not met. As a result, unless further appeals change the outcome, Apotex will have to wait until the FDA approves its application to give the 180-day notice.

Whether the parties agreed on a list of patents to be litigated (as in *Amgen vs. Apotex*) or disagreed and exchanged lists, the sponsor must file an infringement suit within 30 days of completing the applicable process. If the parties agreed on patents to be included, the sponsor's suit must include those patents. If the parties did not agree, the sponsor's suit must include all the patents on the respective lists.

At this point, one of the key ambiguities becomes apparent. Recall that by now two sets of lists have been generated, the first (referred to as a "Paragraph 3" list based on its place in the statute) identifying all patents that either party thought "could reasonably be asserted," and the second (a "Paragraph 5" list) identifying the patents that the parties thought should be involved in litigation. What happens if the Paragraph 3 list is longer than the Paragraph 5 list? Can the sponsor bring suit on patents that were included on the Paragraph 3 lists but not the Paragraph 5 lists? This portion of the statute does not say that the sponsor cannot include such patents, only that the sponsor must include the patents on the agreed list or the patents on both parties' lists if the parties did not agree. Further, Section 271(e)(2) of the Patent Act has been amended to provide that an applicant infringes each patent on a Paragraph 3 list by filing an application for approval of the product (and recall that the lists are not generated until after the application has been filed). If the Paragraph 5 lists are exclusive, Congress would have eliminated key property rights of sponsors in some cases. These considerations make it appear that the statute does not limit patent infringement suits to patents on a Paragraph 5 list.

On the other hand, the provisions governing the creation of the Paragraph 5 lists would arguably be meaningless if they did not limit the potential patents-in-suit. What is the point of allowing the applicant to limit the number of patents on the Paragraph 5 lists if the sponsor can sue on any and all patents it chooses? One principle of statutory construction is that statutes are to be construed as a whole, giving meaning to all the provisions. For example, *TRW Inc. vs. Andrews*, 2001 comments: "It is a cardinal principle of statutory construction that a statute ought, upon the whole, to be so construed that, if it can be prevented, no clause, sentence, or word shall be superfluous, void, or insignificant" (internal quotation marks omitted).

This consideration suggests that the statute does limit litigation to patents on one of the Paragraph 5 lists, at least initially (as will be explained below).

As noted above, an appellate court has already decided that the overall "patent dance" is optional. More recent cases raise a similar issue: namely, once the parties take the initial steps, must they complete the dance? In both cases, Amgen and affiliates sued Sandoz and affiliates (*Amgen, Inc. vs. Sandoz, Inc.*, 2016), in one case over Sandoz's application for a biosimilar to Amgen's Enbrel, the same product for which Sandoz sought a declaratory judgment against Amgen in the first US case discussed above. A second case is based on Sandoz's application for a biosimilar to Neulasta. In both cases, Amgen alleges that Sandoz started the patent dance by providing Amgen with its application and manufacturing information, that Amgen responded with a list of patents of which infringement could reasonably be alleged, and that Sandoz then stopped the music and told Amgen to file suit. Amgen obliged by seeking a declaratory judgment that Sandoz had violated the BPCIA, that there could be no immediate action for patent infringement until the parties had completed the patent dance, and that Amgen would still be entitled to all the remedies for patent infringement, including lost profits and injunctive relief. These cases reinforce the trend that biosimilar applicants view the patent dance procedures as a waste of time that delays the resolution of litigation and their entry into the market. Conversely, the patent owners want the procedures to be completed in full before any litigation begins.

Another section of the statute addresses preliminary injunctions. This section provides that the applicant must give the sponsor 180 days' advance notice of its intention to begin commercial marketing of the biosimilar (42 U.S. Code § 262—Regulation of biological products). Between its receipt of the notice and the expiration of the 180 days, the sponsor can seek a preliminary injunction against sales of the applicant's biosimilar based on any patent that (1) was included on a Paragraph 3 list but (2) was not included on either an agreed list of patents for litigation or a Paragraph 5 list (or, under another section of the statute, based on a patent that issued or was licensed after the sponsor created its Paragraph 3 list).

The 180-day notice of intent to begin commercial marketing does not grant standing to commence a declaratory judgment action before a biosimilar application under 262(k) submitted to the FDA is approved as complying with the requirements for licensing. This arguably provides an additional 6 months exclusivity to the reference product based on the initial court decisions discussed above. The action must meet the requirements of "immediacy and reality." There is no immediacy if the biosimilar manufacturer is still conducting Phase III trials because the product may not yet be final at this point (though it should be near-final for the trial results to be relied upon). It is also unlikely that a court would find immediate and significant impact on the biosimilar company. The biosimilar company may nonetheless try to establish standing at an earlier stage if it can rely on other conventional patent law bases for a declaratory judgment such as having been threatened with a patent infringement suit. It is not clear if the BPCIA could preclude this conventional declaratory judgment jurisdiction until after the biosimilar company had proceeded through the BPCIA framework (*Celltrion Healthcare Co., Ltd. vs. Kennedy*, 2014). The standing issue also has to be reviewed on a case-by-case basis, since merely stating ownership of patents (per *Sandoz*) or having previously asserted patents in the US and elsewhere (per *Celltrion*) against other companies may not provide standing.

An unresolved issue raised by this section is exclusivity again. Can the sponsor only seek a preliminary injunction based on a patent that falls into one of these categories? What if there is pending litigation involving the agreed or Paragraph 5 list patents? Can a patentee seek a preliminary injunction based on those also? The statute does not expressly say that the sponsor cannot. If the answer is that the patentee can seek a preliminary injunction based on such patents, however, what is the point of the provision limiting the requests for preliminary injunctions to patents on Paragraph 3 lists but not the later lists?

The statute also does not address the procedure for seeking a preliminary injunction. Must the sponsor seek the preliminary injunction in any pending case? What if the deadline for amending the pleadings (often fairly short) has passed? Will courts conclude that the applicant's notice of intent to market satisfies the "good cause" required under Federal Rule of Civil Procedure 16 to amend pleadings after any applicable deadline, except in extraordinary circumstances? Can the sponsor start an all new suit on the patent(s) for which it is seeking a preliminary injunction? If so, does it have to be before the same court as the pending suit? If the sponsor chooses to file a new suit before a different court, will the new court transfer the case to the court handling the earlier case? The statute does not address these and other possible issues, so that task will fall to the courts.

Another question is whether the provisions on preliminary injunctive relief provide clues to the earlier question about whether the agreed list or Paragraph 5 lists of patents limit the patents that the sponsor can assert. The preliminary injunction provisions give the applicant an incentive to include on its Paragraph 5 list any and all patents that could reasonably be asserted against the proposed biosimilar. Otherwise, it would expose itself to the risk of being hit with a preliminary injunction at or near the time of its proposed launch of the biosimilar (a time at which considerable investment would have been made). This suggests that Congress may have intended the agreed list or the Paragraph 5 lists to be exclusive for any initial litigation.

On the other hand, an applicant willing to run the risk of a preliminary injunction could force the sponsor to seek such relief, perhaps on an incomplete record with limited time to prepare and/or for the hearing itself. Further, the sponsor would have to satisfy the additional evidentiary burdens of irreparable harm, the balance of the harms, and that the preliminary injunction would not disserve the public interest. The sponsor would have to post a bond, potentially quite large, to enforce the preliminary injunction. What if the court does not rule on the motion for preliminary injunction before the stated date for the applicant to begin commercial marketing?

As a result of these ambiguities, both reference product sponsors and biosimilar applicants can look forward to several years of litigation while these procedural issues are decided, along with issues going to the merits of their cases.

14.4.2 Risk Mitigation: Postgrant Patent Office Challenges in the US

Notwithstanding the risks identified above, there are reported to be many biosimilars in development. In prepared remarks to the US Congress in September 2015, an FDA official stated that "[a]s of July 31, 2015, 57 proposed biosimilar products to 16 different reference products were enrolled in the Biosimilar Product Development (BPD) Program." The same official indicated that "[s]ponsors of an additional 27 proposed biosimilar products have had a Biosimilar Initial Advisory meeting with FDA, but have not joined the BPD program to pursue the development of these products" (Woodcock, 2015).

An obvious way to avoid the uncertainty and expense of litigation under the BPCIA is to wait for any patents covering the reference product to expire. As discussed above, it appears that the patents covering several of the biggest selling biologics have already expired or will expire over the next 5 years.

What if it is undesirable to wait for all the applicable patents to expire? If a good faith invalidity position can be developed for any applicable claims, a biosimilar developer can file a petition with the Patent Trial and Appeal Board (PTAB) pursuant to 35 U.S.C. § 316 for an *inter partes* review or 35 U.S.C. § 326 for a postgrant review in appropriate circumstances. Table 14.1 summarizes features of *inter partes* and postgrant review, such as the timing of when they are available, the types of arguments that may be raised, and other key elements.

Inter partes and postgrant reviews have several advantages over traditional district court litigation, notably:

- Claims receive the broadest reasonable construction, enhancing the prospects that they will be found invalid.
- The challenger bears the burden of proof of invalidity by a preponderance of the evidence, compared to the clear and convincing burden that applies in court.

TABLE 14.1
Features of US *Inter Partes* and Postgrant Reviews

	Postgrant Review	*Inter Partes* Review
Eligible patents	≤9 months of grant filed after March 16, 2013	>9 months from grant and after completion of PGR
Other eligibility requirements	Barred if petitioner has filed civil action challenging validity	Barred if: (1) petitioner has filed civil action challenging validity; or (2) more than 1 year after petitioner served with infringement complaint
Prior art/invalidity grounds considered	All	Patents and printed publications
PTO reviewing body	PTAB	PTAB
Standard for granting request	(1) Information presented in the petition, if not rebutted, demonstrates it is more likely than not that at least one claim is unpatentable; or (2) petition raises novel or unsettled legal question important to other patents or applications	Reasonable likelihood requestor will prevail as to at least one claim
Third-party participation	Yes	Yes
Discovery	Yes	Yes
Impact on pending litigation	Automatic stay of suit by petitioner challenging validity	Automatic stay of suit by petitioner challenging validity
Time to completion (w/o appeal)	12 months from institution, extendable to 18 for good cause	12 months from institution, extendable to 18 for good cause
Appeal to Fed. Cir.	Any party	Any party
Estoppel	Raised or reasonably could have raised	Raised or reasonably could have raised

- The proceedings are less expensive, in part because discovery is limited.
- The proceedings are typically shorter than court litigation, lasting at most 18 months from the filing of a petition to initiate a proceeding.

A reference product sponsor faced with an *inter partes* or postgrant review petition could potentially argue that in the biosimilars context the BPCIA's litigation provisions are exclusive and thus preclude resort to either an *inter partes* review or a postgrant review proceeding. Only time will tell whether such an argument will succeed.

There have been *inter partes* review challenges to biologic patents. The Patent Trial and Appeal Board denied two requests by Amgen for reviews of two AbbVie patents (US Patent Numbers 8,916,157 and 8,916,158) covering the blockbuster biologic, Humira, just months after Amgen had asked the FDA for permission to market a biosimilar version of the autoimmune disease drug (*Amgen Inc. vs. AbbVie Biotechnology Ltd.*, 2015a,b). In both cases, the PTAB found that Amgen had not shown that it was reasonably likely that at least one claim of the patents was invalid. Eli Lilly recently filed an IPR challenging a University of Pennsylvania patent (*Eli Lilly vs. University of Pennsylvania*, 2016).

Even if a patent office challenge is underway, in some cases, where there is an imminent patent threat, it is possible that a declaratory judgment action may be again argued as an available court option to seek to avoid a potentially applicable patent based on noninfringement, invalidity, or unenforceability.

14.5 THE EUROPEAN BIOSIMILAR APPROVAL PROCESS

Biologics are reviewed and approved by the EMA, which has authority to approve drugs for the entire European Union. The EMA implemented a biosimilars approval pathway in March 2004, which took effect the following year (EMA, 2005). Since then, the EU has led the way in first-world biosimilar regulation and approvals. Approved products include somatropins, erythopoietins (epoetins), granulocyte-colony stimulating factors (filgrastims), and a monoclonal antibody (infliximab). The EMA has developed extensive guidance for industry, approved complex biosimilars such as antibodies, and also approved the most biosimilars. The Committee for Medicinal Products for Human Use (CHMP) at the EMA published a series of general and product-specific guidelines in 2005 (European Medicines Agency, 2005).

14.5.1 No Linkage of European Community Patents to Biosimilar Approval Process

There are no European Community (EC) linkage regulations. Certain countries, such as Italy, Hungary, Portugal, and the Slovak Republic do have a national system that, in some way, links patent rights to marketing approval (Bhardwaj et al., 2013) or reimbursement.

It appears that the EC's view is that linkage regulations that delay regulatory review are not permitted at an EC level and may not be permitted at a country level.

In Italy, a local law prevents manufacturers of generic products from submitting a request for marketing authorization prior to the second to last year before patent expiry on a reference drug product. In 2012, the EC called on Italy to comply with the EC view of the European Union rules on marketing authorization of generic drugs (European Commission, 2012). The EC view is that the processing of a submission for marketing authorization can be carried out without being affected by patent rights (European Parliament and Council, 2001), as long as the generic drug company is not allowed to place a product on the market before patent expiry. The request took the form of a "Reasoned Opinion" under EU infringement proceedings. Italy modified its law to remove the law of concern to the EU and then passed another law preventing reimbursement of a drug until the relevant patent expires (Galli, 2012).

14.5.2 Risk Mitigation: Postgrant Patent Office Challenges in Europe

A granted European patent may be opposed by third parties (European Patent Office, 2015). Irrespective of whether the opposition procedure is initiated, it is open to a third party to challenge the validity of a patent in the court of an individual country. An opposition is currently the only way to revoke a patent in all states where the patent is validated. The time limit for filing an opposition is 9 months of the grant being published in the European patent bulletin. The opposition must state the ground of opposition, which can be one or more of the following: lack of novelty or inventiveness, lack of patentable subject matter, insufficient disclosure, or added subject matter beyond the content of the application as filed.

The Opposition Division examines the grounds for opposition and invites the patent owner to respond by setting out why the patent should be maintained. The patent owner can make amendments to the application and claims if it wishes. Typically, an oral hearing is held, after which the Opposition Division will either maintain the patent as is, maintain the patent in amended form, or revoke the patent. An appeal can be filed by either side to the independent Appeal Board. If the Appeal Board maintains the patent as granted or as amended, validity of the patent may still be individually challenged on a country-by-country basis in local courts. If the patent is revoked by the Appeal Board, there is no further appeal. If the Appeal Board overturns the Opposition Division decision to revoke the patent, the opposition may be remitted back to the Opposition Division for further consideration of any outstanding grounds.

14.6 CONCLUSIONS

The biological patent landscape should be proactively assessed by both patent owners having a biologic on the market and subsequent-entry biosimilar manufacturers. Patent owners should consider bolstering patent protection through new filings on improvement inventions and taking advantage of patent term extensions. Biosimilar manufacturers may consider launching early patent challenges in court or at the patent office, particularly in regions where there are linkage regulations that can delay market entry while patent issues are reviewed.

REFERENCES

35 U.S. Code § 284—Damages.

42 U.S. Code § 262(k)(6)—Exclusivity for first interchangeable biological product.

42 U.S.C. § 262(l)—Patents.

42 U.S.C. 262(l)(8)—Guidance documents.

42 U.S.C. § 262(l)(8)(A)—Notice of commercial marketing.

42 U.S.C. § 262(l)(9)(C)—Reasonable cooperation and 35 U.S.C. § 271(e)(2)(C)(ii)—Infringement of patent.

Amgen Canada Inc. vs. Apotex Inc. (2015) FC 1261.

Amgen Canada Inc. et al. vs. Samsung Bioepis Co., Ltd. and the Minister of Health. (2015) T-1283-15.

Amgen Canada Inc. et al. vs. Teva Pharmaceutical Industries Ltd. et al. (2012) T-989-12.

Amgen Inc. vs. AbbVie Biotechnology Ltd. (2015a) Case IPR2015-01514: US Patent No. 8,916,157, Amgen, Inc.

Amgen Inc. vs. AbbVie Biotechnology Ltd. (2015b) Case IPR2015-01517: US Patent No. 8,916,158, Amgen, Inc.

Amgen Inc. et al. vs. Apotex Inc. et al. (2015) U.S. District Court for the Southern District of Florida; Case 0:15-cv-61631-JIC.

Amgen Inc. vs. Sandoz Inc. (2015) 794 F.3d 1347 (Fed. Cir. 2015).

Amgen, Inc. vs. Sandoz, Inc. (2016) Case Nos. 2:16-cv-01118-CCC-JBC and 2:16-cv-01276-SRC-CLW (D.N.J. 2016).

Baker Petrolite Corp. vs. Canwell Enviro-Industries Ltd. (2002) FCA 158.

Bell Helicopter Textron Canada Limitée vs. Eurocopter. (2013) FCA 219.

Bhardwaj R, Raju KD, Padmavati M. (2013) The impact of patent linkage on marketing of generic drugs. *Journal of IP Rights* 18, 316–322.

Celltrion Healthcare Co., Ltd. vs. Kennedy. (2014) No. 14-2256-PAC (S.D.N.Y. March 31, 2014).

Crouch D. (2014) Old applications; new patents. Available from: http://patentlyo.com/patent/2014/01/old-patents.html.

Eli Lilly vs. University of Pennsylvania. (2016) Case IPR 2016-00458 (PTAB).

EMEA. (2005) Guideline on similar biological medicinal products (CHMP/437/04). Available from: http://www.ema.europa.eu/ema/index.jsp?curl=pages/regulation/general/general_content_000408.jsp.

European Commission. (2012) Press release. Available from: http://europa.eu/rapid/press-release_IP-12-48_en.htm?locale=en.

European Medicines Agency. (2005) Scientific guidelines on biosimilar medicines. Available from: http://www.ema.europa.eu/ema/index.jsp?curl=pages/regulation/general/general_content_000408.jsp.

European Parliament and Council. (2001) Directive 2001/83/EC on the Community code relating to medicinal products for human use. November 6, 2001. Available from: http://ec.europa.eu/health/files/eudralex/vol1/dir_2001_83_consol_2012/dir_2001_83_cons_2012_en.pdf.

European Patent Office. (2015) Guide for applicants, Part 1: How to get a European patent. D. (V) Opposition procedure. Available from: http://www.epo.org/applying/european/Guide-for-applicants/html/e/ga_d_v.html.

Food and Drugs Act. (1985) R.S.C., c. F-27.

Food and Drug Regulations. (2016) C.R.C., c. 870.

FTC vs. Actavis, Inc. (2013) 570 U.S. 756, 133 S. Ct. 2223, 186 L. Ed. 2d 343 (2013).

Galli C. (2012) Sometimes they come back: patent linkage for generic drugs reinstated. Available from: http://www.internationallawoffice.com/Newsletters/Intellectual-Property/Italy/IP-Law-Galli/Sometimes-they-come-back-patent-linkage-for-generic-drugs-reinstated#1.

Graeser D. (2012) Free to sell your new biosimilar? The effect of the new FDA rules. OMICS International Conference on Biowaivers and Biosimilars, San Antonio, TX, September 10–12, 2012. Available from: http://www.omicsonline.org/0975-0851/0975-0851-S1.002-004.pdf.

Health Canada. (2009) Summary basis of decision (SBD) PrOmnitrope™. Health Products and Food Branch, Ottawa, ON.

Health Canada. (2010) Guidance for sponsors: Information and submission requirements for subsequent entry biologics (SEBs). Health Products and Food Branch, Ottawa, ON.

Health Canada. (2015) Guidance document: Patented medicines (Notice of Compliance) regulations. Office of Patented Medicines and Liaison, Ottawa, ON. Available from: http://www.hc-sc.gc.ca/dhp-mps/prodpharma/applic-demande/guide-ld/patmedbrev/pmreg3_mbreg3-eng.php.

H.R. 3590, 111th Congress. (2010) Patient Protection and Affordable Care Act.

H.R. 3590, 111th Congress. (2010) § 7002—Approval pathway for biosimilar biological products.

Loftus P. (2016) Panel recommends FDA approval of Remicade knockoff. *The Wall Street Journal*, February 10, 2016.

Patent Act, s. 34. *The Patent Rules*, s. 10 provide for filing of a protest.

Patent Act, s. 48.1–48.5. The Patentee may disclaim anything included in patent by mistake.

Patent Act, RSC 1985, c P-4.

Public Health Service Act, Section 351, (42 U.S.C. § 262).

Sandoz vs. Amgen. (2013) Case No. 3:13-cv-2904 (N.D. Cal. June 25, 2013).

Sandoz Inc. vs. Amgen Inc. (2014) 773 F.3d 1274 (Fed. Cir. 2014).

TRW Inc. vs. Andrews. (2001) 534 U.S. 19, 31.

US FDA. (2015a) Nonproprietary naming of biological products. United States Food and Drug Administration, Silver Spring, MD.

US FDA. (2015b) Guidance for industry: Biosimilars: questions and answers regarding implementation of the Biologics Price Competition and Innovation Act of 2009. United States Food and Drug Administration, Silver Spring, MD.

US FDA. (2015c) Quality considerations in demonstrating biosimilarity of a therapeutic protein product to a reference product. United States Food and Drug Administration, Silver Spring, MD.

US FDA. (2015d) Scientific considerations in demonstrating biosimilarity to a reference product. United States Food and Drug Administration, Silver Spring, MD.

US Patent 8,357,513. (2013) Nucleic acids encoding MPL ligand (thrombopoietin) and fragments thereof. To Genentech.

US Patent 8,399,250. (2013) Anti-MPL ligand (thrombopoietin) antibodies, To Genentech.

US Patent 8,507,196. (2013) Nucleic acid probe of HIV-1, and a method and kit employing this probe for detecting the presence of nucleic acid of HIV-1, To Institut Pasteur.

US Patent 8,557,768. (2013) Human nerve growth factor by recombinant technology, To Genentech.

US Patent 8,603,777. (2013) Expression of Factor VII and IX activities in mammalian cells, To ZymoGenetics.

Woodcock J. (2015) Biosimilar implementation: a progress report from FDA. Available from: http://www.fda.gov/newsevents/testimony/ucm463036.htm.

[illegible]
[illegible] patent [illegible]
H.R. Rep. [illegible]
H.R. [illegible]
[illegible]
[illegible]
[illegible]
[illegible]
Pathways [illegible]
[illegible]
[illegible]
[illegible]
[illegible]
[illegible]
(Grounded)
[illegible] in Patent Cases [illegible]
[illegible]

15 Biosimilars in the EU
Regulatory Guidelines

Sol Ruiz
Spanish Medicines Agency

CONTENTS

15.1 INTRODUCTION

The clinical use of biologics (i.e., medicines obtained from biological material such as animal or human tissues or fluids, e.g., heparin or plasma-derived products or from microorganisms, e.g., some vaccines) and, particularly, biotechnological medicines (obtained using genetic engineering techniques) has been increasing in the last three decades. Insulin from animal origin for the treatment of diabetic patients was already widely used in 1923 (Teuscher, 2007). Its complete amino acid sequence was obtained in 1955; it was the first protein produced through a biotechnology (recombinant) process and was approved by the FDA in 1982 (White Junod, 2009). Similarly, growth hormone from human origin (the pituitary gland from cadavers) was obtained in 1956 and was used in patients with growth hormone deficiency at the end of that decade (Ayyar, 2011). The first recombinant growth hormone was obtained in 1981. The production of these therapeutic proteins from biotechnological processes solved critical problems associated with their equivalents from biological origin, such as immunogenicity or hypersensitivity or infectious disease transmission (e.g., Creutzfeldt-Jakob disease).

Advances in molecular biology and genetic engineering allowed a large and reliable source of these medicines from microbial cell cultures (bacteria, yeast) or mammalian cell lines. Modified versions of these proteins (e.g., pegylated interferon, B-domain deleted factor VIII, insulin glargine) or fusion proteins (e.g., etanercept, abatacept) also became available. Recombinant enzymes and certain monoclonal antibodies are currently the standard of care for many diseases and have also allowed the treatment of rare disorders (e.g., alglucosidase alfa, human C1-inhibitor, eculizumab). Biotechnology drugs have a clearly established efficacy and safety profile as they are the treatment option for many chronic conditions and no transmission of infectious disease has ever been reported even when materials from biological origin are used in their production. More biotechnology-derived medicines are becoming available, and monoclonal antibodies are used in many different therapeutic areas, some providing an entirely novel approach to the treatment of their respective indications (e.g., rheumatoid arthritis, oncology, macular degeneration, or lupus), including rare orphan diseases (e.g., eculizumab for paroxysmal nocturnal hemoglobinuria).

The main drawback of biotechnology-derived medicinal products is possibly their high price, which is posing a challenge for the sustainability of health care systems. Many complex factors contribute to the high cost of biotech drugs (the production process and its control are only a small part of it), but, as with generic medicines, the legislation on similar biological medicinal products (or biosimilars) has the same goal—that is, to try to reduce the cost of clinical treatments based on biological and biotech drugs by introducing market competition. After a decade of experience in the EU with this regulation, more than 20 biosimilar medicines have been approved so far (including the first monoclonal antibodies), some of them widely used. However, some of the initial challenges for their introduction remain and will be discussed here. Guidelines for biosimilar development were first published in 2005 and have been updated according to the experience gained. The main concepts in these guidelines and rationale are also described.

15.2 REGULATORY FRAMEWORK IN THE EUROPEAN UNION

The EU regulatory region has been the pioneer in developing requirements for the approval of biosimilars—that is, those developed as equivalent to other biologics on the market and whose data protection period is over. Given the complexity and heterogeneity of biological medicines, specific legislation has been developed (EC, 2004: Directives 2003/63/EC and 2004/27/EC). Directives are European laws that establish basic principles but leave their implementation to national governments through their respective national laws or regulations. In contrast, Notes for Guidance or guidelines include recommendations based on scientific knowledge in a particular area and, therefore, are subject to periodic review to adapt to scientific progress or changes in a particular regulatory strategy. A different approach from that described in those recommendations may be possible if adequately justified. Guidelines aim to provide a basis for practical harmonization among EU countries on how to interpret and apply the requirements to demonstrate the quality, safety, and efficacy set out in EU directives. Also, they help ensure that applications for marketing authorization are presented in a way that will be recognized as valid by the EMA. Specific Directives and Guidelines applicable to biosimilar medicines in the EU are described briefly in the following.

Directive 2003/63/EC, published on June 27, 2003, amending Directive 2001/83/EC, defines (in *Part II: Specific marketing authorization dossiers and requirements*) the concept of "similar biological medicinal products," commonly called "biosimilars" (EC, 2003). A biosimilar, as has been noted in earlier chapters, is a biological medicinal product that has been developed as equivalent to an other biological medicine already marketed (called "reference product"). The active ingredient of the biosimilar and the reference medicinal product is essentially the same, although there may be slight differences depending on the complexity of its structure and the method of production. Both the reference medicine and the biosimilar have a natural variability inherent to all biological medicinal products. A biosimilar medicinal product is authorized when it has been concluded, based on comparable data available, that those small differences between the two have no significant impact on safety and efficacy.

Biosimilars are normally authorized years after the reference product has been on the market (as the period of data protection has expired) and are used to treat the same disease and use the same dose as the reference medicinal product. Therefore, the large experience in terms of efficacy and safety from the reference product can be used for the development of a biosimilar medicinal product.

This Directive recognizes that, for biological medicines, it is possible that the information required for "essentially similar medicinal products" (generics) does not allow demonstration of the similar nature of two biological medicinal products, and, therefore, additional data should be provided, in particular, the toxicological and clinical profile. The type and amount of additional data (i.e., toxicological and relevant nonclinical and clinical data) shall be determined case by case, taking into account all relevant scientific guidelines and the special characteristics of each drug. If the originally authorized medicinal product has more than one clinical indication, the extrapolation of efficacy and safety of the biosimilar medicine should be justified or, if necessary, demonstrated separately for each of the claimed indications.

Directive 2004/27/EC, published on April 30, 2004, amends Directive 2001/83/EC and also considers that biosimilar medicinal products do not usually meet all the conditions to be considered as generic drugs, mainly due to the characteristics of the manufacturing process, the raw materials used, molecular characteristics, and therapeutic modes of action (EC, 2004). Then the results of the appropriate preclinical testing or clinical trials should be provided to establish a similar efficacy and safety profile for the reference product. The type and amount of additional data should comply with the relevant criteria stated in Annex I and the related detailed guidelines.

15.3 NOTES FOR GUIDANCE OR GUIDELINES

The EMA has developed three main guidelines for biosimilars: a general guide describing the basis for authorization of biosimilars (EMA, 2014); a quality guideline describing the data required on production and control of a biosimilar and emphasizing comparability studies against the reference medicine (EMA, 2012); and guidance on the nonclinical and clinical studies to demonstrate comparability (EMA, 2015). The last-named includes a series of annexes with specific requirements for certain biosimilar medicinal products (e.g., erythropoietin, filgrastim, insulin, monoclonal antibodies). The three general guidelines were initially published in 2005 and have been recently reviewed after a decade of experience in the assessment, marketing, and use of biosimilars. A brief summary of the updated guidelines is provided below.

15.3.1 Guideline on Similar Biological Products

In the updated guideline, the concept of similar biological medicinal products has been introduced, and the basic principles to be applied for the development of biosimilars have been outlined (EMA, 2014).

Although the concept of biosimilarity is applicable to any biological medicinal product, this approach has been successfully used mainly for recombinant proteins as they present less heterogeneity than products obtained by extraction from biological sources (e.g., from tissues or fluids) and can be well characterized. The first biosimilar of enoxaparin sodium has been recently authorized by the EMA and others are under development.

The posology and route of administration of the biosimilar must be the same as those of the reference medicinal product, and for protein active substances the amino acid sequence is expected to be the same. Any differences in strength, pharmaceutical form, or formulation will require justification or additional studies to support comparability. The changes aimed at improving efficacy are not compatible with establishing biosimilarity.

The chosen reference medicinal product must be authorized in the European Economic Area (EEA) (based on a complete dossier under Article 8 of Directive 2001/83/EC as amended) and must be the same for the entire comparability program, that is, quality, safety, and efficacy studies. However, with the aim of facilitating the global development of biosimilars, the updated version of the guideline considers the possibility to compare the biosimilar in certain clinical studies and in some nonclinical studies with a comparator authorized by a regulatory authority with similar scientific and regulatory standards as EMA (e.g., ICH countries), as long as comparability at the quality level (physicochemical structure and biological activity) has been established between the biosimilar and the reference product sourced from both EEA and non-EEA countries.

Briefly, the basic principle of a biosimilar development program is to establish the similarity between the biosimilar and the reference product using the best possible strategy, ensuring that previously demonstrated safety and efficacy for the reference product are also applicable to the biosimilar. The biosimilar must be equivalent to the reference medicine in physicochemical and biological characteristics, and any observed differences will have to be duly justified in relation to their potential impact on safety and efficacy. Any differences that may result in an advantage in terms of safety (e.g., lowest levels of impurities or less immunogenicity) should be explained but are not incompatible with the establishment of biosimilarity.

A step-by-step approach is recommended for the development of a biosimilar, that is, starting with a complete and detailed physicochemical and biological characterization. The scope and amount of the nonclinical and clinical studies will depend on the level of evidence obtained in the previous step, including the robustness of the physicochemical, biological, and nonclinical *in vitro* data and the extent of any structural differences identified between the biosimilar and the reference product.

The ultimate goal of the comparability exercise is to exclude any relevant differences between the biosimilar and the reference product that could have an impact on efficacy and/or safety. Therefore, the clinical studies must be sufficiently sensitive in design, population, endpoints, and implementation to detect such differences if they were present.

Once the marketing authorization is granted, there is no regulatory requirement for redemonstration of comparability because it is acknowledged that the biosimilar will have its own life cycle (in the same way as the reference product has its own). Demonstration of biosimilarity is just a special marketing authorization procedure. Once commercialized, the biosimilar is an alternative product on the market in the relevant therapeutic area, and its production process and control are subject to optimization and evolution, as it is the case with innovative medicines or the reference medicinal product. After changes in the manufacturing process and for demonstration of comparability, the same rules apply for both originators and biosimilars (see also the following section).

15.3.2 Guideline on Similar Biological Medicinal Products Containing Biotechnology-Derived Proteins as Active Substance: Quality Issues

This guideline came into force on June 1, 2006, and has been reviewed recently (EMA, 2012). It describes the requirements for the manufacturing process and control of the biosimilar medicinal product, quality comparability studies, taking into account the choice of the reference medicine, analytical, physicochemical characterization, biological activity, purity, and quality attributes to establish the relevant specifications of the biosimilar. Although the guide refers to biotech drugs, the principles described could be applied to other biological products.

Biosimilar medicines are manufactured and controlled according to their own development and in accordance with relevant guidelines (ICH and CHMP; see the EMA website). The content of Module 3 (production and control data) of the marketing authorization dossier for a biosimilar medicine is the same as for any other biotech drug but should also include all quality comparative studies between the reference product and the biosimilar.

Documentation regarding the development and manufacturing process for biosimilars should cover two distinct but complementary aspects: first, molecular characteristics and quality specifications (comparable to the reference product) and, second, the consistency of the specific production process of the biosimilar.

The comparability exercise should be comprehensive and should show that the biosimilar has a quality profile very similar to the reference medicinal product. It should include side-by-side comparative studies of several batches of the reference and the biosimilar using state-of-the-art, sensitive methods to determine not only the similarities but also the possible differences in quality attributes. The comparability study includes the evaluation of the physicochemical parameters (composition, physical properties, primary, and higher-order structures), identification of variants of the molecule (including product-related substances), and analysis of biological activity. Any differences identified will have to be duly justified in light of their possible impact on the safety and efficacy of the product. To establish comparability, it is necessary to define acceptance criteria (quantitative whenever possible) whose relevance should be discussed, considering the number of batches of reference product analyzed, the quality parameter studied, and the test method used. In principle, these acceptance limits should not be broader than the range of variability of representative batches of the reference medicine.

As described above, the production process and control of a biosimilar are subject to optimization and evolution, as it is the case with any other medicine, including the reference medicinal product. Any subsequent changes in the manufacturing process of the biosimilar (active substance and/or finished product) will have to follow a comparability assessment (as described in Guideline ICH Q5E) (EMA, 2003), and it will be assessed following the same rules as for any other biological medicinal product. If any of those changes result in significant molecular differences that may impact the approved efficacy and safety profile, additional studies may be required.

15.3.3 Guideline on Similar Biological Medicinal Products Containing Biotechnology-Derived Proteins as Active Substance: Nonclinical and Clinical Issues

This guideline came into force on June 1, 2006, and has also been reviewed recently (EMA, 2015). It describes the general principles for the nonclinical and clinical development and evaluation of applications for marketing authorization of biosimilars containing recombinant proteins as active ingredients (although these principles could be applied to other biological products as well).

The nonclinical section addresses the pharmacotoxicological assessment. The clinical section describes the requirements for pharmacokinetic (PK), pharmacodynamic (PD), efficacy studies, and clinical safety and pharmacovigilance studies (including immunogenicity and the risk management plan).

The revised guideline covers several issues, such as a stepwise approach for the design of nonclinical studies, the use of PD markers, study design, choice of appropriate patient population, and selection of surrogate and clinical endpoints in efficacy trials, clinical safety (including the design of immunogenicity studies), the risk management plan, pharmacovigilance, and extrapolation of safety and efficacy. It also recommends a stepwise approach to perform clinical studies.

A key aspect of the nonclinical and clinical studies of biosimilar medicines is the comparative nature of these studies; that is, the objective is not to demonstrate the clinical efficacy and safety of the molecule itself (both of these aspects are assumed to be guaranteed by the initial demonstration of comparability to the reference on quality aspects). Therefore, both nonclinical *in vitro* and *in vivo* studies and clinical studies should be the most sensitive to detect differences between the biosimilar and the reference medicinal product. This is the basic principle on which the specific, product-related annexes of this guideline have been developed and they are briefly described next.

15.3.3.1 Nonclinical Studies

Nonclinical studies should be performed prior to initiating clinical trials using a stepwise approach; that is, conduct, first, comprehensive *in vitro* studies and decide later (based on the results obtained) about the need for *in vivo* studies. Certain factors need special attention to assess the requirement for *in vivo* nonclinical studies (e.g., differences in formulation with new excipients not commonly used for biotechnological products or relevant differences in quality attributes), although they do not necessarily imply that *in vivo* testing is required. If an *in vivo* evaluation is considered necessary, the aim of the studies (PK and/or PD and/or safety) will depend on the additional information required. Animal studies should be designed to maximize the information obtained and considering the principles of the 3Rs (replacement, refinement, reduction) that apply to animal experiments (Directive 2010/63/EU). Toxicity studies in nonrelevant species (i.e., to evaluate only specific toxicity due to the presence of possible impurities) or studies of repeated dose toxicity in nonhuman primates are not recommended.

15.3.3.2 Clinical Studies—Efficacy and Safety

The objective of the comparative clinical trials is to resolve uncertainties that may remain following the quality and nonclinical development of the proposed biosimilar with the reference product. PK and PD studies form the basis of early clinical development and serve to design phase III clinical development. The clinical comparability exercise is normally a stepwise procedure starting with PK and, if feasible, PD studies followed by clinical efficacy and safety trial(s) or, in certain cases, confirmatory PK/PD studies.

To demonstrate PK comparability, the most sensitive model should be selected, which, in some cases, will be healthy volunteers in order to exclude the effect of the disease or concomitant medication and will probably have a lower clearance mediated by the pharmacological target. If the PK study is conducted in patients, a sensitive model/population should be selected—that is, one with fewer factors that cause major interindividual or time-dependent variation. It is recommended that PD markers be added to the PK studies whenever possible. These markers should be selected based on their clinical significance.

Although efficacy comparative studies are usually required, in some cases, PK/PD comparative studies with the biosimilar medicinal product and the reference may be sufficient to demonstrate clinical comparability if there is a clearly demonstrated dose response and the PD marker is an accepted surrogate marker linked to clinical efficacy. Therefore, demonstrating a similar effect on the PD marker ensures a similar effect on clinical outcome. Examples of such markers are the ANC (absolute neutrophil count) to evaluate the effect of granulocyte-colony stimulating factor (G-CSF) or reducing early viral load in chronic hepatitis C to evaluate the effect of alpha interferon.

If surrogate markers for efficacy are not available, it is usually necessary to demonstrate comparable clinical efficacy between the biosimilar and the reference product by adequately powered, randomized, parallel-group comparative clinical trial(s), preferably double-blind, by using efficacy endpoints. The study population should be representative of the therapeutic indication(s) authorized for the reference product and should be sensitive to detect differences between this and the biosimilar. Occasionally, changes in clinical practice may require deviation from an approved therapeutic indication, for example, due to concomitant medication, line of therapy, or the severity of the disease. These deviations should be adequately justified and discussed with regulatory authorities. Generally, the trial design should be performed to demonstrate equivalence. The purpose of the efficacy trials is to detect a clinically significant difference between the reference product and the biosimilar. Clinical comparability margins should be prespecified and justified on both statistical and clinical grounds by using the data from the reference product.

Clinical safety is important throughout the clinical development program of the biosimilar, and comparative data should be obtained from the PK/PD studies and also as part of the clinical efficacy study. Normally, comparative safety data should be collected before the marketing authorization considering the type and severity of the safety problems described for the reference product. The safety assessment continues throughout the commercial life of the biosimilar medicinal product, in the same way as for any other biological medicine.

Evaluation of immunogenicity of the biosimilar in comparison with the reference product should be conducted by using the same assay format and sampling scheme. The duration of immunogenicity study should be justified, depending on the duration of the treatment course, disappearance of the product from the circulation (to avoid interference in the assays), and the time when a humoral immune response develops. The type and amount of immunogenicity data will depend on the experience gained with the reference product and the product class.

15.3.3.3 Extrapolation of Indications

If the reference medicinal product has more than one therapeutic indication, the extrapolation of the efficacy and safety profile of the biosimilar to other indications not studied during the clinical development could be acceptable but needs to be scientifically justified. The justification should consider all the comparability data gathered (i.e., quality, nonclinical, and clinical) and will depend on the clinical experience with the medicinal product, available bibliographic data, knowledge of the mechanism of action of the active ingredient, and the receptors involved for each indication. If there is evidence that different active sites of the reference medicine or different receptors on target cells are involved in the different therapeutic indications, or that the safety profile of the product differs between them, additional data may be required to justify extrapolating the safety and efficacy from the studied indication in pivotal clinical trials. The final decision will be based on the evidence from all the comparability studies and the assessment of any possible uncertainties that may remain.

This still remains one of the most controversial issues regarding the acceptance of biosimilars and is discussed in more detail below.

The extrapolation of indications is discussed in greater detail in Chapter 9 of this book and under 15.5 below.

15.3.3.4 Pharmacovigilance

It is acknowledged that data derived from preauthorization clinical studies with the biosimilar are usually insufficient to identify rare adverse effects. Therefore, clinical safety of biosimilars should continue to be monitored after the marketing authorization.

The applicant should provide a description of the pharmacovigilance system and a risk management plan in accordance with the existing EU legislation. The risk management plan should consider the identified and potential risks associated with use of the reference product and, if applicable, possible additional risks identified during the development program of the biosimilar. The plan should also detail how these issues will be addressed in the postmarketing period. Immunogenicity should also be addressed in this context. Any specific safety measures imposed on the reference medicinal product or therapeutic class should be taken into consideration. Traceability is a critical aspect for the pharmacovigilance of biological medicines in general; therefore, it is not different for authorized biosimilar medicines. The identification and reporting of any adverse reaction require the information reporting the brand name and specific batch number of the concerned product.

Pharmacovigilance for biosimilars is discussed in greater detail in Chapter 13 of this book.

15.3.4 Product-Specific Guidance

In addition to the general considerations previously described for nonclinical and clinical comparability studies, specific additional requirements have been developed for certain biosimilar medicinal products such as human somatropin, erythropoietin, insulin, G-CSF, alpha interferon, low-molecular-weight heparin, monoclonal antibodies, follicle stimulating hormone (FSH), and interferon beta (EMA website). The characteristics and type of nonclinical (toxicological and pharmacological) and clinical studies (PK, PD, safety, and efficacy) are described in these annexes for each drug class considered. Clinical aspects such as the study population, design of clinical studies, and the possibility of extrapolating clinical indications from the reference medicine are discussed. Immunogenicity data needed for the approval of each type of biosimilar are also described. These requirements are based on previous clinical experience with innovative medicines of the same therapeutic class. The complexity of the molecule and the knowledge of its mechanism of action have also been considered. Then the requirements for a biosimilar of a small simple molecule with wide clinical experience such as insulin are simple (since it is a relatively small molecule, nonglycosylated, easy to characterize, and with a well-known mechanism of action), and it is possible that comparability PK/PD studies are sufficient for marketing authorization. Studies required for a biosimilar erythropoietin are more complex and even more so for a monoclonal antibody. Extrapolation of indications is relatively easily done for erythropoietin and G-CSF (as both have a well-known mechanism of action and their effect in different indications is mediated by the same receptor), but it is much more complicated in the case of monoclonal antibodies. These annexes are subject to frequent revision as more experience is gained in the authorization and use of biosimilars.

Other guidelines should also be considered when developing a biosimilar medicinal product, although they are not specific for biosimilars, that is, immunogenicity assessment of biotechnology-derived therapeutic proteins (EMEA, 2006) and of monoclonal antibodies intended for *in vivo* clinical use (EMA, 2010).

15.4 THE FIRST DECADE

To date, more than 20 biosimilars have been approved in Europe, including different active ingredients: somatropin, filgrastim (G-CSF), epoetin, follitropin alfa, insulin, and the first monoclonal antibody, infliximab. A few others have been assessed through the centralized procedure that did not receive a positive opinion. More biosimilar medicines are under development, such as pegylated filgrastim and monoclonal antibodies adalimumab and rituximab.

The availability of biosimilars has resulted in a substantial reduction in the price of biologics in the same therapeutic class, as not only their price is lower than the price of the reference product but also their emergence has resulted in a price reduction from innovators. In Norway, epoetin and filgrastim biosimilars are included in a national tender for drugs used in hospitals, in which prices can be reduced by up to 89% (GaBI online, 2015a).

The pattern of use or market access for biosimilars has been variable in different EU countries, mirroring somehow the countries' previous experience with the use of generics. A wide use of biosimilars in Germany has resulted in significant savings for their health care system (IMS Institute, 2014). In other EU countries (e.g., Spain), the initial market access was low, but, after years of experience and as a consequence of the economic crisis, the use of biosimilars is increasing. Certainly, the Norwegian experience with a recent tender for infliximab is very interesting (GaBI online, 2015b). Orion Pharma had proposed a 72% price reduction for the infliximab biosimilar Remsima (69% lower cost than the price for Remicade), resulting in substantial savings for the hospital budget and the health care system. Patients receiving Remicade are expected to remain on this treatment until more data on switching to the biosimilar can be provided (Stanton, 2013).

It is clear that with more experience in the assessment and clinical use of biosimilars, their uptake is steadily increasing, even in countries initially reluctant to adopt their use. Also, innovator companies have realized the potentially enormous benefit of certain biologics with a very large market and a safe and reduced clinical program. Therefore, they are also getting onboard for biosimilar development.

15.5 REMAINING CHALLENGES

Even after a decade of experience with marketing authorization and clinical experience in the EU with biosimilars, certain challenges still remain in certain countries for a wider use of biosimilars. Among those cultural issues (e.g., previous experience with the use of generics), pricing strategies from originators have played an important role, but the most debated are still related to extrapolation of indications, pharmacovigilance and traceability, and interchangeability and substitution. These strategies are considered further in the following section.

15.5.1 Interpretation of Differences and Extrapolation of Therapeutic Indications

Biosimilars are not identical to their reference medicine due to the complexity of the molecular structure (in contrast to chemicals and their generics). Additionally, the frequently cited paradigm "the process is the product" may leave prescribers and patients concerned about the interpretation of minor differences between the two molecules (e.g., in glycosylation pattern) and any remaining uncertainties with respect to efficacy and safety of the biosimilar as they are approved on a reduced clinical development.

Medicinal products in general are subject to changes in their manufacturing process throughout their life cycle. For biologics, some of these modifications can be relatively minor (e.g., change in a test method, tightening of a specification), and others may result in detectable molecular changes (e.g., elimination of a cell culture reagent can result in a modification of the glycosylation pattern) or in the product purity profile (e.g., different distribution of charge variants, level of aggregates, etc.). The assessment of these modifications is based on comparability studies between batches from the medicinal product before and after the change is introduced, and in many cases these quality data (i.e., analysis of physicochemical structure, biological activity, and purity using state-of-the-art methods) have been considered sufficient to support the change and keep the medicinal product on the market under the same conditions as when it was approved (EMA, 2003). If clear differences are observed, additional *in vitro* characterization or PK/PD data may be required, as will supplementary pharmacovigilance activities to ensure that no new safety signals are identified but in any case a new clinical development (i.e., demonstration of a positive benefit/risk in all approved indications) has been requested. Therefore, regulators have learned from extensive experience in assessing such changes and judging their potential impact on efficacy and safety. Pharmacovigilance of the medicinal product after the change was introduced has also added knowledge from those regulatory decisions. The same tools and rationale are used for biosimilars. That is, assessment of any difference in product quality when compared to the reference medicinal product in the context of the complexity of the molecule and knowledge of its mechanism of action guide the requirements for nonclinical and clinical studies. This whole body of evidence is used to decide on the extrapolation of indications, in the same way as for any biological product on the market for which a change is introduced in its manufacturing process.

For a biosimilar, extrapolation of indications is possible based on the overall evidence of comparability provided from the comparability exercise (quality and nonclinical and clinical data) and with adequate scientific justification. As described in guidelines on biosimilars, extrapolation of efficacy and safety data to all therapeutic indications of the reference medicine is easier when the relevant mechanism of action of the molecule is known and the target receptor(s) involved are the same for all the approved indications (e.g., filgrastim, epoetin). However, it is more challenging when the mode of action is complex or not completely understood and involves multiple receptors or binding sites, the contribution of which may differ between indications (e.g., certain monoclonal antibodies, beta-interferon). In such cases, additional nonclinical or clinical data may be required to ensure that the benefit–risk of the biosimilar is comparable to the reference medicine in the extrapolated indications (Weise et al., 2014).

The approval of the first two biosimilars of infliximab in the EU in all the same indications as the reference medicine (Remicade) raised a lot of controversy, especially after a different regulatory agency decided differently. While the EU, Korea, and Japan granted all licensed indications of the reference product, extrapolation to IBD (inflammatory bowel disease) was not accepted in Canada (Weise et al., 2014).

This different opinion led specialists and some learned societies to doubt the EU's conclusion and state that a specific study in that indication should have been required. The scientific rationale for extrapolation of indications in the EU has been explained in detail (EMA, 2013; Weise et al., 2014). Additionally, at the time of marketing authorization in the EU, preliminary clinical data from a very small cohort of 23 patients with Crohn's disease (15) or ulcerative colitis (8) showed similar response to infliximab biosimilar compared with historical data on Remicade (EMA, 2013). Also, the applicant had extended the enrollment of IBD patients in the postmarketing surveillance study and was going to conduct an additional comparative trial versus Remicade in active Crohn's disease. Clinical data from this trial have been presented recently showing that biosimilar infliximab was comparable to the reference biological Remicade in terms of efficacy and safety (GaBI Online, 2015c).

As other biosimilar monoclonal antibodies are under development, perhaps regulators should make an additional effort to better communicate the rationale for their decision when granting extrapolation of therapeutic indications based on clinical studies conducted on one or two relevant indications only but considering all the comparative data as a whole (quality, nonclinical, and clinical). However, from a scientific point of view, it is difficult to imagine that two molecules (the biosimilar and the reference product) that have been shown to be highly similar in primary and higher-order structure, glycosylation pattern, biological activity, purity profile, and in *in vitro* studies exploring all the potential mechanisms of action of the molecule using different systems (cells and target receptors), will behave differently *in vivo*.

Any remaining concerns about safety (due to the reduced clinical experience with the biosimilar at the time of marketing authorization) and pharmacovigilance are discussed next.

15.5.2 Pharmacovigilance and Traceability

Extrapolation of clinical safety has also been an issue heavily discussed after the marketing authorization of biosimilars. Even though safety is assessed throughout the clinical development program of the biosimilar (PK/PD and pivotal clinical efficacy studies), it is recognized that this safety database is limited (i.e., insufficient to identify rare adverse effects) when they have been approved. Therefore, clinical safety of biosimilars must be monitored closely on an ongoing basis during the post-approval phase, including continued benefit–risk assessment (Ebbers et al., 2012). The objective of the preauthorization comparative safety data is the identification of any new signal not previously identified with the originator medicine, and the extent and type of safety data and duration of the study will depend on the adverse events known for the reference product. Any possible safety concerns that may result from a different manufacturing process from that of the reference product, especially those related to infusion-related reactions and immunogenicity, should be described in the risk management plan and should be monitored closely on an ongoing basis during the postmarketing phase. This approach follows the same rationale as applied from any other biological medicinal product on the market when changes are introduced in its manufacturing process. Long-term postmarketing data are a requirement for both originators and biosimilars and have allowed the identification of rare safety signals—for example, cases of pure red cell aplasia in patients treated with epoetin or progressive multifocal leukoencephalopathy (PML) in patients with multiple sclerosis treated with (natalizumab) or in patients with psoriasis treated with efalizumab (Ruiz and Calvo, 2011).

In general, immunogenicity should be studied before marketing authorization application and using the same principles as described earlier for the clinical safety approach.

After a 10-year experience in the EU no special safety concerns have been identified for biosimilars. However, it is necessary to have a continuous surveillance system as for all biological medicinal products. A good pharmacovigilance system relies on good traceability in order to unequivocally identify the specific medicinal product and precise batch number causing an adverse effect. This equally applies to innovators and biosimilars. Further measures have been proposed recently to improve the traceability of similar biotherapeutic products, such as adding a biological qualifier (BQ) to the international nonproprietary names (INNs), as proposed by the WHO (2014). Adding this four random letter code to the name of the active ingredient would not enhance traceability and could even cause more confusion when reporting adverse events (as it would result in additional information to register and it does not identify the specific batch). The commercial name of the product and batch number, as described in the current legislation, if properly reported, should be enough and any measures to facilitate or improve recording of these critical data would be welcome.

15.5.3 Interchangeability and Substitution

In the EU, recommendations on interchangeability (i.e., the choice of a particular drug between two or more available for the same indication) and substitution (i.e., administrative rules that allow the switch from one medicine to another), either between innovative products or between innovative and biosimilar medicines, are under the remit of national authorities and outside the responsibility of the EMA and, therefore, have never been included in the guidelines on biosimilars.

Automatic substitution (i.e., switch from one medicine to another done at the dispensing level) is possible for the majority of generic medicinal products, therefore, this issue has also been widely raised for biosimilars even before they were available. Unlike generic drugs, the automatic replacement is not considered appropriate for biological medicinal products, and in many EU member states this has been stated through specific regulation (Ruiz and Calvo, 2011). This issue is still highly controversial, and, based on the experience gained in the use of biosimilars, some regulatory authorities have recently expressed opinions in favor of interchangeability and substitution for biosimilars (GaBI Online, 2015d).

The Dutch Medicines Evaluation Board (MEB) has reviewed their position on biosimilars (from 2010) to acknowledge that biosimilars have no relevant differences compared to an innovator biological medicinal product regarding quality, safety, and efficacy. In their initial opinion, patients must be kept on a biological medicinal product as much as possible if their clinical response was good and should not be switched to an equivalent biological medicine. If a switch was considered, it should only occur under strict conditions including the attending physician's approval. MEB's current position on biosimilars is a bit more relaxed, though still cautious about exchange between biologicals (whether originators or biosimilars) and emphasizing critical aspects of adequate clinical monitoring, pharmacovigilance and traceability, and the involvement of physicians and (hospital pharmacists) in any decision on switching between biological treatments.

An interesting experience on switching is currently taking place in Norway where large savings are expected after the acquisition of infliximab biosimilar Remsima through a national tender for biologics for rheumatology, and stomach, intestinal, and skin diseases in February 2015. Although patients being treated with the originator Remicade are expected to continue with their treatment, a clinical trial is under way in which patients will be switched from the originator Remicade to biosimilar Remsima in order to support the uptake of the biosimilar infliximab and to provide reassurance that switching between the originator and biosimilar is safe.

Recently, Australia's Pharmaceutical Benefits Advisory Committee (PBAC) also reviewed its previous position regarding substitution between the originator medicinal product and biosimilars. At its April 2015 meeting, it recommended that biosimilars are suitable for substitution at the pharmacy level where the data are supportive of this conclusion (GaBI Online, 2015e). Relevant considerations for this substitution are no significant differences in clinical efficacy or safety compared with the originator, no identified populations where the risks of using the biosimilar are disproportionately high, data to support switching between the originator and the biosimilar, and data for treatment-naïve patients initiating on the biosimilar.

The FDA has established labeling requirements for biosimilars including a designation as to whether the biosimilar is interchangeable with the reference product (FDA, 2015; GaBI Online, 2015f). However, decisions regarding automatic substitution are left up to the states. At least thirteen states in the US have passed legislation so far allowing substitution of biosimilars for reference products (GaBI Online, 2015g).

As more experience is gained with the use of biosimilars, more regulatory agencies are reconsidering their initial strict approach to interchangeability and substitution toward a more relaxed view. Once the biosimilar is on the market, it becomes an alternative within the same therapeutic group as the reference medicine and others that are equivalent and, therefore, it should be treated as such. When considering switching a patient from a particular biological medicinal product to a biosimilar, the same rules as applied for products within the same therapeutic class should be followed; that is, it is the responsibility of the treating physician and the patient, or it is done following specific local substitution policies or internal protocols at a particular hospital or health care center.

Interchangeability and substitution of biosimilars are discussed in detail in Chapter 10 of this book.

15.6 CONCLUDING REMARKS

In the EU, the regulatory framework for the development of biosimilar medicinal products is well established and provides relatively clear recommendations in terms of quality, nonclinical, and clinical criteria for their development. These recommendations are continuously kept under review as more experience is gained with their use; additional challenges are posed through scientific advice and in accordance with the evolution of scientific knowledge in specific therapeutic areas. More emphasis is given to a solid, robust, and state-of-the-art quality development, as it is the area where differences can be detected easily as characterization techniques are becoming more sensitive. *In vitro* functional studies using different techniques and cell-based systems allow a deeper evaluation of the significance of those differences, while *in vivo* nonclinical studies are being reduced as the information they provide is very limited and there is already wide clinical experience with the reference medicinal product. The requirements for clinical studies are also being refined as it is recognized that in some cases they would not have enough sensitivity to detect differences between two similar products. Extrapolation of indications probably remains the most controversial issue; therefore, regulators should still make an additional effort to better communicate their rationale for the decisions made as more biosimilars are approved. This will contribute to a better understanding of the regulatory requirements and decisions that will be useful for prescribing physicians and learned societies and will certainly support the use of biosimilars. Regulators recognize the limitations of the safety database (including immunogenicity data) at the time of marketing authorization of biosimilars, but the objective of the data gathered preauthorization is just to detect a potentially new safety signal not previously described for the originator. In any case, a robust pharmacovigilance and risk management plan should be in place that will allow continuous monitoring of the biosimilar's safety, in the same way as for any other biological medicine recently approved.

Although the use of biosimilars is still quite heterogeneous across EU countries, their use is increasing as more become available and more experience is acquired with their use. Clearly, biosimilars are being established as a good tool to reduce health care expenditure in the use of expensive biologics for many chronic indications.

REFERENCES

Ayyar VS. (2011) History of growth hormone therapy. *Indian Journal of Endocrinology and Metabolism* **15(Suppl. 3)**, S162–S165. Available from: http://www.ema.europa.eu/ema/ (Accessed July 28, 2015).

Ebbers HC, Crow SA, Vulto AG, et al. (2012) Interchangeability, immunogenicity and biosimilars. *Nature Biotechnology* **30**, 1186–1190.

EC. (2003) Commission Directive 2003/63/EC of 25 June 2003 amending Directive 2001/83/EC of the European Parliament and of the Council on the Community code relating to medicinal products for human use. *Official Journal of the European Union L* **159**, 46–94.

EC. (2004) Directive 2004/27/EC of the European Parliament and of the Council of 31 March 2004 amending Directive 2001/83/EC on the Community code relating to medicinal products for human use. *Official Journal of the European Union L* **136**, 34–57.

EC. (2010) Directive 2010/63/EU of the European Parliament and of the Council of 22 September 2010 on the protection of animals used for scientific purposes. *Official Journal of the European Union L* **276**, 33–79.

EMA. (2003) Note for guidance on biotechnological/biological products subject to changes in their manufacturing process (CPMP/ICH/5721/03). European Medicines Agency, London, UK. Available from: http://www.ema.europa.eu/docs/en_GB/document_library/Scientific_guideline/2009/09/WC500002805.pdf (Accessed July 28, 2015).

EMA. (2010) Guideline on immunogenicity assessment of monoclonal antibodies intended for in vivo clinical use (EMA/CHMP/BMWP/86289/2010). European Medicines Agency, London, UK. Available from: http://www.ema.europa.eu/docs/en_GB/document_library/Scientific_guideline/2012/06/WC500128688.pdf (Accessed July 29, 2015).

EMA. (2012) Guideline on similar biological medicinal products containing biotechnology-derived proteins as active substance: quality issues (revision 1) (EMA/CHMP/BWP/247713/2012). European Medicines Agency, London, UK. Available from: http://www.ema.europa.eu/docs/en_GB/document_library/Scientific_guideline/2014/06/WC500167838.pdf (Accessed July 28, 2015).

EMA. (2013) Assessment report: Remsima. European Medicines Agency, Committee for Medicinal Products for Human Use, London, UK. EMA/CHMP/589317/2013. Available from: http://www.ema.europa.eu/docs/en_GB/document_library/EPAR_-_Public_assessment_report/human/002576/WC500151486.pdf (Accessed July 29, 2015).

EMA. (2014) Guideline on similar biological medicinal products (CHMP/437/04 Rev 1). European Medicines Agency, London, UK. Available from: http://www.ema.europa.eu/docs/en_GB/document_library/Scientific_guideline/2014/10/WC500176768.pdf (Accessed July 28, 2015).

EMA. (2015) Guideline on similar biological medicinal products containing biotechnology-derived proteins as active substance: non-clinical and clinical issues (EMEA/CHMP/BMWP/42832/2005 Rev1). European Medicines Agency, London, UK. Available from: http://www.ema.europa.eu/docs/en_GB/document_library/Scientific_guideline/2015/01/WC500180219.pdf (Accessed July 28, 2015).

EMEA. (2006) Guideline on immunogenicity assessment of biotechnology-derived therapeutic proteins (EMEA/CHMP/BMWP/14327/2006). Available from: http://www.ema.europa.eu/docs/en_GB/document_library/Scientific_guideline/2009/09/WC500003946.pdf (Accessed July 29, 2015).

FDA. (2015) Information on biosimilars. US Food and Drug Administration. Available from: http://www.fda.gov/Drugs/DevelopmentApprovalProcess/HowDrugsareDevelopedandApproved/ApprovalApplications/TherapeuticBiologicApplications/Biosimilars/ (Accessed July 29, 2015).

GaBI online. (2015a) Norway, biosimilars in different funding systems. *Generics and Biosimilars Initiative Journal.* Available from: http://gabionline.net/Biosimilars/Research/Norway-biosimilars-in-different-funding-systems (Accessed July 29, 2015).

GaBI online. (2015b) Huge discount on biosimilar infliximab in Norway. *Generics and Biosimilars Initiative Journal.* Available from: http://gabionline.net/Biosimilars/General/Huge-discount-on-biosimilar-infliximab-in-Norway (Accessed July 29, 2015).

GaBI online. (2015c) Remsima shows comparable safety and efficacy in IBD patients. *Generics and Biosimilars Initiative Journal.* Available from: http://www.gabionline.net/Biosimilars/Research/Remsima-shows-comparable-safety-and-efficacy-in-IBD-patients (Accessed July 29, 2015).

GaBI online. (2015d) Dutch medicines agency says biosimilars "have no relevant differences" to originators. *Generics and Biosimilars Initiative Journal.* Available from: http://www.gabionline.net/Biosimilars/General/Dutch-medicines-agency-says-biosimilars-have-no-relevant-differences-to-originators (Accessed July 29, 2015).

GaBI online. (2015e) Australia's PBAC recommends substitution of biosimilars. *Generics and Biosimilars Initiative Journal.* Available from: http://www.gabionline.net/Biosimilars/General/Australia-s-PBAC-recommends-substitution-of-biosimilars (Accessed July 29, 2015).

GaBI online. (2015f) Biosimilars in the US. *Generics and Biosimilars Initiative Journal.* Available from: http://www.gabionline.net/Biosimilars/Research/Biosimilars-in-the-US (Accessed July 29, 2015).

GaBI online. (2015g) Biosimilars substitution bill become law in Texas. *Generics and Biosimilars Initiative Journal.* Available from: http://gabionline.net/Policies-Legislation/Biosimilars-substitution-bill-become-law-in-Texas (Accessed July 29, 2015).

IMS Institute. (2014) Assessing biosimilar uptake and competition in European markets. IMS Institute for Healthcare Informatics. Available from: www.theimsinstitute.org (Accessed July 29, 2015).

Ruiz S, Calvo G. (2011) Similar biological medicinal products: lessons learned and challenges ahead. *Journal of General Medicine* **8**, 4–13.

Stanton D. (2013) Norway to facilitate switch to biosimilars with $3m Remicade study. *BioPharma.* Available from: http://www.biopharma-reporter.com/Markets-Regulations/Norway-to-facilitate-switch-to-biosimilars-with-3m-Remicade-study (Accessed July 29, 2015).

Teuscher, A. (2007) *Insulin—A Voice for Choice.* Karger, Basel.

Weise M, Kurki P, Wolff-Holz E, et al. (2014) Biosimilars: the science of extrapolation. *Blood* **124**, 3191–3196.

White Junod S. (2009) Celebrating a milestone: FDA's approval of first genetically-engineered product. Available from: http://www.fda.gov/AboutFDA/WhatWeDo/History/ProductRegulation/SelectionsFromFDLIUpdateSeriesonFDAHistory/ucm081964.htm (Accessed July 28, 2015).

WHO. (2014) Biological qualifier—an INN proposal. World Health Organization. Available from: http://www.who.int/medicines/services/inn/bq_innproposal201407.pdf (Accessed July 29, 2015).

16 Biosimilars and Biologics
The Prospects for Competition

Erwin A. Blackstone
Temple University

Joseph P. Fuhr Jr.
Thomas Jefferson University

CONTENTS

16.1 INTRODUCTION

The current US health care system is unsustainable as health care expenditures approach 20% of GDP. One way to cut costs is through competition from biosimilars, which has decreased costs for biological drugs in the EU. Biologics are becoming an even more important part of the pharmaceutical industry. They treat some of the most serious and life-threatening diseases. Their development and production are more costly than the previously dominant small-molecule or chemical drugs. Until passage of the Biological Price Competition and Innovation Act (BPCIA) as part of the Patient Protection and Affordable Care Act of 2010, there was no pathway for biosimilar competition to enter in an expedited and relatively less costly application process. This was similar to the situation for chemical generics before passage of the Hatch–Waxman Act in 1984. Accordingly, even after patents expired on these biologics, lower price biosimilars were not available.

This chapter examines the situation for biologics and biosimilars in light of the new possibilities afforded by the BPCIA. It includes the experience of other countries where biosimilars have been available since at least 2006. Accordingly, we consider the role of patents, provision for data and market exclusivity, entry barriers, competitive considerations, and regulatory matters, among other issues. We also note distinctions between biologics and chemical drugs and their impacts on the market.

16.2 SOME TERMINOLOGY

A biologic is a large-molecule drug produced in living organisms. Unlike a generic in the case of chemical drugs which is equivalent to the originator, biosimilars are highly similar to the originator referred to as a reference product. A reference product is the originally licensed biologic to which the biosimilar is compared to determine if it is highly similar. In both the US and EU, biosimilars must go through the respective regulatory bodies and are rigorously and scientifically assessed. The BPCIA's definition of a biosimilar is "highly similar to the reference product notwithstanding minor differences in clinically inactive components; and … no clinically meaningful differences between the biological product and the reference product in terms of safety, purity, and potency of the product." (42 U.S.C. 2006: § 262(i)(2)(A), (B)). Since they are produced in living organisms and in batches, even for the originator each batch varies and is not identical (McKinnon and Lu, 2009).

Drift occurs frequently among biologics (Schiestl et al., 2011). Given the nature of the production process and the fact that biologics are produced in batches as liquids, no two batches are identical. Thus, each originator biologic varies somewhat from the original batch (drift), and in some cases the biosimilar is more similar to the original biologic than is the biologic itself. The FDA thus allows for a range within which the biosimilar must stay to be considered a biosimilar. Therefore, there is some variability in both the biologics and the biosimilar (Schneider, 2013). Also, there is the issue of the manufacturing process, whereby changes can cause drift in the process. As both biologics and biosimilars change their manufacturing process, variability between the processes can occur. For example, Remicade has had over 35 process changes since the original process. Indeed, Genzyme wanted to produce its originator biologic Myozyme in a larger facility, but the FDA ruled that the product was different from Myozyme, which was produced in a smaller facility. Achieving sufficient product uniformity can sometimes be difficult (Blackstone and Fuhr, 2010).

Since biosimilars are not identical but highly similar, they cannot be automatically substituted at the pharmacy level unless they are interchangeable. According to the FDA, "[a]n 'interchangeable' biological product is biosimilar to the reference product, and can be expected to produce the same clinical result as the reference product in any given patient. In addition, to be deemed an interchangeable biological product, it must be shown that for a biological product that is administered more than once to an individual, the risk in terms of safety or diminished efficacy of alternating or switching between use of the biological product and the reference product is not greater than the risk of using the reference product without such alternation or switch" (FDA, 2014).

Each country has its own unique regulatory requirements for biosimilars. The term *biosimilar* is often misused since many products referred to as biosimilars in some countries with less rigorous regulatory requirements do not meet the standard of being highly similar. Such biosimilars can be referred to as noncomparable biologics. We consider a biosimilar to be one approved in the US, EU, Canada, and Australia and one should be cautious in referring to other countries' noncomparables as biosimilars. Further, biosimilars are sometimes referred to as follow-on biologics.

Biobetters are products that are, as the name suggests, better than originator biologics. They could have greater efficacy, easier or more convenient administration, fewer side-effects, or a lower rate of potential immunogenicity (discussed below) which could lead to greater adherence. A second-generation biologic produced by the reference producer is one example of a biobetter. There has been considerable discussion concerning biobetters, but few are currently on the market except for second-generation originator products. However, many are reportedly being developed. In the US, an alternative to the 351(k) (biosimilars) created by the BPCIA is available, that is, the Biologic Licensing Application 351(a) of the Public Service Act, but it has not been used much for biobetters.

16.3 BENEFITS OF PHARMACEUTICAL INNOVATION

Innovation increases the quality of life and promotes economic growth. Pharmaceutical innovation has led to tremendous advances in the treatment of diseases and has enhanced both the length and quality of life. The National Academy of Sciences states, "Few families know the suffering caused by small-pox, tuberculosis, polio, diphtheria, cholera, typhoid or whooping cough. All those diseases have been greatly suppressed or eliminated by vaccines" (National Academy of Sciences, 2007). Consumer welfare is increased considerably by the replacement of older drugs by newer, more effective drugs (Lichtenberg, 2001). Drug discovery also often reduces medical expenditures. Lichtenberg estimated that the reduction in inpatient spending was four times the prescription costs (Lichtenberg, 2001). Drugs can decrease the need for expensive medical procedures. For example, beta blockers can reduce the need for expensive and risky heart surgery. Antibiotics now treat some ulcers that previously required more extensive and expensive therapies. Drugs, including biologics, can also decrease absenteeism and presenteeism, increasing workers' productivity. Griliches (1992) found that pharmaceutical innovation increases consumer welfare. Thus, public policy should and does encourage innovation.

The primary policy to encourage innovation is that of patents which provide a theoretical 20-year legal monopoly. Also, the government funds some R&D through grants from the National Institutes of Health. However, none of these gains will accrue to people who do not have access to these drugs. The high price of biologics has made access an issue. People without insurance cannot afford the drugs, and even those with insurance often have such high copays that they, too, cannot afford the drugs. However, consideration might be given to some government support for low-income people. This issue is beyond the scope of our chapter. Accordingly, the exception is for communicable diseases whose vaccine usage arguably should be required. In general, once patents expire, more people should have access to these drugs as competition increases and prices decrease. Biosimilars will increase access. We focus on both innovation and competition.

16.4 BIOLOGICS VERSUS CHEMICAL DRUGS

Biologics are large-molecule drugs produced in living systems such as a plant cell or microorganism. Their manufacturing is more complex and costly than chemical drugs (Blackstone and Fuhr, 2012). They are much larger and more complex than chemical drugs (Blackstone and Fuhr, 2015). For example, the biologic Epogen has a molecular weight 168 times that of the small-molecule drug aspirin (Kanter and Feldman, 2012). Their complexity is illustrated by the large molecular drug Remicade, which requires 310 separate production stages (Blackstone and Fuhr, 2007). Biologics' R&D costs are much higher, and their manufacturing is more costly and complex than chemical drugs. Also, small changes in manufacturing can lead to health risks such as immunogenicity. "[U]nique to biologics, is 'immunogenicity': Immunogenicity is a patient's adverse antibody reaction to a drug in which the body perceives a drug to be a foreign microorganism or virus" (Kaldre, 2008). Biologics treat many life-threatening diseases and are generally priced high with few alternative therapies available. Some biologics are priced at more than $500,000 per year per patient. Many biologics are orphan drugs that treat diseases that are relatively rare (Loo, 2015). Finally, biologics are often infused which affects the nature of marketing and advertising and quite possibly even their expected market share.

A major issue initially concerning biosimilars is that of their safety. However, in the EU there have been very few if any safety issues. One safety issue concerning a biosimilar could yield devastating consequences on the biosimilar market. This is why it is especially important to distinguish between a true biosimilar and noncomparable originators. If there is a safety issue with a noncomparable biologic, one must be sure that it is not considered a biosimilar and thus have a negative effect on the biosimilar market. Further, strict regulation of biosimilars to ensure their safety and effectiveness is appropriate, especially in the early period before biosimilars have a track record.

16.5 BIOLOGICS AND INNOVATION: HIGH RISK, HIGH REWARD

Biologic firms spend substantial sums and a high share of their revenues on R&D in the hopes of developing a new drug. These firms are the originator biologic firms. There are also other firms that want to produce the biologic once the patents have expired or are not valid and they can legally do so. The industry thus contains two groups of firms: the originator or reference product producers and the biosimilar producers. Both groups of firms serve an important function. Sometimes the innovator company becomes a biosimilar producer for a particular reference biologic product other than its own biologic. Amgen and Sandoz (Novartis) are examples of originators developing biosimilars. This is somewhat different from the small-molecule or chemical generic drug market where the demarcation is generally stronger.

Innovator firms' spending on R&D as a percentage of biologic sales is among the highest of any industry. Public companies in 2013 spent 32.4% of their revenue on R&D (Loo, 2015). This compares to about 3%, which is typical for US manufacturing companies or industries. Among leading biologic companies, spending in 2014 ranged between 11.5% and 147.5% of revenues (Loo, 2015). Amgen, a leading biological firm, for example, spent 21.4% of its revenues on R&D. Vertex, which spent 147.5% of its revenues on R&D, had only $58 million in sales. Small or startup firms will obviously have such high figures. Spending is so high even though most R&D projects end in failure. Merck estimates that 75% of the R&D it spends is on failures (Blackstone and Fuhr, 2012). There is obviously great risk associated with R&D.

Biologics have greater risk in R&D than chemical drugs. The development of a new biologic is a long and difficult process. On average, a new biologic requires between $1.3 and $2.6 billion (Blackstone and Fuhr, 2012). A new chemical drug involves costs of $500–$800 million (Blackstone and Fuhr, 2012). Further biological development takes on average between 10 and 15 years, with many, if not most, of these efforts ending in failure. Taking failures into account, we find that the cost for an FDA-approved biologic (one able to be marketed) could be as high as $5 billion (Blackstone and Fuhr, 2015).

Moreover, many biologics are developed by small firms that often find it difficult to raise capital. These small firms, often startups, have been responsible for almost 50% of new biologics (Blackstone and Fuhr, 2012). They often partner with a larger firm to complete the development of a new biologic, or sometimes they are acquired by a larger firm. Further, given the increasing merger activities among pharmaceutical firms and even within the biopharma industry, there are fewer potential buyers or partners for these small firms should they encounter problems in the development process or simply wish to sell their project. For example, among the many pharmaceutical mergers in 2014, Actavis acquired Allergan, creating a top 10 company with annual revenues of $23 billion (Loo, 2014). Small firms thus have reduced bargaining power.

An indication of the riskiness of R&D in the drug industry is the fact that only 5%–10% of the drugs that begin clinical trials will receive marketing approval from the FDA (Blackstone and Fuhr, 2012). Further, only 3 out of 10 drugs that receive FDA approval become commercial successes (Loo, 2015). Also indicative of the high risk for originators is the fact that in the past 16 years only 7 out of 103 attempts at developing a melanoma drug resulted in drug approvals. The figures for lung and brain cancer were 10 of 77 and 3 of 78, respectively (Loo, 2015).

Other factors add to the risk and uncertainty for innovators. Biologic marketing involves high cost. In fact, such costs typically rise dramatically soon after a drug is approved as the company attempts to inform potential buyers about the drug. These marketing costs (which normally are included within the sales, general, and administration category) are about equal to R&D costs (Loo, 2015). Should the drug prove unprofitable, these costs cannot normally be recouped. They are what economists refer to as sunk costs.

Also making R&D more risky is the fact that the FDA is increasingly demanding a showing of a statistically significant improvement over the standard therapy. This could increase the required sample size for clinical trials, making these trials more costly, difficult, and time consuming. Already, 85% of clinical trials face delays because of difficulties in obtaining sufficient patient recruitment (Loo, 2015).

16.6 PATENT PROTECTION AND EXCLUSIVITY

Encouraging biological innovation is obviously important. Patents provide 20 years of protection against competition from the date of filing for the patent. The issue is whether such protection is adequate to encourage undertaking the high cost and uncertainty of biological innovation. Since drug approval requires substantial testing and clinical trials, the actual protection is reduced substantially. Given the 8–10 years typically required for approval, actual patent protection may last around 10–12 years. One study reported that actual effective patent life for drugs is 11.7 years (Roth, 2013).

There are numerous issues with the patent system itself that raise questions about its adequacies. Critics contend that too many patents are issued, so that many are flimsy. Defending patents against infringement suits is costly, and the uncertainty about prevailing in court could undermine the incentives for innovation. The result is that the presumption of validity is reduced, encouraging patent challenges. Further adding to the uncertainties of patent litigation is the fact that patent coverage is often unclear (Roth, 2013). The patent system needs more certainty. In the generic market, the producer is making an exact copy. In the case of biosimilars, the producer is making a product that is highly similar to the reference product. In the generic case, infringement would presumably be easier to determine. This adds uncertainty to the market for both the biologic innovator and the copier, whether or not infringement is actually occurring.

This is especially the case for new technologies like biologics since patents tend to be fairly narrow, making them vulnerable to challenges (Sahr, 2009). Also, because of the complexity of large-molecule biologics, proving infringement may be more difficult. Differences in their manufacturing processes and their complexity may combine to yield different properties and structures even though the product is highly similar (Roth, 2013).

On the other hand, biologics tend to be protected by 50–70 patents that cover manufacturing, production, and every aspect of R&D, while chemical or small-molecule drugs are usually covered by 8–10 patents (Blackstone and Fuhr, 2012). Further, manufacturing plays a much more important role in the case of biologics than for chemical drugs, and the manufacturing processes can be kept secret, avoiding the necessity of obtaining patents at all for this aspect. In any event, the extent of effective patent protection for biologics is unclear.

In view of the uncertainties of patent protection for new biologics, it is probably desirable that additional protection be provided. Under the BPCIA, both data and market exclusivity are provided. Four years of data exclusivity begin from the date of FDA approval, which means that no studies or data from the innovator can be used for 4 years. Also, the FDA will not consider any application for a biosimilar during that time. In addition, there is a 12-year market exclusivity from the date of FDA approval of the originator biologic. This means that no biosimilar can be marketed for 12 years from FDA approval for the reference product. Thus, the FDA can approve a biosimilar after 4 years, but the biosimilar cannot be marketed until after the 12-year market exclusivity expires (Gorman et al., 2013).

Also, given that patents are granted for 20 years from the date of application and the 4-year data exclusivity and 12-year market exclusivity are granted from the date of FDA approval, there is a considerable time period, around 8–10 years from patent application through clinical trials to drug approval. In many cases, the patent will probably expire before the end of the exclusivity period.

Interestingly, most commentators discuss data exclusivity without noting the market exclusivity. It is noteworthy that data and market exclusivity cannot be challenged in court (Blackstone and Fuhr, 2013). The exclusivity is automatic once the originator biologic has been approved. Given the uncertainties of patent protection and the long period before drug approval, the market exclusivity included in the BPCIA serves as an important incentive in encouraging innovation. An additional exclusivity of 6 months is available for pediatric applications.

16.7 LENGTH OF EXCLUSIVITY AND TYPE 1 AND TYPE 2 ERROR

There was much debate over the length of market exclusivity that biologics should enjoy. The debate centered around 7 or 12 years. It is obviously difficult to determine the optimal exclusivity time period. It can and almost certainly will differ significantly by drug. Thus, there is almost inevitably going to be some error involved in determining the appropriate exclusivity period. It was eventually decided and implemented that a 12-year exclusivity period was appropriate. This raises the issue of a type 1 and type 2 error. If too short a period were to be chosen, a type 1 error, originator firms would have less time to obtain a return on investment and less incentive to innovate; thus, some beneficial biologics may not be developed. If the period of exclusivity was too long, a type 2 error, then there would be less competition and less access due to higher prices during the exclusivity time period. Thus, given that an error must occur and given the great uncertainty and risk involved in drug innovation, it would make sense to err on the side of too long, since without the drugs there would be no benefit to society; optimal public policy should err on the side of innovation.

16.8 DRUG DEVELOPMENT: A CASE STUDY

A recent blockbuster drug, Sovaldi, illustrates the process of drug discovery and development. Sovaldi is a small-molecule drug, but its discovery and development are common for biologics. The drug was developed by a relatively small firm, Pharamasset, whose senior vice president for chemistry, Michael Sofia, headed a team that discovered the breakthrough therapy for hepatitis C. They surmounted the problem of developing a drug that would pass through the bloodstream and reach the liver where it could act to destroy the virus. The breakthrough occurred by creating a shield that would break down once in the liver to allow the drug to work (Sell, 2015).

Pharmasset was sold in November 2011 for $11.1 billion to Gillead Sciences. Sovaldi was approved in 2013 and achieved revenues of $10.3 billion in 2014, the largest amount ever in so short a time period. This blockbuster drug illustrates many of the elements associated with pharmaceuticals and, to an even greater extent, biological research. In particular, the original developer was not the firm that finally brought the drug to market. A leading figure in the development of Sovaldi, chemical name Sofosbruvir, in referring to biotech R&D, stated "More often than not in the biotech world, the gamble leads to nothing and you are forced to look for a new opportunity" (Sell, 2015).

Sovaldi also illustrates the issues of access. Its price in the US was set at $1000 per day of treatment or $84,000 for the 12-week course of treatment. The World Health Organization would like a lower price, so more of the world's 150 million sufferers of hepatitis C would obtain treatment (Sell, 2015). This concern with pricing is common for insurers, some of which have restricted coverage of Sovaldi to those with advanced liver disease. On the other hand, it is important to provide incentives for R&D so that drugs like Sovaldi are developed. It is also important to note that most drugs are not blockbusters and often do not even cover the cost of their R&D. In effect, the few blockbusters like Sovaldi help cover the costs of developing the low-profit drugs and motivate others to try to duplicate their success. Also, even though its patent has not expired, Sovaldi presently faces competition from another hepatitis C drug, Viekira Pak; this competition and its implications to biosimilar markets will be discussed later in this chapter.

16.9 BIOLOGICS AMONG THE HIGHEST PRICED DRUGS

Biologics are among the highest priced drugs. In particular, the annual price for Soliris in 2015 was $536,529, and for Naglazyme it was $485,747. These are the two most expensive biologics. The 10th most expensive, Revlimid, had an annual price of $128,666 (Loo, 2015). Some of these expensive biologics are so-called orphan drugs; namely, they are used for a relatively small patient population. Soliris is such a drug. To encourage their development, orphan drugs receive a tax credit for 50% of R&D costs, grants to help defray clinical trials, and 7 years of marketing exclusivity. Orphan drugs are typically priced 19 times that of nonorphan drugs. In 2014, of the 41 new molecular entities approved by the FDA, 17 were orphan drugs (Loo, 2015).

16.10 HATCH–WAXMAN AND GENERICS

Thirty years ago, the US developed a regulatory framework for the entry of generic chemical drugs. The resulting Hatch–Waxman Act (H–W) was intended (as is the BPCIA) to balance competition and innovation. The major public policy goal was to enhance competition from generics, which would lead to lower prices but still provide the originator with the incentive to innovate.

The H–W was enacted in 1984 and enabled the entry of generics. Prior to its passage, few generics were available. In fact, only 35% of the drugs that lost patent protection had generic competition (Blackstone and Fuhr, 2012). Further, between 1962 and 1984, 150 drugs lost patent protection but did not encounter generic competition (Behrendt, 2006). There was no abbreviated pathway for generic competition. H–W, also known as the Drug Price Competition Act and Patent Term Restoration, developed such a pathway. The Act created an abbreviated new drug application (ANDA) under which a generic applicant only has to show bioequivalence to the branded product. The generic applicant can claim that no patent is involved, that it will not enter until the relevant patent or patents expire, or make a so-called paragraph 4 certification that the patent is invalid or not infringed. If it succeeds in such a challenge, the first such claimant receives a 6 months exclusivity, which means that no other generic can be marketed during that period. This provision gives a generic entrant the incentive to be the first to file such a paragraph 4 application.

At the same time, to encourage innovation, H–W provided a 5-year market exclusivity for the originator, during which no generic can be marketed. H–W also provided the possibility to extend patent protection for up to 5 years if the FDA approval process took an inordinate time.

H–W seems to have worked quite well in balancing innovation and competition. However, it took some time for consumers and providers to be comfortable with the use of generics. Especially important was the development of automatic substitution, under which the pharmacist can substitute a generic in most states unless the physician specifically prevents that substitution.

H–W has succeeded in its efforts to encourage both competition and innovation. Drug development has continued. Thirty-five new drugs were approved by the FDA through December 3, 2014, compared to a yearly average of 24 for the 2003–2014 period (Loo, 2014). Further, Evaluate Pharma indicates that between 2014 and 2020, "The R&D pipeline is strong" (Loo, 2014). At the same time, generics have increased their share of all prescriptions. Specifically, in 1984 when H–W was enacted, only 19% of all prescriptions were for generics. The percentage increased to 33 in 1990, to 72 in 2008, and to 86 in 2013. Given that generics are sometimes 80% or 90% less expensive than the brand-name drug (Zirkelback, 2014), such a growth in generics is not surprising, and obviously it has increased access. Third-party payers have induced patients to use generics by having lower copays or by not putting the originator product on their formulary. Generic drugs have saved over a trillion dollars in health care costs between 2002 and 2011 (IMS, 2012). Biosimilar competition is also expected to result in substantial benefits.

Thus, there is a strange relationship between the originator and the generic (biosimilar) companies. The generic companies would not exist without the originators since there would be nothing to copy. However, the originators gain from the existence of generics since low-cost generics decrease the pharmaceutical budget costs of insurers and facilitate the ability of insurance companies to use these savings to pay for high-priced new drugs, especially biologics, thus encouraging innovation.

16.11 BARRIERS CONFRONTING BIOSIMILARS

H–W encouraged the development of generics by providing an abbreviated pathway, an ANDA, which required only chemical equivalency for generic approval. The BPCIA has done much the same for biologics that were not included in H–W. Before passage of BPCIA, there was no pathway for biosimilar competition. BPCIA gave the FDA power to develop guidelines for the entry of biosimilars and interchangeable biologics. The FDA approval process requires a stepwise approach, and the FDA will base its decision on a totality of evidence and a case-by-case approach. Among the factors that the FDA will examine are the structural and functional characteristics of the biosimilar and its reference product, animal data, human PK and PD data, clinical data on immunogenicity, safety, and efficacy (Christl, 2015). The FDA has four categories for the proposed biosimilar: not similar, similar, highly similar, and highly similar with finger-like similarity. The last two categories will constitute acceptance. Also, biosimilars can be of two types: biosimilar and interchangeable biologics. The latter must be achieved to be considered for automatic substitution. Biosimilar clinical trials are head-to-head trials with the reference product. Since biosimilars are not exact copies as are generics, whether to require clinical trials and their extent is a major issue. The FDA can waive the need for clinical trials, but this is not expected any time soon.

There are many barriers that make entry of biosimilars more difficult than generics. Biosimilars are much more costly to develop, and the process takes much longer than chemical generics. In particular, one estimate is that biosimilar development takes 8–10 years and involves costs in excess of $100 million compared to under $5 million and 3–5 years for generics (Ramachandra, 2014). The Federal Trade Commission (FTC) has similarly estimated that biosimilar development will cost between $100 million and $150 million and take 8–10 years versus $1–$5 million for a chemical generic (Blackstone and Fuhr, 2012).

In addition, there are likely to be other costs. For example, the cost of establishing a manufacturing facility has been estimated to be around $250 million (Blackstone and Fuhr, 2012). Another barrier making entry difficult and risky for biosimilars is their complexity, which makes expertise in their manufacturing quite important. Companies experienced in biological manufacturing like Amgen and Hospira will have a learning curve advantage that translates into a cost advantage (Blackstone and

Fuhr, 2013). Entrants into biosimilars are likely to be large, biological originators for other reference products.

Marketing costs could be substantial, especially in the early days of biosimilars, as producers have to educate providers and patients about the quality of their products. This means that the biosimilar is likely to have to devote substantial resources to attain buyer acceptance. However, since many biologics are infused, the buyers are physicians or hospitals, so that marketing efforts may be less than expected. On the other hand, originators may have long-term contracts with large buyers that could impede entry of biosimilars.

Moreover, because biosimilars are not exact copies, presently their approval requires clinical trials, whereas for generics or small-molecule drugs, no clinical trials are required. These trials can be quite expensive. A study found that 85% of clinical trials were already being delayed because of difficulties in obtaining sufficient patient recruitment (Loo, 2015). Also, with the large number of biosimilars being developed for each reference product, it will become even more difficult to recruit volunteers for clinical trials, which increases the time and cost to complete clinical trials.

Patients on reference biologics that may be treating life-saving diseases could be understandably reluctant to participate in a clinical trial. However, those who cannot afford the biologics may be willing to participate. Moreover, obtaining the reference product could be quite costly. For example, a study involving 1500 patients receiving a biologic like Soliris would have a $200 million cost in 2015 if the study requires 3 months of treatments. The calculation is only meant to suggest the possible cost of clinical trials for biosimilar entrants. Also, the originators may be reluctant to sell their product to potential competitors.

Another entry barrier factor that makes entry difficult is the uncertainty about the success of the biosimilar. In part, this is because of the large number of companies planning to enter the market with biosimilars or improved versions of the reference product, biobetters. For example, in 2011 it was reported that companies were developing 21 biosimilars and 12 biobetters for Herceptin and 21 biosimilars and 13 biobetters for Rituxan (Blackstone and Fuhr, 2013).

The highly similar but not identical nature of the biosimilar makes obtaining interchangeability status difficult. The lack of interchangeability will preclude automatic substitution at the pharmacy level. Physicians will then have to authorize substitution. Given the possible problems, including the risk of immunogenicity and possible resulting legal actions, physicians may be reluctant to authorize substitutes. Another cost that biosimilars will incur is that of postmarketing surveillance. Increasingly, the FDA is requiring such efforts, and biosimilars would seem to be particularly susceptible to such a requirement.

16.12 PATENT DANCE

Biosimilars also face patent issues. The BPCIA includes detailed requirements for patent disclosures by both the biosimilar applicant and the innovator. This, as noted in an earlier chapter, has been referred to as the patent dance. Within 20 days of its application being accepted by the FDA, the biosimilar shall provide its application to the originator. Then within 60 days after receipt of the biosimilar application, the originator shall provide a list of relevant patents covering the biologic and a list of any patents it is prepared to license to the biosimilar. The biosimilar applicant, within 60 days of receiving the patent list from the originator, shall then explain why it believes these patents are either invalid or will not be infringed by its biosimilar. Alternatively, it may assert that it will not enter until the expiration of the patents.

Next, within 60 days, the originator must explain why it believes certain patents are valid and will be infringed by the biosimilar. The originator and biosimilar applicant are then supposed to negotiate over the list of patents in dispute that may be infringed. The pioneer then has 30 days in which to file an infringement suit. Failure to do so within the 30-day period will result in only reasonable royalties being available in a subsequent suit. An originator's failure to list a patent will preclude it from a subsequent suit on that patent. Finally, the biosimilar firm must notify the originator 180 days before it intends to market the biosimilar (Vatiand et al., 2010).

The patent disclosure and negotiation process is obviously intended to narrow the area of dispute. However, discussions and communications between competitors are fraught with concerns that the public could be harmed. For example, the originator and biosimilar could both potentially gain by delaying entry of the biosimilar. This is the so-called pay for delay that has been a very contentious issue in the generic market (Blackstone and Fuhr, 2006).

Antitrust concerns are likely to ensue from such exchanges of information and communications. Further, the originator could conceivably gain some competitive advantage by the information exchanges, making competition more difficult for the biosimilar. For example, the originator may become aware of the trade secrets of the biosimilar applicant who is probably using the latest technology and gives the originator some important information about this technology.

Concerning the patent dance, which is being contested in *Amgen vs. Sandoz*, the issue revolves around the meaning of the word "shall." The court has to decide if "shall" means that the patent dance is mandatory or if the biosimilar applicant has the option to refuse to dance and thus have the court decide the validity or infringement of any patents. On March 19, 2015, the district court denied Amgen's motion for an injunction to stop the sale of Zarxio. In May 2015, the Federal Circuit gave Amgen a temporary injunction to stop the marketing of Zarxio. The injunction hearing began on June 5, 2015. The district court also concluded that Sandoz had the option to not engage in the patent dance.

Some have questioned why the patent dance even exists since it does not in the generic market. One explanation is that Congress believed that the patent dance could resolve patent issues before the 12-year market exclusivity expired. The law allows a biosimilar to apply after 4 years to the FDA (at the end of data exclusivity), and thus there is an 8-year window to resolve any patent issues. This is the reason for the dance. It is unlikely that an applicant would file as early as 4 years, but even if the applicant were to file in year 9 and receive approval in year 10, there would still be around 2 years to resolve any patent issues through the rules of the so-called patent dance and thus be able theoretically to enter after the 12-year exclusivity expired. Although the goal or intent is admirable, the unintended consequences may make the patent dance problematic.

16.13 SIX-MONTH NOTIFICATION

Another issue has arisen concerning the requirement that a biosimilar firm give the originator 180 days notification before it intends to market the biosimilar. This provision has also led to litigation. The issue revolves around "Subsection 9(k) applicant shall provide notice to the reference product sponsor no later than 180 days before date of the first commercial marketing of the biological product licensed under subsection (42 U.S.C. 262 (1) (8) (A).k)." The issue is whether the 6-month notice of biosimilar entry must be given after approval or whether it can begin after the application is filed, but before approval. In the first case, the originator effectively receives an additional 180 days of market exclusivity. The question is whether Congress intended to give only a 12-year market exclusivity or 12 years plus an additional 6 months.

Sandoz claims that a 180-day notice can be given after an application is accepted but before the application is approved. However, Amgen argues that 180-day approval can only be given after the biosimilar is approved. In *Amgen vs. Sandoz* (2015), Judge Richard Seeborg concluded that if Congress intended for an additional 6 months it would have been "more explicit," and thus he denied Amgen's motion. The court concluded that (1) (A)(8) is satisfied if notice is provided at least 180 days before intent to sell, and thus the notice can be given before approval. In any event, 6 months additional time to earn monopoly profit is highly significant.

16.14 FIRST MOVER DISADVANTAGE

Generally, in economics the first entrant is often thought to have a considerable competitive advantage. This is known as the first mover advantage. This could be true for the originator biologic since it has an established reputation and patients have experience with the product. Patients and physicians need to be educated concerning biosimilars, and even after 30 years there is still some question among consumers about the quality and effectiveness of generics. Thus, the originator is likely to have a competitive advantage over biosimilars. If the originator charges the same price, it should retain all or nearly all of the market. Also, a payer may be willing to pay a premium for the originator's product to avoid the various issues and costs involved with dealing with patients and physicians concerning a switch to biosimilars. This reluctance to switch is because biologics treat some of the most serious diseases. If the reference product is working, patients are reluctant to make the switch.

Interestingly, the first biosimilar entering the market may have a first mover disadvantage. This may be the case for biosimilars due to considerable barriers to entry, which increases costs and uncertainty for the first biosimilar entrant. In the generic or chemical drug market, the first entrant receives a 180-day exclusivity, which means no other generic can enter the market during this time period. During this time period, the generic firm can price considerably higher than after other generics enter and can earn considerable profit (Center for Drug Evaluation and Research, 1998). However, the first biosimilar entrant does not receive any exclusivity; only the first interchangeable biologic does. The first biosimilar still has an advantage since it can enter at a higher price before other biosimilars enter the market, but given the higher cost of R&D and manufacturing and other entry costs, one would not expect much of a return on investment before other biosimilars enter. As other biosimilars enter, price will decrease (Olson and Wendling, 2013). Moreover, in the case of biosimilars, the first mover faces various disadvantages. The first biosimilar in each class will have higher costs of entry than later entrants. FDA approval will be more uncertain, and thus there will be higher costs of preparing for the approval process. Also, the initial entrant may incur higher costs due to legal and patent issues. For example, it may need to resolve any patent issues that could lead to considerable litigation costs and entry delays. There are also the initial costs of educating physicians and patients concerning what a biosimilar is and the quality of biosimilars. The initial entrant will have to overcome the understandable consumer reluctance. Sandoz is facing all of these issues as it attempts to be the first biosimilar in the US market. Later entrants for the same reference product can free-ride on all or at least most of the above costs.

16.15 INTERCHANGEABILITY AND AUTOMATIC SUBSTITUTION

Biosimilars need to be highly similar. However, there is another issue—that of interchangeability. According to the FDA, "[a]n 'interchangeable' biological product is biosimilar to the reference product, and can be expected to produce the same clinical result as the reference product in any given patient" (FDA, 2014). Even though the FDA is still developing guidelines for interchangeability, it is unlikely that an applicant will attempt to get interchangeability in its initial FDA hearing for biosimilar status. In the first biosimilar application before the FDA, Sandoz did not apply for interchangeable status. It is possible that the FDA will not allow a biosimilar to be interchangeable until the biosimilar has a track record and has been demonstrated (through postmarketing studies) to produce results identical to that of the originator product. It is uncertain as to how long the interchangeable process might take, if it ever occurs (Fuhr et al., 2015).

Thus, before a biosimilar can apply for interchangeability, there may be two or three other noninterchangeable biosimilars on the market. The first interchangeable biosimilar receives a 1-year market exclusivity as an interchangeable biologic, but the first interchangeable could still be competing with originator and noninterchangeable biosimilars. An interchangeable biologic has the advantage of automatic substitution at the pharmacy level, depending on state law. However, there would be little competitive advantage for physician-administered biologics since physicians decide directly which product to use. There are also high costs involved in obtaining interchangeability since switching studies will probably be needed for the clinical trials (Shea and Muller, 2010). Payers probably do not care if a biosimilar is interchangeable and are probably not willing to pay a price premium. Noninterchangeable biosimilars may set lower prices, gain formulary status, and obtain most of the market. Finally there is the risk of failure. If a biosimilar tries for interchangeable status and fails, the product may be perceived as not that similar and not of high quality, thus damaging its reputation and market share. A biosimilar firm must weigh the advantages and disadvantages of attempting to achieve interchangeable status for its biosimilar. We suspect that interchangeability status may not be highly desirable and thus not sought after by many firms.

Interchangeability of biosimilars is further discussed in Chapter 10 of this book.

16.16 EU MARKET EXPERIENCE

Biosimilars in the EU are priced on average around 30% lower than their reference products. Some examples of price decreases as a result of biosimilar competition in various EU countries are Orion's Remsima 72%, Sandoz's Binocrut 33%, and Sandoz's Omnitrope 22%.

In terms of market share in the EU, the experience is mixed. In 2011, for example, for granulocyte-colony stimulating factors (G-CSFs) biosimilars had an average market share of 42%. The average market share for the erythropoietin class was 22%, and for human growth hormone it was only 13% (Rompas et al., 2015). With respect to individual countries, market shares for EPO ranged from about 20% in Italy to 80% in the United Kingdom. For G-CSF, the market shares ranged from about 10% in Italy to around 60% in Germany, Greece, Poland, and Sweden (Rompas et al., 2015). Germany had the highest market share for both EPO and G-CSF.

Another estimate of biosimilar market penetration for individual EU countries for 2013 is in terms of patient days of treatment (IMS, 2014). In the case of Epo, the biosimilar share ranged from a low of 0% in Belgium to 66% in Slovenia. Among the larger EU countries, Germany is relatively high, with a 53% share. Norway, Bulgaria, and Sweden are all relatively high, with shares of at least 53%. In the case of G-CSF, biosimilars have a very large share in Sweden (91%), Hungary (99%), and Romania (100%). Germany, Italy, and the UK (73%) have high shares as well. For the human growth hormone (HGH), biosimilar market shares are much lower. For example, Germany had a biosimilar share in 2013 of 9%, the UK 5%, and Sweden 18%. Even Norway, which had 82% for G-CSF and 58% for EPO, had only 2% (IMS, 2014). Again the numbers show a mixed picture but one where biosimilars have begun to capture significant shares of the market.

16.17 SAVINGS FROM BIOSIMILARS

In terms of implications for the US, the experience of the EU and other countries must be reviewed with caution. In the US, generic drug prices generally have been lower and market shares generally higher than in the UK, Germany, Italy, France, and Japan. In Canada, generic drug prices have been lower by about 6% (Danzon and Furukawa, 2003). Further, in terms of biological prices, EU countries have had lower biologics prices than the US. For example, the average price for all biologics in 2005 in the US was $59, and in France it was $44, Germany $14, Italy $19, and the UK $23 (Danzon and Furukawa, 2006). Since prices were already lower in the EU countries, a 20% or 30% discount could be viewed as more significant than in the higher priced US. In other words, one might expect greater market shares and larger price reductions in the US.

One study estimated that biosimilars will have saved between 11.8 billion and 33.4 billion euros in eight EU countries, from 2007 to 2020 (Haustein et al., 2012).

Another study estimated annual savings of 1.6 billion euros across Europe, from a 20% price reduction in the five most popular patent-free biologics (Oldham, 2006). Given the high costs of development and manufacturing, prices are unlikely to decrease by 80% or 90% as in the generics market. Unlike the generics market where the originators generally did not respond to competition, in the biologics market the originators have actively responded in a variety of ways. They have decreased price, developed second-generation biologics (biobetters), attempted patent extensions, developed better devices for injecting the medicine, and reduced the frequency of dosages. These are also gains from competition from biosimilars. On the other hand, given the higher US biologic prices, there would appear to be more ability to decrease prices than in the EU. In any event, each country in the EU has a unique reimbursement system with different incentives for biosimilar use. Over much of Europe, there has been historically little financial incentive for the patient, the physician, or the pharmacists to opt for lower-priced biosimilar products. The major exception is Germany.

16.18 DISINCENTIVES FOR BIOSIMILARS

Some EU reimbursement systems have resulted in incentives that deter biosimilar competition. Some countries have set biosimilar prices at a fixed percentage below the price of the reference biologic. These mandatory discounts can deter competition. If a manufacturer of an originator biologic decreases its price and if the required discounted price of the biosimilar is too low for biosimilar firms to make a profit, then there will be no biosimilar entry. In Ireland, hospital-level tendering had the perverse incentive that could lead hospitals to choose the highest price biologic because the hospital retained the absolute size of the discount (Morton, FM, Stern AD, Stern S., unpublished). Such disincentives make predictions for the US market uptake more uncertain.

16.19 INCENTIVES FOR BIOSIMILARS

Some countries in the EU have begun to restructure incentives. One such mechanism is tendering, which generally leads to lower prices. Under tendering, part or all of a country's product is subject to a competitive bid, with the lowest priced bidder obtaining an exclusive arrangement with that country or region to provide the medicine. In 2014, the Norwegian Medical Agency received a 39% discount for the infliximab biosimilar, Remsima by Orion Pharma (Benassi, 2014). In 2015, the discount was increased to 72% (Generics and Biosimilars Initiative, 2015). Hospira, which offered the same product, had offered a 51% discount. In England, each hospital has a set budget and has the incentive to purchase the lowest priced alternative and can thus use the savings to provide other medical services. As a result, biosimilars constituted 80% of G-CSF sales in the UK. UK hospital physicians moved G-CSF back to first-line treatment when prices decreased and thus increased access.

16.20 GERMANY'S EXPERIENCE

Germany has been successful in terms of usage of biosimilars. The German government has an incentive system that encourages the use of biosimilars. Germany has reference pricing and a rebate system as well as biosimilar quotas for both regional sickness funds and physicians. Under reference pricing, patients must pay out of pocket the difference between the price of the drug chosen and the reference level. Additionally, if physicians exceed 125% of their budget they need to repay the amount above 115% unless the excess can be justified.

The large insurance companies (sickness funds) are paid in a capitated fashion and negotiate for discounts; thus, they have an incentive to choose the biosimilar to be on the formulary. Germany's Federal Healthcare Committee has encouraged the use of biosimilars and is able to bargain for rebates. Sandoz, for example, in 2007 increased its Binocrit discount from 15% to 33% and obtained 30% of the market (IMS, 2014). Another potential reason for high biosimilar penetration may be the fact that several of the biosimilar companies produce in Germany.

16.21 MARKET OPPORTUNITIES

Partly as a result of high development costs, biologics are often among the most expensive drugs. Five of the top ten revenue-producing drugs in 2012 were biologics, an increase from two of the top ten in 2008. Biotech drugs comprised only 7% of the top ten selling treatments in 2001, increasing to 71% in 2013 (Loo, 2015). This represents 18% of the global pharmaceutical industry revenues in 2012. Revenues for biologics are growing at twice the rate of global drug revenues overall. Some estimates have biologics reaching 50% of pharmaceutical sales. US sales in 2014 were around $200 billion and grew over 10% in 2014. Biotechnology products are expected to account for 51% of the top 100 drugs in 2018, an increase from 39% in 2012 and 15% in 2000 (Loo, 2014). The US comprises around 50% of the biologics market. Many biologics have sales of over a billion dollars. Over 30 biologics have lost or will soon lose patent protection. It is estimated that a total of $80 billion in US biological sales are going off patent by 2020. High prices and sales provide an incentive for biosimilar entry once patents expire. This incentive is also indicated by the higher gross margins, which are high because of the great cost of drug development.

The gross margin for the seven biological firms in the S&P 500 index in 2010–2014 ranged from 77.1% to 95.6%, among the highest of any industry (Loo, 2015). The great economic advantage of biosimilars is that a manufacturer only needs to re-create the idea that is already shown to work (Harris, 2011). Biosimilar producers can avoid much of these development costs incurred by originators. Despite the considerable barriers to entry, given the potential market opportunities, an influx of biosimilars into the market is expected. The future for many pharmaceutical firms is in biologics and biosimilars. As S&P states: "With conventional R&D yielding below-par returns, biotechnology is now viewed as the new frontier in breakthrough therapies" (Loo, 2014).

16.22 DUE DILIGENCE

There are tremendous opportunities in the biosimilar market, but there is also tremendous risk. Many companies are developing biosimilars for the same reference product. In 2013, it was reported that 21 firms were developing biosimilars for Herceptin, 27 for Enbrel, and 35 for Rituxan. These figures are not surprising since these drugs were all blockbusters. In particular, 2013 revenues for Herceptin and Rituxan were $6.6 and $7.5 billion, respectively. Humira and Avastin were also blockbuster drugs, with 15 and 13 biosimilar developers, respectively. Patents for these five drugs are to expire between 2018 and 2028 (King, 2014).

As discussed above, not all products labeled biosimilars are truly biosimilars, but if even half of them are true biosimilars, that would still be 17 biosimilar entrants for Rituxan. As in any nascent market, it will take time for the market to adjust to a long-run equilibrium. It is highly unlikely that 17 biosimilar firms can survive in any particular biologic market. The costs of R&D as well as the other costs involved could well exceed $250 million for each biosimilar. Thus, even though the market opportunity exists, firms will presumably understand the complexity of the market and perform due diligence before they decide whether to enter the market.

There is also the issue of which markets to enter. Are there advantages for firms to develop a portfolio of biosimilars? A portfolio of four or five biosimilars will spread the risk but also entail an investment in the area of a billion dollars. Another issue is which markets to enter. Strategies for biosimilar producers involve whether to enter blockbuster or smaller revenue biological markets. There are obviously greater potential profits in blockbusters but more competition, which could result in higher discounts, potentially less profit, and possibly no return on investment. Medium-revenue markets may appear to have less potential profit, but with less entry and lower discounts, profits may actually be higher.

Thus, the issue arises as to whether biosimilars can get a return on their investment. They will face price competition not only from the originator and other biosimilars but also from biobetters, including originators developing second-generation biologics. Many firms are making huge investments to develop biosimilars, including cost of R&D, manufacturing costs, clinical trials, and other costs. How many biosimilars will enter each market is unclear, but one would expect that the mature market could only sustain four or five biosimilar competitors. The biologics market should eventually evolve to be highly competitive as biosimilars enter the market. As in the pharmaceutical market, there will be a few winners and many losers.

16.23 US SUBMISSIONS TO DATE

There have been five applications in the US for biosimilars. However, Sandoz's Zarxio is the only one that has been approved, but it has not yet entered the market because of patent issues. Sandoz submitted considerable evidence of the structure and functionality of its biosimilar. Clinical trials, along with animal studies and the experience in the EU, were provided in support of its application. One clinical trial involved breast cancer patients. Sandoz also noted that its biosimilar had been available in most of the EU, 32 other countries, and has had 7.5 million days of usage (Sandoz, 2015). Also, Sandoz was approved for all five indications of the reference product through extrapolation.

The FDA postponed the hearing on Cellitron's Remsima biosimilar application that was scheduled for March 17, 2015, as the FDA asked for more information.

It was postponed "due to information request pending with the sponsor of the application." The FDA has asked for more details of its statistical analysis (Stanton, 2015).

There are other biosimilar applications under consideration by the FDA. So far, all of the applications have been for biosimilars that have already been approved in the EU, and one would expect this trend to continue. As of the end of February 2015, 50 biosimilar development programs referencing 15 biologics had been submitted to the FDA (Serebrov, 2015). Again, the BPCIA was designed to encourage both competition and innovation, and the former looks promising.

16.24 REIMBURSEMENT ISSUES

The US reimbursement system is much more complex than that of the EU, with both large private and public payers. Biologics are among the most expensive drugs, and they are accounting for an increasing share of drug expenditures. They also treat some of the most serious diseases. The issue of their coverage or reimbursement by insurance companies is likely to become a contentious issue. In some cases, insurance companies already require copayments of $500 per prescription or coinsurance rates of 33% (Blackstone and Fuhr, 2012).

Further, some of these drugs extend life for a short time at a high price. For example, the drug Nexavar extends the life of liver cancer patients on average <3 months (Blackstone and Fuhr, 2012). The National Institute for Clinical Excellence in the UK decided not to pay for this expensive drug, determining that the cost was greater than the benefit. New cancer drugs often cost $10,000 or more per month. Sometimes, the benefit is less than the cost, and physician groups are beginning to question the value of some of these drugs (*Cancer Drug vs. Value*, 2015). The risk for biological producers is likely to increase.

In the US, under the Patient Protection and Affordable Care Act (PPACA) of 2010, a major emphasis was to fund studies on the comparative effectiveness of medical treatments, including drugs. Along with the other factors, this should make drug development riskier as insurance coverage may become more of an issue. Some payers have entered into risk contracts with pharmaceutical firms in which the payer will only reimburse for the drug if it results in positive outcomes. In Germany, in instances when its cancer drug Avastin is not successful, Roche has offered refunds to hospitals and insurers (Blackstone and Fuhr, 2012).

16.25 INCENTIVIZING MARKET ACCEPTANCE

The financial incentives created, particularly through pricing, will determine the development of the biosimilar market. Cost containment is a major issue in the US. Biologics are coming under greater scrutiny because of their high prices. Stakeholders (physicians, patients, and payers) will greatly influence the biosimilar market. In the case of generics, some time was required before they gained acceptance. Biosimilar uptake in the EU has been successful when stakeholders have the right incentives. High biologic prices could lead payers to switch to lower priced biosimilars. For example, Germany has encouraged the use of biosimilars and has experienced some of the highest market shares for biosimilars. Bundling will give physicians the incentive to prescribe the lowest cost-effective alternative. Since under bundling physicians are given a flat rate for each episode, they have the financial incentive to utilize the lower cost drugs, which will increase their profit. Since many biologics are physician administered, bundling would be easily adopted for biologics. Physician confidence in biosimilar products will be determined through experience and use, brand reputation of biosimilars, as well as any educational efforts.

For patients, accountable care organizations (ACOs) may utilize reference pricing. Under reference pricing, the consumer has the incentive to use the lower priced biologics. They are paying out of pocket for the difference between the price of the drug they choose and the reference price. Under such a payment system, physician and patient incentives are aligned.

Similarly, the growth of ACOs, encouraged by the Affordable Care Act, where providers earn higher profits for cutting costs, would seem to encourage the use of biosimilars. Medicare has ACOs and is considering reference pricing and bundling. In the US market, third-party private payers will have the ability to negotiate the best deal for their clients. The biosimilar reimbursement market will likely develop similar to the present generic drug system, with the exception that the reference product will also compete on the basis of price. Manufacturers may have to compete for preferred formulary placement when choosing to enter the market. In the US market, third-party payers will have the ability to negotiate the best deal for their clients and may utilize a tier system. This tier system and copays will drive consumer choice in the biosimilars market. In the US, a bidding process for exclusive arrangements could be utilized to encourage more competition and might lead to more rapid expansion of biosimilars. The use of biosimilars could proceed faster than the experience in the EU and other developed markets. The experience with generics and the higher US prices of biologics would provide support for that possibility.

One model may be evolving from the recent controversy over the price that Gilead has been charging for its hepatitis C drug, Sovaldi. Even though it is not a biologic, Sovaldi was priced at $84,000 for a course of treatment, which raised concern among payers. AbbVie came out with a newly approved hepatitis C drug, Viekira Pak. Express Scripts negotiated a discount for Viekira Pak and entered into an exclusive agreement. It will be the only hepatitis C drug in the Express Scripts formulary. In response, CVS signed an exclusive agreement for a discounted Sovaldi (Harvoni). These arrangements could be employed in the biological markets to encourage greater uptake of biosimilars.

16.26 MEDICARE

Medicare spends billions of dollars on biologics each year, with these expenditures increasing every year. Medicare reimburses biologics under both Parts B (practitioner administered) and D (normal prescribed drugs), but more prominently under B because biologics are usually infused or injected by health practitioners. In 2010, 8 out of the top 10 Medicare Part B drug expenditures were biologics and totaled $8 billion (GAO, 2013). The reimbursement markup for biosimilars is 6% of the selling price of the reference product, so that the physicians receive the same monetary reward for reference products and biosimilars. Medicare has ACOs and is considering reference pricing and bundling. Medicare is currently working on a reimbursement system for biosimilars.

16.27 OTHER ISSUES

16.27.1 State Substitution

There has been considerable debate about state laws concerning the automatic substitution of biosimilars. Many have claimed that such laws are premature and an effort to impede biosimilar competition. Unlike generics which under most state laws can be automatically substituted at the pharmacy level, biosimilars can only be substituted at the pharmacy level if they have been declared interchangeable by the FDA. There are various issues concerning how and under what conditions physicians should be notified about the substitution. Presently, physicians in the generic market can write "do not substitute" and the originator must be dispensed. The same requirement can be established in the biosimilar market. If the physician allows substitution, then for safety reasons and information purposes the pharmacist must notify the physician within a certain time frame specifically what biosimilar was substituted. Prior approval is not required since the physician has already permitted substitution to occur. The notification is required in case there is an adverse event so that the physician knows what drug has been dispensed. Thus, notification of physicians concerning substitution of biosimilars for originator reference products if done correctly should not impede competition and is essentially a nonissue.

Even though the substitution controversy remains, the contending parties currently agree on more aspects than they disagree. Both sides agree that only interchangeable biologics can be substituted; as in generics, the physician can designate do not substitute, the patient should be told by the pharmacist that a biosimilar is being substituted for the originator, the pharmacist should keep records of the substitution, and the physician should have access to dispensing information. The remaining debate is over the last two and how each one should be accomplished.

16.27.2 Naming Controversy

Much controversy has arisen over the naming of biosimilars. The basic issue is whether a biosimilar and reference product should have identical nonproprietary names or different but related nonproprietary names. One side claims that different names will result in a competitive disadvantage for biosimilars. The other side counters that since biosimilars are highly similar products and unlike generics not actually equivalent, different names are appropriate. Also, since most of the biosimilars are being produced by brand-name firms, biosimilars should be at less of a competitive disadvantage than generics since they were produced by relatively unknown firms. In any event, the FDA will have to decide this issue. It does not appear that different names will put biosimilars at a competitive disadvantage (Fuhr et al., 2015).

16.28 POTENTIAL GAINS FROM BIOSIMILAR COMPETITION

Most commentators and experience from Europe suggest that prices of biosimilars will be about 25%–30% less than their reference products. That compares to price reductions on occasion of as much as 80% or 90% for small-molecule or chemical generics, and recall that prices may fall more in the US. In any event, the savings to consumers and society could be much greater in the case of biosimilars because of the higher prices of large-molecule biologics.

Consider the issue by comparing the savings for the tenth most expensive biologic, Revlimid, which treats multiple myeloma and whose annual cost in 2015 was $128,666 (Loo, 2015). A 30% saving on this drug would be about $38,600. Now consider, Lipitor, one of the world's blockbuster drugs which lost patent protection in 2011. The annual cost for a 20 mg regimen of treatment with Lipitor in 2011 was $1939 (Purvis and Schondelmeyer, 2013). Even if the generic price were 90% below that of Lipitor, annual per-patient savings would be $1745. Interestingly, the generic for Lipitor, atorvastatin, may be 90% less than the $1939 figure. In any event, the per-patient saving may be much greater for the expensive biologics than for most chemical drugs.

Biosimilar competition is also expected to result in substantial benefits. A RAND study estimated that savings from biosimilar competition could save $44.2 billion in the US (Mulcahy et al., 2014). However, unlike generics, many biosimilars are being produced by brand-name companies. Because of their reputation, branded biosimilar producers should be at less of a competitive disadvantage than generic entrants initially were.

16.29 CONSUMER WELFARE GAINS

The primary public policy objective is to increase consumer welfare. Thus, the market share of biosimilars is not a fully informative metric. The relevant welfare benchmark is not the price of the biosimilar relative to the reference product, but the comparison price before competition adjusted for inflation. The increase in quantity due to lower prices also increases access.

16.30 IMPLICATIONS AND CONCLUSIONS

The BPCIA was designed to encourage the innovation of biologics but at the same time to allow competition from biosimilars once an originator's legal monopoly, arising from patents or market exclusivity, expired. Similar to Hatch–Waxman for generics but in the case of biologics, biosimilars were effectively denied entry because there was no abbreviated pathway. Hatch–Waxman largely succeeded in that R&D continued, new small-molecule drugs have been developed, and generics comprise about 86% of all small-molecule or chemical drug prescriptions.

Turning to biologics, the question is whether the BPCIA is likely to be similarly successful. Biologics are high-risk and potentially high-reward products. Their R&D is more costly and typically more time consuming to develop than for chemical or small molecule drugs. Their manufacturing is also more complex and costly. A few biologics will become highly profitable, motivating much R&D activity in a kind of spectacular invention argument, much like the prospect of a lottery.

Also, much initial R&D in biologics is done by small firms, which then partner with or are acquired by larger firms that then proceed with the development of the biologic. It is therefore of some concern that the many mergers that have occurred in the pharmaceutical and biologics industries reduce the number of potential partners for small firms.

Effective protection for pioneer or innovator biologics is necessary to encourage R&D. There is some question whether the patent system provides adequate protection. Accordingly, the data and especially the market exclusivity provisions of the BPCIA are helpful in encouraging innovation. The 12-year market exclusivity provision is a kind of insurance policy in case patent protection is inadequate or FDA approval takes an inordinate amount of time. Further, unlike H–W, the FDA does not possess the ability to extend patent protection in cases where approval took excessively long. This gap makes market exclusivity even more desirable.

With respect to biosimilars the BPCIA is intended to encourage their entry when the originator is no longer legally protected but to do so assuring safety and efficacy. Accordingly, the FDA has to determine whether and the extent of required clinical trials. The greater the required clinical trials, the more difficult and costly will entry be and hence the less effective the competitive impact. However, it is important to ensure safety, so that erring on the side of safety could be appropriate. This assurance of safety is especially important in the early days of biosimilars until patients and providers become accustomed to their use. A safety issue with biosimilars could threaten the viability of the market.

In any event, entry barriers for biosimilars are higher than those for chemical drugs. Manufacturing costs, R&D expenses, clinical trials, marketing, and legal and regulation hurdles make entry considerably more difficult than for chemical drugs. The experience in the EU suggests that biosimilar competition will lead to price declines of about 30%. This compares to much greater declines, often 80% or 90%, experienced in the US for generic drugs. However, given the higher biologics prices, savings of even 30% could well be much more significant. Further, since US prices tend to be higher than prices in the EU, potential price reductions might be greater in the US. Given the increasing prevalence and importance of biologics in the overall drug market, such savings will be especially important.

Biosimilars do not enjoy the 6-month exclusivity provided to the first paragraph 4 challenger in the generic market under Hatch–Waxman. Only a biosimilar achieving interchangeable status obtains any exclusivity. In any event, the larger number of firms, including established pharmaceuticals and biological producers, suggests that the industry should enjoy significant competition as the FDA approval process becomes more routine. Also, given the fact that some large established firms are working on biosimilars, the prospects for competition look promising.

The high prices for biologics and the increased concern with their value make reimbursement issues quite important and may lead to success for biosimilars. The high markup for biologics provides some room for competitive response by originators, as has happened in the EU. The fact that these drugs were produced in living organisms and that even batches from the same manufacturer are not necessarily identical may help biosimilars get established in the market. Since many biologics are infused, providers will often determine whether a biosimilar is selected. This may enhance their success or at least contribute to a highly competitive environment. Some concerns arise from the patent dance provisions of the law. Communication between competitors can lead to problems for consumers. Although the idea of resolving patent issues during the period when the originator enjoys monopoly status is admirable, unintended adverse effects may occur.

In any event, the BPCIA seems to have struck a reasonable balance between encouraging innovation and allowing competition. Indeed, the lower prices that competition will provide may well permit more innovation than would otherwise occur and thus increase consumer welfare.

REFERENCES

42 U.S.C. (2006) United States Code, Title 42.

Amgen vs. Sandoz. (2015) Case No. 3: 14-cv-0471 (California Northern District Court).

Behrendt HE. (2006) The Hatch-Waxman Act: balancing competing interests or survival of the fittest? *Food and Drug Law Review* **57(2)**, 247–271.

Benassi F. (2014) Norway's discount on infliximab: a litmus test for biosimilar expansion in Europe? *IHS Life Sciences.* Available from: http://blog.ihs.com/norways-discount-on-infliximab-a-litmus-test-for-biosimilar-expansion-in-europe.

Blackstone EA, Fuhr JP. (2006) Unintended consequences: generic competition in the prescription drugs market. *Medicare Patient Management* **1(2)**, 25–43.

Blackstone EA, Fuhr JP. (2007) Biopharmaceuticals: the economic equation. *Biotechnology Healthcare* **4(6)**, 41–45.

Blackstone EA, Fuhr JP. (2010) Biosimilars and innovation: an analysis of the possibility of increased competition in biopharmaceuticals. *Future Medicinal Chemistry* **2(11)**, 1641–1649.

Blackstone EA, Fuhr JP. (2012) The future of competition in the biologics market. *Temple Journal of Science, Technology and Environment Law* **30(1)**, 1–30.

Blackstone EA, Fuhr JP. (2013) The economics of biosimilars. *American Health and Drug Benefits* **8(8)**, 469–77.

Blackstone EA, Fuhr JP. (2015) Biologics and biosimilars innovation and competition. *SciTech Lawyer* **11(3)**, 4–7.

Cancer Drug vs. Value. (2015) *Philadelphia Inquirer*, June 28 G5.

Center for Drug Evaluation and Research. (1998) Guidance for industry: 180-day generic drug exclusivity under the Hatch-Waxman amendments to the Federal Food, Drug, and Cosmetic Act.

Christl L. (2015) Overview of the regulatory pathway and FDA's guidance for the development and approval of biosimilar products in the US. Food and Drug Administration. Available from: http://www.fda.gov/downloads/AdvisoryCommittees/CommitteesMeetingMaterials/Drugs/OncologicDrugsAdvisoryCommittee/UCM431118.pdf.

Danzon P, Furukawa M. (2003) Analyzing brand-name and generic drug costs in the U.S. and eight other countries. Knowledge, Wharton.

Danzon PM, Furukawa MF. (2006) Prices and availability of biopharmaceuticals: an international comparison. *Health Affairs* **25(5)**, 1353–1362.

FDA. (2014) Information for consumers (biosimilars). Available from: http://www.fda.gov/drugs/developmentapprovalprocess/howdrugsaredevelopedandapproved/approvalapplications/therapeuticbiologicapplications/biosimilars/ucm241718.htm.

Fuhr JP, Chandra A, Romley J, et al. (2015) Product naming, pricing and market uptake of biosimilars. *Generics and Biosimilars Initiative Journal* **4(2)**, 64–71.

GAO. (2013) Medicare information on highest-expenditure Part B drugs. General Accounting Office. Available from: http://www.gao.gov/assets/660/655608.pdf.

Generics and Biosimilars Initiative. (2015) Huge discount on biosimilar infliximab in Norway. March 3. Available from: http://www.gabionline.net/Biosimilars/General/Huge-discount-on-biosimilar-infliximab-in-Norway.

Gorman SP, Pishko A, Iwanicki J, Hutslander JS. (2013) The Biosimilar Act: the United States entering into regulatory biosimilars and its implications. *The John Marshall Review of Intellectual Property Law* **12**, 322–349.

Griliches Z. (1992) The search for R&D and spillovers. *Scandinavian Journal of Economics* **94(Suppl.)**, 27–47.

Harris J. (2011) Promise of biosimilars tempered by complexity, caution. *HemOnc Today*. Available from: http://www.healio.com/hematology-oncology/news/print/hemonc-today/%7B456982d7-b9fe-492d-b72e-827cdd5b0564%7D/promise-of-biosimilars-tempered-by-complexity-caution.

Haustein R, de Millas C, Hoer A, Haussler B. (2012) Saving money in the European healthcare systems with biosimilars. *Generics and Biosimilars Initiative Journal* **1(3–4)**, 120–126.

IMS Institute. (2012) Savings: $1 trillion over 10 years generic drug savings in the U.S. Available from: http://www.ahipcoverage.com/wp-content/uploads/2012/08/2012-GPHA-IMS-GENERIC-SAVINGS-STUDY.pdf.

IMS Institute. (2014) Assessing biosimilar uptake and competition in European markets. Available from: http://www.imshealth.com/imshealth/Global/Content/Corporate/IMS%20Health%20Institute/Insights/Assessing_biosimilar_uptake_and_competition_in_European_markets.pdf.

Kaldre I. (2008) The future of generic biologics: should the United States follow-on the European pathway. *Duke Law and Technology Review* **9**, 1–13.

Kanter J, Feldman R. (2012) Understanding and incentivizing biosimilars. *Hastings Law* **64**, 57–81.

King S. (2014) First World List—the best-selling cancer drugs in 2013–2018. *First World Pharma*. Available from: http://www.firstwordpharma.com/node/1195713#axzz3gX2FGjMu.

Lichtenberg FR. (2001) Are the benefits of newer drugs worth their cost? Evidence from the 1996 MEPS. *Health Affairs* **20(5)**, 241–251.

Loo J. (2014) Standard and Poor's industry surveys healthcare: pharmaceuticals. S&P Capital 1Q, McGraw Hill Financial December.

Loo J. (2015) Standard and Poor's industry surveys: biotechnology. S&P Capital 1Q, McGraw Hill Financial March.

McKinnon RA, Lu CY. (2009) Biosimilars are not (bio)generics. *Australian Prescriber* **32**, 146–147.

Mulcahy A, Predmore W, Mattke S. (2014) The cost savings potential of biosimilar drugs in the United States, Rand Corporation. Available from: http://www.rand.org/content/dam/rand/pubs/perspectives/PE100/PE127/RAND_PE127.pdf.

National Academy of Sciences. (2007) Rising above the gathering storm: energizing and employing America for a brighter economic future. Available from: http://www.nap.edu/catalog/11463/rising-above-the-gathering-storm-energizing-and-employing-america-for.

Oldham T. (2006) Strategies for entering the biosimilar market. In: *Biosimilars—Evolution or Revolution?* Oldham T, ed., Biopharm Knowledge Publishing, London.

Olson LM, Wendling BW. (2013) Estimating the effects of entry on generic drug prices using Hatch-Waxman exclusivity. Bureau of Economics, FTC, working paper no.317, Washington, DC.

Purvis L, Schondelmeyer SW. (2013, June) Rx price watch case study: efforts to reduce the impact of generic competition for Lipitor. AARP Public Policy Institute. Available from: http://www.aarp.org/content/dam/aarp/research/public_policy_institute/health/2013/lipitor-final-report-AARP-ppi-health.pdf.

Ramachandra S. (2014, February 4) Lessons for the United States: biosimilar market. Hospira. Available from: https://www.ftc.gov/system/files/documents/public_events/Follow-On%20Biologics%20Workshop%3A%20Impact%20of%20Recent%20Legislative%20and%20Regulatory%20Naming%20Proposals%20on%20Competition/ramachandra.pdf.

Rompas S, Goss T, Amanuel S, et al. (2015) Demonstrating value for biosimilar: a conceptual framework. *American Health and Drug Benefits* **8(3)**, 129–137.

Roth VJ. (2013) Will FDA data exclusivity make biologic patents pass? *Santa Clara Computer and High Tech Law Journal* **29(2)**, 249–304.

Sahr RN. (2009) The Biologic Price Completion and Innovation Act: innovation must come before competition. *Boston College Intellectual and Technology Forum Current Law Journal*, no pages.

Sandoz. (2015) Zarxio (Filgrastim) FDA Oncologic Drugs Advisory Committee Meeting. http://www.fda.gov/downloads/AdvisoryCommittees/CommitteesMeetingMaterials/Drugs/OncologicDrugsAdvisoryCommittee/UCM428782.pdf.

Schiestl M, Stangler A, Torella A, et al. (2011) Acceptable changes in quality attributes of glycosylated biopharmaceuticals. *Nature Biotechnology* **29**, 310–312.

Schneider CK. (2013) Biosimilars in rheumatology: the wind of change. *Annals of Rheumatic Diseases* **72(3)**, 315–318.

Sell D. (2015) Shepherding a great idea to a blockbuster drug. *Philadelphia Inquirer* May 31, E1 and E7.

Serebrov M. (2015) Building a viable U.S. market is the next big quest. *NovaMedica*, March 10.

Shea TJ, Muller TL. (2010) New U.S. law establishes long awaited abbreviated approval pathway for biosimilars. *National Law Review*, Washington, DC. Available from: http://www.natlawreview.com/article/new-us-law-establishes-long-awaited-abbreviated-approval-pathway-biosimilars.

Stanton D. (2015) Remsima's US FDA review held back by contradictory demands. *Celltrion Marketpharma Reporter.* Available from: http://www.biopharma-reporter.com/Markets-Regulations/Remsima-s-US-FDA-review-held-back-by-data-demands-says-Celltrion.

Vatiand JA, Siekman MS, Loughran CA. (2010) Be prepared for biosimilar. *Managers Intellectual Property* **24**, 24–26.

Zirkelback R. (2014) The reality of prescription medicine cost in three charts. *PHARMA*. Available from: http://www.phrma.org/catalyst/the-reality-of-prescription-medicine-costs-in-three-charts.

17 Plant-Based Production of Biosimilar Drug Products

Kenny K. Y. So
University of Guelph and University of Manitoba

Michael R. Marit and Michael D. McLean
PlantForm Corporation

J. Christopher Hall
University of Guelph and PlantForm Corporation

CONTENTS

17.1 INTRODUCTION

The repertoire of proteins produced and used as biopharmaceuticals is vast, ranging from hormones like insulin to cytokines used for cellular signaling to enzymes for specific replacement therapies to monoclonal antibodies used to treat a wide array of autoimmune and terminal diseases (Rader, 2008). Cumulative sales of biopharmaceutical proteins totaled $140 billion in 2013, of which $63 billion was accounted for by monoclonal antibodies (mAbs) (Walsh, 2014). General conservation of molecular processes such as transcription, translation, and glycosylation means that systems can be developed from organisms as diverse as microbes, animals, and plants for production of proteins for use in diagnostic, industrial, or therapeutic purposes.

In 2012, Elelyso was approved by the FDA as the first plant-cell-produced, recombinant human protein for therapeutic use. This product, taliglucerase alpha, is a carrot-cell-produced version of human β-glucocerebrosidase used for treatment of Gaucher's disease (Aviezer et al., 2009; Zimran et al., 2011). Although numerous other human enzymes for replacement therapy, such as α-galactosidase-A for Fabry's disease (Kizhner et al., 2015), and serum proteins and hormones, such as erythropoietin for stimulation of red blood cell production (Castilho et al., 2013), are under development at plant-based pharmaceutical companies and academic laboratories, research on mAbs dominates the field.

Typically, mAbs are produced using hybridoma technology pioneered by Kohler and Milstein, in which antibody-secreting B cells from the spleens of mice immunized with an antigen of interest are fused to myeloma cells to produce immortal hybridoma cell lines that continually secrete antibodies with affinity for the antigen of interest (Hansel et al., 2010; Kohler and Milstein, 1975). To date, two plant-produced antibody therapeutics have been approved for clinical trials: (1) anti-HIV mAb 2G12, produced by the Pharma-Planta consortium, was approved for a phase I clinical trial completed in 2011 (Ma et al., 2015); and (2) a cocktail of three anti-Ebola virus (Zaire strain) mAbs, produced by Mapp Biopharmaceutical of San Diego, California, entered a phase I clinical trial in 2015 run by the US National Institute of Allergy and Infectious Diseases (NIAID) in collaboration with the government of Liberia (PREVAIL_II, 2016). A phase I clinical trial involving 27 IgG proteins produced in plants for use as idiotype vaccines for non-Hodgkin's lymphoma has also been completed (Anon, 2014; Bendandi et al., 2010). Although plant-based research on just six biosimilar mAb therapeutics has been published—adalimumab, infliximab, and ustekinumab (Westerhof et al., 2014); palivizumab (Hiatt et al., 2014; Zeitlin et al., 2013); nimotuzumab (Rodriguez et al., 2013); and trastuzumab (Garabagi et al., 2012a,b; Grohs et al., 2010; Komarova et al., 2011; McLean et al., 2012; Zeitlin et al., 2013)—at least 21 companies worldwide use plant-based production technologies with many other mAbs and biologics (McLean and Hall, 2012) against human pathogens, autoimmune disorders, cancers, inflammation, and toxins under development.

17.2 CONVENTIONAL mAb PRODUCTION USING MAMMALIAN CELL CULTURE

Most therapeutic antibodies are produced using either Chinese hamster ovary (CHO) or mouse NSO (myeloma) cells. Therapeutic mAbs such as trastuzumab are produced using CHO cells that are transfected with the coding sequences of the desired mAb and are grown to multi-thousand-liter cultures using highly engineered, environmentally controlled culture vessels (Bosch et al., 2013; Butler and Meneses-Acosta, 2012). Once the culture is at a proper cell density, the mAb is harvested, purified, and characterized.

Perhaps the greatest advantages of using mammalian cell cultures are the ease with which the cells can be transfected and the wide array of mAbs that can be produced. Following decades of use for recombinant protein expression, mammalian cell lines have been scrupulously characterized and have been acknowledged by researchers and regulatory agencies alike as viable production platforms for mAb production (Butler and Meneses-Acosta, 2012). Furthermore, the near-human post-translational modification pathways in mammalian cells make them suitable for producing therapeutic mAbs for human use.

Despite their relative ease of use and extensive characterization, the use of mammalian cells for producing mAbs is not without its shortcomings. Because mammalian cell cultures are costly to maintain, their operating costs are transferred to the patient. Large-scale production can be achieved in bioreactors of up to 25,000 L. However, these are costly to operate and maintain due to their high degree of sophistication, and they are poorly suited to meet unexpected high demands for particular mAbs. Also, products produced from mammalian cells must be extensively tested for a larger group of pathogens that are potentially harmful to humans, thereby imposing increased expenditures for rigorous quality control testing and process validation requirements prior to each batch release.

17.3 PLANTS AS BIOREACTORS FOR mAb PRODUCTION

The first demonstration of the successful expression of an IgG in plants was in 1989 (Hiatt et al., 1989); since then, improvements in the production of mAbs by plants have contributed to realistic proposals for the use of plant systems as an alternative platform for the production of therapeutic proteins. This is exemplified by research on plant-based biopharmaceuticals that has been conducted under sponsorship of the US Defense Advanced Research Projects Agency's (DARPAs) "Blue Angel" project, with funding totaling more than \$100 million as of 2013 (Bosch et al., 2013; Sheldon, 2012). Furthermore, although not an antibody, the FDA's approval of Elelyso, which is a recombinant form of taliglucerase alpha produced by Protalix and Pfizer using carrot cell culture, in 2012 for the treatment of Gaucher's disease substantiated the ability of a plant-based platform to produce clinically effective biopharmaceuticals (Maxmen, 2012; Sack et al., 2015a). In addition, the recent USFDA approval of ZMapp, an mAb cocktail produced in *Nicotiana benthamiana* for treatment of Ebola virus disease (Qiu et al., 2014), for use on a provisional emergency basis without clinical trials further confirms the viability of using plant-based systems for producing functional mAbs (McCarthy, 2014).

17.3.1 Agrobacterium and Binary Vectors

Recombinant therapeutic proteins can be produced in plants using either a stable transgenic or transient expression system. Both strategies use *Agrobacterium tumefaciens*, a soil bacterium capable of transferring genetic material to the nuclear genome of the plant hosts it infects (Gelvin, 2003; Smith and Townsend, 1907). *A. tumefaciens* transfers select genetic material, termed transfer-DNA (T-DNA), into plant cells by functions encoded on its tumor-inducing plasmid (pTi): a 200-kbp plasmid that contains an origin of replication, virulence genes that facilitate transfer of the T-DNA region into the plant genome, and the T-DNA region itself (Gelvin, 2003; Lee and Gelvin, 2008).

There are two difficulties that involve the use of wild-type *A. tumefaciens* pTi plasmids for inserting the genes of therapeutic proteins into host plants: cloning these genes into the T-DNA region would be laborious due to the difficulties of working with such a large, low copy-number plasmid; and the presence of oncogenes within wild-type T-DNA regions would hinder growth and affect the morphology of transformed plants. These difficulties have been overcome by the development of the binary vector system, which separated virulence genes and the T-DNA region on two plasmids. The larger pTi plasmid was disarmed by deletion of oncogenes that cause the formation of galls on infected plants while keeping the virulence genes necessary for T-DNA transfer; smaller T-DNA plasmids were developed that contain the T-DNA region itself, delineated by flanking 25 bp left border (LB) and right border (RB) sequences, selectable marker genes, and other genetic elements required for plasmid manipulation in both *A. tumefaciens* and *E. coli* (Gelvin, 2003; Lee and Gelvin, 2008). Binary plasmid systems allow researchers to simply insert genes for therapeutic proteins into T-DNA plasmids, using *E. coli* as an intermediate plasmid host, and then transfer purified T-DNA plasmids to an *A. tumefaciens* strain that contains a disarmed pTi plasmid.

Figure 17.1 illustrates the developmental stages involved in stable transgenic plant production. Sterile leaf discs called explants are cocultivated with an *A. tumefaciens* strain harboring both a disarmed pTi plasmid and a T-DNA plasmid-containing gene(s) encoding therapeutic proteins of interest plus a selectable marker gene. Subsequent development of primary transgenic plants typically requires a 4-month time investment. During this period, the *A. tumefaciens* cells transfer only the T-DNA flanked by the 25 bp LB and RB sequences to the plant genome, and expression of the transferred genes follows. Explants are first treated with medium containing plant growth hormones that direct the formation of callus, which is a mass of unorganized plant cells, and with drug(s) that select for growth of only those cells expressing the selectable marker gene. After adventitious buds have matured to a sufficient size, shoots are excised from the callus and placed on medium lacking growth hormones that induce the formation of roots. Once roots have developed, primary transgenic plantlets can be removed from sterile culture and transplanted into soil for development to mature plants.

Production of a stable transgenic plant line that has a high-level expression level of recombinant therapeutic protein requires many primary transgenic plants to be screened for target gene expression. Screening can be performed when primary transgenic plantlets are judged to be of sufficient size that sampling of tissues will not compromise health and development to mature plants. PCR-based assays can be used for detection of T-DNA incorporation or transcription of the genes of interest (reverse transcription PCR; RT-PCR). However, most useful screening procedures involve assays that measure expression of the therapeutic protein itself. Such assays are usually Western immunoblotting or quantitative enzyme-linked immunosorbent assays (ELISAs) that use an antibody specific for detection of the therapeutic protein. A specific activity assay could also be used for detection of an enzyme intended for use in enzyme replacement therapy (ERT).

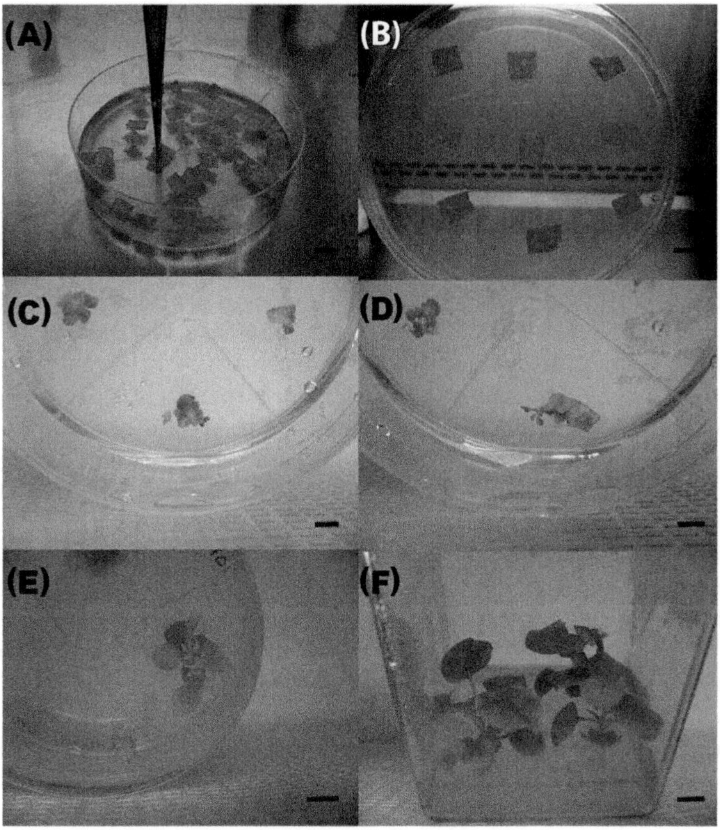

FIGURE 17.1 **(See color insert.)** Agrobacterium-mediated transformation of *Nicotiana benthamiana*. (A) Explants excised from leaf blades of *N. benthamiana*, bruised, and incubated in a culture of *Agrobacterium tumefaciens*; (B) explants plated onto regeneration media; (C) formation of adventitious buds; (D) shoot regeneration; (E) root development; and (F) establishment of primary transgenic plants. Scale bars represent 10 mm.

Any transgenic plant line that is developed should produce seeds for plants that have homogeneous and high-level production of a desired therapeutic protein. Thus, primary transgenic plants (i.e., T_0 plants) with sufficient expression of the protein should be screened for the number of T-DNA loci that have incorporated into their genomes. This requires self-pollination of each T_0 plant and usually involves screening several hundred seedlings of the next generation (i.e., T_1 generation) for drug resistance that results from expression of the selectable marker gene in the T-DNA. For example, kanamycin inhibits development of chloroplasts; therefore, transgenic plantlets expressing a kanamycin resistance gene such as the neomycin phosphotransferase II gene (*npt*II) will develop chloroplasts and healthy green tissues, whereas nontransgenic segregants lacking this gene will become chlorotic, turn white, and die when grown on medium containing kanamycin. Because a resistance phenotype such as this is inherited as a dominant trait, the ratio of drug resistance to sensitivity in the T_1 generation will indicate the number of T-DNA loci in the original T_0 plant according to principles of Mendelian inheritance. A resistance to the sensitivity ratio of 3:1 indicates a single T-DNA locus; 15:1 indicates two T-DNA loci; 63:1 indicates three T-DNA loci; and so on. The complexity of breeding a transgenic plant line to homozygosity increases as the number of T-DNA loci increases; therefore, it is desirable to identify multiple T_0 plants with high-level target protein expression, as this increases the likelihood that some will possess single T-DNA loci. Selection of homozygous plants from the T_1 generation is much easier than from plants that contain multiple T-DNA loci, as T_1 plants must be in turn self-pollinated and their offspring (i.e., T_2 generation) screened for resistance. Successful production of a stable transgenic plant line that produces homogeneous and high-level amounts of a therapeutic protein most easily involves a T_0 plant that has a single T-DNA locus, which will therefore have T_1 progeny that display a 3:1 drug resistance to sensitivity ratio, among which will exist plants homozygous at the T-DNA locus as demonstrated by T_2 progeny that are 100% drug resistant.

Stable transgenic expression of an IgG mAb in a plant host typically involves a T-DNA plasmid that contains two genes encoding both the heavy chain (HC) and light chain (LC) polypeptides; however, individual plasmids that contain either an HC or an LC gene can be used (McLean et al., 2007). Although native mammalian N-terminal signal peptides are sufficient for directing HC and LC polypeptides to the apoplast of plant cells, incorporation of plant secretory signal peptides usually allows for better antibody expression (Grohs et al., 2010). Once the HC and LC polypeptides are expressed, structurally complete IgGs form and accumulate in the apoplast (Vaquero et al., 1999). Plant tissue can then be harvested, and the mAbs can be extracted and purified.

Conventional plant breeding and selection methodologies may be applied to increase transgenic plant biomass and target therapeutic protein yield. Increases in transgene expression, presumably as a result of heterotic vigor, have been reported in progeny arising from conventional crossings between transgenic parental lines. Therefore, conventional breeding efforts may allow for improvements in the yield of a therapeutic protein beyond the improvements that would result from expression vector engineering alone (Chun et al., 2011; Conley et al., 2011; Dong et al., 2007; Guo et al., 2014; Wang, 2012).

Although purification of therapeutic proteins such as mAbs from plant crops involves biomass harvest and several steps to grind, homogenize, and clarify "green juice" primary extracts, downstream purification of target proteins from such feed stocks employs the same biopharmaceutical methodologies used in mammalian cell bioreactor production systems for capture, polish, fill, and finish procedures. The inexpensive upstream production costs for production of a therapeutic protein from a crop of stable transgenic plants initiated from a master seed lot is probably the greatest advantage of using this type of platform. An excellent review of gene-to-harvest GMP (Good Manufacturing Practice) production of anti-HIV mAb 2G12 from stable transgenic tobacco has recently been published in *Plant Biotechnology Journal* (Sack et al., 2015b).

17.3.2 Transient Expression via Agroinfiltration

Transient expression via agroinfiltration refers to the mechanical introduction of *A. tumefaciens* cells containing T-DNA plasmids into the cells of a host plant for short-term expression of genes of interest. In contrast with transient expression using mammalian cells, *Agrobacterium* acts as an intermediate host, introducing T-DNA to the host plant genome (although it is likely that T-DNA genes are also expressed without genomic incorporation because of the high concentrations of *Agrobacterium* cells that are used). Transient expression in mammalian cells typically involves ectopic expression of genes of interest without incorporation into the host cell genome.

Agroinfiltration can be performed at laboratory scale as spot infiltrations by simply forcing a suspension of *Agrobacterium* cells into a leaf through its abaxial (under) side using a needle-less syringe (Garabagi et al., 2012b). The technique was first used as part of a rapid screening method to predict whether T-DNA constructs would be successful at directing expression of proteins of interest in stable transgenic plants (Schob et al., 1997). However, it was soon realized that transient expression could rapidly provide large amounts of proteins of interest (Fischer et al., 1999). In our experience, transient expression generally results in 10 to 100 times better expression of proteins of interest compared with expression of the same proteins by stable transgenic plant lines. In published literature, expression differences upwards of 1000-fold in favor of the transient system have been documented (Nausch et al., 2012).

Agroinfiltration technologies have been developed for commercial-scale production of therapeutic proteins involving thousands of whole plants (Pogue et al., 2010) by immersing them into a liquid *Agrobacterium* suspension, application of a vacuum for a brief period to remove air from leaf tissues, and subsequent release of the vacuum to allow *Agrobacterium* cells to enter the leaf tissues, replace the air spaces, and commence T-DNA transfer to the plant cells. Infiltrated plants are allowed to grow for 2–10 days for the therapeutic protein to be expressed, after which infiltrated tissues are harvested and the protein is purified; in the case of most mAbs, expression typically peaks by 7 days after infiltration. The 1-week time span from infiltration to biomass harvest is much shorter than the many months required either to generate stable transgenic plants (Sparkes et al., 2006) or to grow mammalian cells at the appropriate scale for harvesting product. The increasing adoption of agroinfiltration by industry and the scientific community affirms the ease of use and reliability of this technology.

N. benthamiana, a related but smaller species than smoke tobacco (*N. tabacum*), has become the preferred choice for transient production of therapeutic proteins because of its convenient size to work with, its ability to be grown at high density, its speed of growth, and its capacity for expressing higher concentrations of recombinant proteins (Ahmad et al., 2010; Bombarely et al., 2012; Chen, 2015; Chen et al., 2013; Garabagi et al., 2012b; Goodin et al., 2008; Klimyuk et al., 2014; Kolotilin et al., 2014; Macdonald et al., 2015). *N. tabacum* is commonly used for stable recombinant protein expression due to its larger size ability to produce an abundance of biomass.

17.4 GLYCOSYLATION AND PLANT EXPRESSION SYSTEMS

Notwithstanding the unequivocal advantages of using plants to produce recombinant proteins, differences between plant and mammalian glycosylation had previously limited the potential for exploiting plants for therapeutic protein production.

17.4.1 Glycosylation: General Introduction

Glycosylation is a form of posttranslational modification in which glycans, Y-shaped complexes of various sugar moieties, are covalently linked to proteins as they mature in the Golgi apparatus (Bosch et al., 2013). Although the sugar composition of the core glycan structure is conserved across eukaryotes, the additional sugars that are attached to the core, and their linkages, can be kingdom specific.

Glycosylation begins with protein translation in the endoplasmic reticulum (ER) lumen and the simultaneous assembly of the glycan precursor structure onto a dolichol-phosphate molecule on the cytosolic face of the ER (Stanley et al., 2009). Once assembled, the glycan precursor is flipped across the membrane to the lumen and attached to the glycosylation site of the nascent peptide. The core glycan, which is shown in Figure 17.2, is composed of two *N*-acetylglucosamine (GlcNAc) residues

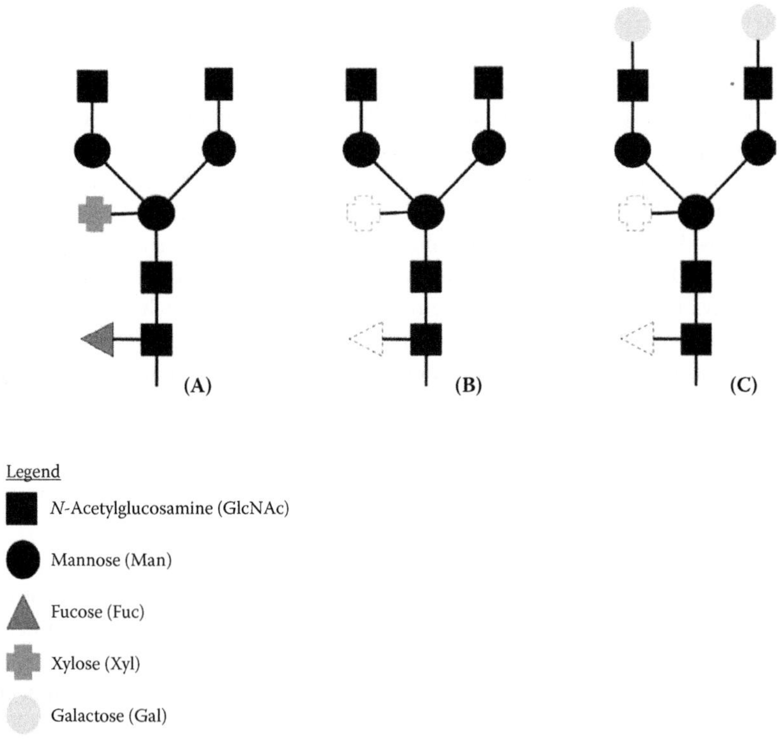

FIGURE 17.2 (See color insert.) Schematic representations of native plant *N*-glycans and their humanized glycoforms. The black core composed of *N*-acetylglucosamine and mannose is conserved across all eukaryotes. (A) A typical native plant *N*-glycan with core fucose and core xylose. (B) A partially humanized *N*-glycan from a glycoengineered plant line where xylosylation and fucosylation are knocked down (Strasser et al., 2008). Note the absence of core fucose and core xylose as indicated by the dotted outline. (C) A target *N*-glycosylation profile for plant-produced therapeutic monoclonal antibodies. Note the presence of galactose at the terminus of each glycan arm and the absence of core fucose and core xylose. (Modified from Bosch, D., et al. *Curr. Pharm. Des.*, 19, 5503–5512, 2013.)

and a mannose (Man) residue from which two arms branch: that is, each arm consists of a Man and GlcNAc (Stanley et al., 2009). The newly formed glycopeptide is subsequently shuttled to the Golgi apparatus where the glycan will mature by the addition of kingdom-specific sugars and linkages to this core structure, after which the completed glycoprotein will be secreted (Stanley et al., 2009).

Two types of glycosylation exist and are differentiated by the amino acid onto which the glycan is attached: *O*-glycosylation, where glycans are attached to serine, threonine, hydroxylysine, or hydroxyproline; and, *N*-glycosylation, where glycans are attached to asparagine (Gomord et al., 2010). The site of *N*-glycosylation is characterized by the amino acid sequence asparagine-*X*-serine/threonine, where *N* is asparagine and *X* may be any amino acid except proline (Stanley et al., 2009). Plants do not contain endogenous glycosyltransferases that perform *O*-glycosylation. Nevertheless, the *O*-glycosylation pathway has been successfully engineered into plants (Castilho et al., 2012, 2013). Because mAbs are not typically *O*-glycosylated, except at *O*-acceptor sites that might serendipitously occur in antibody variable regions, the remainder of this chapter will focus on *N*-glycosylation engineering in plants.

17.4.2 *N*-Glycosylation of mAbs

The constant region in an mAb heavy chain has a single *N*-glycosylation site at asparagine-297. *N*-glycosylation affects various pharmaceutical properties of therapeutic mAbs: structural stability and therapeutic efficacy, as well as its pharmacokinetics and subcellular localization (Arnold et al., 2007; Gomord et al., 2010; Krapp et al., 2003; Wright and Morrison, 1997). The therapeutic efficacy of an mAb partially relies on the ability of its Fc region to bind to the patient's Fcγ receptors, an interaction that is dependent on the quaternary structure of the mAb. Proper *N*-glycosylation reinforces the quaternary structure of the mAb and thus affects its therapeutic efficacy (Forthal et al., 2010). The *N*-glycan of an mAb also facilitates the binding of immune cells to tumors, thereby controlling the growth of malignant cells (Ferrara et al., 2011).

Improper glycosylation, such as the attachment of truncated *N*-glycans, causes mAbs to lose their quaternary open Y-shaped structure. Structure modification leads to a diminished ability to bind to the host's Fcγ receptor, ultimately leading to reduced efficacy of the mAb (Krapp et al., 2003). Interestingly, a lack of core fucose on the plant-produced anti-HIV antibody 2G12 leads to enhanced binding to the natural killer cell FcγRIIIa receptor, and consequently higher antiviral activity as compared to its CHO-produced control (Forthal et al., 2010). Further research demonstrated that the therapeutic mAb rituximab, prescribed for the treatment of lymphoma and leukemia, as well as trastuzumab, a breast cancer therapeutic, were also found to have enhanced binding to FcγRIIIa receptor, possibly due to reduction in steric inhibition arising from the glycan of the receptor (Jefferis, 2009). These results demonstrate that the enhanced binding capacity of nonfucosylated IgGs may reflect the ability of glycosylation to not only alter the quaternary structural conformity of the IgG itself, but also the steric mechanisms with which the mAb Fc region interacts with receptors.

In the absence of *N*-glycosylation, mAbs are more susceptible to thermal and chemical denaturation and to proteolytic degradation. For example, denaturation of three different mAbs due to high temperatures was protected by approximately 10% due to glycosylation (Ionescu et al., 2008; Stanley et al., 2009; Zheng et al., 2011). Furthermore, the susceptibility to chemical denaturation via guanidine hydrochloride of the same three nonglycosylated mAbs was increased by 0.6 M, and their susceptibility to preotolytic degradation by papain was reduced over a 24-h incubation period (Zheng et al., 2011). Thermal and chemical denaturation of mAbs is of particular interest to downstream processing since both parameters may contribute to complications due to aggregation and diminished shelf life (Zheng et al., 2011).

Despite the high degree of conservation in glycosylation patterns in eukaryotes, *N*-glycosylation mutants of *Arabidopsis thaliana* show no abnormal morphology. However, improper *N*-glycosylation in humans has been correlated with the pathogenesis of various acquired and immunological diseases (Nagels et al., 2012; Strasser et al., 2004b; Varki and Freeze, 2009; Xue et al., 2013). With regard to therapeutic proteins, proper *N*-glycosylation is of particular concern as it is required to maintain the pharmaceutical efficacy of the therapeutic; improper *N*-glycosylation may result in diminished therapeutic efficacy and could also lead to immunogenicity in the patient.

17.4.3 *N*-Glycan Profiles of Plants and Mammals

Kingdoms *Plantae* and *Animalia* share core glycan structures but differ in accessory glycan linkages in the following ways: (1) as shown in Figure 17.2C, mammalian *N*-glycans contain α1,6-fucose linked to the proximal *N*-acetylglucosamine; the ends of the branched *N*-glycans are capped with a penultimate β1,4-galactose and a terminal α2,6-*N*-acetylneuraminic acid (sialic acid), though mAbs are not typically sialylated at asparagine-297, and (2) as shown in Figure 17.2A, plant *N*-glycans contain a bisecting β1,2-xylose linked to the core mannose; α1,3-fucose linked to the proximal *N*-acetylglucosamine; and terminal β1,3-galactose (Bosch et al., 2013). The presence of β1,2-xylose and α1,3-fucose is of particular concern for plant-produced pharmaceuticals, as these two sugars are potentially immunogenic to humans.

17.4.4 Immunogenicity of β1,2-Xylose and α1,3-Fucose

Antibodies against biopharmaceuticals, including mAbs, have been detected in the serum of patients in both preclinical and clinical trials (Malucchi and Bertolotto, 2008; Subramanyam, 2008). These antidrug antibodies result in the premature clearance of biopharmaceuticals in patients, thereby reducing the efficacy of the drug, though in some instances no adverse effects on therapeutic efficacy were reported, as in the case of recombinant human growth hormone (Malucchi and Bertolotto, 2008).

In mAbs, two potential structural domains contribute to its potential immunogenicity. Both foreign complementarity-determining regions (CDRs) on humanized antibodies and glycans that contain xenogeneic sugars, such as β1,2-xylose and α1,3-fucose, can elicit a host immune response (van de Weert and Møller, 2008). Although the humanization of murine mAbs by replacement of framework regions can result in clinically safe mAbs, as in the case of trastuzumab, the small proportion of murine sequences in the CDRs that can remain may still be sufficient to induce immunogenicity. Such is the case reported for natalizumab, a humanized mAb of murine origin prescribed for the treatment of multiple sclerosis. Antinatalizumab antibodies with affinity to the CDR regions of natalizumab were discovered in the serum of treated patients, and the levels of circulating natalizumab were depleted (Subramanyam, 2008).

The second potential immunogenic property of mAbs is their glycosylation profile. Plant-produced mAbs bearing nonmammalian *N*-glycans, β1,2-xylose, and α1,3-fucose shown in Figure 17.2A, may be capable of eliciting an immune response in human patients, as these two glymodifications have been reported to be immunogenic in various laboratory animals, including mice and rats (Bardor et al., 2003), rabbits (Faye et al., 1993), and goats (Kurosaka et al., 1991). In all instances, the animals were immunized with horseradish peroxidase, which is heavily glycosylated with both β1,2-xylose and α1,3-fucose, and antibodies with affinity to both sugars were isolated from these immune sera (Bardor et al., 2003; Faye et al., 1993; Kurosaka et al., 1991). Despite the strong evidence of the immunogenicity of β1,2-xylose and α1,3-fucose in mammals, little evidence of their immunogenicity has been reported in humans, although their allergenicity has been well established for more than a decade (Altmann, 2007). Anti-xylose and anti-fucose IgEs, collected from patients with pollen allergies, were found to show affinity toward plant-produced 2G12, which was xylosylated and α1,3-fucosylated, suggesting that patients with pollen allergies may develop an allergic reaction to plant-glycosylated mAbs. However, whether the same reaction would be mounted in nonallergic patients is unclear (Jin et al., 2008). To further complicate matters, Elelyso is both α1,2-xylosylated and α1,3-fucosylated, and was approved for human use by the FDA, suggesting that the presence of these two sugars may not pose a significant immunogenic risk in humans (Shaaltiel et al., 2007).

Although the long-term impact of plant-specific xylose and fucose in plant-produced biopharmaceuticals is unknown, the elimination of these two sugars from the glycan profiles of therapeutic mAbs is a priority for therapeutic mAb producers and regulatory agencies (Bosch et al., 2013). The original draft guidance jointly issued by the FDA and the US Department of Agriculture (USDA), agencies that oversee the approval of pharmaceuticals and their safety, explicitly states that plant-produced therapeutics should be screened for potentially immunogenic xylose and allergenic plant-specific *N*-glycans (FDA, 2002).

17.4.5 Need to Humanize Glycosylation in Plants

The development of biosimilar drugs in plants requires a humanized *N*-glycosylation system. To achieve a CHO-like *N*-glycosylation, for example, xylosylation and α1,3-fucosylation must first be eliminated, as shown in Figure 17.2B, and β1,4-galactosylation must be introduced as shown in Figure 17.2C.

It is worthy to note that the FDA and USDA classify xylose as only *potentially* immunogenic and do not explicitly require the complete elimination of xylose and plant-specific *N*-glycans (USFDA, USDA, 2002). However, to minimize all possible risks of immunogenicity, elimination of these two sugars and their plant-specific linkage is warranted, given that 60% of mAbs produced in wild-type tobacco plants are xylosylated and α1,3-fucosylated (Cabanes-Macheteau et al., 1999). Furthermore, studies indicate that the presence of fucosylated glycans on mAbs may reduce the structural integrity and therapeutic efficiency of the proteins (Woof and Burton, 2004; Wuhrer et al., 2007).

17.4.6 RNAi Knockdown of *XylT* and *FucT*

Transgenic RNAi has been used to silence *FucT* (α1,3-fucosyltransferase) and *XylT* (β1,2-xylosyltransferase) in a variety of model plant species commonly used for recombinant protein expression, including *A. thaliana* (Strasser et al., 2004b), *Lemna minor* (Cox et al., 2006), *Medicago sativa* (Sourrouille et al., 2008), *N. benthamiana* (Strasser et al., 2008), and *N. tabacum* 'BY2' cells (Yin et al., 2011). Single RNAi vectors targeting either *FucT* or *XylT* were individually expressed in *A. thaliana* (Strasser et al., 2004b), *N. benthamiana* (Strasser et al., 2008), and *M. sativa* (Sourrouille et al., 2008), and in all instances, a reduction in the targeted glycotransferase correlated with a reduction in xylosylation or fucosylation of the endogenous protein glycans. Furthermore, the progeny arising from the crossing of *FucT* knockdown and *XylT* knockdown parental lines in *A. thaliana* (Strasser et al., 2004b) and *N. benthamiana* (Strasser et al., 2008) showed significant reductions of both xylosylation and fucosylation. Similarly, *FucT* and *XylT* RNAi knockdown lines of *L. minor* (Cox et al., 2006) and *N. tabacum* 'BY2' cells (Yin et al., 2011) were generated, using a single vector simultaneously targeting both glycotransferases indicating transformation with individual vectors and subsequent crossing is not necessary. In all cases, no aberrant plant morphologies were observed in resultant knockdown lines or their progeny (Cox et al., 2006; Sourrouille et al., 2008; Strasser et al., 2004a; Yin et al., 2011). Collectively, these works corroborate the utility of RNAi to silence native genes and the plasticity of plant glycosylation. Most importantly, these researchers demonstrate the ability to generate plants that lack potentially immunogenic plant-specific *N*-glycosylation and serve as a platform for the use of plants to produce clinically-usable biopharmaceuticals.

17.4.7 Expression of β1,4-Galactosyltransferase in Plants

Human glycoproteins contain galactose in a β1,4-linkage, catalyzed by human β1,4-galactosyltransferase. This enzyme transfers galactose from UDP-galactose, its activated nucleotide sugar form, onto the terminal *N*-acetylglucosamine residue of the glycan arm (Palacpac et al., 1999). Galactosylation affects the conformity of mAbs, such that the absence of terminal β1,4-galactose residues on *N*-glycans results in inappropriate folding of the mAb, diminishing its ability to bind Fc receptors (Krapp et al., 2003).

Functional human β1,4-galactsyltransferase has been successfully expressed in both tobacco cell culture and stable transgenic plants. Constitutive expression of the full β1,4-galactsyltransferase in *N. tabacum* "BY2" cell culture resulted in 47.3% of total endogenous glycans being capped with terminal galactose (Palacpac et al., 1999). Similarly, stable *N. tabacum* "Samsun NN" lines overexpressing β1,4-galactsyltransferase also resulted in the galactosylation of total soluble endogenous proteins, albeit at a lower efficiency of 15% (Bakker et al., 2001). However, upon crossing the primary β1,4-galactsyltransferase transgenic with a stable tobacco line expressing the mouse mAb Mgr-48, 30% of the mAb produced from the hybrid progeny was galactosylated, a level of galactosylation similar to hybridoma-produced Mgr-48 (Bakker et al., 2001; Wright and Morrison, 1997). In a subsequent attempt to improve galactosylation, Bakker et al. (2006) localized the β1,4-galactosyltransferase to the medial Golgi apparatus by replacing its native localization signal with the CTS (cytoplasmic transmembrane and stem region) domain from the *A. thaliana* xylosyltransferase (Fitchette-Laine et al., 1997). The total soluble endogenous protein pool of the resultant stably transformed line showed improved galactosylation, and when crossed with a homozygous Mgr-48 line, the degree of galactosylation of the antibody was again comparable to its hybridoma-produced counterpart (Bakker et al., 2006). The stable expression of β1,4-galactosyltransferase in *N. benthamiana* would result in a line useful for producing humanized recombinant mAbs and other proteins via Agroinfiltration. Furthermore, the stable expression of β1,4-galactsyltransferase in combination with other enzymes in the human *N*-glycosylation pathway would generate a plant host that could produce recombinant proteins with human *N*-glycans suitable for clinical applications. Stable expression of β1,4-galactsyltransferase in a *N. benthamiana* line crossed with a *XylT/FucT* knockdown line (Strasser et al., 2008) produced recombinant HIV antibodies 4E10 and 2G12 with 80% galactosylation and an absence of xylose and fucose, respectively (Strasser et al., 2009).

17.4.8 Expression of the Sialylation Pathway in Plants

Therapeutic mAbs typically do not possess sialic acid on their Fc glycan termini. However, as most human serum proteins have sialylated termini, and some are used for therapeutic applications, work was undertaken to introduce the sialylation pathway into plants (Castilho and Steinkellner, 2012; Castilho et al., 2010, 2014; Schneider et al., 2014). Sialylation requires prior galactosylation of diantennary *N*-glycans (Strasser et al., 2009), CMP-*N*-acetylneuraminic acid (CMP-Neu5Ac), a transporter for delivery of CMP-sialic acid into the Golgi apparatus, and a sialyltransferase (ST) for transfer of sialic acid from CMP-Neu5Ac to galactose termini on *N*-linked glycans. Introduction of the mammalian sialylation pathway has been achieved transiently by coexpressing six genes: *GNE*, *NANS*, *CMAS*, *CST*, *hGalT*, and *ST* from single expression vectors in *N. benthamiana*, along with an expression vector for anti-HIV mAb 2G12 (Castilho et al., 2010). This technique has recently been improved upon by assembly of a single expression vector engineered to contain all six genes encoding the sialylation pathway. Transient coexpression of this multigene vector in *N. benthamiana* with a vector directing expression of human butyrylcholinesterase resulted in efficient sialylation of the recombinant protein (Schneider et al., 2014). The successful demonstration of functionality of this six-gene sialylation vector suggests that the development of stable transgenic host plants capable of performing glycoprotein sialylation is possible.

17.5 CONCLUSION

Plant-based production of recombinant proteins using *N. benthamiana* is a safe, low-cost, and highly efficient alternative production platform to the standard CHO or NS0 cell culture systems for the production of high-value therapeutic proteins. The ability to exploit such a plant-based system was historically hindered by the disparities between human and plant *N*-glycosylation, a crucial posttranslational modification that could have negative consequences for the clinical efficacy of the therapeutic protein being produced. Humanizing *N*-glycosylation in plants by the simultaneous removal of both xylose and α1,3-fucose and by the attachment of galactose will allow a wide variety of therapeutic proteins to be produced. Additional posttranslational modifications, such as sialylation, can be introduced into humanized *N*-glycosylating plant lines to further widen the range of therapeutic proteins that can be produced.

Aside from optimizing the biochemical and molecular aspects of recombinant protein expression and *N*-glycosylation in *N. benthamiana*, the benefit of horticultural research can be applied to further optimize the growth parameters of this species in controlled environmental conditions to facilitate commercial-scale plant production. Production guidelines and recommendations exist for all commercial ornamental and food crops grown under greenhouse conditions; however, none exist for *N. benthamiana*, and therefore must be determined empirically. Fertilizer rate, daily temperature differentials, optimal light intensity, and spectra are among the many crucial aspects of plant production that require investigation. Unlike cell cultures where growth is predictable and consistent year round, the rate of plant growth and development is prone to seasonal fluctuations, especially when grown in a greenhouse setting. Thus, most companies operating in this field are moving toward controlled growth production systems.

Finally, optimization of the growth of *N. benthamiana* can be achieved through breeding means. Growth and developmental characteristics such as leaf area and flowering time have the capacity to be improved through breeding. The two accessions of *N. benthamiana* currently held by the US *Nicotiana* Germplasm collection have been propagated via self-pollination through many decades and can be considered inbred lines (J. Nifong, personal communication, 2015, North Carolina State University). The two accessions may be crossed to produce a population of hybrids from which selections can be made for delayed flowering and improved leaf area among other traits. Furthermore, true wild-type seed may be collected from the field to further increase the genetic diversity of the cultivated population to allow for continued improvement of the desired traits. Although breeding efforts require time and will not be immediately fruitful, the long-term benefits of producing superior plant lines suited for recombinant protein production outweigh the initial time and resource investment required.

ACKNOWLEDGMENTS

This chapter was adapted from Kenny K. Y. So's M. Sc. thesis. J. Christopher Hall is former Canada Research Chair in Recombinant Antibody Technology. The authors thank Drs. Kiva Ferraro and Craig Binnie for critical reading of the manuscript.

REFERENCES

Ahmad A, Pereira EO, Conley AJ, et al. (2010) Green biofactories: recombinant protein production in plants. *Recent Patents in Biotechnology* **4**, 242–259.

Altmann F. (2007) The role of protein glycosylation in allergy. *International Archives of Allergy and Immunology* **142**, 99–115.

Anon. (2014) Icon genetics successfully completes phase I clinical study with personalized vaccines for treatment of non-Hodgkin's lymphoma. *Icon Genetics Company Website.* January 31, 2014. Available from: http://www.icongenetics.com/html/5975.htm.

Arnold JN, Wormald MR, Sim RB, et al. (2007) The impact of glycosylation on the biological function and structure of human immunoglobulins. *Annual Review of Immunology* **25**, 21–50.

Aviezer D, Brill-Almon E, Shaaltiel Y, et al. (2009) A plant-derived recombinant human glucocerebrosidase enzyme—a preclinical and phase I investigation. *PLoS One* **4**, e4792.

Bakker H, Bardor M, Molthoff JW, et al. (2001) Galactose-extended glycans of antibodies produced by transgenic plants. *Proceedings of the National Academy of Sciences USA* **98**, 2899–2904.

Bakker H, Rouwendal GJ, Karnoup AS, et al. (2006) An antibody produced in tobacco expressing a hybrid β-1,4-galactosyltransferase is essentially devoid of plant carbohydrate epitopes. *Proceedings of the National Academy of Sciences USA* **103**, 7577–7582.

Bardor M, Faveeuw C, Fitchette AC, et al. (2003) Immunoreactivity in mammals of two typical plant glyco-epitopes, core α(1,3)-fucose and core xylose. *Glycobiology* **13**, 427–434.

Bendandi M, Marillonnet S, Kandzia R, et al. (2010) Rapid, high-yield production in plants of individualized idiotype vaccines for non-Hodgkin's lymphoma. *Annals of Oncology* **21**, 2420–2427.

Bombarely A, Rosli HG, Vrebalov J, et al. (2012) A draft genome sequence of *Nicotiana benthamiana* to enhance molecular plant-microbe biology research. *Molecular Plant-Microbe Interactions* **25**, 1523–1530.

Bosch D, Castilho A, Loos A, et al. (2013) *N*-glycosylation of plant-produced recombinant proteins. *Current Pharmaceutical Design* **19**, 5503–5512.

Butler M, Meneses-Acosta A. (2012) Recent advances in technology supporting biopharmaceutical production from mammalian cells. *Applied Microbiology and Biotechnology* **96**, 885–894.

Cabanes-Macheteau M, Fitchette-Laine AC, Loutelier-Bourhis C, et al. (1999) N-glycosylation of a mouse IgG expressed in transgenic tobacco plants. *Glycobiology* **9**, 365–372.

Castilho A, Neumann L, Daskalova S, et al. (2012) Engineering of sialylated mucin-type *O*-glycosylation in plants. *Journal of Biology Chemistry* **287**, 36518–36526.

Castilho A, Neumann L, Gattinger P, et al. (2013) Generation of biologically active multisialylated recombinant human EPOFc in plants. *PLoS One* **8**, e54836.

Castilho A, Steinkellner H. (2012) Glyco-engineering in plants to produce human-like *N*-glycan structures. *Biotechnology Journal* **7**, 1088–1098.

Castilho A, Strasser R, Stadlmann J, et al. (2010) In planta protein sialylation through overexpression of the respective mammalian pathway. *Journal of Biological Chemistry* **285**, 15923–15930.

Castilho A, Windwarder M, Gattinger P, et al. (2014) Proteolytic and *N*-glycan processing of human α1-antitrypsin expressed in *Nicotiana benthamiana. Plant Physiology* **166**, 1839–1851.

Chen Q. (2015) Plant-made vaccines against West Nile virus are potent, safe, and economically feasible. *Biotechnology Journal* **10**, 671–680.

Chen Q, Lai H, Hurtado J, et al. (2013) Agroinfiltration as an effective and scalable strategy of gene delivery for production of pharmaceutical proteins. *Advanced Techniques in Biology and Medicine* **1**, 103.

Chun YJ, Kim DI, Park KW, et al. (2011) Gene flow from herbicide-tolerant GM rice and the heterosis of GM rice-weed F2 progeny. *Planta* **233**, 807–815.

Conley AJ, Zhu H, Le LC, et al. (2011) Recombinant protein production in a variety of *Nicotiana* hosts: a comparative analysis. *Plant Biotechnology Journal* **9**, 434–444.

Cox KM, Sterling JD, Regan JT, et al. (2006) Glycan optimization of a human monoclonal antibody in the aquatic plant *Lemna minor. Nature Biotechnology* **24**, 1591–1597.

Dong HZ, Li WJ, Tang W, et al. (2007) Heterosis in yield, endotoxin expression and some physiological parameters in Bt transgenic cotton. *Plant Breeding* **126**, 169–175.

Faye L, Gomord V, Fitchette-Laine AC, Chrispeels MJ. (1993) Affinity purification of antibodies specific for Asn-linked glycans containing α1→3 fucose or β1→2 xylose. *Analytical Biochemistry* **209**, 104–108.

FDA. (2002) Guidance for industry: drugs, biologics, and medical devices derived from bioengineered plants for use in humans and animals. *U.S. Food and Drug Administration*. September 2002. Available from: http://www.fda.gov/OHRMS/DOCKETS/98fr/02d-0324-gdl0001.pdf.

Ferrara C, Grau S, Jager C, et al. (2011) Unique carbohydrate-carbohydrate interactions are required for high affinity binding between FcγRIII and antibodies lacking core fucose. *Proceedings of the National Academy of Sciences USA* **108**, 12669–12674.

Fischer R, Vaquero-Martin C, Sack M, et al. (1999) Towards molecular farming in the future: transient protein expression in plants. *Biotechnology and Applied Biochemistry* **30**, 113–116.

Fitchette-Laine AC, Gomord V, Cabanes M, et al. (1997) *N*-glycans harboring the Lewis a epitope are expressed at the surface of plant cells. *Plant Journal* **12**, 1411–1417.

Forthal DN, Gach JS, Landucci G, et al. (2010) Fc-glycosylation influences Fcγ receptor binding and cell-mediated anti-HIV activity of monoclonal antibody 2G12. *Journal of Immunology* **185**, 6876–6882.

Garabagi F, Gilbert E, Loos A, et al. (2012a) Utility of the P19 suppressor of gene-silencing protein for production of therapeutic antibodies in *Nicotiana* expression hosts. *Plant Biotechnology Journal* **10**, 1118–1128.

Garabagi F, McLean MD, Hall JC. (2012b) Transient and stable expression of antibodies in *Nicotiana* species. *Methods in Molecular Biology* **907**, 389–408.

Gelvin SB. (2003) Agrobacterium-mediated plant transformation: the biology behind the "gene-jockeying" tool. *Microbiology and Molecular Biology Reviews* **67**, 16–37.

Gomord V, Fitchette AC, Menu-Bouaouiche L, et al. (2010) Plant-specific glycosylation patterns in the context of therapeutic protein production. *Plant Biotechnology Journal* **8**, 564–587.

Goodin MM, Zaitlin D, Naidu RA, Lommel SA. (2008) *Nicotiana benthamiana*: its history and future as a model for plant-pathogen interactions. *Molecular Plant-Microbe Interactions* **21**, 1015–1026.

Grohs BM, Niu Y, Veldhuis LJ, et al. (2010) Plant-produced trastuzumab inhibits the growth of HER2 positive cancer cells. *Journal of Agricultural and Food Chemistry* **58**, 10056–10063.

Guo M, Rupe MA, Wei J, et al. (2014) Maize ARGOS1 (ZAR1) transgenic alleles increase hybrid maize yield. *Journal of Experimental Botany* **65**, 249–260.

Hansel TT, Kropshofer H, Singer T, et al. (2010) The safety and side effects of monoclonal antibodies. *Nature Reviews Drug Discovery* **9**, 325–338.

Hiatt A, Bohorova N, Bohorov O, et al. (2014) Glycan variants of a respiratory syncytial virus antibody with enhanced effector function and in vivo efficacy. *Proceedings of the National Academy of Sciences USA* **111**, 5992–5997.

Hiatt A, Cafferkey R, Bowdish K. (1989) Production of antibodies in transgenic plants. *Nature* **342**, 76–78.

Ionescu RM, Vlasak J, Price C, Kirchmeier M. (2008) Contribution of variable domains to the stability of humanized IgG1 monoclonal antibodies. *Journal of Pharmaceutical Science* **97**, 1414–1426.

Jefferis R. (2009) Glycosylation as a strategy to improve antibody-based therapeutics. *Nature Reviews Drug Discovery* **8**, 226–234.

Jin C, Altmann F, Strasser R, et al. (2008) A plant-derived human monoclonal antibody induces an anti-carbohydrate immune response in rabbits. *Glycobiology* **18**, 235–241.

Kizhner T, Azulay Y, Hainrichson M, et al. (2015) Characterization of a chemically modified plant cell culture expressed human α-galactosidase-A enzyme for treatment of Fabry disease. *Molecular Genetics and Metabolism* **114**, 259–267.

Klimyuk V, Pogue G, Herz S, et al. (2014) Production of recombinant antigens and antibodies in *Nicotiana benthamiana* using 'magnifection' technology: GMP-compliant facilities for small- and large-scale manufacturing. *Current Topics in Microbiology and Immunology* **375**, 127–154.

Kohler G, Milstein C. (1975) Continuous cultures of fused cells secreting antibody of predefined specificity. *Nature* **256**, 495–497.

Kolotilin I, Topp E, Cox E, et al. (2014) Plant-based solutions for veterinary immunotherapeutics and prophylactics. *Veterinary Research* **45**, 117.

Komarova TV, Kosorukov VS, Frolova OY, et al. (2011) Plant-made trastuzumab (herceptin) inhibits HER2/Neu+ cell proliferation and retards tumor growth. *PLoS One* **6**, e17541.

Krapp S, Mimura Y, Jefferis R, et al. (2003) Structural analysis of human IgG-Fc glycoforms reveals a correlation between glycosylation and structural integrity. *Journal of Molecular Biology* **325**, 979–989.

Kurosaka A, Yano A, Itoh N, et al. (1991) The structure of a neural specific carbohydrate epitope of horseradish peroxidase recognized by anti-horseradish peroxidase antiserum. *Journal of Biological Chemistry* **266**, 4168–4172.

Lee LY, Gelvin SB. (2008) T-DNA binary vectors and systems. *Plant Physiology* **146**, 325–332.

Ma JK, Drossard J, Lewis D, et al. (2015) Regulatory approval and a first-in-human phase I clinical trial of a monoclonal antibody produced in transgenic tobacco plants. *Plant Biotechnology Journal* **13**, 1106–1120.

MacDonald J, Doshi K, Dussault M, et al. (2015) Bringing plant-based veterinary vaccines to market: managing regulatory and commercial hurdles. *Biotechnology Advances* **33(8)**, 1572–1581.

Malucchi S, Bertolotto A. (2008) Immunogenicity of biopharmaceuticals. In: *Clinical Aspects of Immunogenicity to Biopharmaceuticals*, van de Weert M, Møller EH, eds., 27–56. Springer, New York.

Maxmen A. (2012) Drug-making plant blooms. *Nature* **485**, 160.

McCarthy M. (2014) US signs contract with ZMapp maker to accelerate development of the Ebola drug. *British Medical Journal* **349**, g5488.

McLean MD, Almquist KC, Niu Y, et al. (2007) A human anti-*Pseudomonas aeruginosa* serotype O6ad immunoglobulin G1 expressed in transgenic tobacco is capable of recruiting immune system effector function in vitro. *Antimicrobial Agents and Chemotherapy* **51**, 3322–3328.

McLean MD, Chen R, Yu D, et al. (2012) Purification of the therapeutic antibody trastuzumab from genetically modified plants using safflower protein A-oleosin oilbody technology. *Transgenic Research* **21**, 1291–1301.

McLean MD, Hall JC. (2012) Biologics and biologic products. *Ontario Society for Medical Technologists Advocate* **19**, 5–6.

Nagels B, Weterings K, Callewaert N, Van Damme EJM. (2012) Production of plant made pharmaceuticals: from plant host to functional protein. *Critical Reviews in Plant Sciences* **31**, 148–180.

Nausch H, Mikschofsky H, Koslowski R, et al. (2012) High-level transient expression of ER-targeted human interleukin 6 in *Nicotiana benthamiana*. *PLoS One* **7**, e48938.

Palacpac NQ, Yoshida S, Sakai H, et al. (1999) Stable expression of human β1,4-galactosyltransferase in plant cells modifies N-linked glycosylation patterns. *Proceedings of the National Academy of Sciences USA* **96**, 4692–4697.

Pogue GP, Vojdani F, Palmer KE, et al. (2010) Production of pharmaceutical-grade recombinant aprotinin and a monoclonal antibody product using plant-based transient expression systems. *Plant Biotechnology Journal* **8**, 638–654.

PREVAIL_II. (2016) A randomized, controlled trial of ZMapp for Ebola virus infection. *New England Journal of Medicine* **375**, 1448–1456.

Qiu X, Wong G, Audet J, et al. (2014) Reversion of advanced Ebola virus disease in nonhuman primates with ZMapp. *Nature* **514**, 47–53.

Rader RA. (2008) (Re)defining biopharmaceutical. *Nature Biotechnology* **26**, 743–751.

Rodriguez M, Pérez L, Gavilondo JV, et al. (2013) Comparative *in vitro* and experimental *in vivo* studies of the anti-epidermal growth factor receptor antibody nimotuzumab and its aglycosylated form produced in transgenic tobacco plants. *Plant Biotechnology Journal* **11**, 53–65.

Sack M, Hofbauer A, Fischer R, Stoger E. (2015a) The increasing value of plant-made proteins. *Current Opinion in Biotechnology* **32**, 163–170.

Sack M, Rademacher T, Spiegel H, et al. (2015b) From gene to harvest: insights into upstream process development for the GMP production of a monoclonal antibody in transgenic tobacco plants. *Plant Biotechnology Journal* **13**, 1094–1105.

Schneider JD, Castilho A, Neumann L, et al. (2014) Expression of human butyrylcholinesterase with an engineered glycosylation profile resembling the plasma-derived orthologue. *Biotechnology Journal* **9**, 501–510.

Schob H, Kunz C, Meins F, Jr. (1997) Silencing of transgenes introduced into leaves by agroinfiltration: a simple, rapid method for investigating sequence requirements for gene silencing. *Molecular and General Genetics* **256**, 581–585.

Shaaltiel Y, Bartfeld D, Hashmueli S, et al. (2007) Production of glucocerebrosidase with terminal mannose glycans for enzyme replacement therapy of Gaucher's disease using a plant cell system. *Plant Biotechnology Journal* **5**, 579–590.

Sheldon A. (2012) Growing vaccines. *Chemistry and Industry* **76**, 36–38.

Smith EF, Townsend CO. (1907) A plant-tumor of bacterial origin. *Science* **25**, 671–673.

Sourrouille C, Marquet-Blouin E, D'Aoust MA, et al. (2008) Down-regulated expression of plant-specific glycoepitopes in alfalfa. *Plant Biotechnology Journal* **6**, 702–721.

Sparkes IA, Runions J, Kearns A, Hawes C. (2006) Rapid, transient expression of fluorescent fusion proteins in tobacco plants and generation of stably transformed plants. *Nature Protocols* **1**, 2019–2025.

Stanley P, Schachter H, Taniguchi N. (2009) Essentials of glycobiology. In: *N-Glycans*, 2nd edn., Varki A, Cummings RD, Esko JD, Freeze HH, Stanley P, Bertozzi CR, Hart GW, Etzler ME, eds., Cold Spring Harbor Laboratory Press, CSHL, New York.

Strasser R, Altmann F, Glossl J, Steinkellner H. (2004a) Unaltered complex N-glycan profiles in *Nicotiana benthamiana* despite drastic reduction of β1,2-*N*-acetylglucosaminyltransferase I activity. *Glycoconjugate Journal* **21**, 275–282.

Strasser R, Altmann F, Mach L, et al. (2004b) Generation of *Arabidopsis thaliana* plants with complex *N*-glycans lacking β1,2-linked xylose and core α1,3-linked fucose. *FEBS Letters* **561**, 132–136.

Strasser R, Castilho A, Stadlmann J, et al. (2009) Improved virus neutralization by plant-produced anti-HIV antibodies with a homogeneous β1,4-galactosylated *N*-glycan profile. *Journal of Biological Chemistry* **284**, 20479–20485.

Strasser R, Stadlmann J, Schahs M, et al. (2008) Generation of glyco-engineered *Nicotiana benthamiana* for the production of monoclonal antibodies with a homogeneous human-like *N*-glycan structure. *Plant Biotechnology Journal* **6**, 392–402.

Subramanyam M. (2008) Immunogenicity of biopharmaceuticals. In: *Case Study: Immunogenicity of Natalizumab*, van de Weert M, Møller EH, eds., Springer, New York.

USFDA, USDA. (2002) Guidance for industry: drugs, biologics and medical devices derived from bioengineered plants for use in humans and animals. U.S. Department of Health and Human Services, Food and Drug Administration, Center for Biologics Evaluation and Research, Rockville, MD. Available from: http://www.fda.gov/OHRMS/DOCKETS/98fr/02d-0324-gdl0001.pdf

van de Weert M, Møller E. (2008) Immunogenicity of biopharmaceuticals. In: *Immunogenicity of Biopharmaceuticals: Causes, Methods to Reduce Immunogenicity, and Biosimilars*, van de Weert M, Møller EH, eds., 97–111. Springer, New York.

Vaquero C, Sack M, Chandler J, et al. (1999) Transient expression of a tumor-specific single-chain fragment and a chimeric antibody in tobacco leaves. *Proceedings of the National Academy of Sciences USA* **96**, 11128–11133.

Varki A, Freeze H. (2009) Essentials of glycobiology. In: *Glycans in Acquired Human Diseases*, 2nd edn., Varki A, Cummings R, Esko J, Freeze H, Stanley P, Bertozzi C, Hart G, Etzler M, eds., Cold Spring Harbor Laboratory Press, CSHL, New York.

Walsh G. (2014) Biopharmaceutical benchmarks 2014. *Nature Biotechnology* **32**, 992–1000.

Wang Z. (2012) Segregation and expression of transgenes in the progenies of Bt transgenic rice crossed to conventional rice varieties. *African Journal of Biotechnology* **11**, 7812–7818.

Westerhof LB, Wilbers RH, van Raaij DR, et al. (2014) Monomeric IgA can be produced *in planta* as efficient as IgG, yet receives different *N*-glycans. *Plant Biotechnology Journal* **12**, 1333–1342.

Woof JM, Burton DR. (2004) Human antibody-Fc receptor interactions illuminated by crystal structures. *Nature Reviews Immunology* **4**, 89–99.

Wright A, Morrison SL. (1997) Effect of glycosylation on antibody function: implications for genetic engineering. *Trends in Biotechnology* **15**, 26–32.

Wuhrer M, Stam JC, van de Geijn FE, et al. (2007) Glycosylation profiling of immunoglobulin G (IgG) subclasses from human serum. *Proteomics* **7**, 4070–4081.

Xue J, Zhu LP, Wei Q. (2013) IgG-Fc *N*-glycosylation at Asn297 and IgA *O*-glycosylation in the hinge region in health and disease. *Glycoconjugate Journal* **30**, 735–745.

Yin BJ, Gao T, Zheng NY, et al. (2011) Generation of glyco-engineered BY2 cell lines with decreased expression of plant-specific glycoepitopes. *Protein Cell* **2**, 41–47.

Zeitlin L, Bohorov O, Bohorova N, et al. (2013) Prophylactic and therapeutic testing of Nicotiana-derived RSV-neutralizing human monoclonal antibodies in the cotton rat model. *mAbs* **5**, 263–269.

Zheng K, Bantog C, Bayer R. (2011) The impact of glycosylation on monoclonal antibody conformation and stability. *mAbs* **3**, 568–576.

Zimran A, Brill-Almon E, Chertkoff R, et al. (2011) Pivotal trial with plant cell-expressed recombinant glucocerebrosidase, taliglucerase alfa, a novel enzyme replacement therapy for Gaucher disease. *Blood* **118**, 5767–5773.

Index

C

D

E

Q

R

S

T

U

For Product Safety Concerns and Information please contact our EU representative GPSR@taylorandfrancis.com Taylor & Francis Verlag GmbH, Kaufingerstraße 24, 80331 München, Germany

Printed and bound by CPI Group (UK) Ltd, Croydon, CR0 4YY
08/05/2025
01864438-0001